GRAZING ECOLOGY AND FOREST HISTORY

For Harm van de Veen

'The most important message taught by the history of science is the subtle and inevitable hold that theory exerts upon data and observations.'

'The greatest impediment to scientific innovation is usually a conceptual lack, not a factual lack.'

(Stephen Jay Gould (1989) *Wonderful Life: the Burgher's Shale and the Nature of History*. W.W. Norton & Co., New York, p. 276)

Grazing Ecology and Forest History

F.W.M. Vera
Ministry of Agriculture, Nature Management and Fisheries
Strategic Policies Division
The Hague
The Netherlands

CABI *Publishing*

CABI is a trading name of CAB International

CABI Head Office
Nosworthy Way
Wallingford
Oxfordshire OX10 8DE
UK

CABI North American Office
875 Massachusetts Avenue
7th Floor
Cambridge, MA 02139
USA

Tel: +44 (0)1491 832111
Fax: +44 (0)1491 833508
Email: cabi@cabi.org
Web site: www.cabi.org

Tel: +1 617 395 4056
Fax: +1 617 354 6875
Email: cabi-nao@cabi.org

A catalogue record for this book is available from the British Library, London, UK.

Library of Congress Cataloging-in-Publication Data
Grazing ecology and forest history / [edited by] F.W.M. Vera.
        p. cm.
   Includes bibliographical references.
   ISBN 0-85199-442-3 (alk. paper)
   1. Forest dynamics—Europe. 2. Plant succession—Europe. 3. Range ecology—Europe. 4. Forests and forestry—Europe—History. I. Vera, F.W.M. II. Title.

QK938.F6 G67 2000
577.3'18'094—dc21

                                                                00-029249

ISBN-13: 978-085199-442-0
ISBN-10: 0-85199-442-3

First published 2000
Reprinted 2002,2004
Transferred to print on demand 2009

Printed and bound in the UK by CPI Antony Rowe, Chippenham and Eastbourne.

Contents

About the Author

Franciscus Wilhelmus Maria Vera was born in Amsterdam on 4 June 1949. He gained his HAVO certificate in 1968 and then attended the School of Higher Forestry and Technical Cultivation in Arnhem. He left after 1 year, to take the state HBSb (science) examination in 1970 and to study biology at the Free University of Amsterdam. He completed his studies there in 1978.

In 1979, he started work at the head office of the State Forestry Department in the Inspectorate for Nature Conservation, as an employee with Valuable Agricultural Cultivated Landscapes. His activities concerned the implementation of the Memorandum on the Relationship between Agriculture and Nature, generally known as the Relationship Memorandum. Subsequently he worked on natural development in large natural areas and areas of water; an interest which was stimulated by his involvement in his private activities to safeguard and develop the Oostvaardersplassen in Zuid Flevoland. In his work with the State Forestry Department, this resulted in drawing up a report that played an important part in designating the Oostvaardersplassen as a natural monument and in rerouting the railway line to Lelystad, to safeguard the Oostvaardersplassen. This period was extremely important for the development of his ideas about the relationship between agriculture and nature and the possibilities for natural development.

In 1982, he moved to the Department of Nature, the Environment and the Management of Fauna at the Ministry of Agriculture and Fisheries, where he became the head of the department for Natural Development and Large Natural Areas. In this post, he drew up an Exploratory Study of Natural Development together with F. Baerselman, in which the concept of the ecological main structure (EHS) was developed and presented. This study was published in 1988. The concept of EHS was taken over in the Nature Policy Plan published in 1989 by the Ministry of Agriculture, Nature and Fisheries.

In 1989, he moved to the Faculty of Nature Management in Wageningen to work out questions and ideas related to nature and natural development in more detail. This resulted in a thesis. Since 1996, he has worked as a Senior Policy Employee for the Strategic Policies Division at the Minister's Office in the Hague. In 1997, he received his PhD from the Agriculture University in Wageningen for his thesis entitled 'Metaforen voor de wildernis. Eik, hazelaar, rund en paart'. This book is based on this thesis.

In addition to the above-mentioned activities in his working life, he has also been involved with natural development in his private life on a large scale. The Oostvaardersplassen were mentioned above. He is also one of the founders of the Plan Ooievaar (Stork Plan) which presented the development of nature in river foreland, the area outside the dykes, in addition to the development of agriculture within the dykes of the rivers.

Preface

This book is an updated and expanded version of my thesis on the question of the appearance of the vegetation in the lowlands of Central and Western Europe after the end of the last Ice Age in the Holocene. The thesis (Vera, 1997) was written in Dutch. I publicly defended it in September 1997 at Wageningen University in the Netherlands. During this defence, the comment was made that it was important to publish it in English so that readers outside the Netherlands and Flanders (Belgium) could also become acquainted with its contents. The Dutch Ministry of Agriculture, Nature Management and Fisheries made the translation possible.

During the public defence, the question was also posed how the contents of the thesis were related to the situation in the east of the United States. According to the prevailing opinion, untouched vegetation consisting of a closed forest is a factor which that part of North America has in common with the lowlands of Central and Western Europe. At the time, I had literature containing indications that there was an interesting parallel. Further studies gave sufficient reason to include the area in this book. The literature study for this book on the east of the United States and the lowlands of Central and Western Europe covers the period up to November 1999.

Why study the original natural vegetation in the lowlands of Central and Western Europe? After all, it no longer exists. The entire lowlands have been cultivated. Nevertheless, many nature conservationists, scientists and policy makers have a clear image of what that nature looked like and propagate this image. It is the image of a closed forest. This image influences the policy which determines the future appearance of our (i.e. European) living environment. For example, everywhere in the Netherlands forests are planted, because this supposedly meets the growing need for urban dwellers to have nature close to home. Forest is synonymous with nature. That image also determines the European approach to the disappearance of untouched nature, for example due to cultivation for agriculture. It has been assumed that pre-industrial farmers felled the forest and added open biotopes like grasslands to the original, monotonous, shade-rich virgin forest. According to most nature conservationists – and in their wake politicians and civil servants – European nature has been enriched as a result of this cultivation. The biodiversity would have increased enormously due to pre-industrial agriculture, therefore, agriculture would supposedly be an essential factor for maintaining biodiversity.

This idea influences agricultural politics in the Netherlands, in the rest of Europe and beyond, through the World Trade Organization (WTO). Within this organization, the end of subsidies to agriculture is advocated. The EU supports this, but simultaneously presents the role of agriculture as manager of the landscape and defender of biodiversity as an argument to give farmers in Europe an income allowance after ending the subsidies. In Europe, farmers supposedly do more than simply carry out agriculture, i.e. produce and manage nature. The paradox is that EU agricultural policies themselves are the cause of the disappearance of many species of wild plants and animals and form a threat for many others. The prevailing conception of most nature conservationists in Europe that agriculture is *essential* for maintaining nature, therefore creates an enormous dilemma.

I have often been amazed by the prevailing concept among nature conservationists that agriculture in Europe is the most suitable framework for maintaining biodiversity, since I have never heard them say that the tropical rainforests of Africa, Asia and South America should be reclaimed for agriculture because – like in Europe – it enriches nature. On the contrary, it is regularly stated that cultivation for agriculture leads to loss of nature and thereby to a decrease in biodiversity. Why should the opinion of the European nature conservationists in relation to the rest of the world not be true for Europe?

Particularly during my first position at the State Forestry Service I became entangled in the belief that agriculture was necessary for the maintenance of nature. This was due to the fact that it was my task to carry out policy to ensure that more space was created within agriculture for species of wild plants and animals. Agriculture in turn stated that only that nature which fitted in the farming operations was possible. Usually, this had very few benefits for the species of wild plants and animals. Then the significant question arose: if it is not possible within the framework of agriculture, then how is it possible? Untouched nature was not an option, since it no longer existed.

Then a polder was created in the centre of the Netherlands and a plan was to be made to cultivate it for agriculture. A shallow layer of water remained in the lowest part. This was intended for industry, which did not need the ground as yet. Reclamation was started elsewhere, so that it did not take place in the area intended for industry for the time being. Spontaneously, a large nature area arose, the Oostvaardersplassen. Nature showed a side of itself there that we in the Netherlands no longer knew. A large part of the area consisted of marsh, where greylag geese (*Anser anser*) gathered in the summer to moult. They consumed massive amounts of marsh plants and as a result, the marsh did not close. This had not been expected, based on experience in other small marsh areas of this type where there were no greylag geese. In those smaller nature areas the marsh closed. Nature managers attempted to prevent this by mowing the reeds. This human intervention was unnecessary in the Oostvaardersplassen area, because a species that was part of the natural ecosystem, i.e. the greylag goose, was present and carried out the natural process of grazing. This development not only showed that human management, such as mowing, was unnecessary, but also that there were more species of plants and animals in comparison with those other smaller areas. Through its consumption, the greylag goose created and maintained their living conditions. This taught us that in large nature areas, nature could be given more independence. The greylag goose could in fact establish itself there because of the scale of the area available. The grazing of this species in turn was leading the succession of the vegetation.

The conclusion was that natural processes like grazing could be redeveloped, if the responsible species could live in the nature area involved. This was a new idea, because everyone always assumed that a return to the nature lost to cultivation was no longer possible. Now it appeared that in conditions analogous to untouched situations, nature developed as an analogy of those untouched situations. It was in fact man who had initially *unintentionally* created the conditions for this. The lesson was that it could also be done in a planned way.

The next logical step was from the greylag geese as large grazers in the marsh to the large grazers on land, i.e. the European indigenous fauna of the following species: roe deer (*Capreolus capreolus*), red deer (*Cervus elaphus*), elk (*Alces alces*), European bison (*Bison bonasus*) and the wild ancestors of domesticated cattle and horses: aurochs (*Bos primigenius*) and tarpan (*Equus przewalski gmelini*). The question asked by a number of biologists, including myself, at the time was whether these species have played a role on land that was comparable to the role of the greylag goose in the marsh, in other words did these animals with their consumption as part of nature create the living conditions for the European biodiversity? Would this include the species of plants and animals and biotopes, such as grasslands, that we in Europe only know nowadays from cultivated landscapes? Could we redevelop nature, i.e. create the pre-conditions that lead to modern analogies of that nature which has disappeared through cultivation? If that were possible, then we believed that this was an inviting and challenging perspective for nature conservation. Therefore, we proposed an experiment to reintroduce the disappeared large ungulates.

However, time and again a wall of closed forest, built up in front of us by the nature conservationists confronted our enthusiasm. They said that it could not work, because everything would just become one big closed forest. The large ungulates would not be able to stop the development of the forest. After all, the untouched vegetation has always been forest and the indigenous large ungulates of the time had been unable to change this since it has always been forest.

There was therefore no point in trying; the results of the experiment were known in advance, they continually informed us. The established theory made this perfectly clear. The best thing to do was to manage the area on the basis of agriculture. After all, agriculture has provided open grasslands and great bio-diversity in the lowlands of Europe. In order to maintain nature, the nature area must be converted into a kind of agrarian cultivated landscape.

Years of struggle ensued, particularly with biologists and nature conserva-tionists, to realize the experiment with large ungulates living in the wild. Despite all the opposition and resistance, it has been realized and not only in this one example.

The surprising developments in the Oostvaardersplassen, concern for the European nature of tomorrow, discussions with colleagues, friends, information from literature, and last but not least the arguments used by opponents, were the reasons for me to delve into the question of the appearance of the original vegetation in Europe. It led to a quest that in the first place led to my thesis and now in the second place to this book. One tree, the oak, was my guide through-out.

Frans Vera
Wijk bij Duurstede, 21 March 2000

Acknowledgements

Firstly, I would like to thank my family. In the last few years, Hilda, Thijs and Annemiek have always supported me while they often had to do without me because I spent evenings and weekends in my study working on my thesis, which I completed in 1997. Even when I was with them, I was often absent. Quite rightly, they sometimes dragged me away from my work. Bringing this book up to date and appending it involved more work than I had initially thought, so that I had to isolate myself in my study again for several months.

I would like to thank Dick Hamhuis for suggesting that I concentrate on the scientific approach so that I could explore the facts and ideas on nature and frames of reference. I would like to thank the Wageningen University for giving me the opportunity to carry out this study. I am grateful to Claus Stortebeker for including me in his Faculty of Nature Management at Wageningen University and for supervising the doctoral thesis which resulted in this book. I found a wonderful group of people there and enjoyed working with them very much. I also greatly appreciated my new colleagues when the faculty was reorganized. At the risk of forgetting anyone, I should like to mention two people in particular. These are Stephen de Bie and Chris Geerling, with whom I had many inspirational talks and discussions. Their constructive criticism clarified my ideas about the questions I had, and the answers which I thought I had found. They shared their experiences of intact ecosystems in Africa and I greatly benefited from this. After the completion of my thesis, Chris Geerling made inspiring suggestions about the non-linear character of the succession in my theory. In addition, I learnt a great deal from my talks with Sip van Wieren and Michiel Wallis de Vries about large herbivores, and their experiences with the research they carried out on grazing and large ungulates. In addition, I must mention the background support of an administrative and logistic nature. In this context I would like to mention Gerda Bruinsma and Marjolijn de Brabander, Gerda Westphal and Alie Ormel. I look back with great pleasure to the discussions I had with Herman van Oeveren on etymology. I would also like to thank Mr J. van Wolferen for his explanation of the classical texts and additions to them.

Without the collective memory of the library, my research would not have gone anywhere. Fried Kampes, Elly Boekelman, Jan Roesink and Janny van Keeken were always very helpful with regard to tracing sources. They scoured other libraries for me so that a request for information from abroad which I had given up on months ago was sometimes still discovered. One thing this study has certainly taught me is that there are still many treasures lying in libraries waiting to be discovered and that we cannot value our libraries too highly.

I would like to thank Herbert Prins, who supervised the thesis which resulted in this book, for being prepared to supervise me in research on which little was known at the time. I was and I remain extremely impressed by his broad range of interests and knowledge reaching from the tropics via the Middle Ages to palaeoecology. He was very stimulating. Admittedly, a number of his suggestions led to a great deal of work, but doing this work led to greater clarity and consistency in my thesis. Some of the comments on my ideas were a bit difficult to swallow but, once I had made the necessary adjustments, the compliments which followed were just as direct.

I would like to thank Frank Berendse, who also supervised me, for the fact that he was prepared to supervise me after Claus Stortebeker had indicated that he wished to pass on this task after his retirement. In a sense, he was jumping on to a moving train, which must have been rather difficult. I had heard about his sceptical and critical attitude to my ideas. On the other hand, this seemed all the more reason to ask him. Frank was critical, sharp and constructive. His approval strengthened my confidence in my findings.

I would like to thank Hans van Asté and Koen Peters. As students, they helped me to find pieces of the puzzle with their research, giving greater substance to some vague ideas.

Apart from my two supervisors, others also contributed to rewriting, 'cutting', 'pasting' and 'planing' my thesis and finally this book. At various stages, they worked through and commented on the chapters in draft form. I am very grateful to Leen de Jong, Wim Braakhekke, Roel Janssen, Han Olff, Fred Baerselman, Henk Koop, Gerrit Meester and Dr Blok for this work. I would like to thank Roel Janssen for the fact that he not only commented on the chapter on palynology, but also gave me the opportunity to try out the content with a presentation in his faculty of palaeobotany, and palaeoecology. Dr Blok was prepared to cast a critical eye over my ideas on language and the meaning of words at a late stage. His comments did not leave anything to be desired as regards clarity, but were nevertheless constructive, and Fred Baerselman, Henk Koop and Gerrit Meester went through the entire draft.

I am also very grateful to Gerrit Meester, head of the Bureau for Strategic Policy Formation of the Ministry of Agriculture, Nature Management and

Fisheries, where I am currently employed, for the fact that he gave me the opportunity to complete my thesis and this book and have it translated into English (by V5 Vertalerscombinatie B.V., Amsterdam). I would also like to thank Marja Rotermundt of the secretariat for typing the tables and the extremely detailed bibliography. My thanks go to Rob van Loon of the Repro department of the Ministry for making the drafts of my thesis, which have resulted in this book and Duotone (Wageningen University) for reproducing the figures. In addition, I would like to thank the Nature Directorate of the Ministry of Agriculture, Nature Management and Fisheries, which removed the last financial hurdle with a financial contribution that finally made publication in this form possible.

Finally, the stage of completion was in sight. I had a great deal of support from Fred Baerselman. His understanding, insight and capacity for expressing himself meant that the last hurdles were crossed very quickly at the end.

Last of all, I would like to mention the members of the Critical Forest Management Foundation in the Netherlands, in particular Hans van der Lans, Gerben Poortinga and Harm van der Veen. They put me on the track of the role large ungulates played in the regeneration of trees. They carried out a lot of pioneering work in the Netherlands that I could use as a basis to continue my work. In particular, I would like to thank Harm van de Veen to whom I dedicate this book. For me and for many others looking for new paths in ecology and the conservation of nature in particular, Harm was an extremely inspiring person. His death came much too soon.

General Introduction and Formulation of the Problem

1

1.1 A Closed Forest as Natural Vegetation[1]

It is a generally accepted theory that in the natural state, i.e. if there had been no human intervention, the lowlands of Central and Western Europe, with their temperate climate, would have been covered, in places where trees can grow, with deciduous forest.[2] Figure 1.1 shows this sort of forest. The theories on succession, including that of Clements (1916), form an important basis for this assumption. He formulated a succession theory which states that under certain climatological conditions the bare ground develops into a plant community of a particular type, through a number of successive stages. This is dominated by the largest and tallest plants, which are able to thrive in the prevailing climatological conditions because the tallest plants are always strongest in the competition for light (Clements, 1916, pp. 3, 63, 80, 99, 125; Tilman, 1985). If these are trees, the final stage is forest (Clements, 1916, p. 99). On the basis of this theory and other theories (including Tansley, 1935; Watt, 1947; Whittaker, 1953), it is assumed that every climate zone is naturally characterized by a certain type of vegetation; the *climax* (Clements, 1916, pp. 3, 63) or the *zonal* vegetation (Walter, 1954). In the temperate climate zone in the northern hemisphere, which includes Europe, this is forest. These forest areas are shown in Fig. 1.2.

When the climax forest disappears as the result of some form of intervention, the climax will spontaneously develop again by means of secondary

---

[1] If there is a reference in this chapter or the following chapters to more than five authors, the references are given in a footnote to make the text easier to read.
[2] See, *inter alia*, Tansley (1916), Watt (1947), Walter (1954; 1974, pp.16–17), Ellenberg (1986, pp. 73–74; 1988, pp. 1–2), Harris and Harris (1991, pp. 7–9), Röhrig (1991), Zoller and Haas (1995), Peterken (1996, p. 32 *et seq.*).

Fig. 1.1. Closed deciduous forest in the National Park of Białowieza. The forest in this park is supposed to be the most original vegetation in the lowlands of Central and Western Europe, and therefore most like the original primeval forest (photograph, F.W.M. Vera).

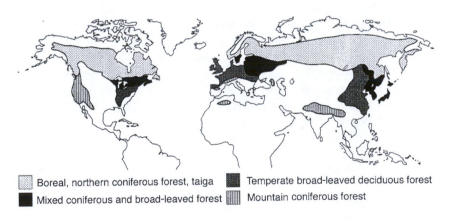

Fig. 1.2. The forest zones in the northern hemisphere. The deciduous forest of the temperate climate occurs in the east of North America and China, as well as Europe. These regions have many genera in common with Europe, including oak (*Quercus*),[*] lime (*Tilia*), beech (*Fagus*), ash (*Fraxinus*), birch (*Betula*) and hazel (*Corylus*) (redrawn from Harris and Harris, 1991, p. 9).

---

[*] In future, I will give the common name followed by the scientific name the first time a species is mentioned in a chapter.

succession, once the intervention has ceased (Clements, 1916, pp. 60, 63, 107, 176; Tansley 1953, pp. 130, 293–295, 487). This theory is based to a significant extent on the observation that abandoned fields and meadows spontaneously develop into forests when man withdraws and the livestock are removed from the fields.[3]

The results of pollen studies also reveal that the vegetation which developed in the lowlands of Central and Western Europe after the last Ice Age ended in approximately 14,000 BP[4] was a closed (deciduous) forest. This conclusion is based on the marked dominance of pollen grains of trees, compared with those of grasses and herbs, which is characteristic of open terrain in the period before the introduction of agriculture. The most important species of trees in this forest, namely oak (*Quercus*), lime (*Tilia*), ash (*Fraxinus*), elm (*Ulmus*), beech (*Fagus*), and hornbeam (*Carpinus*) arrived in the region concerned between 9000 and 1500 BP.[5] Bones that were found show that at that time the fauna in the region included large herbivorous mammals, including aurochs (*Bos primigenius*), tarpan or European wild horse (*Equus przewalski gmelini*), European bison (*Bison bonasus*), red deer (*Cervus elaphus*), elk (*Alces alces*), roe deer (*Capreolus capreolus*), beaver (*Castor fiber*) and the (omnivorous) wild boar (*Sus scrofa*). These species colonized the region approximately 12,000 BP in the Allerød, and were found throughout Europe until the early Middle Ages.[6] Therefore these fauna were present in Central and Western Europe long before the arrival of the most important species of trees which comprised the forests.

---

[3] See Cotta (1865, p. v), Gradmann (1901), Forbes (1902), Warming (1909, p. 326), Moss (1910, p. 36), Moss *et al.* (1910, p. 114), Tansley (1911, pp. 7–8; 1953, pp. 293–294), Clements (1916, pp. 145, 151, 155), Watt (1919; 1947), Spurr (1952), Niering and Goodwin (1974), West *et al.* (1980, p. v), Myster (1993), Kollmann and Schill (1996), Stover and Marks (1998), Bodziarczyk *et al.* (1999).

[4] Coope (1977; 1994), Van Geel *et al.* (1980/81), Lemdahl (1985), Atkinson *et al.* (1987), Berglund *et al.* (1984), Dansgaard *et al.* (1989), Ponel and Coope (1990), Kolstrup (1991). All time indications in this extract for prehistoric times are given in the years of present history. For conformity I therefore use the English abbreviation BP (before present). Present is the year 1950.

[5] See *inter alia* Von Post (1916), Bertsch (1929; 1949), Tschadek (1933), Firbas (1934; 1935; 1949, p. 1), Godwin (1934a,b), Iversen (1941; 1960), Ellenberg (1954), Davis (1967), Manten (1967), Janssen (1974), Huntley and Birks (1983), Birks (1989), Huntley and Webb (1989), Faegri and Iversen (1989, p. 1), Bennett *et al.* (1991), Brown (1991), Real and Brown (1991), Zoller and Haas (1995), Björkman and Bradshaw (1996), Björse and Bradshaw (1998), Björse *et al.* (1996), Bradshaw and Holmqvist (1999), Bradshaw and Mitchell (1999), Mitchell (1998), Mitchell and Cole (1998), Lindbladh (1999), Lindbladh and Bradshaw (1998), Coard and Chamberlain (1999), Tinner *et al.* (1999).

[6] Eichwald (1830, p. 249), Acker-Stratigh (1844), Genthe (1918), Hedemann (1939, p. 310), Degerbøl and Iversen (1945), Degerbøl (1964), Degerbøl and Fredskild (1970), Pruski (1963), Heptner *et al.* (1966, pp. 477–480, 491–499, 861–865), Clason (1967, pp. 31, 60, 76), Evans (1975), Grigson (1978), Jacobi (1978), Volf (1979), Aaris-Sørensen (1980), Simmons *et al.* (1981), Söffner (1982), Stuart (1982), Bosinski (1983), Frenzel (1983, pp. 152–166), Von Koenigswald (1983, pp. 190–214), Louwe-Kooijmans (1985, p. 51; 1987), Birks (1986), Davis (1987, pp. 174–179), Price (1987), Lauwerier (1988, pp. 28–31, 50–51, 56–59, 65, 69, 73, 87, 90, 92–93, 98, 101), Van Alsté (1989), Roberts (1989, pp. 80–83), Aaris-Sørensen *et al.* (1990), Lebreton (1990, pp. 32–44), Cordy (1991), Current (1991), Hously (1991), Street (1991), Stuart (1991), Bell and Walker (1992, pp. 170–173), Litt (1992), Auguste and Patou-Mathis (1994), Chaix (1994), Guintard (1994).

According to the prevailing succession theories, this original fauna did not have any influence on the succession in the forest that was originally present, but followed the developments in the vegetation. (See, *inter alia*, Tansley, 1935; Iversen, 1960; Whittaker, 1977.)

On the basis of pollen studies, Iversen (1941) also formulated the theory that man first started cutting down the closed forest in north-west and Central Europe, as a farmer, to make fields and meadows for cattle approximately 5000 years ago (Iversen, 1941; 1956; 1960, pp. 19–20, 79–80; 1973, pp. 78–92). This so-called 'Landnam' theory has been generally accepted.[7] Because the remaining original forests were also pastured, they disappeared because the seedlings were grazed and trampled, preventing the regeneration of the forest.[8] This means that the original forest degraded into a park-like landscape, and finally into grassland and heathland as a result of retrogressive succession.[9] Historical sources are presented as proof for the damage caused to the forests by the grazing. These include orders and regulations governing grazing in forests. These show that it was known early on that grazing prevented the regeneration of forests, and was therefore regulated.[10] Because of this, grasslands which occur in places where trees can grow, are seen as artefacts created by man. By putting an end to this grazing, these artefacts would spontaneously develop into natural closed forest once again.[11] On the basis of this presumed degradation of the forest to grassland, Watt (1923; 1924; 1925; 1934a,b) put forward his now classical developmental series of the succession from grassland to climax forest, using chronosequences. Based on this, he

[7] See *inter alia* Godwin (1944; 1975, p. 465 *et seq.*), Van Zeist (1959), Sims (1973), O'Sullivan *et al.* (1973), Janssen (1974, p. 80), Rackham (1980, p. 104), Rowley-Conwy (1982), Behre (1988, p. 643), Birks (1986; 1993), Delcourt and Delcourt (1987, p. 374 *et seq.*) Bogucki (1988, p. 33), Huntley (1988, p. 346), Andersen (1989), Faegri and Ivesen (1989, p. 110), Roberts (1989, p. 117 *et seq.*), Jahn (1991, p. 392), Mannion (1992, pp. 64–65), Tallis (1991, pp. 270–280), Bell and Walker (1992, p. 164 *et seq.*), Walker and Singh (1993), Edwards (1993), Waller (1994).

[8] See *inter alia* Cotta (1865, p. 84), Landolt (1866, p. 152), Endres (1888, p. 157), Krause (1892), Forbes (1902), Moss (1910, p. 36; 1913, pp. 91, 111), Moss *et al.* (1910, p. 114), Tansley (1911, pp. 7–8; 1916; 1953, pp. 128–130, 223, 487), Watt (1919), Bühler (1922, p. 259), Vanselow (1926, p. 145), Grossman (1927, p. 33 *et seq.*), Morosow (1928, pp. 279, 318), Endres (1929), Hess (1937), Hedemann (1939), Hausrath (1982, pp. 28, 39), Gothe (1949), Rodenwaldt (1951), Hesmer (1958, p. 454), Krahl-Urban (1959, p. 86), Hesmer and Schroeder (1963, pp. 273–275), Ovington (1965, p. 52), Hart (1966, pp. xix–xxiii), Schubart (1966, p. 95), Streitz (1967, pp. 58, 155), Holmes (1975), Scholz (1973), Bunce (1982), Jahn and Raben (1982), Prusa (1982), Tendron (1983, p. 23), Buis (1985, p. 273), Ellenberg (1954; 1986, pp. 38–49), Mantel (1968; 1990, p. 94 *et seq.*), Holmes (1989), Oldeman (1990, p. 439), Harris and Harris (1991, p. 29), Jahn (1991, p. 395), Pott and Hüppe (1991, p. 23), Harmer (1994), Zoller and Haas (1995).

[9] See Moss (1910, p. 36; 1913, pp. 91, 111), Tansley (1911, pp. 7–8; 1953, pp. 129–130), Watt (1919), Ellenberg (1954; 1986, p. 43), Westhoff (1976), Pott and Hüppe (1991, p. 23).

[10] See *inter alia* Moss (1913, p. 91), Watt (1919), Bühler (1922, p. 259), Vanselow (1926, p. 145), Grossman (1927, p. 33 *et seq.*), Hess (1937), Meyer (1941, pp. 360, 386), Hausrath (1982, pp. 28, 39), Hesmer (1958, pp. 86, 454), Hesmer and Schroeder (1963, pp. 273–275), Hart (1966, pp. xix–xxiii), Schubart (1966, p. 95), Streitz (1967, pp. 58, 155), Wartena (1968), Jansen and van de Westerigh (1983, p. 41), Buis (1985, p. 273), Mantel (1968; 1990, p. 94 *et seq.*), Zoller and Haas (1995).

[11] Cotta (1865, p. v), Gradman (1901), Forbes (1902), Warming (1909, p. 326), Tansley (1911, pp. 7–8; 1953, pp. 128–130, 487), Clements (1916, pp. 102, 107).

constructed a model for the regeneration of the climax forest. He believed that this regeneration took place in the gaps in the canopy which were created when one – or several – trees died or were blown over (Watt, 1925; 1947). This model is known as Watt's 'gap phase' model (1947). In the past few decades it has been generally accepted as the model which explains the mechanism behind the regeneration of natural forests.[12] In addition, the forest ecologist, Leibundgut (1959; 1978) formulated a cyclical model which explains how the primeval forests of Europe were rejuvenated by means of gaps in the canopy; a model which is generally accepted in forestry and forest ecology.[13]

The theory that the natural vegetation in the temperate climate zone of Europe was a closed forest is also generally accepted in European nature conservation circles.[14] Therefore grasslands are seen as vegetation which has evolved from the original primeval forest as the result of human activity. All the other situations, where there are wild species of plants and animals in open terrain, such as wood-pasture, coppices and fields, also evolved from the primeval forest (see Fig. 1.3). These systems, which were partly created by humans when they started farming, are extremely rich in species of plants and animals because they contain many different types of vegetation. According to nature conservationists, the original closed forest will return and the species of plants and animals which are characteristic of grassland will disappear with the end of agriculture. Therefore the original vegetation in Central and Western Europe was thought to contain a relatively limited number of species, in comparison with the situations which developed as a result of human intervention.[15] This means that a large proportion of the great variety of species in Europe are not the result of natural processes, but evolved as a result of the introduction of agriculture, and will disappear when agriculture disappears. Thus, according to nature conservationists, humans have enriched nature in Europe by opening up the natural forest.[16] In their view, human intervention in the form of

---

[12] See *inter alia* Shugart and West (1977; 1980; 1981), Shugart and Seagle (1985), Shugart and Urban (1989), Runkle (1981; 1982; 1985; 1989), Whitmore (1982; 1989), Brokaw (1985), Canham (1989), Lemée (1987), Lemée *et al.* (1992), Peterken (1996, pp. 143–144), Spies and Franklin (1989), Platt and Strong (1989), Prentice and Leemans (1990), Leemans (1991a,b), Dengler (1992, p. 93 *et seq.*), Botkin (1993), Holeska (1993), Van den Berge *et al.* (1993), Koop and Siebel (1993), Abe *et al.* (1995), Tanouchi and Yamamoto (1995).
[13] See Lödl *et al.* (1977), Mayer and Tichy (1979), Mayer (1992, p. 13), Koop (1981, p. 61; 1989 p. 22), Oldeman (1990, pp. 491, 493), Zukrigl (1991), Korpel (1995, p. 14).
[14] See *inter alia* Hampicke (1978), Sissigh (1983), Westhoff (1983), Heybroek (1984), Ellenberg (1986, p. 20), Baldock (1989), Londo (1990, 1991, p. 5), Dolman and Sutherland (1991), Peterken (1991), van der Werf (1991, p. 13), Götmark (1992), Rackham (1992), Burrichter *et al.* (1993), Hondong *et al.* (1993, p. 15), Pott (1993, p. 27 *et seq.*).
[15] See *inter alia* Tüxen (1952; 1956), Sukopp (1972), Scholz (1975), Medwecka-Kornas (1977), Wolkinger and Plank (1981), Sissingh (1983), Westhoff (1983), Heybroek (1984), Ellenberg (1986, p. 20), Dolman and Sutherland (1991), Fry (1991), Götmark (1992), Lindbladh (1999), Lindbladh and Bradshaw (1998), Tipping *et al.* (1999).
[16] See *inter alia* Van Leeuwen (1966), Westhoff (1952; 1971), Westhoff *et al.* (1971; 1973), Bürrichter (1977), Bürrichter *et al.* (1993), Hampicke (1977), Kornas (1983), Heybroek (1984), Ellenberg (1986, p. 20), Green (1989), Berglund *et al.* (1991, p. 421), Dolman and Sutherland (1991), Fry (1991), Götmark (1992), Jennersten *et al.* (1992), McCracken and Bignal (1995).

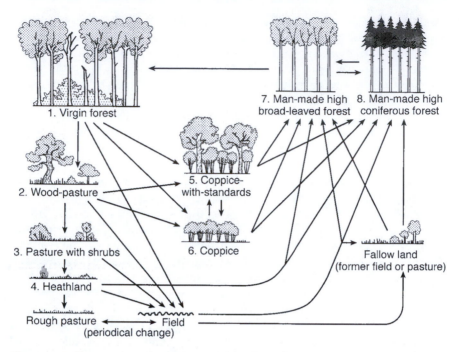

Fig. 1.3. Different types of vegetation and landscapes which have developed from the original primeval forest as a result of human intervention, according to the prevailing theory (redrawn from Ellenberg, 1986, p. 52; 1988, p. 28).

agriculture is necessary to maintain the biodiversity in Europe. For this reason, there is an argument for allowing certain forms of agriculture – in particular, of extensive agriculture – to continue, in order to ensure conservation of nature.[17] Therefore measures are adopted from certain forms of agriculture. These include cutting turf in heathland, mowing and haymaking in grassland, and grazing sheep and cattle.[18]

1.2 Formulation of the Problem

If the above theories are correct, it follows that all the species of trees and shrubs which pollen studies have shown to be present in prehistoric times up to the introduction of agriculture, survived in closed forests and regenerate spontaneously when there are gaps in the canopy. However, this does not apply – or hardly applies – to pedunculate oak (_Quercus robur_), sessile oak (_Q. petraea_) and hazel (_Corylus_

[17] See _inter alia_ Baldock (1989), Bignal and McCracken (1992, Bignal _et al._ (1994), Curtis _et al._ (1991), Goriop _et al._ (1991), Götmark (1992), Hötker (1991), Beaufoy _et al._ (1995), McCracken and Bignal (1995).
[18] See _inter alia_ Westhoff (1952; 1971), Westhoff _et al._ (1970; 1971; 1973), Bakker (1989, pp. 1–5), Londo (1990), Pott and Hüppe (1991, pp. 16–23), Dolman and Sutherland (1991).

avellana).[19] On the other hand, both species of oak and hazel are represented in pollen diagrams taken from Central, Western and north-west Europe,[20] in relatively high percentages, over a period of more than 9000 years. This shows that they were present without interruption throughout this whole period (Faegri and Iversen, 1989, p. 137). Studies of sub-fossil oaks taken from sediments in the flood plains of the Rhine, Danube, Main and Vistula also reveal this sort of discrepancy. Research into the annular rings of more than 5000 of these oaks (pedunculate and sessile oak cannot be distinguished on the basis of the wood) show that oak trees formed part of the vegetation of these floodlands for almost 10,000 years without interruption.[21] However, as far as we know, there is no regeneration of oak trees in forest reserves in these floodlands nowadays. They are displaced by lime (*Tilia* sp.), hornbeam (*Carpinus betulus*), elm (*Ulmus* sp.) and field maple (*Acer campestre*).[22] Therefore the question arises whether the untouched vegetation of Central and Western Europe actually was a closed forest, if it contained oak and hazel, which do not survive in closed forests.

The regeneration of pedunculate and sessile oak and hazel does take place in so-called wood-pasture, actually as the result of grazing by livestock.[23] These

[19] See *inter alia* Gradmann (1901), Forbes (1902), Bernátsky (1905), Watt (1919), Nietsch (1927; 1939, pp. 27–28), Morosow (1928, pp. 269, 279, 311), Vanselow (1926, pp. 9–10; 1929; 1949, p. 252), Meyer (1931, p. 357), Hesmer (1932; 1958, p. 260; 1966), Tüxen (1932), Tschadek (1933), Hedeman (1939), Tansley (1953, pp. 291–293), Reed (1954, pp. 78 and 117), Turbang (1954), Doing-Kraft and Westhoff (1958), Sanderson (1958, pp. 72–87, 128, 153, 253), Krahl-Urban (1959, pp. 146, 191, 212, 214, 216 e.v.), Pockberger (1963), Röhrig (1967), Schubart (1966, p. 168), Pigott (1975), Rackham (1976, p. 33; 1980, p. 327; 1992), Lödl *et al.* (1977), Mayer and Tichy (1979), Malmer *et al.* (1978), Dister (1980, p. 71), Koop (1981, pp. 52–53; 1989, pp. 89, 104, 171), Fricke (1982), Fricke *et al.* (1980), Lüpke (1982; 1987), Röhle (1984), Hytteborn (1986), Jahn (1987a,b), Lemée (1987), Malmer *et al.* (1987), Falinski (1988), Fleder (1988), Lanier (1988), Ebeling and Hanstein (1988), Kwiatkowska and Wyszomirski (1990), Freist-Dorr (1992), Nillson (1992), Le Duc and Havill (1998), Reif *et al.* (1998), Bernadzk *et al.* (1998), Abs *et al.* (1999).

[20] See Von Post (1916), Firbas (1934; 1935; 1949), Godwin (1933; 1934a,b; 1944; 1975), Godwin and Tallantire (1951), Godwin and Deacon (1974), Nietsch (1935; 1939), Iversen (1941; 1960; 1973), Bertsch (1949), Polak (1959), Van Zeist (1959), Van Zeist and van der Spoel Walvius (1980), Andersen (1973; 1976; 1989; 1990), Westhoff *et al.* (1973), Janssen (1974, pp. 55–65), Steel (1974), Planchais (1976), Morzadec-Kerfourn (1976), Huault (1976), Girling and Greig (1977), Ralska-Jaiewiczowa and Van Geel (1992), Waller (1993), Van Geel *et al.* (1980/81), Greig (1982), Perry and Moore (1987), Chen (1988), Huntley and Birks (1986), Huntley (1988; 1989), Kalis (1988), Bennett (1988; 1989), Bartley *et al.* (1990), Day (1991), Bozilova and Beug (1992), Horton *et al.* (1992), Latalowa (1992), Rösch (1992), Caspers (1993), Peglar (1993).

[21] Becker and Schirmer (1977), Becker (1983), Pilcher *et al.* (1984), Becker and Glaser (1991), Becker *et al.* (1991), Becker and Kromer (1993), Kalincki and Krapiec (1995).

[22] See Dister (1980, pp. 65, 66; 1985; 1987), Prusa (1985, pp. 50, 51, 70, 73), Dornbusch (1988), Den Oude (1992, pp. 47–58, 98–99), Bönecke (1993).

[23] See *inter alia* Forbes (1902), Watt (1919; 1924), Grossmann (1927, p. 114), Tüxen (1952), Tansley (1953, pp. 130–133), Peterken and Tubbs (1965), Hart (1966, pp. 180–181, 186, 209, 225, 310), Mellanby (1968), Jakucs (1969; 1972, p. 200 *et seq.*), Musall (1969, p. 95), Ekstam and Sjörgen (1973), Lohmeyer and Bohm (1973), Dierschke (1974), Flower (1977, pp. 28, 32; 1980), Bürrichter *et al.* (1980), Rackham (1980, pp. 173, 293), Addison (1981, pp. 84, 85, 95), Ellenberg (1986, pp. 43–44, 60, 644; 1988, pp. 20–21, 33–34), Pott and Hüppe (1991, p. 23 *et seq.*), Rodwell (1991, pp. 333–351), Kollmann (1992), Oberdorfer (1992, pp. 80–82), Hondong *et al.* (1993), Pietrarke and Roloff (1993).

landscapes consist of a mosaic of grassland, scrub, thickets with trees, solitary trees and forests. Apart from the two species of oak and hazel, all other species of trees that were present in the prehistoric vegetation, according to pollen analyses, such as beech, hornbeam, elm, ash and lime, also formed part of this vegetation and regenerated in this vegetation.[24] This regeneration does not take place in the forest, but in the scrub.[25] The question is whether this type of vegetation could have been the original natural vegetation. In that case, grazing by livestock, i.e. by domestic horses, domestic cattle, sheep, goats and pigs, can be seen as a modern analogy of grazing by the original fauna, the large herbivores.

1.3 The Null Hypothesis and the Alternative Hypothesis

This book is about testing a theory. The theory reads: the lowlands of Western and Central Europe were by nature places where climate, soil and groundwater levels permitted the growth of trees in a closed deciduous forest. The most important genera in this deciduous forest were: oak (*Quercus*), elm (*Ulmus*), lime (*Tilia*), ash (*Fraxinus*), beech (*Fagus*), hornbeam (*Carpinus*) and hazel (*Corylus*). This would still be the case if there had been no human intervention.

Due to the scale and the period in which the phenomena in this theory took place, this theory is tested using the historical method. The events that took place are irreversible. They belong to the realm of contingency, i.e. the outcome of the events depends on what happened beforehand (Gould, 1989, pp. 277–279). The historical method entails that the past is explained from numerous events using induction. In this method, the theory is tested on the basis of the null hypothesis. In the historical method, the null hypothesis is the pattern or the process as it can be predicted based on the theory to be tested (Scheiner, 1993). If the null hypothesis is rejected, this is done in favour of an alternative hypothesis inferred from the same data.

I base the following null hypothesis on the theory that the lowlands of Western and Central Europe were covered by a natural closed forest, where the most important species were oak, elm, lime, ash, beech, hornbeam and hazel, and that this would still be the case if there had been no human intervention:

> Pedunculate and sessile oak and hazel survive in a closed forest, and regenerate in gaps in the canopy, in accordance with Watt's 'gap phase' model (1947) and Leibundgut's so-called cyclical model. The large herbivores which occurred naturally, such as the aurochs, European bison, red deer, elk, roe deer and tarpan or European wild horse, and the omnivorous wild boar followed the developments in the vegetation. They did not have any natural influence on the course of the process of the succession and regeneration of forests.

[24] Kerner (1929, pp. 45–46), Klika (1954), Rackham (1980, pp. 174, 199, 235, 242, 248), Ellenberg (1986, p. 244 *et seq.*), Pott and Hüppe (1991, pp. 289–299) and personal observation.
[25] See *inter alia* Watt (1919; 1924; 1925; 1934a,b), Adamson (1921; 1932), Flower (1977, p. 112), Rackham (1980, pp. 174, 188, 293), Bürrichter *et al.* (1980), Pott and Hüppe (1991, pp. 25–26).

I will compare this null hypothesis with the data in the literature. If this null hypothesis has to be rejected, I propose the following alternative:

> The natural vegetation consists of a mosaic of large and small grasslands, scrub, solitary trees and groups of trees, in which the indigenous fauna of large herbivores is essential for the regeneration of the characteristic trees and shrubs of Europe. The wood-pasture can be seen as the closest modern analogy for this landscape.

1.4 Outline of this Study

This study is based on literature research. Recent publications relating to the formulation of the problem were directly consulted in journals. Other publications were examined on the basis of references in handbooks and articles. The classification and dating of prehistoric eras that are generally used in publications are shown in Fig. 1.4. I use this figure for references to prehistoric eras.

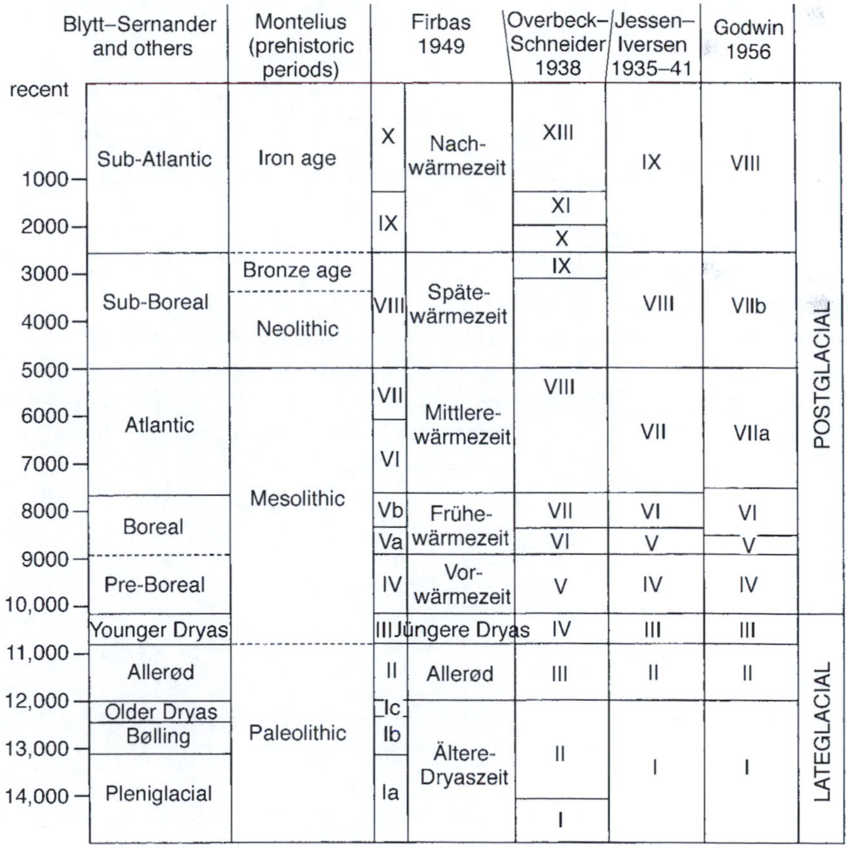

	Blytt–Sernander and others	Montelius (prehistoric periods)	Firbas 1949		Overbeck–Schneider 1938	Jessen–Iversen 1935–41	Godwin 1956	
recent								
	Sub-Atlantic	Iron age	X	Nach-wärmezeit	XIII	IX	VIII	POSTGLACIAL
1000 —								
2000 —			IX		XI			
					X			
3000 —		Bronze age			IX			
4000 —	Sub-Boreal	Neolithic	VIII	Späte-wärmezeit		VIII	VIIb	
5000 —			VII		VIII			
6000 —	Atlantic			Mittlere-wärmezeit		VII	VIIa	
7000 —			VI					
8000 —	Boreal	Mesolithic	Vb	Frühe-wärmezeit	VII	VI	VI	
9000 —			Va		VI	V	V	
	Pre-Boreal		IV	Vor-wärmezeit	V	IV	IV	
10,000 —								
	Younger Dryas		III	Jüngere Dryas	IV	III	III	LATEGLACIAL
11,000 —	Allerød		II	Allerød	III	II	II	
12,000 —	Older Dryas	Paleolithic	Ic					
13,000 —	Bølling		Ib	Ältere-Dryaszeit	II	I	I	
14,000 —	Pleniglacial		Ia					
					I			

Fig. 1.4. The most important pollen zones in Central and Western Europe in the period following the end of the last Ice Age (redrawn from Janssen, 1974, p. 54).

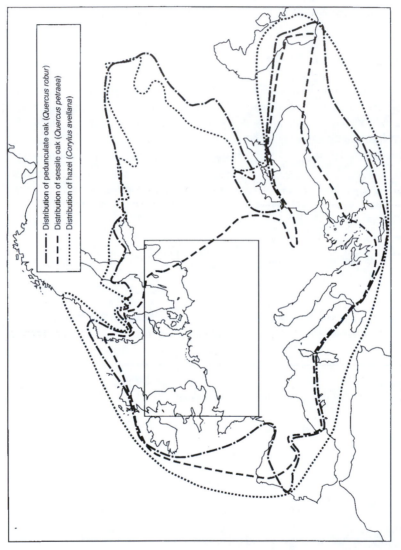

Fig. 1.5. The borders of the distribution in Europe of pedunculate oak, sessile oak and hazel (redrawn from Jahn, 1991, pp. 403, 406). The frame in the centre of the map indicates the area covered by the study. It lies between 45°N and 58°N latitude, and 5°W and 25°E longitude.

This book covers an area in Western and Central Europe between 45°N and 58°N latitude and 5°W and 25°E longitude (see Fig. 1.5). Geographically, this comprises all or parts of England, Scotland, Wales and France, Belgium, Luxembourg, the Netherlands, Germany, Switzerland, Austria, the Czech Republic, Slovakia, Poland, Denmark and Sweden. The area covered by this study lies virtually entirely within the area of distribution of pedunculate and sessile oak and hazel (see Fig. 1.5). The two species of oak are not found in their distribution area at altitudes above 600–700 m (Dengler, 1992, pp. 168–172; Mayer, 1992, p. 111). They are species which grow in the lowlands, hills and mid-mountainous regions. As regards the sessile oak, the eastern border of the area covered by this study coincides approximately with the eastern border of the distribution area of this species. The northern border of the area covered by this study is determined in Scandinavia by the most northern appearance of the pedunculate oak, the species of oak which is found furthest north (see Fig. 1.5).

Chapter 2 describes the development of the theory on succession. It contains an explanation of the development of the climax concept and a description of the models for the regeneration of the climax forest, such as Watt's gap phase model (1947) and Leibundgut's cyclical model (1959; 1978). In addition, it deals with the theories on succession and climax vegetations and the premises on which these are based.

Chapter 3 describes how Europe was covered by a closed forest in its natural state, a conclusion based on pollen studies. It deals with the premises which are used in pollen analysis and describes Iversen's 'Landnam' theory. I also examine whether a pollen collection believed to come from a closed forest could also come from a more open type of vegetation such as a wood-pasture.

Chapter 4 traces the history of what is known as the use of the forest in Western and Central Europe. It examines the premises for the conclusion that humans created pasture in the closed forests, and that this use resulted in the disappearance of the forest.

Chapter 5 examines the theory which states that forests that are no longer exploited, i.e. where the grazing of livestock and cutting wood have ceased, revert to their natural state. This concerns former wood-pasture or scrubland used as pasture. In this respect, I have examined in particular, the way in which the two species of oak and hazel evolve in forests which develop spontaneously without large herbivores.

Chapter 6 deals with the results of an autecological study of the species of trees which are an important component of the deciduous forests of the lowlands of Western and Central Europe, such as pedunculate and sessile oak, broad-leaved lime and small-leaved lime, beech, hornbeam and the hazel bush. These data provide an insight into the mechanisms underlying succession and regeneration in forests, which explain the phenomena in forest reserves. In this chapter, I deal with the autecology of the species concerned in relation to succession in closed forests without large herbivores, and in park-like landscapes, such as wood-pasture grazed by large herbivores.

The vegetation in the east of the United States has been included in these chapters. The reason for this is that the forests in this part of North America are considered an analogy of the untouched vegetation in the lowlands of Central and Western Europe (see Jones, 1945; Whitmore, 1982; Westhoff, 1983; Lemée, 1995; Peterken, 1996, p. 230). From the end of the last Ice Age, during prehistoric to modern times, light-demanding species of oak continuously formed an important component of the vegetation in the east of the United States (see Davis, 1984; Barnes, 1991; Lorimer, 1993; Smith, 1993; Clark, 1997).

Each of these chapters concludes with a paragraph containing a synthesis and conclusions. Chapter 7 synthesizes these results and final conclusions.

Succession, the Climax Forest and the Role of Large Herbivores

2

2.1 Introduction

This chapter deals with the theories of the succession of vegetations and the starting points used in this respect. Succession refers to the succession of stages in the composition of species in vegetation, resulting in a particular final stage (Clements, 1916, pp. 3–4). In this chapter, I attempt to answer the following questions:

- How did we arrive at the conclusion that natural succession in Western and Central Europe resulted in a closed forest?
- What are the premises for this conclusion, and what are they based on?
- Have the theories on succession in vegetation assigned a role to large herbivorous mammals, and if so, what role?

To answer these questions, I deal with the general theories on succession in detail and, in particular, with the theory on the succession from grassland to forest. With respect to the latter, I carry out a detailed examination of Watt's studies (1923; 1924; 1925; 1934a,b), as well as the model for the regeneration of forests drawn up on this basis. The reason for dealing with this in detail is that this so-called 'gap phase' model (Watt, 1947) is generally considered as *the* mechanism underlying the regeneration of natural forests.[1] Therefore Watt's

---

[1] See, *inter alia*, Tansley (1953, p. 405), Peterken and Tubbs (1965), Shugart and West (1977; 1980; 1981), Bormann and Likens (1979a, p. 5), Shugart and Seagle (1985), Borman and Likens (1979a), Flower (1980), Runkle (1981; 1982; 1985; 1989), Whitmore (1982; 1989), Brokaw (1985), Lemée (1987), Lemée *et al.* (1992), Putman *et al.* (1989), Spies and Franklin (1989), Holeksa (1993), Van den Berghe *et al.* (1993), Koop and Siebel (1993), Abe *et al.* (1995), Tanouchi and Yamamoto (1995), Peterken (1996, pp. 91–94, 143–144), Jorritsma *et al.* (1997; 1999), Putman (1996a,b), Pontailler *et al.* (1997), Siebel and Bijlsma (1998), Bradshaw and Mitchel (1999).

publications must contain important convincing elements for the null hypothesis that Europe was covered with a natural closed forest, or should contain points of reference for the alternative hypothesis which postulates that there was a natural park-like landscape.

2.2 History of the Concept of Succession

According to Clements, King's theory on the creation of the 'quaking bog' in Ireland, dating from 1685, is the oldest known study of succession (Clements, 1916, p. 8). He says that it was probably De Luc who first used the term 'succession' in 1806, also in relation to peat bogs (Clements, 1916, pp. 10, 12). Rennie (1810) was the first to write about the succession of forests. His work was based on tree stumps which he found in layers in peat bogs in England. In the 19th century, many tree stumps were found when peat bogs were dug for fuel. According to the Danish writer, Dau, these tree stumps in the peat bogs of Denmark told some of the history of the forest in that country (Iversen, 1973, p. 11). His fellow countryman, Steenstrup (1841), divided the history of the forest into an aspen period, a pine tree period, an oak and an elm period, on the basis of the stratification of tree stumps (Clements 1916, pp. 14–16; Iversen, 1973, p. 12). In 1867, another Dane, Vaupell, established the link between modern forests and the prehistoric vegetation by interpreting the layers of stumps in peat bogs, the so-called 'drowned forests', on the basis of the succession of trees in modern forests. Vaupell (1863) stated that the competition for light was solely responsible for this succession (Clements, 1916, p. 17; Iversen, 1960, p. 7; 1973, p. 12). Vaupell based his interpretation on the book, *Anweisung zum Waldbau*, by the German forester, Cotta (Clements, 1916, p. 17). This book was published in 1816. It is still one of the masterworks in the literature on forestry (Dengler, 1990, p. 16; Mantel, 1990, pp. 141, 173).

Cotta believed that Europe was originally covered by natural forests. In the foreword to his book, he wrote: 'If people leave Germany, it will be completely covered with forest after a hundred years' (Cotta, 1865, p. v). Cotta undoubtedly based this view on the knowledge that cultivated land turns into forest when the land is abandoned. The Roman, Lactanius, had already written about this in AD 300 (Koebner, 1941, p. 24, cited by Darby, 1970). Many of the documents dating from the early Middle Ages up to the 17th century also state that forests take over when fields are abandoned (Gradmann, 1901; Streitz, 1967, p. 40; Darby, 1970; Stamper, 1988). This happened in areas which had become depopulated as a result of epidemics of the plague and war. In the German literature, cultivated land and settlements which had been abandoned and overgrown with trees are referred to as 'Wüstungen' (Hausrath, 1982, pp. 293–297; Rodenwaldt, 1951; Mantel, 1990, p. 64; Jahn, 1991, p. 395). In addition to this anecdotal evidence, in the second half of the 19th century, scientific research into the succession of abandoned fields in Germany, England and the United States also clearly showed that fields and pastures which were abandoned turned into forests (see, *inter alia*, Gradmann, 1901; Spurr, 1952; Tansley, pp. 293–298; West *et al.*, 1980, p. v; Stover and Marks, 1998).

If leaving cultivated land untouched results in a succession to forest, the obvious conclusion, which Cotta also came to (1816), is that there would have been a closed forest if man had not intervened. This conclusion was certainly drawn in the first half of the 19th century, as shown by the words of the Swiss forest ecologist, Landolt. In 1866, he opened his book, *Der Wald. Seine Verjüngung, Pflege und Benutzung* (*The Forest. Its Regeneration, Care and Use*), with the sentence: 'As long as a region remains unpopulated or is only sparsely populated, the majority of the surface area is generally covered with forest, and the first settlers meet their simple needs by hunting the wild animals on land and water' (Landolt, 1866, p. 1). Publications dating from the second half of the 19th century on the history of the use of *uncultivated* land by communities as marks (commons), stated that this uncultivated land was forest (see, *inter alia*, Endres, 1888; Hausrath, 1898). At the beginning of the 20th century, the publications were virtually unanimous in the view that the whole of Europe, including the area of this study, would have been covered with a natural closed forest if there had been no human intervention, and would still be, if humans had not intervened.[2] At that time, descriptions

---

[2] See, *inter alia*, Gradmann (1901), Forbes (1902), Warming (1909, p. 326), Cermak (1910), Moss (1910, p. 36; 1913, pp. 91, 98, 111, 199), Tansley (1911, pp. 7, 65), Dengler (1935, pp. 5–6, 249).

> In Germany, there are no fields which could not easily be turned into forest; if they are allowed to revert to the wilderness, they will automatically become covered with trees as soon as the opportunity for joint approach occurs. This has often been observed, and we can simply assume that the whole of the agricultural land of Germany is destined by nature to be forest, in so far as it is not created by moor or swamps; there is no forester who would doubt this (Gradmann, 1901, p. 363).

> There is little reason to doubt, therefore, what the result of leaving land entirely to Nature would be. So far as indigenous species [of trees] are concerned we have only to fence off a piece of ground from cattle, sheep, and rabbits, and quickly get a sample of indigenous forest of one or other types mentioned above (Forbes, 1902, p. 245).

> These types of forest were forests of oak and forests consisting of Scots pine, beech and oak (Forbes, 1902, pp. 242, 245).

He continued by saying:

> Even when unfenced, thousands of oak, ash, beech, and other seedlings spring up in every pasture after a good seed year, and where seed-bearing trees are within a reasonable distance. Such instances prove the capability of Nature to reassert herself whenever she gets the opportunity, and there is little doubt that this country would regain its original conditions in a hundred years or so if men and domestic animals were to disappear from it (Forbes, 1902, p. 245).

> Pastures in the plains of northern Europe, and other regions that were formerly clothed with forest, are almost without exception artificial products: were the human race to die out they would once more be seized by forest, just as their soil was originally stolen from forest. Exceptions to this rule are provided only by small patches of meadow in old forests, that have been regularly grazed over and manured by wild animals (Warming, 1909, p. 326).

'In the case of established woods, we do not know the progressive associations which culminated in the woodland associations; but we can determine retrogressive stages through scrub to grassland' (Moss, 1910, p. 36). 'Whilst opinions may differ as to whether or not the grassland just described is wholly or only in part due to man's interference, it appears to be generally accepted that such tracts were formerly clothed with forest', ... after which Moss quoted Warming's above-mentioned words (Moss, 1913, p. 111). 'There is no doubt that by far the greater part of the British Isles was originally covered with forest ... ' (Tansley, 1911, p. 65).

by the Romans were also used to support this theory. For example, Caesar and Tacitus are presumed to have written, respectively, in *Bello Gallica* and *Germania* that the uninhabited parts of the lowlands of Europe were covered by extensive dark forests, which they referred to as *silva* (silva Arduenna, silva Caesia) (see Gradmann, 1901, p. 369; Forbes, 1902; Cermak, 1910).[3] In fact, the modern meaning of silva includes forest (Muller and Renkema, 1995, p. 855).

2.3 Clements's Theory of Succession

The development of vegetation from an initial stage of bare soil in a field to a particular end situation, through the invasion of different species of plants, including bushes and trees, has often been observed, and was formalized by Clements in his theory of succession. He believed that one particular type of vegetation which is climatologically determined, develops on the bare soil in a number of successive stages (succession); this is the climax (Clements, 1916, p. 125). The climax is achieved when one species dominates to such an extent that it excludes the introduction of other possibly dominant species (Clements, 1916, pp. 103, 105). According to Clements, this climax continues to exist as long as the climate remains constant (Clements, 1916). He maintains that the succession develops from 'simple' life forms, such as fungi and single-cell organisms, through annual and perennial species of plants, culminating in the dominance of the highest life form that is possible in the prevailing climate. If these are trees, the climax is forest (Clements, 1916, pp. 80, 99, 103, 105, 125).

In his view, animals and people are part of the succession and the climax, and the intervention of humans in particular results in the destruction of the climax. Human intervention can also be responsible for the succession coming to a halt at a particular stage, resulting in a so-called sub-climax; i.e. a situation which precedes the climax (Clements, 1916, p. 107). According to Clements, the sub-climax can have a permanent character if human intervention has resulted in irreversible changes (Clements, 1916, p. 108). Clements made a distinction between primary and secondary succession (Clements, 1916, p. 60). He believed that this distinction was based on whatever had caused the bare ground which

---

[3] Gradmann (1901) wrote that in Roman times there was a clear boundary between densely populated areas on the one hand, and virtually uninhabited landscapes covered with forests, on the other hand. He believed that the Romans described these virtually uninhabited areas in their empire and in Germania with the term silva (silva Arduenna (Ardennes Forest), silva Caesia (forests along the Rhine), Semana silva (Thüringen Forest), Martiana silva (Black Forest). Gradmann quoted Tacitus and his book, *Germania*, in which he described Germania as: 'Terra et si aliquanto specie differat, in universum tamen aut silvis horrida aut paludibus foeda' (Gradmann, 1901, p. 369). Cermak (1910) translated this sentence as follows: 'The land has very different forms in different parts, though in general, it was covered with dark virgin forest or wild bogland' (Cermak, 1910, p. 365). Partly on the basis of this view, he noted that both Caesar and Tacitus 'discovered virgin forest in the heart of Europe which, until they were interrupted by bogs in the lowlands, was virtually unbroken, or at least extended without any major interruptions' (Cermak, 1910, p. 365). Forbes (1902) wrote that the earliest reference to British forests can be found in Book I of Caesar's *Bello Gallica*, where mention is made of the 'vast forests which covered the country at that time' (Forbes, 1902, p. 244) (see Chapter 4 for a further explanation of the term 'silva').

started the succession. When the soil is completely new, without a seed bank, he calls this primary succession. The first seeds to be established are the seeds of pioneering species. He believed that primary successions require an extremely long period of development, for example, because of the lengthy process of creating soil. Secondary succession takes place on bare soil which was covered with a climax, but which was destroyed, for example, by fire, tidal waves, animals or human activity. Generally, these areas contain large numbers of viable seeds and spores of species from several stages of the succession, and therefore they result in relatively rapid series of developments (Clements, 1916, pp. 60, 102, 168–169). If there is a factor which prevents a return to the climax, the result is an apparent climax. This disappears as soon as the interruption ceases.

According to Clements, the destruction of climaxes by humans can result in numerous sub-climaxes; the most striking examples include burning vegetation, felling trees and grazing. This is why he believed that grasslands have developed everywhere in the world that will remain as long as the burning and grazing continue (Clements, 1916, pp. 107, 145). The disappearance of the cause of the destruction will at the same time initiate the development of a series of different types of vegetation which will eventually again result in the climax.

According to Clements, there is no retrogressive development from forest to grassland, e.g. under the influence of grazing by livestock (Clements, 1916, pp. 145–146, 155). In this respect, he disagreed with researchers such as Moss, Rankin and Tansley (Moss *et al.*, 1910), who stated that the forest which originally covered Great Britain was mainly destroyed by livestock (Tansley, 1911, p. 7; Moss, 1913, p. 91). In comparison with countries such as Germany and France, Great Britain had very little forest left in about 1900 (Moss *et al.*, 1910; Tansley, 1911, pp. 64–65). These researchers believed that this destruction was caused by the process of retrogressive succession from forest via scrub to grassland (Moss, 1910, p. 36; 1913, pp. 96–98; Moss *et al.*, 1910; Tansley, 1911, pp. 7, 83–84). Clements did not believe that this had been proved in any way, and he maintained that Moss *et al.* (1910) themselves said that there were no natural forests left in England at all. He considered that the Europeans did have a good working hypothesis, although it would be possible to accept this only after thorough and exacting ecological research (Clements, 1916, pp. 155–156). In England, Salisbury (1918) agreed with Clements's criticisms. He thought that Clements was probably right to say that scrub is not a retrogressive stage of the forest, but a stage in the succession to forest.[4] At that time, succession in

---

[4] Salisbury (1918) started his article with:

Scrub is a particular type of plant-association which is dominated by shrubs, either forming a dense growth or sparsely scattered amongst herbage. The condition has often been regarded as a retrogressive phase of woodland, a view expressed by Tansley, Moss and others (cf. 'Types of British vegetation', pp. 83 and 130), and undoubtedly the scrub-association does often occupy situations formerly held by woodland. But probably Clements is right in regarding such vegetation not as a retrogressive but as the establishment of an earlier phase in a progressive succession (cf. F.E. Clements, 'Plant Succession', Carnegie Institute, Washington, 1916). Nevertheless though scrub, such as the particular examples we shall consider, is often a stage in the passage from pasture, or even arable land, to woodland, there probably are and always have been areas where this condition by some factor or factors, it may be natural or it may be artificial, which prevent the transition to closed woodland (Salisbury, 1918, p. 53).

areas grazed by livestock where seedlings of bushes and trees developed spontaneously was put forward as proof for retrogressive succession (Forbes, 1902; Tansley,
1911, pp. 7–8; 1922; 1953; p. 487). It was seen as the beginning of the return to
forest (Tansley, 1911, pp. 83–84; 1922; 1953, pp. 133, 295, 373, 398, 476, 487).
The succession of grassland to forest was also believed to reveal that cattle formed a
disruptive factor introduced by humans (see Forbes, 1902; Warming, 1909, p. 326;
Tansley, 1911, pp. 8, 66; 1935; Moss, 1913, p. 91).

According to Moss *et al.* (1910), and Tansley (1911, p. 71), British forests
were descended in a direct line from the original forests. They called these forests
semi-natural forests. Although they were exploited, these semi-natural forests, in
contrast with planted forests, were thought to have retained their essential characteristics because of the conservatism of British landowners and the backward
state of British forestry practice.[5] The forests that Moss *et al.* (1910) and Tansley
(1911) were writing about at that time, consisted largely of coppice-with-standards, i.e. coppice with a few trees that were left standing. These were usually
pedunculate or sessile oaks (*Quercus robur* and *Q. petraea*); the scrub consisted
mainly of hazel (*Corylus avellana*) (Moss *et al.*, 1910; Peterken, 1981, p. 108).

2.4 Tansley's Polyclimax Theory

Tansley (1935) defended himself against Clements's criticisms of the concept of
retrogressive succession by stating that livestock are *continuously* present, and
therefore a *continuously* active factor, and that forest can therefore change into
grassland, slowly but surely. He therefore also criticized Clements's mono-
climax concept. He believed that a succession can be interrupted by catastrophes which bear no relation to the laws governing the succession, and which
are responsible for the changes in the vegetation. According to Tansley, such
catastrophes included fire, storm and human influences, such as mowing and
grazing by livestock. He believed that this sort of intervention could result in a
different, stable final situation. In his view, catastrophic factors, such as humans
and their domestic animals, can, under certain climatological conditions, create different climaxes from those which are completely subject to the climate.
On this basis, Tansley defined the climax as 'the highest stage of integration and
the nearest approach to perfect dynamic equilibrium that can be attained in a
system developed under the given conditions and with the available components' (Tansley, 1935, p. 300). The difference between Clements and Tansley is
that Tansley considered climaxes as situations which were almost in balance
with *all* the environmental factors, and not, as Clements believed (1916), only
with the climate. Tansley did agree with Clements that the succession of plant

---

5   One general conclusion at which we have arrived is that the existing English woodlands
     have for the most part been altered in their essential characters to an extent which may
     appear surprisingly slight to those unfamiliar with the actual facts of distribution. This is
     no doubt largely due to the innate conservatism of the English landowner, as well as to
     the backward state of forestry practice in this country (Moss *et al.*, 1910, p. 118).

communities results in a climax community which is dominated by the tallest plants that can survive in a certain area under the prevailing conditions. In temperate climates, these are trees. Therefore he believed that in the temperate latitudes, the climatologically determined climax is forest. If there are also other factors involved, such as humans and livestock, a climax of grassland or heathland can develop (Tansley, 1935).

Tansley rejected Clements's concept of biotic communities, because he believed that plants and animals played a completely different role. In so far as there is a relation between plants and animals, he considered that this was a one-way relationship, because animals are completely dependent on plants. Therefore, in his view, animals follow the development of vegetation and do not play a determining role in the succession. According to Tansley, the term, 'living community', suggests a reciprocal influence between plants and animals which does not exist. In his view, the relationship between the two is no different from that with the abiotic environment.[6] He believed that the system concerned comprises plants and animals, as well as the abiotic environment. On the basis of this view, Tansley is generally seen as the founder of the concept of the ecosystem (Kingsland, 1991, p. 11; Kingsolver and Paine, 1991, p. 310). Nowadays an ecosystem is seen as a system which consists of a *biotic community and its abiotic environment* (see Krebs, 1972, p. 10; Begon *et al.*, 1990, p. 613). It is ironic that the modern meaning of the term *ecosystem* once again includes the concept of a biotic community rejected by Tansley.

Tansley did admit that animals could have an influence on the structure of the vegetation, but that there is no question of this in the usual (i.e. natural) situation.[7] If animals have an effect, he considered that these are exceptional situations, i.e. where there are livestock or a high density of wild animals. According to Tansley, there is no natural equivalent of the effect of livestock.[8] Red deer (*Cervus elaphus*) and roe deer (*Capreolus capreolus*) are woodland animals which occur in such great numbers in the natural situation that there is an equilibrium between the seedlings that are eaten and the regeneration of the woodland, so that the survival of the forest is not jeopardized (Tansley, 1953, p. 143). If red deer and roe deer do prevent the regeneration of the forest, he believed this is an unnatural situation.

---

[6] 'But is it really necessary to formulate the unnatural conception of biotic community to get such co-operative work carried out? I think not. What we have to deal with is a system, of which plants and animals are components, though not the only components' (Tansley, 1935, p. 335).
[7] 'This is not to say that animals may not have important effects on the vegetation and thus on the whole organism-complex. They may even alter the primary structure of the climax vegetation, but usually they certainly do not' (Tansley, 1935, p. 335).
[8] 'grassland or heathland have no doubt originated mainly from the clearing of the woodland, and the pasturing of sheep and cattle. ... In some cases where grassland is not pastured, the shrubs and the trees of the formation recolonize the open land, and woodland is regenerated' (Tansley, 1911, p. 7). '... if pasturing were withdrawn their areas would be invaded and occupied, as they were *originally* [italics by the author] occupied, by shrubs and trees' (Tansley, 1953, p. 487).

2.5 The Lack of Regeneration in the Climax Forest

From 1908 to 1920, Tansley studied a number of permanent quadrats in limestone grasslands bordering on a beech forest (Tansley, 1922). Tansley wished to examine the effect of the end of grazing, particularly by rabbits, on the proliferation of woody species of plants and, in particular, the beech forest. He found that when the grazing ended, this resulted in the spreading of the scrubland. After a number of years, young beech trees (*Fagus sylvatica*) appeared. It was partly on the basis of his findings in that study that he suggested to Watt that a study be carried out into the regeneration of oak and beech forests which were considered to be climaxes (see Forbes, 1902; Moss *et al.*, 1910; Tansley, 1911, p. 79; 1922). Because they were climaxes, these forests had to survive by means of regeneration. However, there was no regeneration of oak trees and beech trees in these forests, and the question arose as to why this was the case.

According to Watt (1919), the failure of the oak forests to regenerate was due to the general degeneration and disappearance of woodlands. In his opinion, this could be attributed to the shipbuilding industry and to farmers turning woodland into pasture for their livestock in the past. He supported the theory that forests had been destroyed by livestock grazing, which appeared in a publication by the German author, Krause (1892a) (a source which Watt in his turn borrowed from Moss (1913, p. 91)), as well as the fact that the forests in England, like those in Germany, were used for grazing cattle and pigs.[9] Watt (1919) was surprised that in certain years the floor of the oak forests was strewn with acorns, and that millions of seedlings appeared the following spring, while the spontaneous regeneration of oak trees was rare in those forests. He studied the predation of acorns and seedlings by animals, and the lack of germination of the acorns because they dried out. As a result, he concluded that apart from diseases, it was mainly the animals that eat acorns that are responsible for the failure of the regeneration of oak forests. Because these herbivorous animals, both wild and domesticated, were no longer kept under control by their natural enemies, he believed they were increasing proportionately, and therefore destroy any vegetation which could serve as a source of food. Thus he saw humans as being punished for killing and suppressing the predators which would normally have kept the number of herbivores under control.

Apart from the failure of the regeneration of oak trees *in* the forest, Watt (1919) also referred to the successful regeneration of oak trees *outside* the

---

[9] Moss wrote:

> The inability of certain forests to rejuvenate per se has been pointed out by many foresters and plant geographers. In discussing the causes of the succession of forest to heath in north Germany, Krause (1892) emphasized the view that the narrowing of the forest area has been largely due to errors in sylviculture, especially to the grazing of cattle in the forest. That such a factor is a causa vera in the degeneration of forests is undisputable (Moss (1913, p. 91).

Watt (1919) subsequently wrote:

> Krause (1892) attributes the limiting of the forest area in North Germany to 'errors in sylviculture, especially to the grazing of cattle in the forest,' and this may be legitimately applied to this country as cattle and pigs were formerly driven into the woods for pannage. This is generally concluded as a factor responsible (Watt, 1919, p. 174).

forest, i.e. in grassland. In his opinion, acorns not only germinate there in large numbers, but the seedlings thrive there when the grazing stops. He believed that there were no problems for oak trees to recolonize grasslands and scrub in grasslands. He supported this observation with a quotation from Forbes (1902), which showed that grassland almost seems to be the natural seed bed for oak.[10] According to Watt (1919), oaks also germinate successfully in grassland which is lightly grazed by cattle during the summer. Therefore Watt assumed that there was regeneration of trees, including oaks, *with* grazing. He also studied the New Forest where he noted that young oaks developed among thorny and spiny shrubs, such as hawthorn and blackthorn, which protected them against cattle, ponies, deer and sheep roaming through the forest. In addition, he mentioned that many others had also observed this.[11]

Watt did not devote any further attention to this phenomenon with regard to the question: why oaks do not regenerate. In this respect, he looked only at regeneration *in* the forest. The reason for this must be that in his view, oaks should regenerate *in* and not *outside* the forest, because the climax is forest. With regard to the failure of the regeneration of oaks in the forest, this put an end for Watt to the conclusion that the cause is the disturbed equilibrium between herbivores browsing on acorns and oak seedlings, and the predators which prey on these herbivores; an explanation which was supported by Tansley (see Tansley, 1953, pp. 141, 291–293).[12]

2.6 Secondary Succession from Grassland to Forest

Watt studied the failure of beech woods to regenerate in the South Downs in the south of England (Watt, 1923; 1924; 1925) and in the Chilterns, west of London

---

[10] Forbes (1902) said: 'A grassy surface seems the natural seed bed of oak, for very successful examples may often be seen on rough pasture adjoining woods which for some reason or other has been allowed to lie waste or is only slightly stocked with cattle during the summer' (Watt, 1919, p. 175).

[11] Supplementary evidence as to the value of protection [by thorns] was patent from observations made in the New Forest towards the end of August. Among such spiny plants as Ilex aquifolium, Prunus spinosa, Crataegus monogyna, etc., usually near the periphery of clumps of these, saplings [of oaks] of various heights were found growing up among the protecting branches. Their demands for light led them to incline their stems to the outside of this protection but rarely did I find any protruding twigs which would be liable to be nibbled by the cattle, ponies, sheep, deer etc., which roam through the Forest, or if they do project they are promptly eaten back. Once these branches emerge from this protection sufficiently high up to escape the browsing animals, the future of the tree is assured. It is no uncommon thing to find a large oak standing in the centre of such a clump, an oak which has grown up with the thorny species, the latter affording the necessary protection. This phenomenon has been recorded by numerous observers for trees in general and undoubtedly the protection thus afforded was the salvation of the oaks in question (Watt, 1919, pp. 196–197).

[12] On this subject, Tansley wrote:

particularly of the widespread mice and voles, are one of the main cause of general failure of oakwood to regenerate: and there can be no doubt that the constant war carried out by gamekeepers against the carnivorous birds and mammals which prey upon the small rodents has contributed in an important degree to preventing regeneration in the existing English oakwoods (Tansley, 1953, p. 293).

(Watt, 1934a,b). As for his study of the oak forests, Watt (1923) looked at all the factors which he considered could be responsible for the failure of the beech forests to regenerate. He believed that under a closed canopy, these included the eating of beech nuts, particularly by mice, and the destruction of seedlings by a combination of being eaten by insects, and the reduced vitality of the seedlings because of lack of light. Although all the seedlings died in some years, Watt believed (1923) that there *must* be some natural regeneration, because the beech forests in that area were characterized by a great degree of continuity.

Watt was able to explain successful regeneration by years where there had been a full mast. He believed that in those years there was so much seed that even after the mice and birds had eaten their fill, there was still plenty left for seedlings to become established. He noted that the regeneration of beech trees took place in gaps in the canopy. According to Watt, this regeneration was seriously hampered by the fact that as a result of the increased amount of light in the gap, there was a great proliferation of species such as dog's mercury (*Mercurialis perennis*), common enchanter's nightshade (*Circaea lutetiana*), bracken (*Pteridium aquilinum*) and brambles (*Rubus* sp.). In addition, trees such as the ash (*Fraxinus excelsior*), birch (*Betula* sp.) and sycamore (*Acer pseudoplatanus*) became established. According to Watt, their canopy allowed only a limited amount of light to penetrate, which is far below the minimum that seedlings require to thrive (Watt, 1923). In grassland scrub, he believed that the canopy is so dense that the beech has no chance of becoming established.

Watt then turned his attention from woodland to grassland (Watt, 1924). In his opinion, the climax forest returns there when grazing stops. He believed that this secondary succession could therefore probably shed some light on the question of how the climax is regenerated. He then deduced the recovery of the climax from chronosequences which reflected the secondary succession from grassland to beech forest. He believed that the succession started with the establishment of scrub in grassland, followed by the establishment of oaks and ash. Eventually these form an oak–ash forest in which beech appears next. Beech trees can become established because the oak and ash grow taller, their crowns overlap, and the shade of the closed canopy makes the scrub more open. Because of their shape, Watt called the oldest, and therefore the first beech trees to appear, pioneer beeches. Their crown started low down on the trunk, and developed as a broad crown because the trees had grown in light conditions. According to Watt, these pioneer beeches were then surrounded by clusters of young beech trees. In his opinion, these 'beech families' formed the start of the beech climax in which beech trees are replaced only by other beech trees.

Watt (1924) found trees everywhere in the scrub. In his view, oak and ash are constantly being established on the outer edge of the scrub extending into the grassland. These young trees are protected against being eaten by herbivores by the thorny scrub. He considered that this protection is essential for the establishment of these species of trees. He believed that the rate at which oak and ash advance into the grassland is equal to the rate at which the outer edge of the scrub, the fringe, extends into the grassland. As this moves further into the grassland, trees which have been established are found further and further *within* the scrub. When their crowns have formed a continuous canopy, this is a forest. Eventually, the trees dominate the vegetation in this way in places where there was grassland before. As these forests and scrub become older, Watt believed that the crowns of the trees become thinner. Then beech seedlings appear. Because they grow more rapidly, the beech trees then overtake the oak and ash, and eliminate them. Therefore, according to Watt, the succession series results in the complete dominance of beech, both in limestone soil and in the leached soil of the South Downs. He described this development as: 'Grassland → scrub associes → ash-oak associes → beech associes → beech consociation' (Watt, 1924, p. 149). Thus, according to Watt, the beech forest is the climax vegetation and *the oak does not form part of this* (author's emphasis).

According to Watt (1925), virtually the only seedlings to survive in the forest in the gaps in the canopy are beech seedlings, when this gap is first colonized by ash, sycamore or beech. The few oaks which appear between the ash and beech disappear very quickly; the ash survive a little longer. However, after a few years, they are also overshadowed by young beech trees from the periphery of the gap (see Figs 2.1 and 2.2). In the end, only the odd ash could survive. In his view, no seedlings survive under the closed canopy of the climax vegetation itself. He found only beech

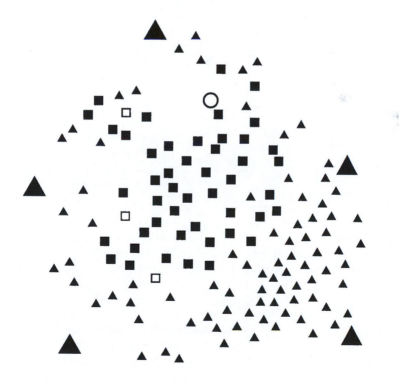

Fig. 2.1. The regeneration of trees in a gap in the canopy of a beech forest which is seen as the climax. Young beech trees appear on the periphery, and young ash grow in the centre. See Fig. 2.2 for key to symbols.

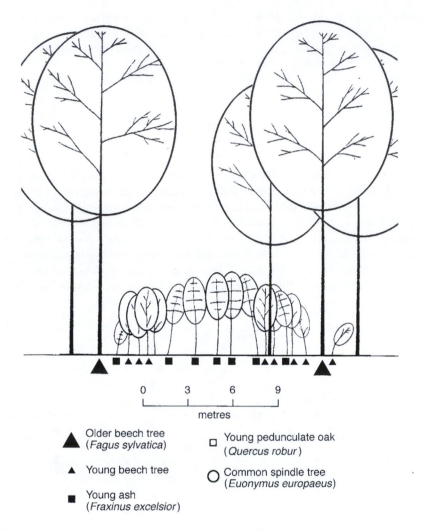

0   3   6   9

metres

▲ Older beech tree   ▢ Young pedunculate oak
   (*Fagus sylvatica*)   (*Quercus robur*)

▲ Young beech tree   ◯ Common spindle tree
                                (*Euonymus europaeus*)

■ Young ash
   (*Fraxinus excelsior*)

Fig. 2.2. The regeneration of trees in a gap in the canopy of a beech forest which is seen as the climax. Young beech trees grow directly under the old beech trees. As there is more shadow, the young beech trees do not thrive as well (redrawn from Watt, 1925, pp. 31, 32).

seedlings up to 5 years old. In some places, he did find numerous young oaks in the gaps in the canopy. He saw these as paving the way for beech because the beech win the competition against the oaks. Thus, at the climax stage, the forest consists of 90% beech trees with only a sporadic oak or ash (Watt, 1925).

According to Watt (1924), the so-called pre-climax stages, which always consist of thorny and spiny vegetation, form the start of the climax. The most important species are: (thorny) one-styled hawthorn (*Crataegus monogyna*), blackthorn (*Prunus spinosa*) and juniper (*Juniperus communis*), and (non-thorny) yew (*Taxus baccata*) and common dogwood (*Cornus sanguinea*) (Watt, 1924). According to Watt (1924), these are the pioneers. He thought that the spiny and thorny species became established in grasslands when grazing by sheep came to an end, was greatly reduced, or the sheep were replaced by cattle. By postulating that cattle are less choosy than sheep, Watt explained the establishment of shrubs which takes place when the area is grazed by cattle. However, he believed that the effect of the cattle could be undone by rabbits.[13]

According to Watt (1924), grazing prevents the development of both scrub and oak–ash forests. Therefore he saw the evolution of scrub and woodland in grassland as phenomena which appear when land is no longer cultivated.[14] Nevertheless, Watt observed that when the land was grazed, trees were regenerated in thorny scrub. In this respect, he noted that this grazing slowed down the spread of the scrub, and thus the colonization of grassland by forest. He considered that, as a result, the colonization is limited to the periphery of oak–ash forests.[15] In view of this, the regeneration of trees in scrub would not be hampered by grazing, but merely slowed down, in comparison with grassland which was not grazed. He interpreted the emergence of thorny shrub and forest in grassland grazed by cattle as a stage in the succession from grassland to the climax forest, which would be completed when the grazing came to an end. He did this because there seemed to be a reduction in the pressure of grazing in comparison with grazing by sheep. Therefore Watt included grasslands which

[13] And there is evidence that colonisation of grassland by woody plants has in the immediate past been due to sudden releases of biotic pressure, e.g. the abandonment of cultivated ground, and the reduction in the number of sheep and their partial replacement by the more fastidious cattle, although an excessive number of rabbits tends to minimise these effects (Watt, 1924, p. 182).

[14] How far man directly affected the vegetation of the S. Downs in the past awaits the unveiling of historical records for an answer, but it is quite evident that enclosure and cultivation have occured in the past where there now is high forest, scrub and grassland. [...]. In the eastern half, however, [...] most of the land was brought under cultivation during the Napoleonic war and the abandonment of some of it has occured within living memory. Ground was given up during the period of agricultural depression following 1870 is now covered with scrub and is being invaded by trees (Watt, 1924, p. 147).

[15] Watt (1924) remarked:

The scrub associes with certain noteworthy exceptions forms a marginal zone of varying width to the woodlands of ash-oak and beech. In certain localities the scrub community exists surrounded by grassland, but the number of such in which trees have not gained a footing is very small and when they occur it is at some distance from the nearest woodland, where withdrawal of grazing animals, whether due to economic conditions or simply to the advance of the scrub itself has permitted more luxuriant growth of the latter, this community occupies considerable areas (Watt, 1924, p. 154).

Another citation which illustrates this is:

The influence of the animal factor upon the limitation of this community has been described by Tansley (1922) and between animal pressure on grassland and beech pressure from behind scrub is eliminated or reduced to the merest fringe, capable as such of affording protection to young trees but necessarily retarding the rate of the invasion and succession of grassland by woodland (Watt, 1924, p. 155).

contained thorny shrub and trees while being grazed by livestock, in the chronosequences which he used to reconstruct the succession from grassland to forest after the grazing stopped. He interpreted the phenomena in these grasslands on the basis of the theory prevailing at the time, i.e. that grassland evolved when the forest that was originally present was destroyed by livestock, and that livestock prevent the return of the forest. In his opinion, the regeneration of trees where grazing took place was the beginning of the return of the forest which existed before man introduced livestock. Therefore he interpreted the regeneration of trees in thorny scrub where there is grazing by livestock as an analogy of the return of the original forest after grazing has stopped in the grassland, because what happened when livestock were grazing should, according to the prevailing theory, take place after the end of grazing by livestock, i.e. the establishment of thorny scrub and the regeneration of trees there. In my opinion, this is why he included these grasslands in his chronosequences, which show the succession from grassland to forest after the end of grazing.

The different effects of grazing by sheep and cattle as observed by Watt (1924) can be explained by the different effects both species of large herbivores have on woody plant species. Sheep browse and peel bushes and trees much more than cattle (Buttenschøn and Buttenschøn, 1978, 1985; Mitchell and Kirby, 1990). For this reason, there were no wild shoots of young trees at all (beech and Scots pine) on dry poor sand ground with a sheep density of 22 kg ha^{-1}, while at a density of horses and cattle of 100 kg ha^{-1}, this took place without problems (Van Wieren and Borgesius, 1988, quoted by Van Wieren, 1996). Sheep also browse prickly species such as blackthorn and juniper much more than cattle. They eat a larger percentage of the annual shoot growth on the blackthorn and are even able to completely eliminate wild shoots from rhizomes. An explanation for this difference could be that it is easier for sheep to stick their narrower snout and incisors between twigs and branches than it is for cattle (Buttenschøn and Buttenschøn, 1978; Coops, 1988; Mitchell and Kirby, 1990). The increased effect on woody plant species will have been the reason why during the 17th century sheep pasturage was strictly limited on common ground from which firewood and timberwood were also extracted (see Chapter 4). In these situations, sheep were often treated like goats (Endres, 1888, p. 113; Mantel, 1980, pp. 134, 439; 1990, p. 97; Buis, 1985, pp. 130, 353).

Watt (1934a,b) verified the results of his study of the South Downs in grasslands, scrub and woodland, on the slopes and on the plateau of the Chilterns. On the basis of his own observations, he concluded that in principle the succession of open grassland to beech forest took place in the same way in both places, i.e. in stages from spiny and thorny scrub of hawthorn, blackthorn and juniper. In shallow, drier soils, richer in limestone, with little humus and a short carpet of grass, he believed that the succession took place via the scrub, which consisted largely of juniper, and in deeper, damper soil which was richer in clay, via scrub consisting mainly of hawthorn. When it is intensely grazed, the dominance of hawthorn in the scrub tends to be replaced by juniper. According to

Watt, blackthorn was very common in both types of scrub, and even suppressed the juniper in some places (Watt, 1934a).

As in the South Downs, Watt found that in the Chilterns, there were only beech seedlings in the gaps in the beech forest. He did not find any influence of grazing animals on the vegetation of the forest floor, which he attributed to the fact that the plants which grew there were inedible because they contained toxic elements for the herbivores (Watt, 1934a). Thus, here too, Watt found that scrubland became established where there was grazing with the regeneration of trees including oaks. According to Watt (1934b), oaks were found mainly on the so-called 'commons' on the plateau of the Chilterns. 'Commons' are traditional grazing grounds for common use in the vicinity of settlements. According to Watt, the oak was the most successful colonizer of grassland there. He believed that the importance of scrubland on this common grazing land for the development of oak forests was revealed by the fact that five of the six forests which had developed there had evolved from scrubland. Watt saw the dead vestiges of hawthorn and blackthorn under the old oak trees as proof of this, as well as the fact that young trees, particularly oaks, grew on the periphery of the scrubland which was advancing into the grasslands. Although the oak was the most successful colonizer of grassland on the plateau, Watt (1934b) claimed that the beech dominated everywhere. He based this conclusion on the observation that there were young beech trees growing under the oaks in a planted oak forest, and he believed that they would eventually suppress the oaks. Apart from the fact that the oak would be the loser in the competition in the beech climax, Watt believed that spiny and thorny scrub with hawthorn, blackthorn and dog rose (*Rosa canina*) would also disappear under the canopy of the beech forest. Thus he believed that all these species occur exclusively in the pre-climax, which has the greatest wealth of species. This applies both for shrubs and trees, and for grasses and plants. Therefore in the climax stage, there are no new species of trees or shrub; in his opinion, the number of species only declines.[16]

Watt (1947) synthesized the results of his study in the South Downs and the Chilterns in his article, 'Pattern and process in the plant community', which is seen as a milestone in ecology (Bormann and Likens, 1979, p. 5; Greig-Smith, 1982; Begon *et al.*, 1990, p. 647). In his opinion, the vegetation is constantly subject to a cyclical process of a pioneering stage, a developmental stage, a maturation stage and a degeneration stage. He called this total cycle, 'the regeneration complex'. In his opinion, there is no rejuvenation in these stages. This occurs only in gaps in the vegetation. In the beech forests, these are gaps in the canopy which occur when one or more trees die or are blown over. In his opinion, this stage should be added to the regeneration complex. He called it the 'gap phase'. According to Watt, it is the last stage. He believed that the different

---

[16] Watt (1934b) remarked in this respect: 'No tree or shrub makes its first appearance either in the beech associes or in the beechwood' (Watt, 1934b, p. 490). For the individual species, see the tables in Watt (1924, pp. 156–157, 169, 172–175, 192–193; 1925, pp. 33–35, 50–51; 1934a, pp. 241–242; 1934b, pp. 455–457, 463, 471, 476, 480).

stages of the regeneration complex are not synchronized, so that a natural forest consists of a mosaic of stages. He thought that only the species which can survive as seedlings in the gaps in the canopy determine the species diversity of the climax forest (Watt, 1947).

Apart from Watt, Adamson (1921; 1932) also noted that in grasslands where animals grazed, hawthorn and blackthorn spread and trees regenerated there. He noted the regeneration of field maple (*Acer campestre*), pedunculate oak, beech, ash, common dogwood, hazel, birch, sallow (*Salix caprea*), guelder rose (*Viburnum opulus*), privet (*Ligustrum vulgare*) and wild pear (*Pyrus pyraster*). Fenton (1948) noted the regeneration of trees and shrubs in grazed areas in gorse (*Ulex europaeus*). Tansley (1953) also wrote that where there is grazing, thorny and spiny scrub develops with hawthorn, blackthorn, gorse, brambles and all sorts of roses, and young trees grow there. In his opinion, this scrub is not a thicket which covers everything, but consists of vegetation with alternating grassland and scrub, where trees can grow. Tansley described the grazing which occurs here as rough grazing (Tansley, 1953, pp. 130–133, 489).

2.7 Regeneration of the European Primeval Forest

As I indicated earlier, and illustrated with quotations from Cotta (1865) and Landolt (1866), the forest ecologists of Central Europe at the beginning of the 19th century certainly supported the currently prevailing theory that the original natural vegetation in Europe was a closed forest. Dengler (1935) referred explicitly to what he called Cotta's famous assertion (1816) that if Germany were to be abandoned, it would be covered by forest within 100 years. In his opinion, this prediction still applied.[17] In the first half of the 20th century, forest ecologists studied the regeneration of what they considered to be the last European primeval forests. These were the forests in the mountains of southeast Europe, particularly in the Balkans, covered by beech, Norway spruce (*Picea abies*) and silver fir (*Abies alba*). One of these forest ecologists was Cermak (1910), referred to above. He saw the primeval forests in the Balkans as being representative of the original vegetation of the lowlands of Central and Western Europe, because he maintained that Caesar and Tacitus had described the lowlands of Central Europe, stating that they were covered with virtually uninterrupted, dark primeval forests.

---

[17] Dengler (1935) said the following about Cotta (1816):

> Das dritte der älteren Hauptwerke, das nun auch zuerst den Namen 'Waldbau' trägt, ist das von Heinrich Cotta, dem ersten Direktor der 'Sächsischen Forstakademie Tharandt. [. ...] Es beginnt mit dem berümt gewordenen Wort: 'Wenn die Menschen Deutschland verließen, so würde dieses nach 100 Jahren ganz mit Holz bewachsen sein', eine Gedanke, den auch das vorliegende Buch [Waldbau auf ökologischer Grundlage] zum Ausgangspunkt für die Stellung des Waldes in der Natur genommen hat (Dengler, 1935, p. 249).

The motivation for this study was to answer the question: whether and to what extent the techniques used for the regeneration of productive forests in the European lowlands can be considered analogous to the processes of regeneration in natural, virgin forests. These techniques for regeneration were developed in Germany at the beginning of the 18th century. They consisted of allowing a new generation of trees to grow from the seeds which had been spontaneously shed by the trees which remained standing in productive forests. In forestry circles, these techniques are known as 'natural regeneration'. They were introduced in France and England in the 19th century (see Forbes, 1902; Reed, 1954, p. 48). Gaps are made in the forest canopy to initiate this regeneration.[18] In Chapter 4, I will return to this technique in detail.

Cermak (1910), Müller (1929), Hesmer (1930), Fröhlich (1930), Markgraf and Dengler (1931) and Rubner (1934) found that the regeneration of trees in primeval forests took place where there were gaps in the canopy. They felt that there was no regeneration in these primeval forests, because there were virtually never any younger trees. They appeared only when the forest canopy became thinner as the trees grew older, and gaps were created when individual trees died or fell over (Cermak, 1910; Fröhlich, 1930; Markgraf and Dengler, 1931; Rubner, 1934).

The Swiss forest ecologist, Leibundgut (1959), was the first to describe in detail how the primeval forests in the mountains went through a number of stages which can be clearly distinguished in terms of structure (Zukrigl, 1991). Leibundgut described his theory in two articles (1959; 1978). He made a distinction between the regeneration of the forest, with and without catastrophes, such as fire, storm and plagues of insects. In the case of catastrophes, he thought that the succession started on the bare surface with the establishment of less light-demanding species like birch, aspen and willow. He called this stage 're-establishment and colonisation' ('Vorwaldstadium'). This stage progresses through a number of transitional stages, where increasing numbers of species which tolerate shade become established, into the closed forest, which he called the 'Schlußwald'. In the terms of Clements (1916), Tansley (1935) and Watt (1947), this is the climax (see Dengler, 1935, p. 5). By this stage the forest consists only of shade-tolerant species. Leibundgut called this stage the 'optimal phase' ('Optimalphase'). In his opinion, this is followed by the 'ageing phase' ('Alterphase'), when the trees lose their canopy and move into a phase of degeneration which Leibundgut called the 'breakdown and die-back phase' ('Zerfallphase'). As the canopy becomes thinner, more light penetrates to the forest floor and this is when the regeneration begins, partly because of the gaps which appear in the canopy when trees die. During the next, so-called 'regeneration phase' ('Verjüngungsphase'), the new generation of trees is established. This is again followed by the 'optimal phase', the 'ageing phase', the 'breakdown and die-back phase', and back to the 'regeneration phase'.

---

[18] See, *inter alia*, Cotta (1865, p. 2), Landolt (1866, p. 197), Gayer (1886, pp. 32, 43, 45, 68), Bühler (1922, pp. 302–303), Vanselow (1926, pp. 222–227), Hausrath (1982, p. 64), Schubart (1966, pp. 100–103, 125–127) and Mantel (1990, p. 357).

According to Leibundgut (1978), all these stages occurred side by side in the European primeval forests. He believes that this cycle continues unless it is interrupted by a catastrophe. In that case, the succession begins with the re-establishment and colonization phase and ends with the climax. According to Leibundgut, this model applies to the original primeval forest in the lowlands of Europe, as well as to the primeval forest in the mountains. Therefore, oaks, which require light, also regenerate in gaps in the canopy with a diameter of 20–40 m.[19]

According to forest ecologists, wild animals have no effect on the regeneration of the primeval forests because they occur naturally in very low numbers.[20] Furthermore, they maintained that grazing livestock is bad for the forest (Cotta, 1865, p. 84; Landolt, 1866, p. 152; Rubner, 1920; Fröhlich, 1930). Watt (1919) based his view on the consequences of grazing for the forest partly on the opinions of German and Swiss foresters.

2.8 The East of North America as an Analogy for Europe

Based on his research, the British scientist Jones (1945) stated that the forest is probably the most common and most complex type of climax vegetation. He considered it feasible that the 'climax forest' is merely a concept, which never existed in practice as a result of the initiation of young stages resulting from continually changing surroundings. With respect to the facts on the structure and regeneration of forests, he felt that 'it must be noted that most of the evidence relates to forests of tolerant ("shade-bearing") genera such as *Fagus*, *Acer*, *Abies* and *Tsuga* spp., which are undoubtedly the ultimate dominants in mesic sites with adequate rain' (Jones, 1945, p. 145). He postulated that shade-tolerant species of trees formed the climax. In his opinion, this could be the beech in Europe. He felt that the position of so-called 'intolerant genera', i.e. those which require light, such as *Pinus*, *Larix* and *Quercus*, is not clear. He considered that they form a developmental stage in the succession towards forests of species which tolerate shade. According to Jones, these intolerant species are found particularly in extreme climatic conditions, for example, in places where fire may also play an important role, or in places where animals graze. With regard to the oak, he considered that oak forests in lowland areas, where excessive

---

[19] Leibundgut talked about gaps the size of a 'horst'. According to Dengler (1992, p. 26) a 'Horst' has a diameter of 20–40 m, which covers a surface of 0.1–0.5 ha.

[20] Fröhlich wrote on this subject: 'Schäden durch Wild sind nicht zu verzeichnen, weil das Wild (Hirsch- und Rehwild) in den fraglichen Urwäldern derart rar ist, daß ein Wildschaden hier eine große Seltenheit zu sein pflegt' (Fröhlich, 1930, p. 59). In a later publication he is even more explicit when he writes:

> Daß sich die Urwälder Europas in einigen Gegenden bis auf den heutigen Tag in ihrer ursprünglichen Verfassung erhalten und sich immer wieder auf natürlichem Wege verjüngen konnten, ist nicht in letzter Linie darauf zurückzuführen, daß diese Wälder von jeher nur einen *sehr geringen Stand an Hochwild bargen* [italics by Fröhlich] (Fröhlich, 1954, pp. 124–125).

numbers of animals graze beneath its light canopy, cannot be seen as proof for the structure and behaviour of climax forests. Therefore he implied that the grazing of livestock in oak forests is an anthropogenic artefact. According to Jones, it is possible to study climax forests in European mountainous forests dominated by beech, Norway spruce and silver fir. This view is questionable, because these forests are found at higher altitudes than those at which oaks grow. In this region, oaks are found only in areas where cattle grazed and their leaf-fodder was cut down.[21]

Jones (1945) maintained that the primeval forests of Europe and North America did not differ from each other with regard to the mechanism of regeneration. This belief is currently generally accepted among ecologists and foresters (see Whitmore, 1982; Westhoff, 1983; Lemée, 1985; Peterken, 1991; 1996, p. 230; Holeksa, 1993). Before Jones (1945), German foresters like Schenck (1924) and Dengler (1935, pp. 22–25, 78) pointed out the fact that the forests considered to be untouched in the east of the United States could be seen as an analogy for the untouched forests in Europe. Watt (1947) referred to Jones's article (1945) and realized that it confirmed his 'gap phase' model and his theory about the irregularity of the climax vegetation. Jones based his work on both American publications, and particularly on the above-mentioned and other German publications about the virgin primeval forests of south-east Europe.

Jones (1945) reviewed the publications on the primeval forests of Europe and North America and linked them; whereby he indicated that the mechanisms of regeneration in the zone of the moderate forests did not essentially differ. Jones (1945) refers to the above-mentioned German-language publications (Cermak, 1910; Fröhlich, 1930; Hesmer, 1930; Markgraf and Dengler, 1931) on the structure and the regeneration of the forests considered to be the last untouched primeval forests in south-east Europe. He referred to American publications (including Lutz, 1930; Hough and Forbes, 1943) on original forests particularly in the east of the United States, which discussed the structure and regeneration of these forests. These and other publications of the time (Bromley, 1935; Morey, 1936; Marks, 1942) show that species requiring light such as the white oak (*Quercus alba*), black oak (*Q. velutina*), scarlet oak (*Q. coccinea*), northern red oak (*Q. rubra*) and white pine (*Pinus strobus*) are suppressed by shade-tolerant species such as eastern hemlock (*Tsuga canadensis*), American beech (*Fagus grandifolia*), red maple (*Acer rubrum*) and sugar maple (*Acer saccharum*). Here Jones quoted the German forester Schenck (1924) who stated that he had never seen regeneration of oak, hickory or chestnut in the east of the United States. Even a financial incentive for finding a seedling of these species in the forests did not provide Schenck with any evidence (Schenck, 1924). According to Jones (1945), light-demanding trees such as oak and pine occur only in early developmental stages, so-called early seral stages, of forests consisting of

---

[21] See Hoffmann (1895), Cermak (1910, p. 360), Markgraf (1927, pp. 53–54, 138; 1931), Müller (1929, pp. 10, 209, 289), Nietsch (1935, pp. 58–59) and Fröhlich (1954, p. 126).

shade-tolerant species of trees. Fire was considered an important factor for the
continuance of these developmental stages.

In America, as in Europe, grazing is considered to be a factor that prevented
the regeneration of trees and degraded forests to park landscapes and finally to open
grassland. Livestock are considered to be an artefact introduced by the Europeans
that for precisely this reason caused a great deal of damage to the original,
untouched, so-called pre-settlement forest.[22] However, regeneration of trees where
livestock graze does take place in America, as in Europe. Oak (*Quercus* spp.), elm
(*Ulmus* spp.), hickory (*Carya*), apple (*Malus* spp.) and hawthorn (*Crataegus* spp.)
became established in open grassland where livestock graze. Scot (1915) described
how, in the state of New York, hawthorns were particularly noticeable in terms of
both number and shape in pastured grassland. Owing to pruning by livestock, they
were round, conical or shaped like an hourglass. Species without thorns, such as
oak and elm, survived browsing by livestock well. However, they only grew above
the browse line of livestock if they were protected by thorny bushes, such as
hawthorns or roses (*Rosa* spp.). Nut-bearing species could grow when unprotected,
since the livestock avoided them, probably due to their taste (Scot, 1915).

Analogous to what Watt and others observed in England, oaks, elms and
other species of tree grew among thorny bushes in grazed grasslands in the east
of North America as well. The following statement by Scot (1915, p. 461) is evidence that this phenomenon was not a rarity: 'Doubtless many others observed
related phenomena in pastoral regions'. In fact, other authors (Bromley, 1935;
Marks, 1942; Stover and Marks, 1998) observed this phenomenon in the east of
the United States, although they did not publish about it in as much detail as Scot.
For example, Marks (1942) described how thorny species, such as hawthorns,
gooseberry (*Ribes cynosbati* and *R. missouriense*) and prickly ash (*Zanthoxylum
americanum*) remained intact in grazed grassland. Furthermore, the white pine
(*Pinus strobus*) grew there unprotected, because it was avoided by the livestock (see
also Bromley, 1935). Grazing by livestock even promoted the regeneration of the
ponderosa pine (*Pinus ponderosa*) (Covington and Moore, 1994). Further, when
there was grazing, zones emerged from grassland via grassland scrub to forest.
The grassland scrub consisted in part of thorny bushes and sumach (*Rhus glabra*
and *R. typhina*) bushes, which were avoided by the livestock due to the excess tan-
nin and which grew clonally. Furthermore, dogwood (*Cornus femina*) and
American hazel (*Corylus americana*) occurred there. Analogous to the mantle and
fringe vegetation in Europe, light-demanding trees grew among these bushes,
including aspen (*Populus tremula*), oak (white oak, northern red oak and black
oak) and hickory (*Carya* spp.). Stover and Marks (1998) reported that many
thorny species and apple (*Malus* spp.) became established in grazed grassland.
Analogous to the situation in Europe, we can speak in ecological terms of the
establishment of thorny bushes and species of trees requiring light and the growth
of young trees among thorny bushes in grazed grassland. Consumption by the

---

[22] DenUyl (1945; 1962), DenUyl *et al.* (1938), Marks (1942), Steinbrenner (1951), Curtis
(1970), Parker *et al.* (1985), Whitney and Somerlot (1985) and Peterken (1996, pp. 237–238).

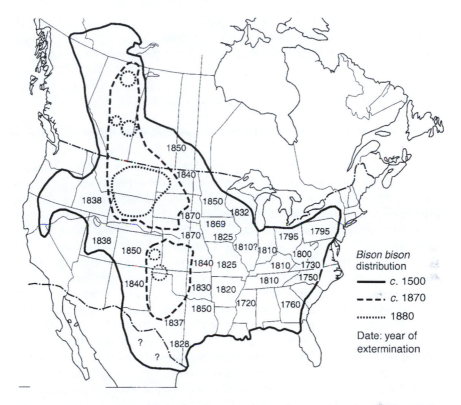

Fig. 2.3. Distribution map of the bison (*Bison bison*) from AD 1500 in the east of the United States. The years given are the years in which the species became extinct in the various regions (based on McDonald, 1981, p. 104; Semken, 1983, p. 186; Hodgson, 1994, p. 70).

animals reduced the amount of flammable plant material, which reduced the frequency of fires. This promotes the establishment of young bushes and trees (Archer, 1989; Covington and Moore, 1994; Fuhlendorf and Smeins, 1997). In this way, savannas have changed into closed forests (Archer, 1989).

In the east of America, grazing was by a non-indigenous species, namely cattle. Although this should be considered as introduced by man, this is not true for cattle's food strategy, i.e. that of a specific bulk and roughage eater, since this is also true for the indigenous bison (Hofmann, 1973; Van de Veen, 1979, p. 132; Van Soest, 1982, pp. 7, 338; see Fig. 2.6). This grazer was found in historical times throughout the east of the United States in places where the climax vegetation of a closed forest is postulated[23] (see Fig. 2.3). In terms of diet, bison and cattle are more or less similar species. Therefore, cattle are sometimes seen

[23] Branch (1962, pp. 52–65), Smith (1962), Thompson and Smith (1970), McHugh (1972), Matthiae and Stearns (1981), McDonald (1981), Semken (1983, p. 186), Jokela and Sawtelle (1985), cited in Crow *et al.* (1994), Day (1989, pp. 183–186), Hodgson (1995).

as surrogates for the bison in America (Reynolds *et al.*, 1982; Milchunas *et al.*, 1998: Knapp *et al.*, 1999). Many descriptions from the period when the bison was found in the east of the United States give an impression of open forests and park-like landscapes with grasslands[24] (see Chapter 4), meaning that regeneration of trees could have taken place where bison grazed, analogous to the regeneration of trees in grasslands grazed by cattle.

In general, Watt (1925, 1947) is seen as the founder of the 'gap' or 'gap mosaic' concept. He identified the gap as the mechanism for the regeneration of the plant community in general, and the forest in particular.[25] Furthermore, Jones (1945) and Aubréville (1933, 1938) are seen as experts who also made an important contribution to the development of the 'gap' concept (Picket and White, 1985; Shugart and Urban, 1989). Aubréville formulated his concept with reference to the tropical rain forests of West Africa. However, his view ignores the concepts of the regeneration of forests which were developed in Germany and Switzerland on the basis of forestry. The common element of those concepts is the fact that the regeneration of the species of trees which form the climax takes place in gaps which are created in the canopy when trees die or are blown over. Nowadays the 'gap' concept is generally accepted as the mechanism which explains the regeneration of all species of trees in the natural forests in temperate zones throughout the northern hemisphere, as well as in the tropics.[26] The concept has become a paradigm (Platt and Strong, 1989; Whitmore, 1989), which means that it has become an a priori starting point for other theories on the regeneration of forests. Moreover, it is widely accepted that the grazing of cattle and high concentrations of wild animals inhibit the regeneration of forests.[27]

---

[24] Bromley (1935), Day (1953), Gordon (1969), Thompson and Smith (1970), Russel (1981), Cronon (1983, pp. 4–5, 10, 25–31), Whitney and Somerlot (1985), Whitney and Davis (1986), Kjidwell (1992), Nabokov and Snow (1992), Williams (1992, pp. 32–33, 42–44), Covington and Moore (1994).

[25] Greig-Smith (1982), Shugart (1984, pp. 48 and 214), Runkle (1985), Shugart and Urban (1989), Whitmore (1989), Coffin and Urban (1993).

[26] See, *inter alia*, Leibundgut (1959; 1978), Grime (1979, p. 140), Mayer *et al.* (1980), Mayer and Neumann (1981), Koop (1981; 1982; 1989, pp. 90–91), Koop and Siebel (1993), Runkle (1981; 1982), Korpel (1982; 1995, pp. 18–22), Whitmore (1982; 1989), Shugart (1984), Shugart and Urban (1989), Collins *et al.* (1985), Ellenberg (1986, pp. 119–120), Falinski (1986), Lemée *et al.* (1986; 1992), Oldeman (1990), Leemans (1991a,b), Remmert (1991), Röhrig (1991), Zukrigl (1991), Dengler (1992, pp. 95–98), Holeksa (1993) and Peterken (1996, pp. 141–143).

[27] DenUyl (1945; 1962), Mayer (1975; 1976; 1981; 1992, pp. 407–408), Curtis (1970), Adams and Anderson (1980), Pigott (1983), Parker *et al.* (1985), Whitney and Somerlot (1985), Ellenberg (1986, p. 41), Putman *et al.* (1989), Dengler (1992, pp. 42–46, 122), Harmer (1994), Peterken (1996, pp. 237–238), Peterken and Tubbs (1965), Putman (1996a,b), Jorritsma *et al.* (1999).

2.9 Pioneer and Climax Species

The observations of the succession from open terrain to forest, and succession where there are gaps in the canopy, has resulted in a classification of species of trees based on the light required by the seedlings in pioneer and climax species. A pioneer species is intolerant, a climax species tolerant of shade (Bormann and Likens, 1979a, p. 106; Swaine and Whitmore, 1988; Whitmore, 1982, 1989; Ellenberg, 1986, p. 82; Harris and Harris, 1991, p. 51). The pioneer species produce small seeds which are easily dispersed, and can therefore colonize rapidly where there are large gaps (Swaine and Whitmore, 1988; Whitmore, 1989; Harris and Harris, 1991, p. 51). The seeds of this group germinate only in full daylight. They grow rapidly, and quickly fill up open terrain and gaps in the canopy (Whitmore, 1989; Harris and Harris, 1991, p. 51). On the other hand, the climax species produce fewer seeds. They are large and contain sufficient reserves for germination and to become established with low levels of daylight. They grow slowly, the seeds can germinate in the shade, and the seedlings tolerate the shade for a relatively long time. They are able to regenerate in small gaps in the closed canopy, and under the canopy of the closed forest. Species which are intolerant to shade and require light can do so only in open terrain or in gaps in the canopy which are so large that there is actually a clearing in the forest (Swaine and Whitmore, 1982; Whitmore, 1982; 1989). The category of species which tolerate shade includes the beech, and that which includes pioneer species intolerant of shade, the birch (Whitmore, 1982). Table 2.1 shows the tolerance of shade of species which are found in Central Europe. Whitmore (1989) considers that the distribution of trees in these two groups corresponds to our present knowledge about the dynamics of forests. Despite some people's criticism of this rigid classification of trees (see Canham, 1989), it is used in the models which were developed to simulate the succession in forests in gaps in the canopy, and the establishment of species diversity in climax forests (see Botkin *et al.*, 1972; Shugart and West, 1980, 1981; Botkin, 1993, p. 68). Because of the dominance of shade-tolerant species, these 'gap' models are sometimes described as shade models (Shugart and Seagle, 1985).

The seedlings and young trees of pedunculate and sessile oak require a great amount of light, and because they easily colonize open terrain, they are classified as pioneer species (Koop, 1981, p. 46; Tubbs, 1988, p. 142). On the other hand, neither of them complies with the other criteria which are used for pioneer species. In fact, they have many characteristics which are ascribed to climax species tolerant of shade, such as slow growth and the propensity to grow very old and produce large seeds.

Table 2.1. The tolerance to decreasing amounts of daylight for different species of trees indigenous to Central and Western Europe, as a tree (T) and as a germinating plant and growing seedling (S) (redrawn from Ellenberg, 1986, Table 9).

Name of species	S	T
Silver fir (*Abies alba*)	00	00
Norway spruce (*Picea abies*)	xx	0
Scots pine (*Pinus sylvestris*)	()()	()()
Yew (*Taxus baccata*)	0	0
Sessile oak (*Quercus petraea*)	x	x
Pedunculate oak (*Q. robur*)	()()	x
Turkey oak (*Q. cerris*)	x	x
Beech (*Fagus sylvatica*)	00	00
Small-leaved lime (*Tilia cordata*)	xx	0
Large-leaved lime (*T. platyphyllos*)	0	0
Smooth-leaved elm (*Ulmus minor*)	xx	xx
Wych elm (*U. glabra*)	0	0
Fluttering elm (*U. laevis*)	0	0
Sycamore (*Acer pseudoplatanus*)	0	0
Norway maple (*A. platanoides*)	0	0
Field maple (*A. campestre*)	xx	xx
Ash (*Fraxinus excelsior*)	0	xx
Hornbeam (*Carpinus betulus*)	0	00
Native alder (*Alnus glutinosa*)	xx	xx
Wild cherry (*Prunus avium*)	0	xx
Bird cherry (*P. padus*)	xx	xx
Crab apple (*Malus sylvestris*)	()()	x
Wild pear (*Pyrus pyraster*)	xx	xx
Black poplar (*Populus nigra*)	xx	x
White poplar (*P. alba*)	xx	x
Aspen (*P. tremula*)	x	x
Birch (*Betula pubescens*)	()()	()()
Silver birch (*B. pendula*)	()()	()()
True service tree (*Sorbus domestica*)	0	0
Wild service tree (*S. torminalis*)	0	xx
Whitebeam (*S. aria*)	()()	xx
Rowan (*S. aucuparia*)	x	x
White willow (*Salix alba*)	xx	x
Crack willow (*S. fragilis*)	xx	x

Tolerance to decreasing amounts of daylight: 00 = very high; 0 = high; xx = fair; x = low; ()() = very low; () = extremely low.

2.10 Variations on the Succession Theme and the Gap Phase Model

Whittaker (1953) drew up the so-called 'climax pattern hypothesis'. This means that climax vegetation is a collection of stages containing populations of plant species. Whittaker approached the climax on the basis of population dynamics. He believed that the climax is determined by all the factors which are part of the system, and which have a constant or regularly recurring effect on populations. The effect is such that the populations are not destroyed, so that a new succession is initiated. Whittaker defined the climax as the fluctuations of species around a particular equilibrium. He saw his theory as being in between, or a synthesis of Clements's mono-climax theory (1916) and Tansley's polyclimax theory (1935), because his theory contained elements of both. In his theory, the polyclimax theory is expressed in the diversity of climax vegetation as a result of populations of species which are at different stages of development. The mono-climax theory is expressed because different climax vegetations cannot be seen as two or more clearly distinct associations, but are part of a single climax pattern.

Following a visit to Israel, Whittaker noted that Mediterranean landscapes have been misused for thousands of years because of overgrazing by goats, camels, donkeys, sheep and cattle. He wrote that these heavily overburdened communities were very rich in different species of plants. They were among the richest in species that he had studied (Whittaker, 1977). In his opinion, the species diversity in these meadows was probably the product of a rapid evolution of the species of plants concerned over 10,000–12,000 years which had occurred because of chronic interference by livestock. He believed that the effect of goats and other animals is primarily to constantly suppress the scrub. When this returns, there is a great decline in the variety of species. The scrub that becomes established and develops when the influence of the herbivores disappears and the climax returns therefore has a much poorer variety of species. Hence, the disruption caused by grazing and browsing herbivores makes an important contribution to maintaining the diversity, according to Whittaker.[28]

In the Mediterranean area, the domestication of wild herbivores as cattle took place approximately 10,000 years ago. The wild varieties of livestock are found there naturally (see Davis, 1987, pp. 127–133; Bell and Walker, 1992, pp. 112–114; Mannion, 1992, pp. 83–90). Whittaker's argument (1977) shows that he believes that the natural vegetation in the presence of wild herbivores consisted of a closed forest, which he assumed would develop once again with an intermediary stage of scrub, once the influence of cattle disappeared. Thus there could be no grasslands with a rich variety of species when there were wild herbivores.

---

[28] 'Through time under grazing – decades, centuries, millenia – species accumulate in the disturbed communities, enriching these. Given sufficient time, as in Israel, the disturbed communities may become very rich indeed compared with the climax' (Whittaker, 1977, p. 418).

Remmert (1991) formulated the so-called 'mosaic cycle concept'. He postulated that the climax consists of a mosaic of very different plant communities which all go through their own cycle independently of each other. In his opinion, these stages take place in accordance with Watts' gap phase model (1947). Some of the stages in this cycle have a large species diversity; others do not. He believes that the regeneration cycle of forest also includes a stage of grassland. He based this theory, *inter alia*, on a particular type of forest in northern Botswana. This consists of one species, namely, mopane trees (*Colophospermum mopane*). In his view, this type of forest can be seen as an analogy for the beech forests in the European temperate climate. He found dead mopane trees in large areas of the grasslands. There were also large areas of bare grassland, where there were many stumps of dead mopane trees, as well as many seedlings of this species. From this, Remmert deduced that large areas of grassland develop in this type of forest as they grow older and degenerate. Therefore he believed that a grassland stage is normal in forests which consist virtually of one species, such as mopane forests and beech forests. Dying trees are not replaced by a new tree, but by a vegetation of plants where trees subsequently become established.

He considered that the appearance of grasslands in forests in Europe is incorrectly interpreted as an indication that the forest is starting to die off. Remmert developed his mosaic cycle concept on the basis of these findings. He believed that storms and diseases perpetuate the system, as well as the dying trees. He thought that gaps in the canopy can result in a domino effect, because more and more trees can be blown over or die off, one by one, on the perimeter of the gap. The reason for this is that the trunk of a tree on the periphery of the gap is exposed to direct sunlight and dies. This happens with beech trees. Remmert considered that his hypothesis also explains why large herbivores such as aurochs, wild boar, European bison, horse and red deer were found in the virgin forest. In fact, the grassland stage allows for much higher densities of these animals than when there is no grassland stage, and the forest is not damaged, because the grasslands provide the food that is needed. Although Remmert clearly placed the large herbivores in the natural system, their role is essentially no different from that assigned them, for example, by Tansley (1935). The herbivores *follow* the developments in the vegetation, a view shared with Whittaker.

2.11 Catastrophes as the Mechanism of Succession, Regeneration and Diversity

The deciduous forests of Europe and in the east of North America are considered comparable because of the similarity of the genera, and because forests are believed to be fundamentally the same throughout the world (Whitmore, 1982, p. 45). The results of studies and theories on the forests in the east of the United States are implicitly and explicitly considered to be applicable to the forests of

Europe on the basis of this assumed similarity.[29] This includes the role of cata-strophes in the succession of these forests.

In Watt's gap model (1947), catastrophes such as fire and storm do not play a significant role. It is only the gaps which develop when one or more trees die that serve as the mechanism for the regeneration of forests. According to many authors,[30] fire did not play a significant role in the succession of the original deciduous woodlands of the European lowlands, particularly where the soil was rich, because the forest in those areas would have been virtually unsusceptible to fire. Only fire caused by people could have led to changes in the forest naturally present there (Mellars, 1975, 1976; Bogucki, 1988, pp. 34, 38–40; Clark *et al.*, 1989; Roberts, 1989, p. 83; Tinner *et al.*, 1998; 1999). According to Müller (1929), fire does play a role in the regeneration of natural coniferous forests in Europe, or in forests dominated by conifers. He believed that large-scale forest fires in the primeval forests of Bulgaria and the Rodopi and Rila mountains are respon-sible for large-scale regeneration and the establishment of the Scots pine, more or less at the same time (Müller, 1929, pp. 49, 268, 288). In his opinion, forest fires in areas with a climate of summer drought are 'a special kind of catastrophe which, we observe, affects the primeval forest as an inevitable event, as the great "matador" of the primeval forest' (Müller, 1929, p. 317).

Jones (1945) pointed out a dispute in America regarding the degree to which fire should be seen as a natural or man-made phenomenon. This dispute resulted from what Raup (1964) pointed out, namely that throughout America, forests whose history could be traced back to the period before European colo-nization did not meet the conditions of a self-perpetuating 'climax'. Raup (1964) believed that disruptions, which destroyed or decimated the forests, must have been at work. These disruptions would have ensured that light-demanding genera such as oak (*Quercus*) and pine (*Pinus*) were present in the degree shown by historical descriptions. These genera are nowadays suppressed in the forests by more shade-tolerant species such as red maple (*Acer rubrum*), sugar maple (*Acer saccharum*), American beech (*Fagus grandifolia*), white ash (*Fraxinus americana*) and American elm (*Ulmus americana*). That process is ascribed to the absence of fire.[31] The forests which formed the climax vegetation in the east of the United States would be an invention by the Europeans, which occurred due to exclusion of fire (Pyne, 1982). Fire, either due to lightning or burning by Native Americans, would always have had an effect on the vegetation.[32]

---

[29] See, *inter alia*, Dengler (1935, pp. 22–25), Jones (1945), Whitmore (1982), Westhoff (1983), Lemée (1985), Peterken (1991; 1996, pp. 58, 230), Peters (1992), and Holeksa (1993).
[30] See, *inter alia*, Müller (1929, p. 3), Falinski (1976), Mellars (1976), Lemée *et al.* (1986), Koop (1981; 1982; 1989), Röhrig (1991), Remmert (1991, p. 9), Zukrigl (1991), Dengler (1992, pp. 93–98), Rackham (1992) and Holeksa (1993).
[31] Lutz (1930), Bromley (1935), Morey (1936), Hough and Forbes (1943), Raup (1964), Ware (1970), Leitner and Jackson (1981), Pyne (1982), Hibbs (1983), Adams and Anderson (1980), Barnes (1991), Abrams (1992; 1996), Abrams and Downs (1990), Abrams and McCay (1996), Covington and Moore (1994).
[32] Day (1953), Botkin (1979), Host *et al.* (1987), Abrams (1992; 1996), Abrams and McCay (1996), Covington and Moore (1994), Barnes and Van Lear (1998), Ruffner and Abrams (1998).

In recent decades, fire has emerged more and more as a disruptive factor, which would have determined the structure and regeneration of forests in the east of the United States. This is particularly true for the regeneration of oak.[33] The subsistence of oak would have been guaranteed in the first instance by fire as a result of lightning and in the second instance by burning by Native Americans.[34] Ground fire would have promoted the oak compared with the above-mentioned shade-tolerant species of tree and in certain types of forest it would have ensured the dominance of oaks.[35] The oak would be better adapted to fire than other species of tree, because this species has deep roots and a strong ability to shoot from the stump after fire (Whittaker and Woodwell, 1969; Russell, 1983; Whitney, 1987).

Abrams (1992) formulated the so-called fire and oak hypothesis. According to this hypothesis, fire eliminated the regeneration of shade-tolerant species, such as sugar maple, red maple, white ash, American beech and American elm and the light-demanding and fast growing yellow poplar (*Liriodendron tulipifera*), which can grow very old, that suppress oaks in spontaneously developing forests (Peterken, 1996, p. 65; Brose *et al.*, 1999a,b). According to Abrams and Seischab (1997), fire explains why oaks could continue to exist in the presence of maple and ash for thousands of years at a time. 'What disturbance factor other than fire could historically have prevented these species from replacing oak?' (Abrams and Seischab, 1997, p. 374). Partly on this basis, there have been calls to reintroduce fire as a disruptive factor (Abrams and Ruffner, 1995). Other authors point out that fire is not as selective with respect to oak as is assumed, and that it is not clear at all whether oak does benefit from fire (Whitney and Davis, 1986; Huddle and Pallardy, 1996; Arthur *et al.*, 1998).

Fire is equally bad for all young trees, including young oaks (Korstian, 1927; Whitney and Davis, 1986; Huddle and Pallardy, 1996; Barnes and Van Lear, 1998). Only an incidental fire followed by a long period without fire leads to an increase of red maple, for example, as a result of a strong increase in the number of shoots springing up from the remaining stumps. A species requiring a large amount of light like the scarlet oak did not benefit at all from two fires in quick succession; there was no regeneration (Arthur *et al.*, 1998). In relation to the presence of the oak in the original vegetation in the east of the United States, fire would not have played the role that others have assigned to it (see

[33] See, *inter alia*, Lutz (1930), Monk (1961b), Pyne (1982, pp. 27, 42, 75), Thompson and Smith (1970), Whittaker and Woodwell (1969), Lorimer (1984), Clark (1993), Van Lear and Watt (1993), Crow *et al.* (1994), Orwig and Abrams (1994), Arthur *et al.* (1998) and Barnes and Van Lear (1998).

[34] Monk (1961b), Pyne (1982, pp. 27, 42, 75), Jokela and Sawtelle (1985), cited in Crow *et al.* (1994), McCune and Cottam (1985), Crow *et al.* (1994), Abrams (1996), Abrams and Nowacki (1992), Abrams *et al.* (1995), Ruffner and Abrams (1998).

[35] Monk (1961b), Ogden (1961), Pyne (1982, p. 38), Abrams (1992), Abrams *et al.* (1995), Mikan *et al.* (1994), Orwig and Abrams (1994).

Clark and Royall, 1995; Clark *et al.*, 1996; Clark, 1997). Based on analyses of pollen and charcoal in sediments, Clark *et al.* (1996) questioned the role of fire in maintaining the composition of species in forests in the east of the United States. Here it appears that fire cannot have played the role that has been ascribed to it (Clark and Royall, 1995; Clark *et al.*, 1996). According to Clark (1997), fire can have played a role in the subsistence of the oak, but in most of the east of the United States, the oak appears to have maintained itself for thousands of years without clear palaeoevidence of fire. These findings make it necessary, according to Clark (1997), to look for other factors, which allowed for the subsistence of oak in the long term, without fire. As such, he names windthrow. Other authors are also of the opinion that fire did not play a significant role in the east of the United States.[36] In that region, only occasional storms ensured that there is an increase in the relative proportion of species which do not tolerate shade (Bormann and Likens, 1979a; Runkle, 1981, 1982). However, this is not true for the oak. More light on the ground in the form of gaps in the canopy, either in the form of large open areas caused by storms or heavy tree felling, lead, in contrast, to a fast growth of shade-tolerant species that were already present as advance regeneration under the canopy of the forest.[37]

According to Loucks (1970), who carried out research in forests in the east of the United States, a certain diversity is the result of succession following a disruption which created open terrain. He reached this conclusion on the basis of the results of research into the diversity of seedlings in forests which grew on abandoned agricultural land. In the initial stage, species became established which he believed to be adapted in a unique way to colonize open terrain. In his opinion, the large-scale regeneration of these species of trees which require light shows that disruption is essential for the survival of certain species. Therefore the species diversity is not a static condition in the form of a stable climax, but a function of time in relation to disruptions. Following such disruptions, Loucks believed that there is a strong increase in the species diversity, because more and more seedlings of species which tolerate shade are established. Subsequently, the diversity declines again as the first colonists grow older, because these colonists are no longer replaced. The diversity declines further because eventually only the seedlings of a few species which tolerate a great deal of shade are able to survive under the canopy, such as sugar maple (*Acer saccharum*). After about 100 years, the only seedlings are those of sugar maple, while the canopy

---

[36] See Runkle (1981; 1982), Whitmore (1982; 1989), Shugart (1984), Shugart and Urban (1989), Collins *et al.* (1985), Oldeman (1990), Röhrig (1991) and Peters (1992).
[37] Spurr (1956), Gammon *et al.* (1960), Skeen (1976), Ehrenfeld (1980), Barden (1981), Hibbs (1983), Loftis (1983), Lorimer (1984), Lorimer *et al.* (1994), McGee (1984), Parker *et al.* (1985), Beck and Hooper (1986), Ross *et al.* (1986), Host *et al.* (1987), Ward and Parker (1989), Abrams and Downs (1990), Cho and Boerner (1991), Nowacki and Abrams (1992), Peterson and Picket (1995), Norland and Hix (1996), Cook *et al.* (1998).

consists of species which require light, dating from the initial stage. Some seedlings of species from the initial stage may appear among the seedlings of the species which tolerate shade, but they never thrive. Eventually the species which tolerate shade penetrate the canopy and the species from the initial stage disappear. Therefore the decline in the diversity of seedlings is followed by a decline in the diversity of the canopy. In the end, the canopy consists only of the climax species which tolerate shade (Loucks, 1970).

Loucks (1970) considered that species from the initial stage which require light only become established after a catastrophe. In his view, the diversity is greatest when there are also species which tolerate shade from the later stage, sometime after a disturbance. Therefore the diversity of communities can only be maintained by disturbances which occur with a certain regularity (once every 30–200 years), destroy the climax and therefore ensure that the initial stage starts again.

Connell's 'intermediate disturbance hypothesis' (1978) supports Loucks' hypothesis (1970). Connell (1978) referred to the extremely high species diversity in tropical rain forests and coral reefs. He maintained that studies of succession in these systems indicate that the species diversity is a function of the intervals between disturbances. He also stated that the climax has a low diversity. As proof for this thesis, he referred to areas where there are disturbed forests next to undisturbed forests. The undisturbed forests consisted virtually of a single species of tree, while parts of the same forest which were exploited by man contained a variety of species. According to Connell, old secondary forests actually have the greatest variety of species because they contain species from the climax stage, as well as the species which became established immediately after the disturbance. Connell believed that the occurrence of species which require light and species which tolerate shade at the same time confirms his hypothesis. In coral reefs, he maintained that disturbances contribute to the diversity because the highest diversity of species occurs in the places where reefs are most exposed to the forces released by storms.

Huston (1979) also believed that disturbances play an essential role for ensuring diversity. He approached the effects of disturbances from the aspect relating to the composition of the population of species. He thought that strongly competitive species exclude many others when they have the opportunity to develop up to a particular population level without disturbance. Although certain species have the capacity to exclude others by means of competition, this does not happen because the populations of all the species, including the most competitive species, are decimated by catastrophes. According to Huston, the species diversity is the result of the inability of the competing species to interact in a balanced way, because regular catastrophes are constantly affecting the composition of the population of all species.

Loucks' theory (1970), as well as Connell's theory (1978) and Huston's theory (1979) therefore all state that a high species diversity is a state of imbalance which develops into a balanced community that does not have many different species, when there are no disturbances. The duration of the intervals between the disturbances actually determines the diversity to a great extent. When the disturbances are very frequent, a community of species develops which colonize rapidly or quickly build up a population; when there are few disturbances, a small number of highly competitive species become dominant. Therefore the greatest diversity occurs when disturbances take place between these two extremes. Pickett (1980) and Vogl (1980) arrived at the same conclusion with regard to the effect of catastrophes on diversity.

Although Bormann and Likens (1979b) also considered that disturbance is an important factor with regard to maintaining diversity in forests, they did not believe that it is necessary for the survival of all the species of trees that are found there. On the basis of the number of trees struck by lightning, and storms which raged in the north and north-east of the United States between 1945 and 1976, they ascertained that in certain parts of this region, forest fires and storms are regular events. This means that the natural forest vegetation consists of strips of vegetation of different sizes and ages. They believed that every system encounters disturbances, i.e., there is never a so-called 'steady state' (= climax in the sense used by Clements). However, they did add that relatively large areas can remain unaffected by catastrophes. In their opinion, the regeneration in the climax stage takes place only in the 'steady state' in the gaps in the canopy, in accordance with Watt's model (1947). If gaps form in the canopy because one or more trees die, Bormann and Likens (1979a, p. 194; 1979b) refer to this as an endogenous disturbance. They describe factors such as fire or wind as exogenous disturbances. They maintain that the climax stage is characterized by alternating sizes of gaps, so that the species which appear in the succession after a disturbance also appear in the climax without such a disturbance (Bormann and Likens 1979a, p. 5; see Fig. 2.4). They described the pattern of vegetation which develops as a result of endogenous disturbances as the 'shifting mosaic steady state' (Bormann and Likens, 1979a, p. 5). In this, the total biomass oscillates around a certain average (Bormann and Likens, 1979b). In their view, the 'shifting mosaic steady state' is almost a replica of the model proposed by Watt (1947) for the succession in an initially mature beech forest with trees of the same age (Bormann and Likens, 1979a, p. 5).[38]

To summarize, it can be noted that the theories and models on disturbance described in the preceding paragraphs are characterized by the fact that they superimpose factors on the gap phase model from outside the system, which initiate certain processes and conditions that are necessary for the survival of certain groups of species. Large herbivorous mammals are not taken into consideration in this respect.

[38] 'Many aspects of the reproductive behaviour of the forest associated with later phases of our biomass-accumulation model closely approximate those set forth for forest ecosystems more than a quarter of a century ago in the landmark paper by A.S. Watt (1947)' (Bormann and Likens, 1979a, p. 5).

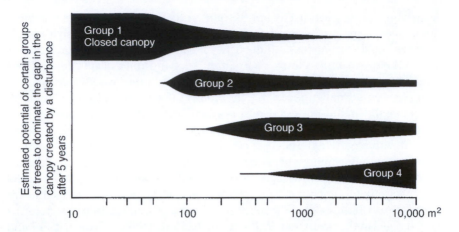

Fig. 2.4. Schematic representation of the potential of certain groups of trees to dominate gaps within 5 years after the gaps appear in the canopy. These are forests in the east of North America. Group 1 contains species which form part of the canopy and close the gap by developing sideways. These are the climax species. Group 2 consists of species which tolerate shade, including beech (*Fagus*). Group 3 contains so-called intermediary species, including ash (*Fraxinus*). Group 4 are species which do not tolerate shade, such as birch (*Betula*) and poplar (*Populus*). The observations concerned gaps in the canopy varying from 50 to 500 m² which developed when several trees died or fell over. In addition, data from the literature have been used on tree felling over 400–10,000 m² (redrawn from Bormann and Likens, 1979a, p. 132).

2.12 Establishment Factors in Relation to Succession and Diversity

Apart from disturbances, factors relating to the establishment of species may also have contributed to maintaining diversity in natural vegetation. According to Eglar (1954), data obtained from the study of old fields showed that 95% of the species of trees which developed there had already been present during the initial stage. He maintained that species of which there were no seeds during that stage were excluded for a very long time. This explains, for example, why grasslands persist for a long time in open areas in forests. He thought that when trees eventually grew up from the shrubs in old fields, this was not the result of the establishment of these species at the end of the shrub stage. They were already present during the initial stage, and therefore grew up above the other surrounding species during the shrub stage. Thus Eglar considered that the succession and the climax were mainly determined by the species which were already present from the very beginning. He called this phenomenon the 'initial floristic composition' of a vegetation. According to Eglar, this consists of the species of trees which survive in the shrub stage and then outlive all the other species. They eventually form the climax. He made a distinction between this model and the customary model in which the different species appear one after the other, which Eglar called the 'relay floristic model' (1954).

Horn (1975) compared the diversity of virgin forests in New Jersey in the United States with the succession in old fields to see whether all the species survived there. On this basis, he developed a predictive model which he validated using chronosequences in the old fields. He made two assumptions regarding this model. The first is that every seedling under the canopy has an equal chance of growing into a tree which will form part of the canopy. The second is that analogous to the 'gap phase' concept: the replacement of trees in the canopy occurs one tree at a time.

The results of his model correspond closely to his observations in the old fields. Species which require light colonized the old fields. Then they created an environment in which only species which tolerate shade could survive. Eventually, a stable distribution developed between the different species. According to Horn, his observations and his model also showed that it was possible to predict the progress of the succession from the architecture of the crowns of the different species of trees. In light conditions, trees with a crown consisting of several layers of leaves, such as red oak (*Quercus rubra*) and hickory (*Carya* sp.), grew at a faster rate and had a superior biomass development; in shady conditions, these are species of trees with a crown consisting of a single layer of leaves, such as American beech (*Fagus grandifolia*) and silver maple (*Acer saccharinum*). The humidity of the soil also plays a role with regard to the establishment of the two types of trees. Species with a crown which consists of several layers of leaves evaporate relatively less water per unit of surface area under the tree than species with a crown consisting of one layer. According to Horn (1975), the reason for this is that the heat radiated in a crown consisting of several layers is distributed over a larger total surface area of leaves than in species with a crown consisting of one layer of leaves. Therefore the latter evaporate more water per unit of surface area under the tree, and consequently require moister soil. In his opinion, the most aggressive species consisting of one layer of leaves eventually win the competition against all other species in favourable conditions of humidity, because their crown casts such a large shadow that there can be no regeneration underneath. Therefore Horn (1975) believed that in virgin forests, species with a single layer of leaves predominate when the humidity is favourable.

Establishment in relation to diversity is the core of the 'regeneration niche' concept formulated by Grubb (1977). He emphasized the fact that the competitive relationship between species depended to a large extent on the stage in the life cycle of the plant. He believed that many species require different conditions to become established from those which are required once they have become established. He considered the process of establishment to be very important for maintaining diversity in a plant community. In one particular place, different conditions may be present at different times, and therefore different species of plants. According to Grubb (1977), this is clearly illustrated by the fact that species of trees which require light, such as the silver birch, and species of trees which tolerate shade, such as the beech, are found in the same forests side by side. He referred to Watt (1934; 1947), who described this phenomenon. The amount of light penetrating through the gap in the canopy to the forest floor where the regeneration takes place determines which of the two types of tree will become established. When the gap is large, he thought species which require light would predominate; when it is small, the species which tolerate shade would predominate.

According to Grubb, when you add up the specific characteristics of species of plants for regeneration, on the one hand, and the capacity for survival once they are established, on the other hand, this produces an almost infinite number of possibilities of species which can survive.[39] In his opinion, the gaps in the vegetation cover, whether in woodland or in grassland, form the niche which determines the conditions for the regeneration of species. To find out more about the mechanisms of species diversity and maintaining this, he considered that it is important to know the conditions under which species regenerate. Disturbances by outside forces (Grubb, 1987, p. 131) such as moles, earthworms, ants, rabbits and ungulates which disturb the soil (by trampling and interfering with it), and mowing or breaking up the soil with tractor wheels, result in a great diversity of 'regeneration niches', according to Grubb (Grubb et al., 1982; Grubb 1985; 1987). He made a distinction between two types of disturbance, namely, periodic and continuous disturbances (Grubb, 1985) and considered that grazing fell under the latter category.

Grubb (1985) referred to a particular category of species, the 'edge' species, which are found in the mantle and fringe vegetation of forests. He raised the question of whether these species can be accommodated in the 'gap phase' model (Grubb, 1985). He based this question on an observation which surprised him in New Guinea, that certain species of trees in the secondary forest on abandoned agricultural land, or where trees have been felled are not found in the largest gaps which occur naturally in the canopy of the primary forest. These species grow naturally on the edges of forest grasslands, gravel banks, and areas of subsidence in and along rivers (Grubb, 1985). He claims that this phenomenon can also be observed in old fields in North America. Therefore, in his view, disturbances show that species of plants react individually. In a secondary succession induced by man, this means that species can appear which come from a broad range of natural habitats, and have therefore not evolved together (Grubb, 1987). Thus he indicated that communities of plants can be loose collections of species for which there are no historical analogies; a view put forward by Gleason (1926) at a much earlier date.

---

[39] According to Grubb (1977), it is even the case that 'species diversity has more to do with requirements for regeneration than with partitioning of the habitat niche of the adult' (Grubb, 1977, p. 133).

According to Fenner (1987), the characteristics of plant seeds also play an important role in the establishment of species and therefore the diversity of vegetation. A 'gap' creates a change in the physical environment which can remove obstacles to the germination of seeds in the seed bank. These obstacles serve to prevent the seeds from germinating in a situation that is not favourable to them. Mechanisms which are related to gap creation help the germination. These mechanisms include, for example, sensitivity to light and the nature of the light, such as the changing ratio of red/far-red (R/FR) and the sensitivity to fluctuations in temperature. In addition, Fenner pointed out the importance of the dispersal of seeds by birds. One example he gave was jays (*Garrulus glandarius*) hiding acorns in the ground, and the fact that certain seeds will only germinate properly once they have passed through the intestinal tract of a bird. The dispersal of this last group of seeds is, in turn, dependent on the fact that there are places for the animals to rest.

Apart from remaining in the soil as seeds, species of plants can also remain as seedlings under the canopy for many years, until the conditions are favourable for growth. This happens when more light penetrates because the canopy grows thinner, or if a gap develops. The seedlings can then shoot up (Canham, 1985, quoted by Fenner, 1987). Canham and Marks (1985) also pointed out the importance of the right seeds being in the right place for the succession. Eglar's 'initial floristic composition' concept (1954) also emphasizes this.

To summarize, it may be said that the theories and concepts described in this section endeavour to explain natural phenomena which have been observed, without involving the large herbivores. These are not discussed, except in terms of the disturbance they cause.

2.13 Nutrients in Relation to Establishment and Succession

According to Grime (1974), there are three main factors which determine what the vegetation looks like as regards species diversity, namely, competition, stress and disturbance. Grime defines competition as the attempts of plants growing next to each other to secure the same units of light, water, mineral nutrients or space; stress is the presence of a factor which inhibits growth, such as a low pH or a shortage of nutrients, light or water; and disturbance is the loss of biomass as a result of processes such as grazing, trampling, ploughing, mowing, burning, agents which start diseases and erosion (Grime, 1974; 1977; 1979, pp. 152, 159).

Grime (1974) maintained that both stress and disturbance prevent competition, because they restrict the development of plants. Stress restricts the competition of plants because it restricts the primary production of the plants. Disturbance restricts the competition of plants because species are damaged causing a decline in their capacity to compete (Grime, 1974). During the succession the tolerance to stress, which includes shade, becomes progressively

more important in the competition between plants. According to Grime (1977),
a reduction in the amount of light because of the shade created by competitive
species coincides with a decline in the amount of mineral nutrients available for
the plant, because the vegetation is dominated at the end of the succession by a
large biomass of big, long-living forest trees. He believed that succession eventually results in the dominance of species of plants with two strategies, i.e. competitive species and species which are tolerant of stress. According to Grime,
every species has developed particular autecological characteristics which allow
the species concerned to flourish under a particular combination of circumstances. In this way, every species has acquired a fixed place within the force
field of the main factors of competition, stress and disturbance. The species concerned can survive in this place, but not outside.

Grime (1977) asserted that many plant communities contain species which
may monopolize the available natural resources essential for growth. In the
temperate zones, these are bushes and trees. On the other hand, many species
of plants can survive side by side when disturbance or stress, or a combination
of both, restrict the competitive capacity of these potentially dominant species
(Grime 1977; 1979, pp. 157, 159). In his opinion, this is proved by experiments
and forms of management such as mowing and burning, as well as grazing and
trampling by animals. The diversity of the vegetation is maintained because the
potentially dominant species are weakened by the management (Grime, 1979,
p. 159). For example, in unfertilized limestone grassland, shrubs and species of
trees such as beech trees become dominant when the grazing by sheep and rabbits stops, according to Grime (1979, p. 128), on the basis of the results of
Watt's research (1957).

Connell and Slatyer (1977) pointed out that the research and theories on
succession and climaxes are concerned mainly with the competition for light
and nutrients. They believed that certain species of plants and animals not only
tolerate the presence of others, but even need them to pave the way. They called
this model the 'facilitation model'. They collected proof above all from het-
erotrophic organisms, such as animals which live on carcasses, manure and
humus. Another example of facilitation which they give is the primary succession which is found on soil where glaciers are receding. The first plants to colonize this soil fix the nitrogen, which enters the ecosystem by means of
atmospheric N-deposits and N_2 fixation in the form of a store of humus, which
is the initial stage in forming the soil. Other species of plants can then obtain
nitrogen from this. Crocker and Major (1955) had already ascertained that
stores of nitrogen formed in soil where glaciers had receded. Berendse (1990)
described this phenomenon for the Dutch heathlands. He found that the N-min-
eralization could increase by more than a factor of ten over a period of 50 years,
as a result of humus accumulation. Connell and Slatyer (1977) pointed out that
the appearance of a particular species later on in the succession could be the
reason why these species need other species in some way to become established,
because the first colonists made the environment more suitable for species from
the later stages of the succession. Therefore the first species of plants to appear

have a facilitating effect for the later species. In this context, Connell and Slatyer (1977) referred to the distinction which Eglar made (1954) in the succession in old fields, between the 'initial floristic composition', on the one hand, and the 'relay floristic', on the other hand. In the 'relay floristic' model, they believed that the succession of species can involve facilitation.

In addition to the facilitation model, they also distinguish two other models, namely, the tolerance model and the inhibition model. According to Connell and Slatyer, these two models apply particularly to secondary succession. The tolerance model implies that succession results in a community consisting of species which exploit the natural resources in the most efficient way. This model predicts that in regions with a temperate climate, the species which are most tolerant of shade will dominate the community, if the community is in balance with the environmental factors, i.e. the climax has been achieved. In contrast, in the inhibition model, no one species is superior to any other. Once a species has become established, it inhibits the arrival of others. The species diversity then changes only as a result of the death or destruction of one species which occupies a particular place. For example, the prevention of the establishment of plants and grasses by trees and shrubs can be removed by grasshoppers, other nibbling insects and grazing vertebrates, or by fire. If these are regularly recurring phenomena, Connell and Slatyer believe that the climax can even consist of herb-like vegetation in places where trees and shrubs could grow.

Connell and Slatyer referred in particular to the effect of grazing animals as a possible facilitating mechanism. They felt that this effect is ignored. In their view, interactions between plants, herbivores, predators and organisms which spread disease should be added to the processes which play a role in the succession and the climax. They believed that the only people who had studied succession were people whose main interest was plants. The mechanisms which are considered to be responsible for succession have therefore remained limited to the interactions of plants with their abiotic environment and interactions between plants. They agreed that some attention is devoted to the consumers of plants, but this is particularly with regard to the role which they play in the cycle of nutrients, and not the role they play in the succession. Shugart (1984) and Finegan (1984) also referred to the importance of facilitation, as did Edwards and Gillman (1987), who also mentioned mammals as facilitating factors. They referred to Watt's research (1926), which describes how yew developed in juniper bushes in areas where animals grazed. In addition, they ascribed a facilitating role for plants to large mammals, because these disperse seeds in their fur and droppings.

Tilman's 'resource ratio' hypothesis (1985) follows directly from Connell and Slatyer's facilitation model (1977). According to this theory, there is a relationship between the succession on the one hand, and the build up in the soil of a store of nitrogen available for plants, on the other hand. Tilman (1985) stated that light and nutrients are the most important determining factors for succession. At the beginning of the primary succession there is plenty of light, but a shortage of nutrients, particularly nitrogen. Species which are able to deal with these shortages most efficiently become established and survive. As more nutrients become available as a result of the formation of a blanket of vegetation and the subsequent creation of humus, more productive species of plants appear and the competition for light becomes the determining factor. According to Tilman, the relative availability of limited resources forms a range over time. The difference between Grime and Tilman is that Tilman believes that species can live side by side because they are restricted in their development to a different extent by the individual available resources, while Grime maintains that species of plants have a particular place within the whole force field of competition. Therefore Grime says that there are 'strong' and 'poor' competitors, which Tilman denies. According to Tilman, competition is important in all the stages of a primary succession. Only the object of the competition changes; in the first instance, the competition is for nutrients, later it is for light. A particular ratio in the availability of resources causes one species to be a formidable competitor, while the opposite applies for a different ratio. Therefore the absolute supply is not important; rather, what is important is the ratio in which the different natural resources are available. Cajander (1909, pp. 8–18) was the first to put forward this theory. He came to this conclusion on the basis of different zones in the vegetation. He believed that there are shifts in the competitive relationships because of the different ways in which species make use of the available natural resources, and this results in a change in the dominance of particular species in vegetations. Subsequently this leads to the different zones that can be observed. This theory is sometimes referred to as the Cajander–Tilman school (Oksanen, 1990).

According to Tilman, the evolution of terrestrial species of plants has taken place on ranges of habitats from soils poor in natural resources but rich in light, to soils rich in natural resources but poor in light. Plants which colonize new, bare soil are believed to be restricted by a shortage of available nitrogen. Therefore the first colonists include many species which fix nitrogen, and over the course of time they form a store of nitrogen which can be assimilated by other species from the organic material they formed in the soil. As a result, larger species can eventually become established, such as trees, so that subsequently light becomes a restricting factor.

Tilman (1985) maintained that plants in the later stages of the succession are larger, grow more slowly, and become fertile later. As an example, he referred to studies of the primary succession in soils which had appeared in Alaska, where glaciers were receding. The first plants to become established included many species which fix nitrogen, such as *Dryas drummondii*, followed by trees which fix nitrogen, such as *Alnus crispa*. Eventually these became overgrown by trees established even later, such as *Populus* sp. (cottonwood trees). On the basis of his resource ratio hypothesis, Tilman (1990) developed the theory that species which invest mainly in roots are strong competitors where nutrients are concerned, but poor with respect to light.

Like Tilman, Vitousek and Walker (1987) stated that there is a difference in the management of nutrients between the primary and the secondary succession. In new, bare, volcanic soil, where a primary succession starts, they found that there was an abundance of phosphates at the beginning, but virtually no nitrogen. After a time, they believe the opposite applies, because more and more nitrogen accumulates in the system and the phosphates slowly but surely disappear. In the primary succession, Vitousek and Walker believe that species which fix nitrogen have a competitive advantage, although they are not always the first colonists by any means, for example, in areas where glaciers have receded. By fixing atmospheric nitrogen, they have a facilitating effect for species which do not fix N_2.

The theories and concepts described in this section explain phenomena such as succession and diversity in vegetation on the basis of the interactions between species of plants and their abiotic environment. Large herbivorous mammals are also mentioned as factors which can have an influence on the vegetation.

2.14 The Role of Large Mammals in Succession

One of the first striking works on vegetation and succession in a system which ascribes an important role to large herbivorous mammals is the book, *Serengeti. Dynamics of an Ecosystem* (Sinclair and Norton-Griffith, 1979). According to these authors, an experiment on the scale of a complete ecosystem took place in the Serengeti. The parameters in the experiment were the changes in the number of animals of particular species and the changes in the distribution of rainfall throughout the year.

Following the occurrence of rinderpest in 1890, 95% of the wildebeest (*Connochaetus taurinus*) and buffalo (*Syncerus caffer*) population died in East Africa in the space of 2 years. As a result of rinderpest, the populations of these two species in the Serengeti in 1961 were 250,000 and 30,000, respectively (Sinclair, 1979a). The disease disappeared from the Serengeti in 1962–1963 because the livestock around the National Park was vaccinated. As a result, the population of wildebeest grew from 250,000 to 500,000, and the number of buffalo increased from 30,000 to 50,000 between 1961 and 1967. This was followed by more rain in the dry season and less rain in the wet season, while the total rainfall for the year remained the same. The higher rainfall in the normally dry season meant that there was more green grass in that period, and therefore less starvation and death among the herbivores. As a result, the wildebeest population increased to 1.3 million in 1977. With a delay of 2 years, the number of buffalo also increased to approximately 70,000 in 1975 (Sinclair, 1979b). Because of the large numbers of wildebeest grazing on the grass, the plants were in a physiologically young stage for a longer time, so that there was green grass for a longer time. Therefore grazing by the wildebeest had a facilitating effect for Thomson's gazelle (*Gazella thomsoni*) (McNaughton, 1979). The latter preferred to forage in places where wildebeest had grazed. This facilitation was also observed with regard to large predators, such as the hyena

(*Crocuta crocuta*) and the lion (*Panthera leo*). They increased as a result of the increased number of potential prey (Hanby and Bygott, 1979). The lion increased because of the increase in the number of prey, such as the buffalo, which are more or less sedentary. On the other hand, the increase in the number of nomadic herbivores such as the wildebeest, resulted in an increase in the hyena population (Hanby and Bygott, 1979).

Another consequence of the large number of wildebeest grazing was that fewer plants went on growing, and therefore fewer plants grew old and withered. This meant there was less dead plant material, less fuel for fire, and therefore fewer fires. Because there were fewer fires, a larger number of acacia seedlings (*Acacia* sp.) survived. The giraffe (*Giraffe camelopardalis*), in particular, feeds on acacia seedlings. Therefore there was an increase in the number of giraffes, and consequently in the intensity with which the acacias were cut back. As a result, they were limited to the size of shrubs. There was a decline in the number of tree-shaped acacias; on the one hand, the seedlings did not reach tree size because they were eaten by giraffes, and on the other hand, the tree-shaped acacias were trampled by elephants (*Loxodonta africana*). The results of different parts of this study are summarized in Fig. 2.5 in a 'feedback' diagram.

Because of the changing conditions, the Serengeti system goes through various stages in which it can become stuck for a longer or shorter time. Therefore the system has different balances (Sinclair, 1979a). Another important conclusion of the results of the study is that although the complex of herbivores is dominated by wildebeest and buffalo (Sinclair, 1979b; Jarman and

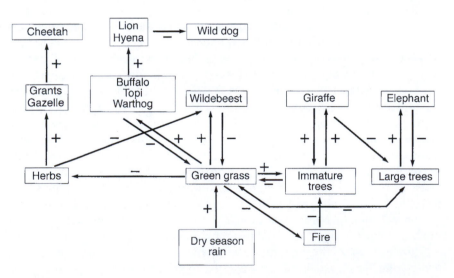

Fig. 2.5. Summary of the interrelationships and their effects on the ecosystem in the Serengeti. Minus signs show that higher values in the previous box result in lower values in the box the arrow points to. Plus signs result in higher values (redrawn from Sinclair, 1979a, p. 23).

Sinclair, 1979), the disappearance of the wildebeest will lead to great changes in the system, while the disappearance of the buffalo will result in very few changes (Sinclair, 1979a). Therefore some animal species fulfil a key function in an ecosystem, such as the wildebeest, while others, such as the buffalo, do not (Sinclair, 1979a). Furthermore, the conclusion is that although there is some facilitation, competition is still the most important process (Sinclair, 1979a). It is the propelling force behind the distribution of the resources (Jarman and Sinclair, 1979). In addition, it is striking that the grazing of wild herbivores is able to prevent the grasses and plants in the vegetation from flowering, to such an extent that the grasses now flower only sporadically (McNaughton, 1979).

In the Netherlands, the results of this study led to a great deal of discussion about the effect of large herbivorous mammals on the vegetation, and whether or not these herbivores had a facilitating effect for species of plants and animals (see, *inter alia*, Van de Veen and Van Wieren, 1980; Van de Veen, 1985; Van der Lans and Poortinga, 1985; Vera, 1986; 1988; 1989). In Germany, a discussion on the so-called megaherbivore theory has developed around the effect of large herbivores on the vegetation.[40] The fact that large herbivores can clearly halt plants' development at a particular physiological stage and bring about a particular succession in the vegetation greatly appealed to the imagination (see Sinclair, 1979a; McNaughton, 1979; Van de Veen and Van Wieren, 1980). Van de Veen (1979) drew the parallel between the situation in Africa and that in Europe as regards the role of large herbivores in ecosystems in the European lowlands, including the Netherlands; others[41] subsequently followed.

These authors believed that it is curious to assume a priori that all sorts of phenomena and processes in vegetations take place without the intervention of large herbivorous mammals, while it has been shown that there are systems in which this group of mammals have a determining effect on the development of the vegetation. This applies not only for the herbivorous fauna of Africa, but also for the domesticated species of Europe, such as cows, horses and pigs. These authors argue that if livestock can determine the succession, the wild forms of livestock, the aurochs (*Bos primigenius*), the tarpan or European wild horse (*Equus przewalski gmelini*) and the wild boar (*Sus scrofa*), which were present in Europe naturally before the introduction of livestock, should, in principle, also have been able to do so. Moreover, during the interglacial periods preceding the Holocene era, the European fauna did not fundamentally differ from the fauna of, for example, Africa and Asia, while with temperatures virtually the same as those of today, modern species of trees, such as the oak, hornbeam and lime, survived in the presence of the wild forms of cattle, the aurochs, wild boar, wild

[40] See Geiser (1992), Beutler (1996), Gerken and Meyer (1996; 1997), Bunzel-Drüke (1997), Bunzel-Drüke *et al.* (1994; 1995) and Müller-Kroehling and Schmidt (1999).
[41] Van de Veen and Van Wieren (1980), Geiser (1983; 1992), De Bruin *et al.* (1986), Vera (1988), Overmars *et al.* (1991), Beutler (1992), May (1993), Bunzel-Drüke (1997), Bunzel-Drüke *et al.* (1994), Beutler (1996), Gerken and Meyer (1996; 1997), Van Wieren (1998), Wallis de Vries (1998), Olff *et al.* (1999).

horse, as well as a species of elephant (*Palaeoloxodon antiquus*) and a species of rhinoceros (*Dicerorhinus kirchbergensis*) (Geiser, 1983; 1992; Beutler, 1992; May, 1993; Bunzel-Drüke *et al.*, 1994; Kahlke, 1994, pp. 98–99). The question which arises is why these large herbivores in Europe could not have an influence on the succession, while species of trees survive.

Van de Veen and Van Wieren (1980) considered that the processes which took place in the Serengeti had analogies in the natural world of Europe, because the feeding strategies of European herbivores are basically no different from those of the herbivores of Africa and the other continents. They based this analogy on research by Hofmann (1973) into the structure of the digestive system of ruminants in relation to the digestive physiology and choice of food eaten by different species of herbivores. Within the ruminants, Hofmann (1973; 1976) distinguished three main strategies, namely, browsers, intermediate feeders and grazers, which are found both in Africa and North America, as well as in Europe (see Fig. 2.6). The browsers feed mainly on leaf buds, leaves, twigs and the bark of trees and shrubs, as well as herbs. The grazers are specialized grass eaters, while the intermediate feeders take up an intermediate position; their choice of food alternates between grasses and woody plants (Van de Veen, 1979, p. 225; Van de Veen and Van Wieren, 1980). In the European lowlands, the elk and roe deer are browsers; the red deer and European bison are intermediate feeders, and the auroch is a grazer (Van de Veen, 1979; Van de Veen and Van Wieren, 1980; Hofmann, 1973; 1976; 1985; Van Wieren, 1996). The horse is also a specialized grass eater, but is not included in this classification as it is not a ruminant (see Fig. 6.25).

According to Van de Veen and Van Wieren (1980), the European bison (*Bison bonasus*) and aurochs probably had a complementary effect on the vegetation because of their feeding strategy and their size. They believed that the European bison played an important role with regard to inhibiting the succession in the gaps in forests created by storms, and by creating small open spaces in the forest by systematically removing the bark from trees in winter. This means that some of the tree trunks died and consequently more light penetrated to the forest floor, resulting in natural regeneration. Because the aurochs will have grazed on the old grass in these open spaces, at least locally, Van de Veen and Van Wieren (1980) believed that permanent grazed areas developed there, where plants with good defensive qualities could become established, such as broom (*Genista anglica*), gorse (*Ulex europaeus*), hawthorn (*Crataegus monogyna* and *C. laevigata*) and juniper (*Juniperus communis*). Subsequently, oak, beech, hazel, aspen, ash and crab apple will have become established in this scrub, a process which they maintain is still taking place in the lowlands of Central and Western Europe (see Figs 3.6, 4.2, 4.3 and 4.5). In the forest itself, the regeneration is restricted to holly and yew, while there is bush encroachment with thorny scrub in the open areas. Eventually a savanna-like landscape develops as a result of the interaction between plants and animals, with bushes, solitary trees and groups of trees. In order to restart these processes, which Van de Veen and Van Wieren described as natural processes, the original indigenous large

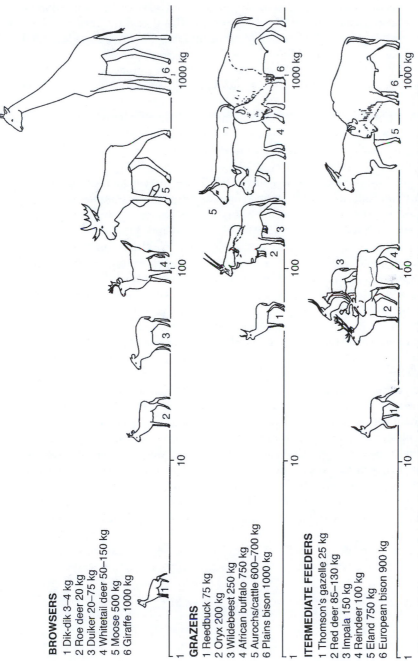

BROWSERS

1 Dik-dik 3–4 kg
2 Roe deer 20 kg
3 Duiker 20–75 kg
4 Whitetail deer 50–150 kg
5 Moose 500 kg
6 Giraffe 1000 kg

GRAZERS

1 Reedbuck 75 kg
2 Oryx 200 kg
3 Wildebeest 250 kg
4 African buffalo 750 kg
5 Aurochs/cattle 600–700 kg
6 Plains bison 1000 kg

ITERMEDIATE FEEDERS

1 Thomson's gazelle 25 kg
2 Red deer 85–130 kg
3 Impala 150 kg
4 Reindeer 100 kg
5 Eland 750 kg
6 European bison 900 kg

Fig. 2.6. Hofmann's classification (1973) of large ruminant herbivores according to their feeding strategy and the structure of the stomach, originating from North America, Africa and Europe (redrawn from Van de Veen, 1979, p. 132).

herbivores will have to be reintroduced into natural areas in the Netherlands. They considered that these herbivores are essential and integral elements of the indigenous ecosystems. Because they create and maintain a landscape which in its totality contains the biotopes of many species of indigenous flora and fauna, their grazing therefore has a facilitating effect for many species of plants and animals. Although Van de Veen and Van Wieren (1980), as well as the other above-mentioned authors, cast doubt on the theory of the closed forest as being the natural vegetation, their criticism of this theory is based, above all, on the data relating to fauna. The above considerations have led in the Netherlands to experiments with grazing by large herbivores living in the wild, including horses and cattle, with related research.[42]

Apart from the study carried out in the Serengeti, the fact that large herbivorous mammals have an influence on succession in vegetation and on the climax as integral components of the ecosystems is also shown by a study in the Lake Manyara National Park in Tanzania (see Drent and Prins, 1987). The herbivorous fauna in that area consists, *inter alia*, of elephant, buffalo, giraffe, wildebeest, black rhinoceros (*Diceros bicornis*), hippopotamus (*Hippopotamus amphibus*), impala (*Aepyceros melampus*) and Burchell's zebra (*Equus burchelli*). Drent and Prins (1987) found that the herbivorous fauna consumed at least 80% of the total annual production of grasses, plants and sedge every year. Elephant and buffalo account for the majority of this, i.e. 43% and 35% of the total production, respectively. With the high consumption that was observed, the number of herbivores fluctuated for about 25 years around a particular equilibrium with a biomass of approximately 175 kg ha^{-1}.

According to Drent and Prins (1987), restricting the animals' freedom of movement can result in dramatic changes, such as localized overgrazing. This does not automatically mean that there is an unnatural situation in terms of disturbance; in fact, the herbivore is a prisoner of its food supply. Therefore, under natural conditions, the vegetation cannot by definition be disturbed by indigenous herbivores, even when the pressure of grazing is extremely high, as in Lake Manyara, because the herbivores and all their effects on the vegetation are integral components of the system. Van Andel and Van de Bergh (1987) also share this view. With regard to the grasslands of Europe, they maintain that when the grasslands are a result of grazing, and if the species of herbivores responsible for these grasslands by their grazing are indigenous species, the grasslands must be considered as a natural phenomenon.

---

[42] See, *inter alia*, Vulink and Drost (1991a,b), Vulink and van Eerden (1998), Wallis de Vries (1994; 1998), Baerselman and Vera (1995), Vera (1988; 1998), Cornelissen and Vulink (1995; 1996a,b), Helmer *et al.* (1995), Huijser *et al.* (1996), van Wieren (1996), van Wieren and Kuiters (1997), Anonymous (1999), Groot Bruinderink *et al.* (1999), Jorritsma *et al.* (1999) and Krüger (1999).

2.15 Conclusions and Synthesis

This chapter looked at the prevailing theories and the way in which they were developed. The most important findings and comments on these are presented below.

- In the 19th century and the first half of the 20th century, there was a growing conviction, and theories were formulated, that in places where the composition of the soil and the climate would allow trees to grow in the lowlands of Europe and in the east of the United States, the original vegetation consisted of a closed forest.

The basis for this view is the observed succession of vegetation in old fields and pastures towards forests in Europe and North America. The fact that the last vegetation considered to be virgin vegetation, the mountainous forests of south-east Europe, consisted of closed forest, was seen as supporting this theory. Historical sources which described the last wildernesses as 'silva', a term that now means 'forest', were also seen as confirmation of the theory. These theories then formed the basis for the models developed in plant ecology and forestry for the regeneration of primeval (deciduous) forests in Central and Western Europe.

- This theory does not take account of large herbivores, because it is assumed that these do not play a meaningful role in the original vegetation in the process of the regeneration of trees, or in the succession in these vegetations in general.

The fact that large herbivores do not play a significant natural role as regards the vegetation is deduced from historical sources. These are used to show that livestock destroy the germinated seedlings of trees in the forest by eating them and trampling on them. On the basis of these historical data, researchers such as Moss, Tansley and Watt concluded that the grazing of livestock resulted in a retrogressive succession from forest to grassland or heathland. They believed that this conclusion is supported by the fact that in woodland and meadows grazed by livestock, where parts are fenced off, trees and shrubs spring up spontaneously. When the grazing stops in grasslands, they believe that the original forest vegetation will be restored. These researchers did not think that there is an equivalent in Europe and in the east of the United States, in nature, of the effect which livestock have on the natural vegetation. Wild herbivores, such as the red deer and roe deer, were thought to have occurred in such low concentrations that they did not have a significant influence on the natural vegetation. The low concentrations of red deer and roe deer in the last primeval forests, the mountainous forests of the

Balkans, were seen as the natural situation, which confirmed that wild herbivores could not have a determining influence on the natural vegetation. The bison is not considered to be a herbivore that can have had any influence on the vegetation in the east of the United States analogous to livestock.

> - Watt and Leibundgut both developed a model for the regeneration of the original primeval forest. Watt's model was based on the succession of grasslands to forest, after the end of grazing by livestock, and Leibundgut's model was based on the primeval forests in the mountains. In both models the regeneration takes place in gaps in the canopy (gap phase model), and large herbivores do not play a significant role.

Subsequent researchers did not question these models, but simply built on them. In order to explain certain phenomena with regard to species diversity and regeneration, these researchers superimposed catastrophes such as fire and storm. Aspects such as nutrients and light were also taken into account. Large herbivores were sometimes mentioned to explain the diversity in the vegetation, but without looking at the role which they played in natural systems.

> - Research in Africa indicates that indigenous large herbivores certainly do determine the succession as integral components of the ecosystem. In Europe and the east of the United States, they are generally not considered able to do this in natural conditions. An exception are the authors who believed that, on the basis of the similarities in the fauna, the European situation can be considered as being analogous.

The research in the Serengeti in particular revealed that large herbivores, combined with climatological conditions (rainfall), can be an important determining factor in the ecosystem, and have a great influence on the dynamics in the vegetation. Some species of herbivores appear to be more important than others (key species). The feeding strategies appear to be very important for the effect on the vegetation. In addition, certain species of herbivores facilitate the regeneration of certain species of plants. On the basis of the analogy between the feeding strategies of large herbivores in Africa and Europe, it is also possible to make an analogy regarding the facilitating and monitoring effect in the European situation.

> - The working hypothesis adopted by Watt *et al.*, that the grazing of livestock results in a retrogressive succession, has never been tested. However, many of the observations made by the researchers of this work allow for a retrospective evaluation of this working hypothesis.

Where the succession of grassland to forest is presented as a succession with-out grazing, it appears that in the grasslands on which this view is based, there is regeneration of trees where there is grazing by cattle. Therefore Watt inter-prets something which takes place *with* grazing as a phenomenon which takes place *without* grazing. He does *not* show that the grazing of cattle results in a ret-rogressive succession from forest to grassland; on the contrary, he demonstrates that *all* species of trees and shrubs, including beech, as well as pedunculate and sessile oak, *do* regenerate in the presence of large herbivores. Therefore he inter-prets his observations in line with the present theory. Moreover, his observations show that oak and hazel regenerate only in thorny scrub in an open landscape where animals graze.

To summarize, the above findings lead to the conclusion that the working hypothesis that the grazing of livestock results in a retrogressive succession from forest to grassland does not apply as a general hypothesis. On the basis of the results of Watt's study, it can even be rejected. His research shows that the regeneration of trees certainly does take place where there is grazing of live-stock, though it is *outside*, rather than *in* the forest. This is also true for the east of the United States. There as well, regeneration of trees occurs outside the for-est in grassland grazed by livestock. The assumption that wild herbivores do not have a determining effect on the succession in the original vegetation is based on observations of the succession *without* large, wild herbivores. Large herbi-vores were present in the original system. Therefore the assumption is based on a circular argument. The 'gap phase' model is based on the same assumption and therefore on the same circular argument.

The assumption that wild herbivores do not have a determining influence on the succession in the original vegetation leads to the view that where wild animals are found, they only occur in low concentrations, because the wild ani-mals do not have an influence on the vegetation. The starting point in this is that *because* the natural vegetation is a closed forest, the wild animals must have existed in low concentrations because they would have had an influence on the forest if they had existed in high concentrations.

The presumed analogy between the mountain forests in the Balkans and the original primeval forest in the lowlands of Central and Western Europe is put forward as proof to support these assumptions. However, these mountain forests cannot be seen as being analogous to the vegetation in the European lowlands, because they are situated at altitudes where the two species of oak cannot grow, and where certain species of large herbivores (aurochs and tarpan), which were present originally in the lowlands, are missing. The bison, which traditionally appeared in the vegetation zone of the deciduous forest, is not considered at all in the development of the vegetation in the east of the United States. The closed forest is assumed to be the natural vegetation. This large herbivore, analogous to large herbivores in the lowlands of Europe, is, in natural conditions, not seen as a determining factor in the succession in the original vegetation. It is therefore very questionable whether the results of studies of these forests can serve as a guideline for what took place in the untouched vegetation in the lowlands of Europe, given the preconditions. Within the context used for theory forming, the lack of fire has been presented as an explanation for the suppression of the oak genus by shade-tolerant species such as maple, ash, elm and beech. This hypothesis was also explicitly defined in the context that grazing by indigenous large grazers, the bison in the natural conditions, does not play an important role in the succession of the original vegetation considered to be a forest.

Palynology, the Forest as Climax in Prehistoric Times and the Effects of Humans

3.1 Introduction

The results of the search into fossil pollen collections have contributed to the theory that Europe was originally covered by a natural closed forest.[1] In this chapter I will address the following questions:

- What are the interpretations of fossil pollen collections based on?
- What are the starting points and facts which support the conclusion that the virgin vegetation in the prehistoric Holocene era was a closed forest?
- Can fossil pollen material also be interpreted as indicating a more open landscape, consisting of alternating meadows, scrub, trees and groves, as postulated by the alternative hypothesis?

At the end of this chapter, I will discuss the situation in the east of the United States. Based on pollen analyses, it was assumed for this as well that it was originally covered with a closed forest.[2] As mentioned in the previous chapter, this forest vegetation is considered comparable to the forest vegetation originally present in Europe.

As virtually all the fossil pollen material from Western and Central Europe is dominated by species of trees (see, *inter alia*, Bertsch, 1949; Firbas, 1949; 1952; Godwin, 1975; Huntley and Birks, 1983), a park-like landscape would have deposited pollen material in the sediment, dominated by the pollen grains of trees. To find out whether this is actually the case, I have made use of

---

[1] See, *inter alia*, Nietsche (1927), Firbas (1935), Tansley (1953, pp. 154–169), Ellenberg (1954), Rackham (1980, pp. 97–109), Peterken (1996, pp. 33–45), Prins (1998) and Bradshaw and Mitchell (1999).

[2] Watts (1979), Davis (1984), Delcourt and Delcourt (1987; 1991, pp. 90–91), Webb (1988), Roberts (1989, pp. 72–74), Barnes (1991), Tallis (1991, pp. 235–263), Peterken (1996, pp. 45–53).

studies of modern pollen material in wood-pastures. Despite the objections inherent in the use of modern pollen spectra (pollen analyses of modern vegetations), they can serve as an important key to shed light on prehistoric times (see Grimm, 1988; Faegri and Iversen, 1989, p. 122). Up to now, the research has been limited, and only recently publications have tackled the problem of the representation of tree pollen in the pollen spectra of park-like landscapes.[3] Therefore there are still only a few publications available to give access to these data. In addition, I have looked at species in pollen collections which are known to vegetation experts as being species found in open grassland, and in mantle and fringe vegetation; types of vegetation which are known to be characteristic of landscapes where large herbivores graze. These include grasses (*Gramineae*), species of the parsley family *Umbelliferae* and goosefoot family *Chenopodiaceae*, nettle (*Urtica* sp.), sorrel (*Rumex* sp.), and hazel (*Corylus avellana*).[4]

I have based my interpretation of the data on two synonymous principles, both used in palynology and palaeoecology for reconstructing the past: methodological uniformitarianism and actualism. These two principles form the only key to the past for which there are no written sources or illustrations. They are the foundation of the picture which we have built up of the past. This means that the past can be interpreted on the basis of modern phenomena. To do this, it is assumed that the ecological characteristics of species have not changed through the ages. At the same time, this means that there are limitations on the use, namely, that an understanding of the past can never extend beyond the knowledge and understanding of the present.

For the reconstruction of the past with the help of the present, analogies are also often used as an *inductive* argument. This reasoning is based on the assumption that when similar matters have a few common characteristics, they will also share other characteristics.[5] I will also make use of this method.

3.2 The Reconstruction of Prehistoric Vegetation

In 1916, the Swedish geologist, Von Post, laid the basis for modern palynology when he presented the first 'modern' pollen diagram (Davis, 1967; Manten, 1967; Delcourt and Delcourt, 1987, p. 29; Real and Brown, 1991). Before that time, attempts had been made to reconstruct the vegetation of the past on the basis of macro-fossils of plants, such as leaves, fruit and woody parts found in peat (Iversen, 1973). The Danes, Dau (1829) and Steenstrup (1841), who were mentioned in Chapter 2, used this method. In 1829, Dau suggested that

---

[3] Gaillard *et al.* (1992), Jackson and Wong (1994), Sugita (1994), Sugita *et al.* (1999), Calcote (1995; 1998), Hjelle (1998), Broström *et al.* (1998).
[4] See *inter alia* Watt (1923; 1924; 1925; 1934a,b), Tüxen (1952), Tansley (1959, pp. 259–266), Müller (1962), Dierschke (1974), Westhoff and Den Held (1975, pp. 115–125, 239–241), Behre (1981), Rodwell (1991, pp. 333–351), Oberdorfer (1992, pp. 81–106) and Pott (1992, pp. 270–341).
[5] See Scott (1963), Gould (1965), Birks (1973; 1981; 1986), Rymer (1978), Shea (1983), Roberts (1989, p. 21), Dodd and Stanton (1990, pp. 5–12, 15), Delcourt and Delcourt (1991, p. 2) and Bell and Walker (1992, p. 15).

the widespread occurrence of tree stumps of conifers no longer found in Denmark, in the peat bogs of the Danish island of Zealand, told part of the history of the forest in that country (Iversen, 1973, p. 11).

On the basis of the layers of peat and the subfossil stumps and leaves of particular species of trees, Steenstrup concluded in 1841 that there had been a succession of tree species in the prehistoric forests of Denmark. He divided what he called the history of the forest into four periods, an aspen, pine, oak and alder period (Clements, 1916, pp. 14–16; Iversen, 1973, p. 12). The Swedish botanist, Nathorst, was the first to find remnants of arctic flora with dwarf shrubs such as mountain avens (*Dryas octopetala*), least willow (*Salix herbaceae*), (*S. polaris*) and reticle willow (*S. reticulata*) in the clay under the peat (Nathorst, 1870, 1873, in Clements, 1916, p. 21; Iversen, 1973). In the layers above this, he found the fossil remains of species which require more warmth, such as birch (*Betula* sp.), pine (*Pinus* sp.) and willow (*Salix* sp.).

On the basis of these layers, which he described as 'forest beds', he assumed that temperatures had gradually changed since the Ice Age up to the present time. The Norwegian botanist, Blytt (1876), elaborated on this and concluded from the appearance of tree stumps above a particular level that there were alternating dry and wet periods. He believed that trees germinated in the peat during dry episodes, because it dried out superficially (Clements, 1916, pp. 21–22; Iversen, 1973). The Swedish botanist, Sernander, elaborated further on this theory, and interpreted the difference in the structure of the layers in the peat as reflecting climate fluctuations. On this basis, he divided the period following the Ice Age into four climatological episodes, the Boreal, Atlantic, Sub-Boreal and the Sub-Atlantic episodes (Sernander, 1881, 1894, 1895, 1899 etc., in Clements, 1916, pp. 22–24; Sernander, 1908, in Janssen, 1974, p. 52). He believed that the Boreal period was hot, the Atlantic period was hot and damp, the Sub-Boreal period was hot and dry, and the Sub-Atlantic period was cool and damp. Nowadays, this classification is known as the theory of Blytt and Sernander (Janssen, 1974, p. 52 and see Fig. 1.4).

As stated above, Von Post was the first person to present a 'modern' pollen diagram. He noted that it was not possible to make any quantitative decisions about species of trees on the basis of macro-fossils, because coincidences in their dispersal played too large a role. However, he did think it was possible to do so using tree pollen.[6] He said that with the exception of *Populus*, there are

---

[6] In his presentation in 1916, Von Post posed the following rhetorical question:

> is it possible to use fossil material preserved in peat deposits for tracing the postglacial history of a terrestrial plant at all? Yes, it is, but only if a plant is represented by fossil remains evenly, and in great quantity, uninfluenced by the genesis of the enclosing peat type. Only under these conditions could the absence of a fossil have the indicator value, and only under these conditions might the possibility exist for reliable evaluation of any changes that may have occurred in the plant's frequency.
>
> Every observer in nature is well aware that pollen of certain anemophilous plants, especially trees, is scattered over the terrain in enormous quantities during the flowering period. In the forests and in their vicinity this pollen can form deposits several millimeters thick, covering the ground evenly, and in lakes and rivers the pollen rain even causes a sort of water bloom (Von Post, 1916, p. 38).

well-preserved grains of pollen present in the peat of all the trees which he knew
to exist in the Swedish forests. It was possible to identify them, in terms of the
genus to which they belonged. If the sediment where the pollen is found had not
been disturbed, he believed that the pollen flora in these sediments reflected the
average composition of the pollen rain at that time. In its turn, the pollen rain
reflects the type of forest which produced this pollen rain at the time.

In his analyses of pollen flora of the period from the end of the Ice Age up
to modern times, Von Post found that there was a certain change in the order
of dominance of the genera. He interpreted these in the light of changes in the
climate which Sernander had deduced from the different layers of peat, and
concluded that the Blytt–Sernander climatological periods were synchronized
with the changes in the pollen flora throughout the world. In his opinion, this
demonstrated that the changes which occur in the pollen flora in the different
peat profiles can be correlated without looking separately at the stratification
of the peat bogs.

Von Post examined the pollen of tree species only, because like all the
other early palynologists, he was not yet familiar with pollen from any herbs
(C.R. Janssen, Utrecht, 1997, personal communication). He calculated the
percentages of pollen on the basis of the total sum of pollen from trees of the
species in the forest. He interpreted the diagrams on the basis of the structure
and the species diversity of modern forest vegetation, because he saw these as
modern analogies of prehistoric vegetation. Therefore he assumed that the
virgin prehistoric vegetation was forest. In this respect, he was no different
from Dau, Steenstrup, Nathorst, Blytt and Sernander. They all published work
on what they called the history of the forest.

As we noted in the previous chapter, there was absolutely no doubt at the
beginning of the 20th century that the original vegetation of Europe was forest. Wholly in line with this, Von Post classified the genera *Quercus* (oak), *Tilia*
(lime) and *Ulmus* (elm), in a category which he called the *Quercetum mixtum*,
the mixed oak forest, a type of forest found in southern Sweden in which oak
(*Quercus* sp.), lime (*Tilia* sp.) and elm (*Ulmus*) formed the tree layer, while hazel
formed a shrub or lower layer. This type of forest is the so-called coppice-with-
standards (see Fig. 4.14). Because the hazel forms the shrub layer in these
modern forests, Von Post kept the sum of the pollen of the hazel outside the
total sum of the pollen from trees.[7]

---

[7] As I wanted to be able to read the relative displacements among the competing forest types
– not only among the individual tree species – I have treated the mixed-oak-forest pollen,
that is, the sum of pollen percentages of oak, lime and elm … as a unit. For the same reason
I have not included hazel pollen in the sum, but instead have only given its frequency as
percent of the sum. This is because hazel occurs mostly as a shrub layer in mixed oak-
forest, and forms only exceptionally a separate community competitive with other forest
types (Von Post, 1916, p. 466).

Although I do not know, the fact that Steenstrup (1841) also found large quantities of hazel
nuts in the layers of peat where he found the oak stumps, possibly contributed to Von Post's
view that hazel grew underneath the oaks in prehistoric times.

Von Post distinguished three groups of pollen: a group of birch (*Betula* sp.) and pine (*Pinus* sp.), a group of the mixed oak forest, and a group of beech (*Fagus*), Norway spruce (*Picea*) and hornbeam (*Carpinus*). He believed that the period of the mixed oak forest coincided with the post-glacial warmer period which comprises the Boreal, the Atlanticum and the Sub-Boreal periods. He thought this came to an end at the same time as the transition from the Bronze Age to the Iron Age (see Fig. 1.4).

3.3 The Forest in Palynology

Palynologists who published research after Von Post into the pollen flora dating from after the Ice Age throughout Europe, always wrote that their publications were concerned with the reconstruction of the history of the forest (see Bertsch, 1929; 1932; 1949, p. 4; Tschadek, 1933; Firbas, 1934; 1935; 1949, p. 1; Godwin, 1934a,b). In their diagrams they also gave the total sum of tree pollen as 100%, and for the reasons given by Von Post (1916), they excluded the levels of hazel pollen from the total sum of tree pollen (see Fig. 3.1). For them, it was self-evident that the closed forest was the natural vegetation, just as it had been for Von Post, an a priori and not an a posteriori view. As we remarked before, the theory of succession and the discovery of stumps of trees in the peat contributed to this.[8] The pollen of grasses, heather and herb species were also excluded from the total sum of pollen. The reason was that the experts wished to make a diagram in which it was certain that the pollen did not come from the peat, but from the vegetation around the peat, and because grasses and herbs could not be identified from the pollen (C.R. Janssen, 1997, Utrecht, personal communication). The first person to examine types of pollen from herbs was Firbas (1934; 1935).

Firbas (1934) wondered how the Ice Age vegetation with arctic species such as mountain avens (*D. octopetala*) was replaced by trees at the end of the Ice Age.[9] According to Firbas, very little was known about the length of time

---

[8] The fact that the discovery of tree stumps in peat bogs contributed to the view that the prehistoric vegetation was a closed forest, in addition to the theory of succession, is shown by comments made by Godwin (1934b) and Nietsch (1939). Godwin (1934b) said: 'A very large component of the pollen rain incorporated in these deposits has always been the pollen of the forest trees, for these are anemophilous and over large areas are the natural climatic dominants' (Godwin, 1934b, p. 278). He added that the history of the forest had actually been written before the development of pollen analysis by the research of Blytt and Sernander (also see Clements, 1916, pp. 8–32; Iversen, 1973). Concerning the results of the pollen study, Nietsch (1939) wrote: 'Es konnte auf diesem Wege vieles bestätigt und weiter ausgeführt werden, was schon früher die allgemeine Auswertung des Handhaften Torffunde, auch die Bestimmung von Holzstücken, Holzkohle und Samen aus vorgeschichtliche Funden an Einblicken in den Einstigen Landschaftszustand ergeben hatte' (Nietsch, 1939, p. 6).

[9] He wrote:

Je eingehender wir aber die nacheiszeitliche Waldentwicklung kennen lernen, und je dringender wir hierbei nach ihren Ursachen fragen, um so mehr wendet sich das Enteresse jenen frühen, hoch und spätglazialen Zeitabschnitten zu, in denen der Wald in Mitteleuropa offenbar noch eine sehr ungeordnete Rolle spielte (Firbas, 1934, p. 109).

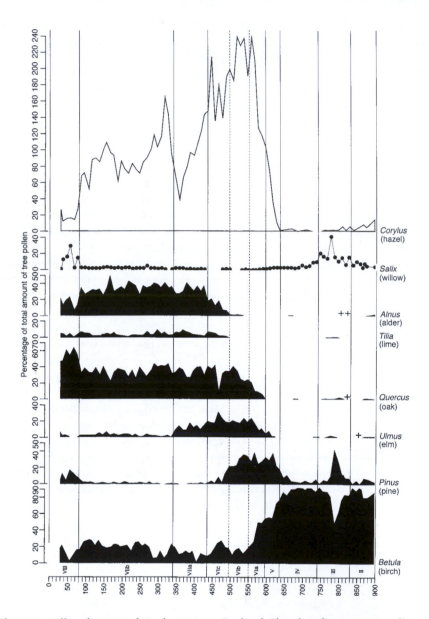

Fig. 3.1. Pollen diagram of Hockam Mere, England. The classification into pollen zones used is that of Godwin. The high percentage of hazel pollen (*Corylus*) compared with oak (*Quercus*), lime (*Tilia*) and elm (*Ulmus*) is striking. The percentage of hazel pollen was kept outside the sum of pollen of trees, and is therefore shown in white. During a period when it was assumed that there was a virgin closed forest, i.e. during the Atlantic period (zone VII), this percentage was just as high as that for all of the species of trees together. Relative percentages which fall below 0.5% in this calculation are shown with a + sign (redrawn from Godwin and Tallantire, 1951, p. 292).

during which there was no forest, where the tree line lay in Europe during the Ice Age, and how the forest developed. He wondered whether it would be possible to plot the areas where there were no forests or sparse forests, with the help of pollen analyses, and whether it was possible to determine the density and the species diversity of the forest from pollen in sediments. If this were done, he considered that it would be possible to determine how the tree line moved in treeless vegetation in Europe. In order to find the answers to the questions raised above, Firbas (1934; 1935) was the first to point out the importance of types of pollen which did not come from trees as the perfect indicators for treeless vegetation. He believed that a pollen analysis of the treeless plant communities could also provide some insight regarding any layers of plants that might have been present in the forests.

He answered the above questions with the help of the transition from the tundra to the forest in Lapland. In this region, he took surface samples of pollen over an area of open tundra up to the Boreal birch–pine forest (see Firbas, 1934). He considered sampling the open terrain towards the forest and into the forest, because it was analogous to the advance of the forest front in open regions with arctic species at the end of the Ice Age. He applied a time/space substitution. In addition, he took pollen samples in Germany in peat moors and in fields surrounded by forests to find out whether the pollen collections formed a reflection of the forest. He also sampled forests with a rich undergrowth of heather to examine whether the plant layer could be found in the pollen samples. The most important conclusions of Firbas's research are given below:

- The succession of the arctic tundra and sub-arctic (Boreal) bands of vegetation can easily be identified in the pollen spectra, even without looking at the pollen which is not from trees.
- The combined ratio of pollen from trees and pollen that is not from trees demonstrates the density of the forest with an adequate degree of reliability. The forestation of open areas and the density of forests can therefore be determined with the help of pollen analyses. The absence of trees or sparse density of forest is clear from relatively low percentages of pollen that is not from trees, compared with percentages of tree pollen. (Converted into the total pollen sum of tree pollen and pollen that is not from trees, as used nowadays, the percentages of pollen not from trees accounts for between 20 and 26%.)
- It is only in areas with few or no forests that the percentage of pollen that is not from trees exceeds that of tree pollen.

The possibility of identifying small open areas by means of pollen analyses is fairly limited. In the pollen spectrum, the tree pollen exceeds the pollen that is not from trees. On the basis of the analytical research into pollen in Central Europe, Firbas (1935) drew up the following classification of vegetation after the Ice Age (see also Fig. 3.1):

- A first, treeless period following the high point of the Ice Age.
- A second, sub-arctic period with pure birch–pine forests.

- A third, pre-Boreal period in which species of trees appeared for the first time, and are characteristic of higher temperatures and more fertile soil, i.e. hazel, oak, elm and (later) lime.

Firbas (1949, p. 49) called the last period the start of the 'Wärmezeit', when he believed the so-called mixed hazel–oak forest evolved. To summarize, Firbas (1934; 1935) concluded that the European lowlands north of the Alps became covered after the Ice Age with a thick forest, so that the sub-arctic steppe which had been there initially disappeared. During the sub-arctic period, birch and Scots pine became established. The percentage of pollen which was not from trees then reached levels of 10% and lower of the total pollen sum. He believed that this should be interpreted as the establishment of a closed forest. In his opinion, mountain slopes covered with small and large rocks, sandy deposits, chalky soil and dunes, which developed during the Ice Age along the ice cap are all places where species of plants from the vegetation of the treeless period have survived. Because of the composition of the soil, the forest would have remained sparse there quite naturally, so that the species of plants of open terrain could survive. Apart from the treeless areas above the tree line in the high mountains, Firbas believed that these open areas had too small a surface area to be identified by means of pollen analyses. He did not think that their absence detracted from the general picture of the prehistoric landscape; this was without any doubt a closed forest.

Firbas's conclusions (1934; 1935) are still supported in palynology and palaeoecology today.[10] On the basis of pollen studies, it is now thought that the closed forest developed more or less as follows in north-west and Central Europe at the end of the Ice Age (see Figs 3.1 and 3.2).[11] Initially, the forest consisted of birch and pine, and subsequently hazel became established. Then oak, lime, elm and ash developed as well as alder (*Alnus*) in the marshlands. These species largely or completely suppressed the birch and pine. The forest consisting of oak, elm, lime and ash is known as the Von Post mixed oak forest. Hazel, oak, elm, lime and ash came from their Ice Age refuges in the south or south-east of Europe to the north-western part of Europe, sometimes at long intervals. It was only in southern Germany, which is close to the Ice Age refuges of these species,

[10] See *inter alia* Iversen (1941), Roberts (1989, p. 71), Bell and Walker (1992, p. 156), Bradshaw (1993), Edwards (1993), Simmons (1993), Walker (1993), Waller (1993; 1994) and Bradshaw and Mitchell (1999).
[11] See Von Post (1916), Godwin (1934a,b; 1944; 1975), Firbas (1949, pp. 49–53; 1952), Nietsch (1935; 1939), Iversen (1973, pp. 54–55), Bertsch (1949), Polak (1959), Van Zeist (1959), Van Zeist and van der Spoel-Walvius (1980), Andersen (1973; 1976; 1989; 1990), Janssen (1974, pp. 55–65), Steel (1974), Planchais (1976), Morzadec-Kerfourn (1976), Huault (1976), Girling and Greig (1977), Van Geel *et al.* (1980/81), Greig (1982), Huntley and Birks (1983), Huntley (1988; 1989), Birks (1986; 1989), Delcourt and Delcourt (1987), Perry and Moore (1987), Chen (1988), Kalis (1988), Bennett (1988a,b,c), Bartley *et al.* (1990), Bennett *et al.* (1991), Day (1991), Bozilova and Beug (1992), Horton *et al.* (1992), Latalowa (1992), Ralska-Jaiewiczowa and Van Geel (1992), Rösch (1992), Caspers (1993), Peglar (1993), Waller (1993), Björkman (1997; 1999), Björkman and Bradshaw (1996) and Bradshaw and Holmqvist (1999).

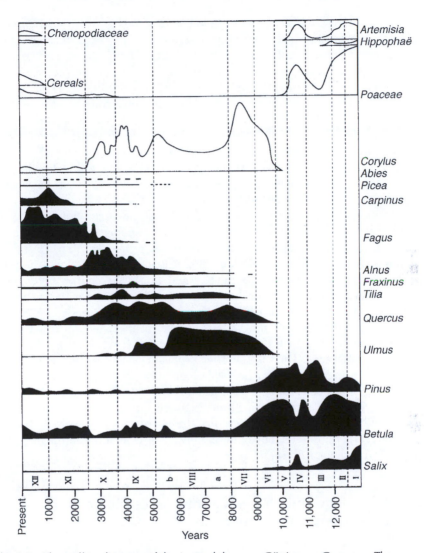

Fig. 3.2. The pollen diagram of the Lutter lake near Göttingen, Germany. The classification into pollen zones is in accordance with Overbeck's classification. The frequency of the pollen of hazel (*Corylus*) was excluded from the sum of the tree pollen. The diagram shows how the various species of trees and hazel shrubs colonized Central Europe after the Ice Age, one after the other. It is striking to see how high the percentage of hazel pollen remains during the period when there was a virgin primeval closed forest, according to the current theory, i.e. in period VII, the Atlantic period (8000–5000 BP) (redrawn from Jahn, 1991, p. 392).

that they appeared at more or less the same time (Firbas, 1949, quoted by Godwin, 1975b). Oak had refuges throughout the south and south-east of Europe (Huntley and Birks, 1983, p. 355; Bennett *et al.*, 1991). This species colonized progressively across the whole width of the continent at the rate of approximately 500 m year^{-1} (Delcourt and Delcourt, 1987). At the start of the Boreal period (9000 BP), the oak was present throughout Central and Western Europe. This did not apply to the lime, which had a refuge in the Ice Age in south-eastern Europe (Huntley and Birks, 1983, p. 395; Bennet *et al.*, 1991). This species migrated northwards fastest in the east of Europe. The lime arrived only 2000 years after the oak in the south and south-east of England, i.e. in approximately 7500 BP (Birks, 1989). The British islands which were furthest removed from the Ice Age refuges for the species of deciduous trees, with the exception of the hazel, were respectively colonized in 10,000 BP by birch, in 9500 BP by pine, hazel, oak and elm, in 8000 BP by alder (*Alnus*), in 7500 BP by lime, and in 6000 BP by ash (Birks, 1989).

According to this view, the mixed oak forest developed fully during the Atlantic period (8000–5000 BP). Because of the dominance of oak pollen, it was even known as an oak age (Nietsch, 1935, p. 56; Firbas, 1949, p. 169) or an Atlantic oak forest (Tansley, 1953, p. 163). Floodplain forest along the great rivers in Europe such as the Saale, the Elbe and the Sava which were dominated by oak, were considered by some authors as a modern analogy of the prehistoric mixed oak forest (see Nietsch, 1927; 1939, pp. 102 and 105; Iversen, 1973). The beech appeared in Central and Western Europe relatively late. In the south of England, which is the furthest away from the Ice Age refuges of this species, the beech eventually arrived in 3000 BP. The hornbeam came even later. In the south of Sweden, the beech arrived as the last deciduous species around 1450 BP.

The appearance of these two species was restricted in Great Britain to the south of England. Central Europe is distinguished from the north-west by the arrival of Norway spruce and silver fir. These two species also arrived there fairly late. The high frequency of hazel pollen[12] is evident in all the diagrams of Western and north-west Europe.

3.4 Palynology and Theories about Succession and the Forest as the Climax Vegetation

Firbas considered the order in the dominance of different tree species during the colonization following the last Ice Age as an analogy of the secondary succession in open spaces in the forest (Firbas, 1949, pp. 275–277). This analogy is still used in palynology. It is also assumed that the regeneration of forests in prehistoric times took place in the way that it does in modern forests, according to the prevailing theory, namely, in gaps in the canopy (see Iversen, 1960, p. 7; 1973 p. 72; Birks, 1986; 1993; Delcourt and Delcourt, 1987, p. 378;

---

[12] See *inter alia* Erdtman (1931), Bertsch (1949, pp. 26–31), Firbas (1949, pp. 147–165), Godwin and Tallantire (1960), Iversen (1973, pp. 54–55), Janssen (1974, pp. 55–63), Rackham (1980, p. 100), Huntley and Birks (1983, pp. 167–184), Kalis (1983; 1988), Kalis und Meurers-Balke (1988), Kalis und Zimmermann (1988), Kalis und Bunnik (1990) and Berglund (1991, pp. 110, 170, 222, 248).

Clark *et al.*, 1989). A distinction is also made between pioneer species of trees which require light, and the climax species which tolerate shade (see, *inter alia*, Iversen, 1960; 1973; Birks, 1986).

It was, above all, Iversen (1941; 1960) who took an interest in the ecology of trees and forests in order to ascertain whether the reconstruction of the pre-historic forest corresponds with the ecology of the species of trees concerned in the modern forest. He devised starting points for the development of forests in prehistoric times from the knowledge and theories on succession in forests and the regeneration of the climax vegetation which was available at the time. These are described below. The starting points formulated by Iversen still form the basis for the interpretation of pollen diagrams and are widely used in palynology.[13]

Iversen borrowed arguments for the interpretation of pollen diagrams from the above-mentioned researchers such as Moss, Clements, Tansley, Dengler, Rubner and Morosow (see Iversen, 1941, pp. 21, 36, 44, 47; 1960, pp. 6, 26–27). He used Clements's succession theory (1916) and Tansley's modifica-tions of this theory as the starting point for his research and theories. He described Tansley's book *The British Isles and their Vegetation* (Tansley (1949), the first edition appeared in 1939) as an excellent work. He believed that the concepts which Tansley presented, including that of retrogressive succession, should be used in the pollen analyses.[14] In addition, he based his work on 20th century

---

[13] See Iversen (1960; 1973, p. 63), Janssen (1974, pp. 46–47), Birks (1986), Jenik (1986), Bottema (1987), Faegri and Iversen (1989), Delcourt and Delcourt (1991, p. 61 *et seq.*) and Mannion (1992, p. 4).

[14] Iversen commented on Clements's succession theory as follows:

A great deal of critisism may be levelled at the climax theory in its original and rather dogmatic presentation by Clements (1916). However, the concepts of succession and climax are sound provided that they are used as designation for actual conditions, and they are indispensable in vegetational history. Indeed if these terms had not already been in existence, it would have been necessary to introduce them to meet the requirements of research in that field (Iversen, 1960, p. 6).

In the notes, he then continued with:

The ideas and the system of Clements (1916) have stimulated to fruitful research, and also given rise to much sophisticated dispute. The latter may be due to the fact that Clements' terminology was not always kept clear of his theory. A sound modification of Clements' system is presented by Tansley in his excellent book; 'The British Islands and their Vegetation'. We feel that Tansley's conception should be adopted in pollen analysis, which, in its turn, may serve as an inductive method for dynamic plant geography, when properly adapted to this partcular purpose. In this way a sound factual basis may be established for the study of succession and the assessment of climax vegetation, and there is no need for resorting to problematic analogisms from vegetational zonations. Since Neolithic time successions are often greatly modified by human activities ('deflected successions' *sensu* Godwin, 1929), and even Tansley's climax communities are, no doubt, modified by man. It might perhaps be better to replace the word climax by a more neutral term when dealing with regions or periods where deflected successions prevail (cf. Walter, 1954). When climate is stable or improving most successions are progressive, while retrogressive successions (Moss) normally occur in response to climate decline. Soil degradation too may without man's interference bring about retrogressive succession in a climax forest, as suspected by Tansley (l.c. p. 26) and clearly demonstrated by pollen analysis (see, e.g. the present paper p. 12). Through retrogressive succession forest may develop into moorland (disclimax), as described by Peasall (1950) (Iversen, 1960, p. 26).

forestry science, as shown, *inter alia*, by the fact that he regularly referred to Dengler's book (1935) *Waldbau auf ökologischer Grundlage* for information and views on forests (see Iversen, 1941, pp. 24, 28, 30, 36, 44–47; 1960, p. 27). In his turn, Dengler based his work on the climax concept and the theory that the climax in the area being studied was a closed forest.[15] Therefore Iversen (1960) based his work on theories on succession and climax vegetation which were current in the first half of the 20th century in plant ecology and forestry.

In order to acquire knowledge about the natural succession of species of trees, Iversen (1960) considered that more research should be carried out in forests which were no longer being exploited. For this purpose the Geological Survey in Denmark selected the Draved skov (Draved Forest) in 1947. A comparison of the fossil pollen flora with the present forest resulted in the conclusion that the species diversity there had remained virtually unchanged since the Atlantic period (8000 BP) (Iversen, 1958; 1973, p. 72). According to historical sources, it was used as a wood-pasture until 1785 and then was used until at least 1955 for the production of timber using rational forestry techniques (Iversen, 1958; Andersen, 1970, p. 17; Aaby, 1983, pp. 21–25). Systematic studies were carried out in the Draved Forest into competition and the succession of tree species (Iversen, 1960). It was used as the reference forest for palynological research (Iversen, 1958; Andersen, 1970). Iversen (1960) stated that apart from forests no longer in use, the few remaining virgin forests of Europe were also of great importance, in particular those in the mountains of south-east Europe, in the Balkans.[16]

---

[15] Dengler stated on the wood

Der Wald als Schlußformation. Wir können das auch heute noch gelegentlich hier und da beobachten, wo einmal Neuland durch natürliche Ereignisse (An- oder Abschwemmungen, Erdrutschungen u. dgl.) entsteht oder wo der Mensch derartiges Neuland künstlich schafft (wie auf alten Kiesgruben, Steinbruchshalden, Wegeböschungen, auch auf aufgegebenen Weide, Wiesen, und Äckern, sog. Ödland). Meist bilden sich hier zuerst andere Vegetationstypen aus wie Grasfluren, Zwergstrauchheiden und Buschwerk. Aber schließlich findet sich ein Bäumchen nach dem andern ein, diese wachsen empor, schließen sich zusammen und verdrängen die Waldfremden Elementen in den Unterstufen, während andere zum Walde gehörende sich ansiedeln. Am Ende dieser Reihenfolge, die man Sukzession genannt hat, steht als schlußglied (Klimax) immer der Wald! [This is a description which fits the ones about secondary succession of grassland to wood as described by Watt (1924; 1925; 1934a,b)]. Das geht bald rascher, bald langsamer, es braucht manchmal nur Jahrzehnte, oft aber auch ein Jahrhundert und mehr. Aber es geht, wenn keine gewaltsamen Störungen eintreten, unaufhaltsam und stetig immer dem Endziel, dem Wald, entgegen. Ein Beispiel siegreichen Vordringen von Wald in gewaltigem Umfang auf öde gewordenen Ackerland haben wir in Deutschland nach dem Verwüstungen und der Entvölkerung des Dreißigjährigen Krieges gehabt, wo der Wald ganze Dorfstätten mit ihren Feldfluren wieder vollstandig überzog und der Spruch entstand: 'Wo der Wald dem Ritter reicht bis an den Sporn, da hat der Bauer sein Recht verlor'n!'

Das Wort unseres forstlichen Altmeisters H. Cotta, daß Deutschland, wenn es von allen Menschen verlassen würde, in 100 Jahren wieder ganz von Wald bedeckt sein würde, gilt sicher auch heute noch zu Recht!' (Dengler, 1935, pp. 5–6).

[16] 'Of supreme interest are, of course, the few remaining specimens of virgin forest in our temperate zone, especially in the mountains of southeastern Europe. A number of valuable studies of the virgin forests and their regeneration in southeastern Europe are available' (Iversen, 1960, p. 7). Iversen refers to authors such as Fröhlich (1930), Mauve (1931), Markgraf and Dengler (1931) and Rubner (1935).

Therefore, on the basis of information from research into succession in general and into the last virgin forests of Europe in particular, Iversen (1960) tried to answer the question of *what* the virgin climax forest in Europe looked like in prehistoric times and how it functioned as an ecosystem. On the basis of data from research by Fröhlich (1930), Markgraf and Dengler (1931), Mauve (1931) and Rubner (1935) into the primeval forests of south-east Europe, which consisted mainly of species of trees which tolerate shade, like beech (*F. sylvatica*), Norway spruce (*P. abies*) and silver fir (*A. alba*),[17] Iversen (1960) concluded that stable climax forests cannot comprise pioneer species which require light, such as willow (*Salix*), poplars (*Populus*), birch (*Betula*) and rowan species (*Sorbus*), but only species which tolerate shade, such as beech and lime (Iversen, 1960). Together with his experience of forestry, from which he knew that the lime which tolerates shade is a very strongly competitive species, Iversen came to the conclusion that *all* the climax primeval forests in Europe, i.e. also those which he believed had covered the lowlands in Europe in prehistoric times, must have consisted of species of trees which tolerate shade.[18]

He also considered that a prehistoric primeval forest in the lowlands regenerated itself in the same way as the primeval forests in southern Europe, i.e. in the gaps in the canopy when one or more trees die or fall over. According to Iversen, the hazel, which is strongly represented in the pollen diagrams of the lowlands of Central, Western and north-west Europe, was able to prevent the establishment of trees for a long time in certain places because of the shade cast by that shrub.[19]

Iversen (1960) described from the experience of forestry that the shade-loving lime and the elm suppress the oak, and that oaks hardly or never regenerate in forests. In this respect, he referred to Morosow (1928) and Tansley

---

[17] These primeval forests were therefore sometimes described as 'Schattholzwälder' or 'Schattholzurwälder' (Nietsch, 1939, p. 83).

[18] Iversen (1960) wrote about the relationship between the primeval forest in the mountains of south-east Europe and that in the lowlands of Denmark: 'Pollenstatistical evidence demonstrates that the Atlantic virgin forest in Denmark was of a similar stable type (climax forest)' (Iversen, 1960, p. 27). In a later publication (Iversen, 1973), he wrote about the regeneration of the climax forest in the Atlantic period under the title 'Renewal of the primeval forest':

There has been much discussion over the process of renewal in the primeval forest. Pollen analysis shows that the Danish Atlantic forest must have renewed itself in the same manner as that described by experts from the last South-East European primeval forests. It was not a catastrofic renewal with a change in tree species, as is known to occure in coniferous regions and in cultivated forests. The forest was stable, and new growth of the same tree species appeared upon the fall of old trees: pioneer tree species no longer appeared (Iversen, 1973, p. 72).

[19] 'Hazel appears to have been an exception: it is a dogged shade-tree which could well have held the larger trees' reproduction in check for some time, but by virtue of its shade it only assisted in keeping light-loving trees out of the dark of the primeval forest' (Iversen, 1973, p. 72). He probably concluded this from the practice of coppicing hazel shrubs in coppice-with-standards, where it is well known that trees cannot establish themselves in the hazel scrub in the lower levels, or find it very difficult to do so (see Lindquist, 1938, p. 26; Rackham, 1980, p. 297; Watkins, 1990, p. 95).

(1953) who devoted a great deal of attention to this.[20] Iversen realized that there was a discrepancy between the postulated view of the closed oak forest in prehistory, and the autecology of the oak. Consequently, he formulated the theory that the oak grew in places where the lime could not grow because of the composition of the soil. In his view, this geographical division remained until humans intervened in the forest. He believed that this explained the dominance of oaks and the limited presence of lime trees in certain 'modern' forests, and the fact that both species are nevertheless found together in modern forests, despite the competitive strength of the lime. He did not specify what he meant by human intervention (see Iversen, 1960: pp. 10, 28). The high values of oak pollen in pollen diagrams from prehistory still presented a problem. In this respect, Iversen remarked that Andersen's research (1970), described in the next section, had given him an adequate explanation of this. In his view, the lime is under-represented in the pollen diagrams for reasons which are also explained in the next section.

According to Iversen, Andersen made corrective factors which corrected the distorted picture created by the pollen diagrams. He believed that the strongly competitive line was much more widespread in the prehistoric virgin forest, and the oak much rarer (Iversen, 1973, p. 65). Iversen suggested that, the oak is a climax species mainly in wet soil and as an example, he gave the floodplain forest dominated by oaks along the River Sava in Croatia (Iversen, 1973, p. 68). These forests consist mainly of pedunculate oak. Up to 100 years

---

[20] See Morosow (1928, pp. 269, 279–283, 298–299, 318–321) and Tansley (1953, pp. 291–293). With regard to the discussion in Russia about the failure of oaks to regenerate in oak forests, Morosow wrote:

> Bei seinen Wanderungen durch die Eichenwälder des Gouvernements Kasan fiel Korshinski 'das Fehlen von Eichenjungwuchs in ihnen auf, was ihn nicht wunderte, da er es für normal hielt. 'Es ist bekannt', schreibt er, 'daß die Eiche eine äußerst lichtbedürftige Holzart ist, die sich im schatten überhaupt nicht entwickeln kann, daß sogar Keimlinge unter ihrem Kronendach in 2 bis 3 Jahren verschwinden' (Morosow, 1928, p. 279).

Morosow does not agree with this observation according to his statement:

> Vor allem scheint es ganz unwarscheinlich, daß die Natur eine Holzart mit derartigen Lichtbedürfnis geschaffet hätte, daß es ihr unmöglich wäre, sich unter dem Schirme der Mutterbäume zu verjüngen. Die Birke, Kiefer, Eiche, jede beliebige Holzart ist imstande, sich unter dem Schirm reiner Bestände der eigenen Holzart zu verjüngen; nicht immer volzieht sich die Verjüngung mit genügendem Erfolg, doch diese Bewertung des Erfolges oder der unbefriedigenden Verjügung ist Sache des Wirtschafters. Wäre es denn denkbar, daß bei natürlicher Auslese eine Holzart in der Natur enstehen könnte, die den Schatten ihres lichten mütterlichen Schirmes nicht ertragen könnte? ... Die Kulturen erforderten Pflege und vor allem Schutz vor dem Vieh; es genügte eine zuverlässige Einzäunung, um den Eichenaufschlag, der unter dem Altholzschirm vorhanden war und ständig durch das Vieh verbissen wurde, zum Leben zu erwecken (Morosow, 1928, p. 280).

Tansley (1953, pp. 291–293) referred to the disturbed equilibrium in nature, which was Watt's conclusion (1919) from his research into the causes of the failure of oaks to regenerate in oak forests (see Chapter 2).

ago these forests were regenerated with the help of the so-called shelterwood system, a regeneration technique which is incorrectly described as 'natural regeneration'. It is a technique which is accompanied by a great deal of human intervention, particularly where the regeneration of the oak is concerned (see Chapter 4 and Appendix 5). For the past 100 years, the oak has been regenerated above all with planting programmes (Glavac, 1969; Klepac, 1981; Raus, 1986). In contrast with what Iversen had assumed (1960), these forests are therefore by no means natural, and the oak does not generate spontaneously.

According to Iversen, the hazel also grew in the closed forests. He called this the first shade-tolerant species to reach Denmark (Iversen, 1973, p. 53). He did not explain the basis for the classification of shade species. Perhaps it was taken from the practice of coppicing hazel, leaving coppice-with-standards. Another possible reason why he called the hazel a shade-loving species could be that it is the only way to explain how the hazel became established in the Boreal birch/pine forest. Iversen maintained that this forest was present when the hazel arrived. He does not describe how it became established in that forest. He believed that the hazel survived in the dark Atlantic forest by forming shoots 10–12 m tall which were replaced when they grew old and died by shoots which grew rapidly from the roots. In this way, he thought that the hazel could prevent light-tolerant species from becoming established, although the hazel eventually lost out against species of trees which tolerate shade, such as the lime (Iversen, 1973, pp. 64, 72). According to Iversen, the hazel then continued to play an important role on the fringes of forests by lakes and fjords, and as a pioneer in gaps in the canopy in the primeval forest (Iversen, 1973, pp. 64–65). He believed that this explains the high percentage of pollen in the sediments from the Atlantic period (Iversen, 1973, pp. 54–55).

Regarding the role of the large herbivores in the prehistoric primeval forests, Iversen supported the views of forest experts, that deer are scarce in the primeval forests of south-east Europe, and did not have any natural influence on the regeneration of the forest (Iversen, 1960, p. 26). In his view, wild animals are found there in such low densities because the forests are so shady. Because the virgin forest in the Atlantic period was also shady, he believed that there were very few wild animals living there, such as deer, aurochs, tarpan and European bison. According to Iversen, there was no undergrowth for the animals to live on, and consequently there were very few people because they would have had to live from hunting these animals (Iversen, 1973, pp. 72 and 73, also see Aaris-Sørensen, 1980; Bottema, 1987). Therefore Iversen supported the prevailing theory at the time that large herbivores did not have an influence on the climax vegetation forest in the sense that they had influence on the regeneration of the forest.

3.5 Drawing up Corrective Factors for Pollen Frequencies

On the basis of the theory that Europe was naturally covered by a closed forest, a number of palynologists[21] calculated correction factors for the values given in pollen diagrams for different species. The aim of these was to modify the picture produced by the sampling of pollen in peat, in terms of the relative frequency of pollen of different species of trees. As a result of the long distance travelled by pollen before forming a sediment in the peat bogs, Andersen believed that a selection took place so that some species were over-represented, while other species were under-represented in the peat bogs (Andersen, 1970; 1973). Because of the lack of winds in the (postulated) original primeval forest there was no selection there of the pollen by weight and the rate at which it descended. Therefore it could have sedimented by vir-tually falling straight from the trees. This is why there might be a direct rela-tionship in some of the small depressions in the forest between the amount of pollen dispersed by the trees and the percentage of the species concerned in the sediment in the depression (see Andersen, 1970; 1973; Bradshaw, 1981a,b; Prentice, 1986). By taking surface samples in modern forests where the species diversity and the relative abundance of the various species of trees in the direct vicinity of the sample is known, it is possible to deter-mine to what extent the pollen flora in the sample deviate from the composi-tion of the forest. This produces a ratio, and this ratio gives the correction factors with which the relative frequency of pollen in the sediment in small depressions can be converted, so that the actual presence of the different species of trees in the (alleged) forest can be calculated. Because the sedimen-tation of pollen took place in the original primeval forest only in small depres-sions without selection, the correction factors can therefore be applied only to samples from small depressions.

Andersen drew up the relative frequency of beech at one, divided the levels found for oak by a factor of four, and multiplied that for lime with a factor of two. This means that when the correction factors are applied, the relative frequency of lime pollen compared with that of oak pollen is increased by a factor of eight. Therefore the application of the correction factors pro-duces a very different picture of the vegetation in prehistoric times from the picture produced when this is not done. Forests which would be dominated by oak without applying this correction factor are changed into forests domi-nated by lime trees tolerant of shade (see Fig. 3.3). The values of pollen of species such as *Juniperus* (juniper), *Poaceae* (grasses), *Empetrum* (crowberry), *Artemisia* (mugwort), *Rumex* (sorrel), and *Filipendula* (dropwort) were divided by a factor of four by Andersen (1976), and grasses by a further factor of two (Andersen, 1990). The reason for this was that they are characteristic species

[21] Andersen (1973; 1990), Girling and Greig (1977), Bradshaw (1981a,b), Greig (1982), Prentice (1986), Bennett (1988c), Mitchell and Cole (1998).

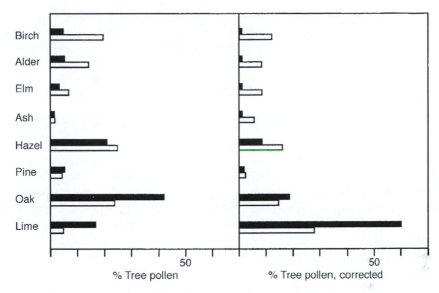

Fig. 3.3. The average percentages of pollen in a peat bog in Elsborg (white bars) and the forest at Eldrup (black bars) in Denmark, from the Atlantic period (8000–5000 BP). According to the theory prevailing at the time, there was a virgin closed forest. The uncorrected frequencies and the frequencies adjusted using Andersen's correction factors are shown side by side (redrawn from Bradshaw, 1988, p. 739, taken from Andersen, 1984).

in open terrain, which would therefore be over-represented in pollen diagrams.

Andersen (1970; 1973) considered that the application of correction factors gives a more exact picture of the Atlantic and Sub-Atlantic climax forest as a forest of species which tolerate shade. Furthermore, he believed that the accuracy of the results of his correction method confirmed the fact that the last primeval forests in Europe consisted mainly of species of trees which tolerate shade. Iversen supported Andersen's argument, and thenceforth no longer referred to the Atlantic period as the period of the mixed oak forest, but as the lime period (Iversen, 1973, p. 62).

On the basis of the correction factors, the assumption of a mixed oak forest was replaced by a number of researchers[22] with forests dominated by lime. Nevertheless, in these corrected pollen diaspectra, oak still accounts for a share of 20–30%, compared, for example, with 60% for lime,[23] which is a

---

[22] See *inter alia* Andersen (1973; 1976; 1989; 1990), Girling and Greig (1977), Rackham (1980, p. 237), Greig (1982) and Bennett (1988c).
[23] See *inter alia* Andersen (1973; 1976; 1989; 1990), Perry and Moore (1987), Bennett (1988c), Chen (1988), Huntley (1988), Kalis (1988), Bozilova and Beug (1992), Horton *et al.* (1992), Latalowa (1992) and Caspers (1993).

significant percentage. If the correction factors are not applied, the oak was most strongly represented among all the terrestrial species of trees in the Atlantic period (8000–5000 BP) and the Sub-Boreal period (5000–2500 BP) in most of Central, Western and north-west Europe (see Figs 3.1 and 3.2). In many regions in Europe, beech pollen dominated the common flora only in the Sub-Atlantic period, but even then, oak pollen is still clearly present (for example, see Firbas, 1949, pp. 50–51). Either with or without the application of the correction factors, oak therefore counts for an important share in the pollen diagrams. With regard to the question of the presence or absence of oak in prehistoric vegetation, this is an important fact. Pollen analyses actually reveal only one thing, that if the fossilized pollen of a species is found, and is a clearly demonstrable trend in the sediment, an example or a whole population of this species was present in the near or distant vicinity when this sediment was formed (Faegri and Iversen, 1989, p. 137). All the pollen diagrams of Western and Central Europe show that there has been oak there since the Boreal period. This means that oak has survived for many thousands of years and has regenerated. This also applies for hazel, which is represented by even higher relative frequencies in the period when there was a closed forest according to the prevailing theories.

As remarked above, the starting points and theories which Iversen used for the development of forests in prehistoric times still form the basis for the interpretation of pollen diagrams. In view of the data presented by Iversen, taken from forestry and forest ecology, it is improbable that oak and hazel were present in a closed forest together with lime in prehistoric times. Therefore the question is whether pollen analyses support the theory that the virgin vegetation in Central and Western Europe was a closed forest.

3.6 The 'Landnam' Theory

In all the Danish pollen diagrams Iversen found that there was a striking and significant decline in the relative frequency of pollen from the mixed oak forest just above the boundary of the Atlantic and the Sub-Boreal period, which was caused mainly by a decline in the relative frequency of the elm (Iversen, 1941, pp. 21–22). In north-west Europe the date of this decline coincided with archaeological discoveries which indicate the first presence of agricultural activities there (Hilf, 1938, pp. 72–73; Nietsch, 1939, p. 23). Iversen assumed that before this time there had been a closed forest. In the same place in the diagrams where the sharp decline of the relative frequency of the pollen of the mixed oak forest was shown by a sharp decline in the relative frequency of the elm, he found some pollen of greater plantain (*Plantago major*) and narrow-leaved plaintain (*P. lanceolata*) for the first time in Denmark. According to Iversen, there is no doubt that these two species, which are common in meadows, are clearly indicative of cultivation. If humans disappeared, Iversen believed that these species would also disappear from Denmark.[24] Therefore he considered the pollen of greater plantain and narrow-leaved plantain as indicating the presence of man as a farmer. In his opinion, the beginning of the *Plantago* curve in pollen diagrams is a good guide for determining the establishment of agriculture in Europe (Iversen, 1941, pp. 25, 29–30). Iversen (1941) pointed out that the cattle of these farmers could not find food in the closed forest, so every settlement started by creating a clearing in the forest. He believed that this resulted in the decline of the frequency of pollen of the mixed oak forest.[25] This felling of trees is described in Danish as 'Landnam' and the theory developed by Iversen on this is generally known as the 'Landnam' theory (Iversen, 1941).[26]

The herbs and grasses which grew, once the forest had been cut down, could serve as food for cattle, as did the young shoots which sprang up from the stumps of the deciduous trees. In this respect, he referred to Dengler who described how man changed the original closed forest (see Dengler, 1935, p. 39 etc.). Dengler described, among other things, that the felling of trees in

---

[24] With regard to the connection between plantain and farming, Iversen wrote (1941):

Apparently Plantago came to Denmark together with the first farmers in the same manner as it 'has since followed the European all over the world, wherever he has settled. Plantago has been called 'the white man's trail' by the American Indians; the trail of the Neolithic conquerors is the Plantago pollen in our diagrams

(Iversen, 1941, p. 25). He commented further on greater plaintain:

Plantago major is so closely associated with places with direct cultural influences that it would doubtless disappear from our flora if the culture ceased. True, it is encountered now and then by the shore; but, in all the cases I have been able to observe, it would no doubt disappear together with the cultural interferences in nature (grazing, mowing etc.) which keep the vegetation low or open. As regards Pl. lanceolata the position is less clear, as this plant can also grow in well-lighted oak forests. Here again, however, we find that it is dependent on culture. The only occasion on which I have seen Plantago occuring in abundance in a wood was in an open oak scrub at Froslev in South Jutland, but the scrub was browsed by cattle [*by which he indicated that these light oak forests developed as a result of human intervention*]. Untouched oak forest is no place for Plantago lanceolata as appears clearly e.g. from Olsen's (1938) comprehensive statistical investigations into the ground flora in Danish oak forests and oak scrubs (Iversen, 1941, pp. 40–41).

Therefore Iversen still supported the theory that the Atlantic primeval forest was a mixed oak forest, the theory to which he returned later, assigning a more important place to the lime in the Atlantic primeval forest.

[25] Accordingly, a number of people settled in an area and built a settlement in a suitable spot. Before moving their domestic animals to the new home they all embarked on a wide clearance of the forest; the cattle had to have food, but the forest as it stood could not provide it. That forest was nothing like the oak forest of today; innumarable dead, fallen trees lay in every direction, making access to it difficult and hindering the growth of grass and leafy bushes. Light and air had to be provided for herbs and low bushes to grow (Iversen, 1941, p. 29).

[26] See *inter alia* Godwin (1944; 1975, p. 465 *et seq.*), Van Zeist (1959), Janssen (1974, p. 80); Birks (1986; 1993), Delcourt and Delcourt (1987, p. 374 *et seq.*), Behre (1988, p. 643), Bogucki (1988, p. 33), Huntley (1988, p. 346 *et seq.*), Roberts (1989, p. 117 *et seq.*), Tallis (1991, pp. 270–280), Bell and Walker (1992, p. 164 *et seq.*), Walker and Singh (1993), Edwards (1993), Waller (1994), Lindbladh (1999), Lindbladh and Bradshaw (1998), Bradshaw and Mitchell (1999).

the original forest for establishing settlements and creating fields and mead-ows, must have been resulted in coppices. These developed when the stumps of the trees started to grow again. Apart from the grasses and herbs which appeared, Dengler thought that the coppices also provided food for cattle in the form of young shoots which grew from the stumps (Dengler, 1935, p. 79). He based this on written records dated from the 13th and 14th centuries, which referred to a prohibition on grazing in coppices to protect the young shoots from being eaten by cattle. According to Iversen, the farmers also took over the surrounding forest for grazing their cattle after they had cleared open spaces in the forest for them. In his view, the composition of the forest changed so little as a result, in comparison with the original forest, that there is no evi-dence of this grazing in the pollen diagrams. Therefore he believes that cutting down the forest and the appearance of plaintain (*Plantago* sp.) are the only demonstrable effects of the introduction of agriculture. On the basis of a bone of a house cow (a tibia) found by Degerbøl, and pollen which he found in the hollow of the bone, dating from the time of the minimum relative frequency of pollen of the mixed oak forest, Iversen concluded that this was the palynologi-cal proof of the appearance of the first farmers.[27]

With regard to the structure of the prehistoric forest where the first farm-ers settled, Iversen noted that it must have been an impenetrable forest because of the trees that had fallen down, and that grass could barely grow there. Therefore the shrubs and herbage first had to be allowed light and air so that they could grow and serve as food for the cattle (Iversen, 1941, p. 29). In his view, the use of fire was at least as effective as the use of the axe for this purpose. According to Iversen, it may be true that it is difficult to set fire to present-day cultivated deciduous forests, but because of the large amount of dead wood on the floor of the prehistoric natural forest, it burnt very easily. He came to this conclusion about the use of fire by the first farmers from the dis-covery of a small layer of charcoal in one place. He also based this view on the way in which forests were turned into agricultural land in Karelia, formerly part of Finland, which became part of Russia in the 20th century. The trees were cleared, the wood was burnt and then the ash was ploughed in and used as fertilizer. In addition, Iversen referred to the 'clearance fire' method ('slash and burn'), which was the earliest method of farming throughout the world. Consequently, he felt it could be assumed a priori that this form of farming also took place in Denmark in prehistoric times. When the soil was exhausted after a number of harvests of cereals, the farmers left the area and the forest recov-ered. Iversen believed that this is shown in the pollen diagrams by the increase of pioneer species such as birch and hazel. Therefore, according to Iversen, the first farmers practised 'shifting cultivation'. However, in more recent layers,

---

[27] 'we arrive at a natural and satisfactory explanation of the courses of the various pollen curves if we assume that the pollenfloristic changes express the vegetation developments in a region where land-tilling people have occupied the land and cleared this dense primeval forest with axe and fire' (Iversen, 1941, p. 23).

Iversen did not find any charcoal and consequently he did not believe that the secondary forest was cleared by fire. He explained this by stating that as a result of the first cultivation, the forest never recovered its character of a primeval forest, and it was therefore not necessary to burn it down a second time.

In order to test his theory, Iversen carried out a practical experiment by clearing an open area in the Draved Skov with a stone axe, burning the chopped wood and ploughing in the ash and growing grain. He believed that this experiment showed that it was possible to clear the forest with a stone axe, and that it could therefore serve as an explanation of phenomena in the pollen diagrams (Iversen, 1956; 1973, pp. 88–92).

Initially, Iversen (1941) saw the appearance of plantain as the indication for the arrival of the first Neolithic farmers. In the pollen diagrams, this comes after the time at which the levels of elm pollen declined. He explained this difference with reference to the results of research by Troels-Smith (1954; 1955; 1960). According to Troels-Smith (1960), the decline in the levels of elm pollen and the isolated occurrence of low levels of plantain pollen were the first signs of agriculture. He pointed out that in virtually every part of Europe it is common for livestock to be fed with leaves and twigs cut from trees. As regards the nutritional value of this foliage, he considered that elm takes first place and ash, second. Therefore he ascribed the decline of the elm pollen to the cutting of elm foliage by farmers as fodder for their livestock. Cutting elm twigs prevents the tree from flowering, which explained why there was such a strong sudden decline in the frequency of elm pollen at that time.

Troels-Smith (1960) also based his interpretation on the discovery of wooden bowls in Denmark and Switzerland, which date from the time of the decline in levels of elm pollen, and which were made from the roots of elm and ash. He believed that this showed that man was familiar with cutting down elm and ash trees for use as fodder at this time. On the basis of Troels-Smith's research, Iversen (1960) assumed that the decline of elm pollen could be attributed entirely to the first farmers cutting elm foliage as fodder for their livestock. He considered the decline in elm pollen as an indication for the appearance of the first farmers, and *not* the higher levels of the pollen curve of greater plantain and narrow-leaved plantain, although he did think these were indicative of agriculture (Iversen, 1960; 1973, pp. 79–80).[28] Figure 3.4a and b indicates the scale on which the relative frequency of elm pollen declined in Western and Central Europe. On the basis of research by Troels-Smith, Iversen also concluded that the livestock must have been kept in the barn all year round and fed on foliage (Iversen, 1960, p. 20; 1973, pp. 79–80).

---

[28] Initially, Iversen (1960) was still cautious, as is shown by his question: 'Is it really possible that the number of flowering elms in a region could be reduced to such an extent and in such a short time only because of the use of their folliage for fodder?' (Iversen, 1960, p. 19). He answered this question further in his article on the basis of modern analogies, though he implicitly answered it affirmatively (Iversen, 1960, p. 20).

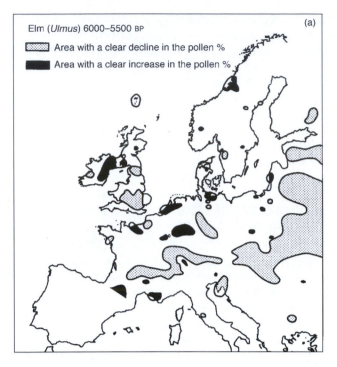

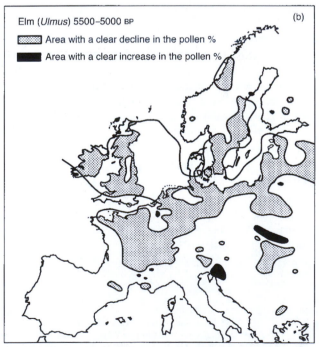

After the decline in the levels of elm pollen, Iversen noted an increase in the relative frequency of pollen of species which were recorded as being pioneers in open areas in the forest, such as hazel and birch. He considered that this increase showed that the forest grew back after it had been cleared for the cultivation of cereals, once the farmers had abandoned the fields. Because the decline in the relative frequency of elm pollen does not coincide with the appearance of plantain pollen in the pollen diagrams, Iversen (1973) considered that initially there must have been two cultures side by side, i.e. a primitive culture which fed the cattle in barns with foliage, and a later culture which grazed the cattle outside. As a result of the fairly heavy grazing of the cattle, grassland was thought to have developed from the original forest. In his opinion, the later appearance of plantain pollen indicates this (Iversen, 1973, p. 92). On the basis of the above-mentioned practical experiment, in which cattle were also allowed to graze in the forest when it had been cleared, he also came to the conclusion that cattle were kept outside (Iversen, 1956; 1973 pp. 88–92). According to Iversen, cattle in prehistoric times ate leaves and mast (a term which comprises acorns, beech nuts, the fruit of wild fruit trees, and berries) as well as grass, while they were fed on dried foliage in the barn in winter (Iversen, 1973, p. 92). As noted earlier, the growing forest was transformed by this into coppice, where it was not only easy to pick the foliage, but where the cattle could also easily browse on the leaves themselves (Iversen, 1973, pp. 79–80).

Nevertheless, some doubts have arisen as to whether human activity *alone* can have been the cause of the sudden decline in the relative levels of elm pollen in the so-called 'Landnam level' in Central and Western Europe.[29] Increasingly, data indicate that elm disease could have played an important

---

Fig. 3.4. The areas where the frequency of elm pollen declined significantly between 6000 and 5500 BP (a), and 5500 and 5000 BP (b) (redrawn from Huntley and Birks, 1983, pp. 435 and 436). Note that the decline in Central and Western Europe which was ascribed to the cutting of elm foliage by humans for fodder for cattle (a) was preceded by a decline in an area to the east of this (b). Considering the enormous surface area in which the relative frequency of elm pollen declined in Central and Western Europe of over approximately 500 years, the possibility of the outbreak of an elm disease cannot be excluded.[30] Elm disease as an explanation becomes even more probable when one considers that the decline in Central and Western Europe was preceded by a strong decline in Eastern Europe, though there are no indications that agriculture was being practised there before it appeared in Western and Central Europe. Therefore the cutting of foliage for fodder cannot be an explanation there, while an outbreak of elm disease could be. In that case the disease spread from the east to the west.

[29] See *inter alia* Evans *et al.* (1975), Groenman-van Waateringe (1983), Barker (1985, p. 234), Edwards (1985), Watts (1985), Zvelebil (1986), Price (1987), Simmons and Innes (1987).
[30] Rackham (1980, p. 266), Rowley-Conwy (1982), Groenman-van Waateringe (1983; 1988), Edwards (1985), Girling and Greig (1985), Watts (1985), Perry and Moore (1987), Moe and Rackham (1992), Graumlich (1993), Peglar (1993), Peglar and Birks (1993).

role in this.[31] In that case, the decline in elm pollen would have been caused by an outbreak of elm disease in combination with the cutting of elm foliage (Moe and Rackham, 1992; Graumlich 1993; Peglar, 1993; Peglar and Birks, 1993). Discoveries of pollen of cereals (*Cereale* type) from before the Landnam and the frequent appearance of pollen of narrow-leaved plantain and of species of plants which are characteristic of agriculture, such as sorrel (*Rumex* sp.), goosefoot (*Chenopodiaceae*) and mugwort (*Artemisia* sp.), have given rise to the question whether the Landnam level was actually the first sign of agriculture (see Groenman-van Waateringe, 1983; Dimbleby, 1984; Görannson, 1986; 1988; Price, 1987). According to a number of authors (Görannson 1986; 1988; Zvelebil, 1986a; Louwe-Kooijmans, 1987; Price, 1987), there was a much more gradual transition from hunting and foraging to agriculture. In that case, there were initially two cultures side by side; not, as Iversen supposed, two agricultural cultures, but one of hunting and foraging, and one of agriculture.

According to some researchers (Evans, 1975; Mellars, 1975), hunter-foragers, as well as farmers, could have had an influence on the vegetation. Hunter-foragers could have destroyed the forest vegetation with fire in order to create meadows. The grass, the herbs and the ash (minerals) would have attracted the species of animals that were hunted, such as auroch and red deer. Analagous to modern situations, these burnt areas could have led to a tenfold increase in the biomass of wild animals (Mellars, 1975). Hunter-foragers could also have burnt the vegetation, so that it would have been more accessible for finding fruits, hazel nuts and acorns. This would have meant that the hazel increased as a pioneer, and man then profited from it (Regnell *et al.*, 1995).

Several authors[32] pointed out that the role of the large herbivores in palynology has continued to be underestimated. They said that large wild herbivores were responsible for grasslands, although they did not consider that these were more than large open areas in the forest with the size of a gap in the canopy arising as a result of one or several trees dying or falling over. The only species which Coles and Orme (1983) considered capable of creating open areas in the closed forest is the beaver (*Castor fiber*). Schott (1934) used the consequences on the vegetation of the presence of beavers in Canada as a point of reference for the prehistoric vegetation of Europe. When beavers build dams in streams and in small rivers, the trees drown in the adjacent forest and then disappear. When the beavers themselves depart, the dam breaks after a while because it is no longer maintained, and the pools empty. Subsequently, a meadow of grass and sedge develops in the drained pool, which attracts

---

[31] Rackham (1980, p 266), Rowley-Conwy (1982), Groenman-van Waateringe (1983; 1988), Girling and Greig (1985), Perry and Moore (1987), Moe and Rackham (1992), Graumlich (1993), Peglar (1993), Peglar and Birks (1993).
[32] Smith (1958; 1970), Birks (1981), Coles and Orme (1983), Buckland and Edwards (1984), Barker (1985, pp. 18–19).

European bison, elk and deer. Analogous to this, Schott (1934) and Coles and Orme (1983) considered that as a result of the activities of beavers in prehistoric times, meadows developed which attracted aurochs, European bison, deer and roe deer. These beaver meadows could have covered large areas. Coles and Orme referred to the beaver meadows in the province of Ontario in Canada, where areas of 40 ha were not exceptional. The colonists used these meadows for making hay. Enormous lakes had been created by the millions of beavers which built their dams in families in streams and small rivers throughout the province of Ontario. According to Coles and Orme (1983), this could also have happened in large areas of the European lowlands. The meadows created by the beavers could have resulted in the same palynological picture in pollen diagrams as these open areas created in the forest by humans. Therefore they considered that it was not justified to assume a priori that changes in pollen diagrams are the result of human activity.

Despite the above-mentioned criticism, many authors still base their interpretation of pollen diagrams on the concepts and theories of Firbas, Godwin and Iversen (Edwards, 1983; Walker, 1990; Edwards and McDonald, 1991; Birks, 1993), who, in turn, drew on the theories of succession and climax vegetation of Moss, Clements and Tansley, Morosow, Dengler and Rubner. Iversen's Landnam theory is more or less supported up to the present both inside and outside palynological circles as the theory which indicates how humans cultivated the primeval forest and how this is expressed in pollen diagrams.[33] Every change in pollen diagrams which cannot be explained on the basis of climate changes is ascribed to human invention.[34] Only the beaver is considered to be able to affect the succession in a natural vegetation. The large herbivores, such as aurochs, elk, European bison and red deer, are not considered, or when attention is devoted to them, they are not deemed to play a significant role.

3.7 The Closed Forest versus the Half-open Park Landscape

I have described the current theory about the original closed forest as supported by palynology and palaeoecology. In this section, I would like to postulate the alternative hypothesis which is based on a more open, park-like landscape. I would like to suggest the wood-pasture as a modern analogy for

[33] See *inter alia* Godwin (1944; 1975, p. 465 *et seq.*), Van Zeist (1959), O'Sullivan *et al.* (1973), Simms (1973), Janssen (1974, p. 80); Rackham (1980, p. 104), Rowley-Conwy (1982), Birks (1986), Delcourt and Delcourt (1987, p. 374 *et seq.*); Behre (1988, p. 643), Bogucki (1988, p. 33), Huntley (1988, p. 346 *et seq.*), Andersen (1989), Roberts (1989, p. 117 *et seq.*), Jahn (1991, p. 392), Tallis (1991, pp. 270–280), Bell and Walker (1992, p. 164 *et seq.*), Walker and Singh (1993); Edwards (1993), Waller (1994), Lindbladh (1999), Lindbladh and Bradshaw (1999), Bradshaw and Mitchell (1999).
[34] See *inter alia* Turner (1962), Smith (1970), Jacobi (1978), Birks (1981), Huntley (1986), Zvelebil (1986b), Zvelebil and Rowley-Conwy (1986), Simmons and Innes (1987), Price (1987), Kalis (1988), Faegri and Iversen (1989, p. 175 *et seq.*), Berglund *et al.* (1991), Evans (1993), Simmons (1993), Wiltshire and Edwards (1993), Magri (1995).

this: a mosaic of grassland, scrub, trees and groves. If there was such natural vegetation, this means that this landscape produced the same pollen diagrams as those which led to the conclusion that there was a closed forest. After all, pollen diagrams constitute the factual information. Another fact is that the large herbivores which grazed the land and maintained the landscape, according to the alternative hypothesis, had occurred throughout the lowlands of Western and Central Europe since the Allerød (12,000 BP). They were present there, at least locally, until the early Middle Ages.[35] Therefore there was a large degree of continuity in their presence in virgin vegetation. In so far as it is possible to identify patterns in pollen diagrams, they can therefore be seen as an interpretation of facts; the patterns in themselves are not facts. This section therefore mainly concerns a different interpretation of the facts than that which was described in the previous sections.

Grasslands and mantle and fringe vegetation (see Figs 3.5 and 3.6) are characteristic of park-like landscapes grazed by large herbivores. The fringe vegetation comprises, *inter alia*, Umbelliferae, nettle (*Urtica* sp.), mugwort (*Artemisia vulgaris*), garlic (*Allium* sp.), sorrel and goosefoot (Müller, 1962; Dierschke, 1974; Westhoff and Den Held, 1975, pp. 115–125; Behre, 1981). The mantle vegetation consists mainly of shrubs which require light, such as blackthorn (*Prunus spinosa*), common hawthorn and English hawthorn (*Crataegus monogyna* and *C. laevigata*), guelder rose (*Viburnum opulus*), common privet (*Ligustrum vulgare*), dogwood (*Cornus sanguinea*), wild apple (*Malus sylvestris*), wild pear (*Pyrus pyraster*), wild cherry (*Prunus avium*), rowan (*Sorbus aucuparia*), many species of roses (*Rosaceae*) and a great deal of hazel.[36] With the exception of hazel, these species are pollinated by insects. The species which are pollinated by insects emit little or no pollen into the atmosphere and therefore remain entirely or almost entirely invisible, from a palynological point of view, in contrast with hazel. This means that the presence of the species in prehistoric vegetation is difficult or impossible to ascertain using pollen analyses (Davis, 1963; Faegri and Iversen, 1989, pp. 4, 12). They form the so-called blind spots in the landscape (Davis, 1963, p. 905).[37] In grazed,

---

[35] See *inter alia* Degerbøl and Iversen (1945), Pruski (1963), Heptner *et al.* (1966, pp. 477–480, 491–499, 861–865), Clason (1967, pp. 31, 60, 76), Szafer (1968), Degerbøl and Fredskild (1970), Iversen (1973), Evans (1975), Jacobi (1978), Volf (1979), Aaris-Sørensen (1980), Simmons *et al.* (1984), Von Koenigswald (1983, pp. 190–214), Söffner (1982), Frenzel (1983, pp. 152–166), Louwe Kooijmans (1985; 1987), Davis (1987, pp. 174–179), Price (1987), Lauwerier (1988, pp. 28–31, 50–51, 56–57, 58–59, 65, 69, 73, 87, 90, 92–93, 98, 101), Van Alsté (1989), Roberts (1989, pp. 80–83), Lebreton (1990, pp. 23–44), Bell and Walker (1992, pp. 170–173), Litt (1992), Auguste and Patou-Mathis (1994), Chaix (1994), Guintard (1994), Coard and Chamberlain (1999).

[36] See Watt (1924; 1925; 1934a,b), Tüxen (1952), Tansley (1953, pp. 259–266, 473–475), Müller (1962), Dierschke (1974), Westhoff and Den Held (1975, pp. 239–241), Rodwell (1991, pp. 333–351), Oberdorfer (1992, pp. 81–106).

[37] This is revealed, for example, by the fact that Bertsch (1929) did not find a sample of wild apple or wild cherry in the pollen archives, though he did find the remains of fruits of these species in the same place. Greig (1992) found the thorns and seeds of hawthorn and blackthorn, but no pollen in sediment in West London.

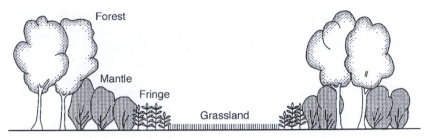

Fig. 3.5. Schematic transverse section of the structure of a grazed park-like landscape with mantle and fringe vegetation (redrawn from Pott, 1993, p. 222).

Fig. 3.6. Mantle and fringe vegetation with blackthorn, hawthorn, birch and hazel in the western Jura, France. The mantle and fringe vegetation form the transition between woodland and grassland (photograph, F.W.M. Vera).

park-like landscapes there are different species of trees which emit a great deal of pollen into the atmosphere, such as oak, elm, lime, beech and hornbeam. However, the low levels of grass pollen compared with tree pollen in pollen diagrams do not suggest that there were such grasslands in the prehistoric vegetation. On the other hand, a number of factors can be mentioned that can describe why, in park-like landscapes grazed by large ungulates, these landscapes are characterized by a low percentage of non-arboreal pollen (NAP) in pollen diagrams. What is relevant here is not so much what the factors do separately but rather their cumulative effect. They follow below.

The low values of grass pollen in the pollen diagrams may be caused in these cases by the natural presence of the wild forms of domestic cows and horses, aurochs and tarpan, which also eat grass. By grazing, they may have greatly limited the flowering of the grasses, and thus the pollen emitted by grass into the atmosphere. Apart from these specialized grass eaters, red deer, the European wild horse and wild boar also eat a great deal of grass.[38] The fact that eating grass in these natural systems can result in the reduction of flowering is apparent from a study carried out in the Serengeti, where grasses flower only sporadically as a result of grazing by wild herbivorous fauna (McNaughton, 1979). Groenman-van Waateringe (1993) demonstrated with surface pollen samples that an area of forest and open terrain populated all year round by wild cattle which were not given any extra fodder, results in a pollen spectrum which looks like one which would normally be ascribed to a closed forest. Grazing by wild cattle clearly reduced the flowering of grasses to such an extent that the paradoxical situation arises in which an intensely grazed open meadow bordering on the forest produces a high percentage of tree pollen.

According to the prevailing view, this is interpreted as a closed forest with limited undergrowth where no grazing animals can live (Groenman-van Waateringe, 1993). Grass-eating aurochs, tarpan, red deer, European bison and wild boar, which are found in the natural state, may therefore change the ratio of tree pollen to non-tree pollen in favour of tree pollen. Comparisons of modern surface samples of pollen from pastured grassland and grassland where hay is made with forest in the background show that this type of shift can occur. Arboreal pollen occurs more on the foreground as a result of reduced blooming in the grazed grassland. As a result, the percentage of NAP drops and the percentage of arboreal pollen (AP) rises. In a heavily grazed vegetation, the influence of forest background pollen is relatively high, due to the low pollen production in the pasture (Hjelle, 1998).

The ratio between tree pollen and non-tree pollen may have moved even further in the direction of tree pollen because mantle vegetation does not allow – or hardly allows – wind to pass through (Jakucs, 1959; 1972, pp. 47–49; Dierschke, 1974). In a park-like landscape there is an enormous peripheral length of fringes which constantly move in relation to each other, so that there is a virtually impenetrable barrier for grass pollen. The fact that grasses do not flower until late in the season, when all the trees and shrubs in the mantle and fringe vegetation are in leaf, also contributes to this barrier (Rogers, 1993). In addition, pollen which moves low above the ground is not dispersed very easily (see Tauber, 1965; Jacobson and Bradshaw, 1981; Edwards, 1982). As a result of the presence of mantle vegetation, this will occur to an even lesser extent.

In the modern pollen rain of forest meadows that are grazed, pollen, particularly of species of sorrel (*Rumex acetosa/acetosella*), appears to be a particularly important indicator of grasslands (Berglund and Persson, 1986; Gaillard *et al.*, 1992; 1994). Narrow-leaved plantain is significantly under-represented in the modern pollen rain of park-like landscapes (Berglund and Persson,

---

[38] Borowski and Kossak (1972), Van der Veen (1979, pp. 212–213, 218), Van Soest (1982, pp. 202, 208, 338), Falinski (1986, pp. 163–170), Hofmann (1986), Groot-Bruinderink *et al.* (1992; 1997; 1999), Cornelissen and Vullink (1996a,b), Van Wieren (1996).

1986). Both greater plantain and narrow-leaved plantain are characteristic of grasslands with slight to heavy grazing (Weeda *et al.*, 1988, pp. 252–256).

Greater plantain and narrow-leaved plantain are admittedly considered in palynology as characteristic species of meadows created by humans (Iversen, 1941; Behre, 1981; 1988), but both species are also found in natural meadows with wild herbivores (Groenman-van Waateringe, 1968). This means that the pollen of both varieties of plantain from the period before the establishment of agriculture could be important indicators of grasslands, and therefore of park-like landscapes grazed by wild herbivores. The pollen of both species is regularly found in the period preceding the establishment of agriculture, namely, from the beginning of the Bølling (13,000 BP) up to the end of the Atlantic period (5000 BP).[39] This indicates that before there was agriculture, there may have been grasslands.

As we remarked above, there is a great deal of mantle and fringe vegetation in park-like landscapes which consists of shrubs and young trees pollinated by insects, in addition to hazel. We also established that the presence of these species and therefore of this vegetation is very difficult to demonstrate with the help of pollen analyses. The only exception to this is the hazel. Despite the unlikeliness that the pollen of species pollinated by insects could have been found in this type of vegetation, they are regularly found in pollen diagrams relating to the post-glacial and the whole of the Holocene period.[40] This is of great importance for the reconstruction of the prehistoric landscape (Faegri and Iversen, 1989, pp. 4, 12). Pollen of the *Prunus* type, including the blackthorn, has been found throughout the Holocene period (Godwin, 1975, pp. 195–196; Day, 1991; Peglar, 1993). In settlements from the Mesolithic and Neolithic eras, seeds were also found of blackthorn, hawthorn, rowan, wild cherry, apple and pear.[41] In addition, pollen was regularly found of species which are known to vegetation experts as fringe species such as mugwort, goosefoot, nettle and sorrel.[42] As noted above, in the modern pollen rain of park-like landscapes, sorrel (*R. acetosa/acetosella*) appears to be an important

[39] See Müller (1947; 1953), Godwin and Tallantire (1951), Godwin (1975, pp. 326–330), Van der Hammen (1952), Smith (1958; 1970), Polak (1959), Seagrief (1960), Van Zeist (1964), Königsson (1968), Pennington (1970; 1975), Smith (1970), Van Geel *et al.* (1980/81; 1989), Barber (1975), Scaife (1982), Bennett (1983), Groenman-van Waateringe (1983; 1988), Wiegers and Van Geel (1983), Buckland and Edwards (1984), Bohncke *et al.* (1987), Van Leeuwaarden and Janssen (1987), Huntley (1989), Bartley *et al.* (1990), Edwards and McDonald (1991), Bush (1993), Peglar (1993), Waller (1994), Linbladh and Bradshaw (1995), Regnell *et al.* (1995).

[40] See Polak (1959), Smith (1970), Godwin (1975a, pp. 183, 197, 199, 200, 462 *et seq.*, 471), Moore and Webb (1978), Bennett (1983), Van Leeuwaarden and Janssen (1987), Day (1991), Peglar (1993), Waller (1993; 1994), Regnell *et al.* (1995).

[41] Bertsch (1929; 1932; 1949, pp. 62–63), Firbas (1949, pp. 188–189), Smith (1958), Tüxen (1974), Godwin (1975, p. 196), Louwe Kooijmans (1987), Regnell *et al.* (1995), Dietsch (1996), Marziani and Tachini (1996).

[42] See *inter alia* Müller (1947; 1953), Smith (1958; 1970), Polak (1959), Casparie and Van Zeist (1960), Seagrief (1960), Van Zeist (1964), Godwin (1975, p. 345), Pennington (1975; 1977), Van Geel *et al.* (1980/81), Bennett (1983; 1986), Clark *et al.* (1989), Van Leeuwaarden and Janssen (1987), Day (1991), Greig (1992), Peglar (1993).

indicator for grasslands where there is grazing. *Artemisia* in particular is represented by low percentages from the post-glacial and throughout the Holocene period and is therefore an important indication for the permanent presence of mantle and fringe vegetation, and therefore grassland.

In my opinion, very important convincing evidence for the presence of mantle and fringe vegetation, and therefore of park-like landscapes, in prehistoric times, is the high percentage of hazel pollen (Erdtman, 1931; Huntley and Birks, 1983, pp. 167–184; see Figs 3.1 and 3.2). In the total pollen sum throughout Western and Central Europe this species was represented by very high percentages, from 20 to 40%.[43] These high percentages of pollen were found in collection basins of the regional pollen rain, such as relatively large peat bogs and lakes (see Tauber, 1965; Tamboer-van den Heuvel and Janssen, 1976; Jacobson and Bradshaw, 1981; Lutgerink *et al.*, 1989), and in smaller collection basins where the extra-local and local pollen rain sedimented (see, *inter alia*, Iversen, 1958; Peglar, 1993; Mitchell, 1998; Mitchell and Cole, 1998).

On the basis of the prevailing theory of the closed forest, the high levels of pollen in the deposits of the regional pollen rain, such as in peat bogs and lakes, cannot come from hazel in a shrub layer of the forest, because hazel does not flower – or hardly flowers – under the canopy of trees.[44] When hazel grows in the gaps in the forest canopy, this cannot result in such high percentages in the regional pollen rain either, because hazel pollen rarely moves outside the forest (Faegri and Iversen, 1989, p. 5). One possible explanation could be that hazel shrubs grew in the transitional stage of the forest, which is assumed to have occurred, according to the current theory, in places where no trees could grow as a result of abiotic circumstances such as flowing or stagnant open water, peat bogs or soil covered with small or large rocks. The appearance of the hazel in the lowlands of Europe would then have remained virtually restricted to the banks of lakes and rivers and the edges of peat bogs. This seems improbable in the light of the high relative percentages in the local pollen rain, which is apparent from the high percentages in small depressions. In fact, these actually indicate the universal presence of this species.

Modern pollen rain has shown that in areas where hazel grows, the shrub is represented in the regional pollen rain by significant percentages (Tamboer-van den Heuvel and Janssen, 1976; Lutgerink *et al.*, 1989). On the other hand, it was found that hazel pollen is dispersed locally over only a relatively small distance (Jonassen, 1950, pp. 26–27, 63; Linnman, 1978; Faegri and Iversen, 1989, p. 5). Only a few metres away from a hazel scrub, Jonassen found a very strong decline in the amount of pollen. In scrub consisting of

---

[43] See *inter alia* Erdtman (1931), Bertsch (1949, pp. 26–31), Firbas (1949, pp. 147–165), Godwin and Tallantire (1951), Janssen (1974, pp. 56–63), Rackham (1980, p. 100), Huntley and Birks (1983, pp. 167–184), Kalis (1983; 1988), Kalis and Meurers-Balke (1988), Kalis and Zimmermann (1988), Kalis und Bunnik (1990), Berglund (1991, pp. 110, 170, 222, 248).
[44] Bertsch (1929), Borse (1939), Breitenfeld and Mothes (1940, cited by Firbas, 1949, p. 151), Andersen (1970, p. 68), Bunce (1982), Rackham (1992).

hazel, the relative percentage declined by more than 1.5 times to 1% of the percentage of tree pollen in the direct vicinity of the scrub (Jonassen, 1950, p. 63). Linnman found a similar decline over a comparable area (Linnman, 1978). This low level of horizontal dispersal and the high percentages in the scrub could explain the high concentrations in small depressions, but not those in the collection basins of the regional pollen rain, such as peat bogs and lakes. Air currents can draw the pollen of hazel woods into the regional pollen rain as has been shown in the mountains with the help of samples of modern pollen rain (Tamboer-van den Heuvel and Janssen, 1976; Lutgerink *et al.*, 1989). In that case, the high percentages of hazel pollen in prehistoric times could have been from hazel woods, which alternated with forests of other trees (Erdtman, 1931; Rackham, 1992). However, it is not probable that this situation occurred in prehistoric times, as hazel does not regenerate in scrub or in forests (Sanderson, 1958, pp. 74–79, 83–87; Rackham, 1980, p. 220, and see Chapter 6). Therefore the shrub would not have survived.

The high levels of pollen could be explained by a different interpretation than the current theory, on the basis of park landscapes (with grazing). In these landscapes, hazel is found in the mantle and fringe vegetation (Fig. 3.6), as well as in the form of scrub consisting virtually exclusively of hazel, next to forests and trees (see Fig. 3.7). This is the result of the fact that the species is regenerated in mantle and fringe vegetation and in grasslands (where there is grazing) (Sanderson, 1958, pp. 74, 80, 82, 85–87, 113; Dierschke, 1974;

Fig. 3.7. Park-like landscape with grazing in Slovenski Kras, Slovakia. The tall pyramid-shaped crowns are those of free-standing large-leaved lime. In addition there is pedunculate oak, hawthorn, blackthorn and hazel. The last species is there as bush as well as closed scrub (photograph, F.W.M. Vera).

Oberdorfer, 1992, pp. 82–106). As we pointed out earlier, hazel pollen moves low above the ground over a relatively small distance (Jonassen, 1950, pp. 26–27, 63; Linnman, 1978; Faegri and Iversen, 1989, p. 5). In that case, the high relative percentages in small basins must have been from local shrubs, i.e. shrubs in the direct vicinity of the depression. In a park-like landscape there is a long periphery of mantle and fringe vegetation with hazel shrubs, so if the original landscape was a park-like landscape, there would have been local sources of hazel pollen in the form of flowering hazel shrubs in many places in the mantle and fringe vegetation and in the grassland (Fig. 3.6). In addition, the pollen could have been picked up by rising air currents in open spaces in this sort of landscape and transported over a greater distance. This would explain why hazel is represented with such high percentages in the collection basins of regional pollen rain, such as peat bogs and lakes. As we noted earlier, modern pollen rain has established (Tamboer-van den Heuvel and Janssen, 1976; Lutgerink *et al.*, 1989) that hazel pollen is deposited at a large distance from the place where the hazel grows. In the model of a park-like landscape for prehistoric vegetation, the high levels in the frequency of hazel pollen in the local and regional pollen rain can therefore be easily explained, in contrast with the situation for the vegetation of a closed forest.

The combination of high percentages of hazel and oak pollen in all the pollen diagrams also supports the interpretation of a park-like landscape. As we saw in Chapter 2, the oak also regenerates in the fringes of scrub but not (or hardly) in a mature forest. Except in the situations described by Watt (1919; 1924; 1925; 1934a,b), regeneration of oak in the fringes or scrub was found in all the wood-pasture in England (Rackham, 1980, pp. 293–297, 300). In the pollen spectra of limestone grasslands where there are high percentages of grasses, which is interpreted as the presence of open grassland, there are relatively high percentages of hazel and oak pollen (see Bush and Flenley, 1987; Moore, 1987; Bush, 1993). This reveals the presence of grassland in combination with hazel and oak, as is the case in wood-pasture and in limestone grasslands, as we noted in Chapter 2.

Apart from the combination of high levels of hazel and oak pollen, another combination also indicates that there was a prehistoric vegetation consisting of a semi-open landscape. This is the combination of hazel and oak pollen, together with high relative frequencies of lime pollen, as found in sediments in loess soil (see Kalis, 1988a,b). Hazel cannot have flowered in lime forests because of the heavy shade provided by this species. On the other hand, oak, hazel and lime are found side by side and in combination in park-like landscapes grazed by large herbivores. Lime trees, like oaks, grow as free-standing trees and in the form of groves. The hazel grows in free-standing shrubs and in closed scrub (see Figs 3.6 and 3.7). These landscapes, with many free-standing trees or stands of trees of broad-leaved lime (*Tilia platyphyllos*), pedunculate oak (*Quercus robur*) and hazel are found in Slovakia in the Slovenski Kras area (Jakucs and Jurko, 1967; author's observation). Figure 3.7 illustrates this. In contrast with lime trees in a closed forest, free-standing or groups of lime flower profusely, particularly in the middle and lower parts of their crowns, and therefore emit a relatively large amount of pollen into the atmosphere (Rempe, 1937; Hyde, 1945; Eisenhut, 1957, pp. 24–25, 27, 35, 37, 94). In lime trees in a closed forest, only the top of the crown flowers (Borse, 1939). Admittedly, insects pollinate the lime blossom, but the wind also does this. In contrast with the assumption of palynologists (e.g. Iversen, 1973, p. 65; Kalis, 1988a), a lime tree cannot be described as a tree pollinated by insects, as regards emission of pollen into the atmosphere. In this respect, the lime is hardly any different from species of trees pollinated by the wind (Eisenhut, 1957, pp. 40–41). Iversen's argument (1973, p. 65) that the lime is under-represented in pollen diagrams because this species is pollinated by insects and therefore emits little pollen into the atmosphere, is therefore incorrect. In the case of lime, pollination by insects is in itself not a sufficient reason for applying a correction factor to the levels reached by lime pollen in pollen diagrams. Even uncorrected realistic high percentages of lime pollen, such as those found in the vicinity in the loess soil, are not necessarily from lime forests. As remarked above, they could also have originated from park-like landscapes such as that in Slovenski Kras (see Fig. 3.7).

Like the lime, other species of trees also emit much more pollen when they are free-standing or grow on the periphery of a grove than when they are growing in a closed stand (Borse, 1939; Faegri and Iversen, 1989, p. 14). In open terrain they develop a broad crown, which starts low down on the trunk and flowers profusely. Combined with a long periphery in park-like landscapes, this could explain why the tree pollen, as opposed, for example, to grass pollen, is so highly represented in the pollen floras. The trees protrude above the scrub and grassland, and consequently their pollen is easily transported by air currents and moved over long distances. This could have contributed to tree pollen being relatively over-represented in comparison with the species in the grasslands, including grasses themselves in the regional pollen rain, which sedimented in the regional collection basins, such as lakes and peat bogs, even when there are relatively few trees (see Tauber, 1967; Janssen, 1973; 1981; Jacobson and Bradshaw, 1981). Firbas (1934) had already indicated that pollen from forests which is collected in the surface samples in open terrain was over-represented as a result of the transportation over long distances; a phenomenon that was subsequently confirmed by many other researchers (see, *inter alia*, Berglund *et al.*, 1986; Faegri and Iversen, 1989, pp. 21, 27; Lutgerink *et al.*, 1989).

To summarize, it may therefore be concluded that in a park landscape of a certain area, a smaller number of trees may emit an equal amount of pollen (or even a larger amount) into the atmosphere than a closed forest of the same area, while in addition, factors such as the presence of mantle vegetation that is impenetrable for wind and grazed by large herbivores, can also affect the ratio between tree pollen and non-tree pollen in favour of that of tree pollen.

Finally, it has been determined based on modern pollen samples of park-like landscapes which have been grazed, that they give a pollen spectrum with a very high dominance of tree pollen (Gaillard *et al.*, 1994); a spectrum which is interpreted according to the prevailing theory as originating from a closed forest. Fossil pollen collections from landscapes, which according to historical sources were grazed and are characterized as very open forests (so-called out-fields) give a pollen image that is interpreted as a closed forest (see Lindbladh and Bradshaw 1998; Lindbladh, 1999). In addition, a landscape with grass-lands where hay was made, alternated with scrub and bushes with a lot of hazel in the south of Sweden, a so-called lövang (Heybroek, 1984), showed in a modern pollen assemblage a pollen spectrum that corresponds most with that of the Atlanticum (8000–5000 BP), the period during which the virgin closed forest was best developed in Europe, according to the prevailing theory (Gaillard *et al.*, 1994). Therefore it would seem that grasslands remain practi-cally invisible with respect to pollen from grasses in the pollen spectra. Furthermore, modern pollen assemblages from sediments and model calcula-tions from lakes of several hectares show that the traditional use of the per-centage of non-arboreal pollen (NAP) is insufficient as a direct measure for the percentage of open terrain. In those cases, no direct relationship appeared to exist between the percentage of NAP and the percentage of open terrain (Broström *et al.*, 1998; Gaillard *et al.*, 1998; Sugita *et al.*, 1999).

Trees and forests in the background give a high regional pollen load that lowered the percentage of NAP, even in the case where open areas were located close to the sample site, the lake (Davis and Goodlett, 1960; Sugita *et al.*, 1999). This background load could include 25–90% of the pollen in sedi-mentation basins (Jackson and Wong, 1994; Sugita, 1994; 1998; Calcote, 1995; 1998). Research in the south of Sweden in 22 lakes of 0.1–19.2 ha sur-rounded in a radius of 500–1000 m by landscapes with *Pinus* and *Betula* as dominant trees and with an openness varying from 20 to 80% showed a vari-ation in NAP of 10–20%. It was striking that even in the most open land-scapes the percentage of NAP was never above the 20% mark (Broström *et al.*, 1998). A situation where a closed zone of trees was situated along the bank of the sampled lake (0.2 ha) within (radius up to 1000 m) a landscape that con-sisted of more than 50% open cultivated land and of 30% semi-open land (20–50% tree coverage) showed a pollen image of 90% tree pollen (AP) (Broström *et al.*, 1998). In other situations, 45% openness showed a percent-age of NAP of 6–16% (Davis and Goodlett, 1960) and elsewhere a variation of open land between 0 and 75% showed a variation of NAP of 2–23%. Particularly in semi-open landscapes, the percentage of NAP proved insuffi-cient as a measure for the openness (Gaillard *et al.*, 1998; Sugita *et al.*, 1999). A semi-open landscape covered with forest with 30% open terrain with grasses and grains showed in two samples a percentage of NAP of 4.5% and 2.0% respectively (Sugita *et al.*, 1999). Closed landscapes with a percentage of forest of 60–75% showed a percentage of tree pollen of 90% (Gaillard *et al.*, 1998).

Both empirical and model-based research show that the size of the basins also plays an important role in differentiating open areas in the surroundings of the sedimentation basins (Sugita, 1994; 1998; Hicks, 1998). Due to the enormous load of background pollen, there were no differences shown in the pollen assemblages with respect to forest composition or the openness of the vegetation. The landscape appears to be homogeneous (Gaillard *et al.*, 1998; Sugita *et al.*, 1997). A simulation with a model shows that disruptions bordering on a lake of more than 3 ha can only be found in the pollen records if the surface of the disruption is at least eight times that of the lake itself and is not located further than a few hundred metres from the edge of the lake (Sugita *et al.*, 1997). This suggests that the openness in a landscape consisting of grazed grasslands alternated with bushes in a park landscape (see Figs 3.7 and 5.47) is difficult or even impossible to discover in the pollen assemblages (Sugita *et al.*, 1997; 1999). Empirical research confirms this. In the case of a boreal forest, where 300–400 m of forest was present between open areas and the pollen sample site, the percentage of NAP dropped to a general background level of 5–6%. Based on these findings, it was concluded that it is not at all possible to trace open areas in pollen assemblages based on sedimentation basins with a diameter smaller than 200 m, if these are located at a distance of more than 300–400 m and there is forest between them (Hicks, 1998).

3.8. The Situation in the East of the United States

In the east of the United States, like in Europe, a closed forest has been assumed to be the natural vegetation. Again like in Europe, the ratio of AP to NAP of 90% to about 10%, respectively, is considered evidence for this.[45] As remarked earlier, this forest vegetation is considered to be a vegetation comparable with the original vegetation forest in Europe. However, in the east of the United States, the percentage of pollen from oak forms a much larger portion in the pollen diagrams than in Europe, namely 40 to up to 70% (Davis, 1967; Wright, 1971; Watts, 1979; Delcourt and Delcourt, 1987; 1991, pp. 90–91; Clark, 1997), although based on modern pollen assemblages it is assumed that oak could be over-represented (Davis and Goodlett, 1960). In the east of the United States, there are 30 species of oak (Smith, 1993; Abrams, 1996). A large majority of these species is light demanding, including the most widespread species, i.e. white oak (*Quercus alba*). Only two species, namely coast live oak (*Q. agrifolia*) and canyon live oak (*Q. chrysolepis*) are shade tolerant (see for example Ross *et al.*, 1986; Smith, 1993; Abrams, 1996; Arthur *et al.*, 1998). The oak, together with thermophile genera such as *Acer, Fagus, Ulmus* and *Castanea*, would have taken its Ice Age refuge in the south-east of the United States. After the end of the Ice Age, these species would have migrated to the north (Davis, 1984; Watts, 1984; Webb *et al.*, 1984; Delcourt and Delcourt, 1987; Webb, 1988). There is a noticeable difference here in that the

---

[45] Watts (1979), Davis (1967, 1984); Wright (1984); Delcourt and Delcourt (1987; 1991, pp. 90–91), Webb (1988), Roberts (1989, pp. 72–74), Barnes (1991), Tallis (1991, pp. 235–263), Peterken (1996, pp. 45–53).

beech moved much faster than in Europe. An explanation given for this is that the blue jay (*Cyanocitta cristata*) buried both beechnuts and acorns as food-stock far away from the tree where they collected them (Darley-Hill and Johnson, 1981; Johnson and Adkinson, 1985; Johnson and Webb, 1989; Delcourt and Delcourt, 1991, pp. 27–28). In Europe, the beech lacked this type of vector. The oak did have one, namely the jay (*Garrulus glandarius*). This species hardly collected beechnuts (Schuster, 1950; Chettleburgh, 1952; Bossema, 1968; 1979, pp. 16, 20–26, 29, 32–33; Nilsson, 1985).

It has been stated that oak would be limited to edaphic extreme climatic conditions (Abrams, 1992), but pollen analyses show that oak also occurred in more moist regions on soils with good moisture levels in the company of shade-tolerant genera such as beech, etc. (see *inter alia* Gordon, 1969; Delcourt and Delcourt, 1991, pp. 90–91; Abrams, 1996; Ruffner and Abrams, 1998). This is particularly true for the white oak (*Q. alba*) which requires light. In historical times, this species occurred in all types of forest. In the vegetation from before European colonization, the so-called pre-settlement forests, it was a common species (see Chapter 4).[46] According to the pollen diagrams, oak survived next to shade-tolerant species, where this genus is now suppressed by these shade-tolerant genera such as *Acer*, *Fagus*, *Ulmus* and *Fraxinus* in forests. As noted in the previous chapter, lack of fire in histori-cal times could be the cause of this (see *inter alia* Crow *et al.*, 1994; Abrams, 1996; Abrams and Seischab, 1997; Clark, 1997). However, as Clark and Royall (1995) and Clark (1997) point out, there are no indications that the phenomenon of fire played a major role in the east of the United States during prehistoric times. This makes the comment by Clark (1997), that an explana-tion for the subsistence of oak must be sought after in places where fire is rare, topical. In a closed forest, this has not been successful as yet. It cannot be excluded in advance that the presence of oak could be under the influence of grazing by bison in park-like landscapes, given the fact that oak was able to regenerate in the east of North America in the ecologically analogous grass-land grazed by cattle. As for the possible interpretations of the pollen analyses, it is true that for North America a more open, park-like landscape in the Holocene than assumed until now on the basis of the percentage of NAP, can-not be excluded.

3.9 Conclusions and Synthesis

This chapter examined the contribution of palynology to theories on succes-sion, climax vegetation and the influence of humans on the original vegeta-tion. The most important findings and comments on this are presented below.

---

[46] See *inter alia* Smith (1962), Gordon (1969), Barnes (1991), Abrams (1996), Abrams and Downs (1990), Abrams and Ruffner (1995), Abrams and McCay (1996), Cho and Boerner (1991).

- When Von Post presented the first 'modern' pollen diagram in 1916, there was absolutely no doubt in scientific circles at the time that the lowlands of Western and Central Europe had been covered by a natural closed forest.

When the pollen diagrams were interpreted, this was done with an a priori point of view based on the notion of a closed forest as the original vegetation. The pollen study served, in the first place, to reconstruct the history of the forest, and the only questions that were asked were about the species diversity of the forest and the arrival of the forest in an open landscape.

- In palynology, the theory that the original vegetation was forest was never seen as a working hypothesis and was therefore never evaluated.

The theories and concepts which were used by plant geographers and forest researchers at the beginning of the 20th century are still used today. Palynology has not contributed any new building blocks to theories relating to the succession and composition of the original vegetation.

- Palynologists extrapolated the prevailing theories and concepts back to prehistoric times. Because of this process, data from prehistory were then interpreted as confirming the prevailing theories. This added another circular argument to the circular argument mentioned in Chapter 2.

The data obtained by palynology about prehistoric times reveal that trees, grasses and plants did exist in the original vegetation. In the interpretation of the pollen diagrams, the trees are seen as being highly dominant. This is viewed as confirming the prevailing theory, which entails that the original vegetation in the lowlands of Europe was a closed forest and that large herbivores did not have any influence on succession in it.

- The sudden strong decline in the relative frequency of elm pollen between 5000 and 5500 BP (the Landnam level) is attributed, above all, to the introduction and development of agriculture.

Together with archaeological discoveries the decline of pollen in the pollen diagrams in Central and Western Europe forms the basis for the Landnam theory. This decline was preceded by a decline in the east of Europe before the first farmers had arrived there. The decline of the relative frequency spread towards the west (see Fig. 3.4a and b). In view of the fact that it spread in this direction, and in view of the enormous area where the frequency of elm pollen declined in Central and north-west Europe over approximately 500 years, an outbreak of elm disease is a more plausible explanation for the sudden decline in the frequency of elm pollen than the cutting of foliage by the recently arrived farmers as fodder for their cattle. The course of the decline on both maps suggests that the infection of the elm disease spread to the west.

> • Palynology explains that the decline in the relative frequency of tree
>   pollen in the pollen diagrams is the result of the establishment of
>   agriculture (the Landnam theory). When trees were cut down in the
>   forest and cattle were grazed, the forest was believed to become more
>   and more open. For the palynologists, this showed that grazing cattle
>   results in a retrogressive succession from forest to grassland and
>   heath, once again confirming the prevailing theory.

The Landnam theory is also based on the circular argument raised in Chapter 2. Thus, one assumption follows another, and consequently it becomes increasingly difficult to establish where the theory started, and what are the causal relations between observations and theorization.

> • When the research data do not simply corroborate the prevailing
>   theory, for example, in the case of species which require light such
>   as oak, grasses and hazel, correction factors are applied to ensure
>   that observations correspond with the theory. The theory itself is
>   no longer a matter of discussion.

Even with the correction factors, discrepancies continue to exist between the theory, on the one hand, and the actual content of the pollen spectra and the observations in the field, on the other hand. Pollen of species that require light, such as oak and hazel, are found in high frequencies in the pollen diagrams, while empirical evidence from forestry shows that the oak does not regenerate in the closed forest, and that the hazel does not flower under a canopy, and therefore does not produce pollen.

> • The high relative percentages of hazel pollen in sediments in Central and Western Europe are consistently left out of or added to the total sum of tree pollen by palynologists. This selection or addition contributes to the view that the original vegetation was a closed forest.

Palynologists are wrong to exclude hazel pollen from or add it to the total sum including tree pollen. The assumption on which this is based, namely, that the hazel is part of plant communities which do not compete with the forest, is incorrect. Hazel should be seen as a species which likes open, unshaded situations. It does not flower and does not regenerate in the shade. Together with grasses and herbs, hazel is part of plant communities found in open grassland. Therefore hazel pollen should be presented in the pollen spectrum, together with the pollen of trees and the pollen of grasses and herbs. Toegether with grasses and herbs, the hazel represents open terrain. Here the hazel indicates mantle vegetations of bushes at the transition from grassland to groves. If the hazel pollen is presented in one sum together with the tree pollen in the total pollen sum, but as a representative of a seperate vegetation category, namely the mantle vegetation of groves at the transition from grassland to groves, the pollen diagrams immediately reveal a different picture. The species characteristic of forests then do not dominate the total pollen sum to such an extent at all (see Fig. 3.8). The pollen spectrum of hazel pollen, once it is included in the total sum as a representative of a mantle vegetation of groves at the transition from grassland to groves, can easily be explained by a park-like landscape where there is grazing.

> • Modern pollen spectra of park-like landscapes grazed by large herbivores reveal great similarities, in terms of species diversity and relative representation, to the pollen spectra of prehistoric times, which are interpreted as being of a closed forest.

The pollen diagrams, which are interpreted nowadays as being of a closed forest, could also come from a park-like landscape. Apart from the high levels of hazel pollen, the appearance of species which tolerate shade, such as the elm, lime, beech and hornbeam, at the same time as the oak which does require light, also indicates this. The low relative percentage of pollen grains of grass can be explained by grazing, which stops grasses from flowering so readily, and by the thick mantle and fringe vegetation which is present in park-like landscapes and acts as a virtually impenetrable barrier for the pollen of species which flower lower down, such as grasses and plants which grow in the grasslands and fringe vegetation.

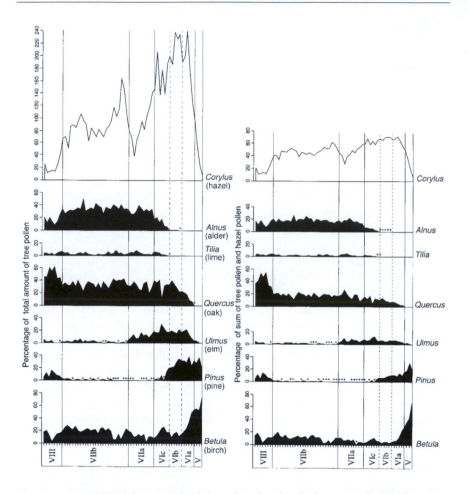

Fig. 3.8. Two pollen diagrams, both based on levels which were calculated on the basis of Fig. 3.3. Both diagrams cover the beginning of the Boreal period (V), which began in 9000 BP, up to the Sub-Atlantic period (VIII) which began in 2500 BP, and continue up to the present. Period VIIIa is the Atlantic period, the period in prehistoric times when the primeval forest was fully developed in Central and Western Europe, according to the prevailing theory. The left-hand diagram presents the amount of tree pollen as set at 100% in Fig. 3.3. The percentage of hazel pollen is expressed in that of the sum of tree pollen. In the right-hand diagram, the amounts of tree and hazel pollen are added together and percentages of both the individual tree species and hazel are expressed as percentages of this sum. Relative percentages which fall below 0.5% in this calculation are shown with a + sign. Note that in the right-hand diagram, trees dominate the diagram far less than they do in the left-hand diagram (redrawn from Godwin and Tallantire, 1951, p. 292).

> • Pollen diagrams from the original vegetation in the east of the United States from the Holocene can, like those in Europe, be interpreted as those of a much more open growth pattern than the closed forest that is currently postulated.

Like in Europe, it is true that for the east of the United States it can be concluded only on the grounds of the division of the percentage of AP and NAP of 90% to 10% that the original vegetation in the Holocene was a closed forest. With respect to the *Quercus* genus, this is an important fact. During the entire Holocene, this genus is very heavily represented in the pollen diagrams. It includes species requiring light, such as white oak (*Q. alba*), black oak *(Q. velutina)* and scarlet oak (*Q. coccinea*), which by all appearances cannot survive in closed forests. An interpretation of a more open vegetation with grasslands in prehistoric times offers an opening to the possibility of regeneration in grassland with grazing, as discussed in Chapter 2.

To summarize, it can be concluded that palynology has not added more to the theory of the original vegetation than was presented in Chapter 2 as premises and theories. The only new fact is that the dominance of tree pollen in sediments has been shown. This makes it plausible to assume that the untouched, prehistoric original vegetation was a closed forest. However, based on empirical forestry data and data from plant ecology, questions have been raised on the prevailing theory based on the permanent presence of oak and hazel in the prehistoric vegetation. Instead of drawing up a working hypothesis, the data are interpreted based on the prevailing theory, on the grounds of which plant ecologists and foresters see confirmation of their theory of the closed forest as original vegetation. An important finding from pollen analyses of modern pollen assemblages with respect to the situation in the lowlands of Central and Western Europe is that pollen spectra from park-like landscapes show great similarities with pollen floras from prehistoric times. Moreover, research into the relationship between modern pollen spectra and landscapes show that pollen spectra interpreted as closed forest according to the prevailing theory, were actually semi-open to very open landscapes. These analyses show that the percentage of NAP, particularly in semi-open landscapes, is an unreliable measure for the openness of an area. These and other data show that pollen diagrams that are traditionally interpreted as a closed forest, could also have been a grazed park-like landscape. As in Europe, it holds that it cannot be concluded based on pollen diagrams of the Holocene that the untouched, original vegetation in the east of the United States was a closed forest.

The Use of the Wilderness from the Middle Ages up to 1900

4.1 Introduction

As I indicated in Chapter 2, the theories on succession and the natural climax vegetation in the lowlands of Western and Central Europe are based on the assumption that the grazing of livestock resulted in the disappearance of the original forest.[1] According to this view, livestock prevented the regeneration of the original forest because they trampled and ate the seedlings. The grazing of livestock and the exploitation of timber were believed to have resulted in more and more openings in the original forest so that the wilderness, the closed forest, eventually changed into a park-like wood-pasture and ultimately into grassland or heath.[2] This is described as retrogressive succession, according to the theories which were dealt with in Chapter 2.

As we found in Chapters 2 and 3, the above was based, on the one hand, on tests in which areas of grasslands or forest were fenced off from livestock, and on the other hand, on historic sources. The original forest, which was believed to have been present over considerable areas up to the early Middle Ages, was also thought to have disappeared during the course of the Middle Ages, partly as the result of grazing livestock (Schubart, 1966, p. 145; Darby, 1970; Mantel, 1980, p. 116; 1990, p. 55). In order to prevent livestock from trampling seedlings of trees in the forest, regulations were issued in the Middle Ages and

---

[1] See: Moss (1910, p. 36; 1913, pp. 96–98), Moss *et al.* (1910), Moss (1913, pp. 96–98), Tansley (1911, pp. 7–8; 1935; 1953, pp. 128–130), Watt (1919; 1923; 1924; 1925; 1947).
[2] Kerner (1929, pp. 44–46), Nietsch (1927; 1935, pp. 47, 55; 1939, pp. 109, 116, 117), Hilf (1938, p. 91), Hart (1966, p. xix), Clason (1977, pp. 114–115), Ellenberg (1954; 1986, p. 36; 1988, pp. 17–21 and 28), Mantel (1990, pp. 95, 423).

later, to regulate the grazing of livestock in forests.[3] The oldest known regulations date from the 6th century. A fairly large number of records and other texts have survived from a later period containing similar edicts. They were discovered by researchers working on what is known as the use of the forest in history.[4] In the light of the problem which I formulated, I would like to address the following questions, using these publications:

- Do the historic written sources on the use of land that has not been cultivated show that it originally consisted of forest?
- Can the edicts on the use of the last areas of wilderness in the lowlands of Central and Western Europe explain a priori what the nature of the vegetation there was?
- Do edicts on the grazing of livestock show that grazing resulted in retrogressive succession of the forest to grassland, i.e. were the seedlings trampled and eaten by livestock?
- Do edicts on the use of the last wildernesses of Europe explain how the regeneration of trees and shrubs in general, and of oak and hazel in particular, took place?

4.2. The Wilderness and the Concept of 'Forestis'

The German term 'Forst', the French term 'forêt' and the English term 'forest' nowadays mean a large interconnected forest complex.[5] They are derived from the term 'forestis', which appeared in the early Middle Ages in deeds of donation of the Merovingian and Frankish kings (Kaspers, 1957, pp. 23–26; Gilbert, 1979, p. 10; Mantel, 1980, pp. 63–65; 1990, p. 36; Buis, 1985, p. 6). In Central and Western Europe the term 'forestis' in the High German language area and in lower Saxony, became closely linked in the Middle Ages to the term 'Wald'.[6] According to Mantel, not every 'Wald' was a 'forestis' in the Middle Ages, but the term 'forestis' did relate to large areas of 'Wald' which were brought under their jurisdiction by the kings (Mantel, 1980, pp. 63–65; 1990,

---

[3] Hausrath (1898, p. 101; 1982, pp. 39, 206–209), Bühler (1922, pp. 300–301, 339, 610), Meyer (1931, pp. 345, 386), Rodenwaldt (1951), Hesmer (1958, p. 454), Hesmer and Schroeder (1963, pp. 151–153), Streitz (1967, pp. 53–54), Buis (1985, pp. 40, 50), Mantel (1990, pp. 95, 358).

[4] See for example: Endres (1888), Hausrath (1898; 1982), Bühler (1922), Vanselov (1926), Grossmann (1927), Meyer (1931), Hilf (1938), Rodenwaldt (1951), Reed (1954), Kaspers (1957), Hesmer (1958), Hesmer and Schroeder (1963), Hart (1966), Schubart (1966), Streitz (1967), Mantel (1968; 1990), Rackham (1980; 1993), Buis (1985), Perlin (1991), Tack *et al.* (1993).

[5] See: Kaspers (1957, p. 18), Rackham (1980, p. 175), Buis (1985, pp. 6, 25), Herckenrath and Dory (1990, p. 400), Mantel (1990, pp. 38, 153–154), Ten Bruggencate (1990, p. 347), Stoks (1994, p. 233).

[6] See Habets (1891, pp. 351–360), Vanselow (1926, pp. 171–213), Kaspers (1957, pp. 18, 40–50, 143, 151, 154–157, 168), Hesmer (1958, pp. 408–423), Streitz (1967, p. 39), Buis (1985, pp. 235–236, 304–307), Mantel (1980, pp. 883–890, 925–996, 1026; 1990, p. 36).

pp. 36–37).[7] During the course of the Middle Ages, he did think that every
'Forst' was a 'Wald' in the sense of the word 'forest'. Mantel therefore considered the term 'forestis' as a new term for a forest (Mantel, 1980, p. 1005; 1990,
pp. 36, 38). This argument is based on the present meaning of the term 'Wald',
which is forest (Stoks, 1994, p. 760).

The term 'forestis' appeared in the 7th century in the deeds of donation of
Merovingian and Frankish kings in the form of the use of 'forestis nostra'. This
was a legal concept which described or confirmed the royal rights in certain
areas (Kaspers, 1957, pp. 23–25; Mantel, 1980, pp. 63–67; Buis, 1985, p. 26;
Tack *et al.*, 1993, p. 96). There is a consensus that the term 'forestis' related to
the wilderness which had not been cultivated and which had no clear owner.
(Kasper, 1957, p. 25; Hesmer, 1958, p. 408; Mantel, 1980, p. 1005; Buis,
1985, p. 223; D.P. Blok, Nederhorst den Bergh, 1997, personal communication). This land lay beyond the land which had been cultivated and therefore
had a clear owner. These were the fields where crops were cultivated, and the
hay fields and settlements (Kasper, 1957, pp. 25, 95, 238; see Appendix 1).

In the Frankish kingdoms, the ownership of uncultivated lands without a
clear owner went to the king or lord according to Gallic-Roman law.[8] This so-
called royal prerogative to land or wilderness was based on Roman law, the
Codex Iustinianus X, 10. This stated that so-called 'bona vacantis', goods without a clear owner, belonged to the 'government' (De Monté Verloren and Spruit,
1982, p. 123). This law produced the Frankish-Latin legal concept 'forestis'. It
was a new concept. There is no unanimous view about the origin of this concept (Mantel, 1980, pp. 59–65, 1000–1006; D.P. Blok, Nederhorst den Bergh,
1997, personal communication). The theory which is most widely supported is
that 'forestis' was derived from the Latin 'foris' or 'foras', which means 'outside',
'outside it', and 'outside the settlement' (see Kaspers, 1957, p. 24; Hesmer,
1958 p. 408; Buis, 1985, p. 26 *et seq.*; Muller and Renkema, 1995, p. 363).
There is some agreement that the concept 'forestis' applies to the wilderness in
general, and to trees, forest, shrubs, wild animals, water and fish in particular

---

[7] On this subject, Mantel wrote:

Seit dem 7. Jahrhundert erscheint der Begriff 'forestis nostra' in den fränkischen
Königsurkunden. Zumeist wurde Wald ausgeschieden, es wurden aber auch andere
Ländereien, insbesondere Wildland und sogar Fischwasser 'eingeforstet', wie alte Urkunden
vom 6.Jahrhundert zeigen. Die erste Erwähnung einer forestis findet sich in einer fränkische
Urkunde von 648, durch die der König einen Besitz in den Ardennen – in foreste nostra
nuncupante Aduinna – einem Kloster gab. Die Einbeziehung von Wald in eine forestis wird
auch Einforstung oder Forestifikation genannt. Manchmal wurde nicht der gesamte Wald
eingeforstet, sondern nur ein Teil enes gröberen Waldgebietes als königlicher – oder später
herrschaftlichre – Forst ausgeschieden.

As an example of the 'forestis' or 'forst' relating mainly to a forest, Mantel quoted a text dating
from 1324, which states that the Archbishop of Mainz 'sin abgescheiden wald hait, mit namen
der forst' (Mantel, 1980, p. 64).

[8] For example, one text in which King Chilperich II granted an area of land to the abbey of St
Denis in 717 described this as: 'foreste nostra Roveito'. Also see Grossmann (1927, p. 14),
Kaspers (1957, pp. 23–25), Hesmer (1958, pp. 12, 408), Schubart (1966, pp. 40–41), Buis
(1985, p. 26), Mantel (1990, p. 153).

(Kaspers, 1957, pp. 24–30; Hesmer, 1958, p. 408; Mantel, 1980, p. 1005; Buis, 1985, pp. 25–26, 223 *et seq.*; D.P. Blok, Nederhorst den Bergh, 1997, personal communication). In other words, in a 'forestis', every individual tree, as well as every wild animal, belonged to the king. The concept 'forestis' comprised everything which lives and grows in the wilderness or the land which has not been cultivated, i.e. the land, water, wind and every animal that lives and every plant that grows there.[9] The term 'forestis nostra' not only indicated that the land concerned ('forestis'; plural, 'forestes'), and all the wild plants and animals living there belonged to the king; it also meant that only the king had the right to make use of these. Without his express consent, others were not permitted to graze their livestock, cut down trees, collect firewood or create fields for crops (Kaspers, 1957, pp. 23–26, 39–40). To emphasize that these were areas where no one else had any rights, the words 'eremus', 'solitudo' or 'in deserto' were often added to the 'forestis nostra'. This clearly showed that it concerned an unpopulated or abandoned wilderness (Kaspers, 1957, p. 23; Buis, 1985, p. 26).

From the 9th century, the term 'forestis', 'vorst' or 'Forst' was found in combination with the term 'Bann', e.g. as 'Forstbann' (Kaspers, 1957, pp. 48–56). This emphasized that a 'forestis' fell under royal jurisdiction and was protected as such (' ... forestum etiam cum nostro banna ... ') (Kaspers, 1957, p. 52). The 'bannum' was the royal power to prohibit and command (Van Caenegem, 1967, p. 64). From the 11th century, the term 'wiltban', ('wiltbant', 'wiltpant', 'wiltbanck', 'wiltbahn', 'wiltbane', 'wildbaenne') was equated with the term 'forestis' (' ... forestum vulgariter dictum wiltban ... ius forestale quod Wiltban dicitur ... ') (Weimann, 1911, p. 61; Kaspers, 1957, pp. 52–53; Kiess, 1998). According to Kaspers (1957), this was a popular interpretation of what was meant by 'forestis' as a legal concept, i.e. the wild uncultivated land which fell under royal jurisdiction.[10] The term 'wild' referred to all animals and plants which existed without being cared for by man, and which had no clear owner (Kaspers, 1957, p. 53). According to Trier (1963), a 'wild' thing was anything which was not obtained by cultivation from seed and tending, but from collecting, breaking off, pulling, picking, shaking, seizing or catching for use by humans. For example, this included wild oats, wild grass ('Wildheu') and wild animals. According to Trier, anything that was used meant that it was called

[9] For example, the Act in which the Frankish king Sigibert III gave permission for establishing the Stablo-Malmedy monastery in the Ardennes in Belgium in the middle of the 7th century states: 'in locis vastae solitudinis, in quibus caterva bestiarum germinar' (in places of emptiness and wilderness where herds of animals live) (Kaspers, 1957 p. 23; Buis, 1985, p. 26). The legal term 'forestis' was used in relation to hunting ('omnesque venationes vel foresta') fishing (' ... in opsa aqua forestem piscationis'), water ('necnon foresten aquaticam in fluvio ... '), trees (' ... silvae vel forestes nostrae') and wild animals ('et feramina nostra intra forestes bene custodiant') (Kaspers, 1957, pp. 25–26, 30).

[10] The fact that the term 'wild' in this context was not limited to wild in the sense of animals that were hunted, as had always been assumed up to that time by researchers, according to Kaspers (1957), is deduced from the existence of terms, such as, for example, 'wilthube' and 'wilthubener', which respectively meant farm and owner of a farm in a 'forestis' in the 'wild' (for further explanation, see Kaspers, 1957, p. 54).

'wild'. Anything which was not tended by man, but which was not taken for use, was not 'wild'. Bats and moles were not 'wild', but aurochs, red deer and wild boar were, because they were hunted for human use (Trier, 1963, pp. 48–49).

Because wild animals and wild plants did not have a clear owner, they were the property of the local lord. Therefore, the so-called 'ius forestis' applied to these animals and plants. The express consent of the local lord was needed for collecting these 'wild' animals, 'wild' plants, or 'wild' fruits, and this consent might or might not be granted on the basis of the 'ius forestis'. Because livestock in the wilderness fed on the 'wild' grass and 'wild' plants, it was necessary to obtain the lord's consent to graze livestock in a 'forestis'. The livestock as such did not fall under the 'ius forestis' because it had a clear owner. But this did apply for anything the livestock ate, like the 'wild' grasses and herbs. The 'bannum', the royal prerogative to prohibit and command, applied to anything that was 'wild'. Because these 'wild' things lived in the 'forestis', the meaning of the term 'wiltban' became synonymous with the concept 'forestis' in the course of the Middle Ages. Because the 'bannum' applied in the 'forestis', during the Middle Ages the 'forestis' also came to be referred to as 'banvorst', 'Forstbann', 'wiltforst' or simply as 'Bann', as well as 'Wiltban' (Kaspers, 1957, pp. 53–54, 154; Mantel, 1980, p. 989).[11]

The administration and management of the 'forestes' was passed by the king to so-called 'forestarii'. The first reference to 'forestarii' dates from 670 (Buis, 1985, p. 223). These officials had to oppose every infringement by anyone of the rights arising from the concept of 'forestis' (Kaspers, 1957, pp. 32–39; Hesmer, 1958, p. 408; Buis, 1985, pp. 223–225). These 'forestarii' were often given a farm ('mansus'; plural, 'mansi'). The 'forestarii' were also given the rights by the king to pass judgement on matters which were related to respecting and exercising the rights covered by the legal concept, 'forestes'. This included regulating the use of the 'forestis' (in Dutch: 'foreest', 'forest', 'voorst' or 'vorst') for grazing livestock and cutting wood. The 'forestarii' would appoint a 'magister/minister forestariorum'. The latter chaired the legal forum as *primus inter pares* amongst the 'forestarii', the court in which the use of the 'forestis' was arranged in accordance with the 'ius forestis', the 'iura forestarorium' or the 'ius nemoris', as well as dealing with the infringements of the regulations (Kaspers, 1957, pp. 32–39, 50; Buis, 1985, pp. 223, 225; Buis, 1993, p. 41). In the records written in Latin, this 'magister forestarorium' was also called the 'comes nemoris' ('comicia lignorum', 'cometia nemorum', 'comes sylvestris' or 'silvae comes'), in (lower) German, 'Waldgraf', 'Wildgraf',

---

[11] In the Dutch language, the terms 'wildernisse' and 'Wyldnisse' were used respectively in the Middle Ages and subsequently, for land that was not cultivated. A letter donating land dating from 1328 states: ' … dat alinge broeck und alle die wildernisse, die wij voortmeer dit Niebroeck geheyten willen hebben … ' (Blink, 1929, p. 27). In 1571, the Duke of Gelre issued the 'Placeat ende ordonnantie op het Nederrijckse-Walt', and in 1605, The Count of Holland issued the 'Placaten ende Ordonnancyen op 't stuck van de wildernissen' (Buis, 1985, pp. 296, 345). Also see: Wartena (1968), Buis (1985, pp. 238–240, 294, 304–306).

'waltgreve', 'holtgreve' ('silvae comes, qui dictur holtgreve'), 'marckgreve', 'vorstmeister', 'waldmeister' or 'Dingherr', and in Dutch, 'waldgraaf', 'woudgraaf', 'wautmeester', 'woudmeester', 'wautmaire', 'vorster' or 'houtvester'.[12] At the end of the 12th century and the beginning of the 13th century, the terms 'vorsthinc', 'holzgedinghe' and 'waltgedinghe' appeared in north-west German records for this jurisdiction (Kaspers, 1957, p. 40; Buis, 1985, p. 226). The 'ius forestis' or 'ius nemoris' was also referred to as 'forstrecht', 'Recht der Forste', 'altes Recht der Wälder', 'recht der Welde', 'reicht der welde', 'waltrecht', 'recht des walts', 'des walts recht' and 'woudtrecht' (Habets, 1891, p. 352; Kaspers, 1957, pp. 41, 50, 113, 150, 222, 230; Wartena, 1968). Licences for the use of the 'forestis' were granted on the basis of the 'Waldrecht' (after 'waltrecht') or the 'Woudt-recht' (Kaspers, 1957, p. 166; Wartena, 1968, pp. 36, 38; Ten Cate, 1972, p. 201). The records which contained these licences were known in the 13th century as 'waltbuch', 'waldtforstbuch' or 'vorsterboych' (Kaspers, 1957, pp. 103, 106).

'Forestes' could contain settlements with cultivated land. In this case, these were traditionally royal possessions such as farms ('curtes' or 'fisci'), or new settlements established in the 'forestis' with royal consent, or so-called 'villae' (Kaspers, 1957, pp. 28–29, 93). The 'forestarii' supervised these cultivated areas (Buis, 1985, pp. 224–225). They made sure that the colonists carried out the obligations which they had *vis-à-vis* the local lord. These obligations arose from the concession which had been granted to them by the lord for establishing a farm. This entailed that they could create a field to cultivate crops of the size which they needed to maintain a household, and they were granted the rights to use the remaining wilderness to graze their livestock and use as much firewood and timber for building as they needed to meet their own needs (Blink, 1929, p. 16; Kaspers, 1957, pp. 124, 149, 185, 205, 214). The obligation to do something in return meant that the colonists were in servitude to the lord (De Monté Verloren and Spruit, 1982, pp. 83–85). Therefore, the 'forestarii' collected from the colonists the quota of the harvest, the 'medem', to which the lord was entitled according to the 'ius nemoris', the so-called 'Holzkorn', 'Forstkorn' or 'Wildbannkorn' (Kaspers, 1957, pp. 29, 38, 126, 137, 185, 205, 220–224, 236, 238; Buis, 1985, pp. 223–224).

The king could also loan parts of a 'forestis' or 'wiltban' to vassels, or establishments such as monasteries. This right became known in the records as the 'ius eremi' or 'ius nemoris' (Kaspers, 1957, pp. 23, 50). The royal prerogative ('sub banno nostro' or 'in regio banno') still applied to a 'forestis' that was given in loan. Thus the 'forestes' were loaned 'cum banno' (Kaspers, 1957, p. 52), which meant that the king also transferred the power to prohibit and command. For the use of this land, permission was therefore needed from the new owner; the person or establishment who had been granted the loan by the king. In his turn, he was entitled to grant this consent (Kaspers, 1957, pp. 25, 39,

---

[12] Kaspers (1957, pp. 42–43), Mantel (1968), Wartena (1968), Ten Cate (1972, p. 201), Buis (1985, pp. 238–240, 294), Tack *et al.* (1993, p. 204).

40). If the person or establishment who had been granted the loan gave consent for the establishment of a farm on the loaned land, this was in servitude to the person or establishment to whom the land had been loaned (De Monté Verloren and Spruit, 1982, pp. 83–85). If the king loaned 'forestes' (with 'villae') to people or church establishments, the organization of the 'forestes' was also transferred to the person who had been granted the loan by the king (Buis, 1985, p. 224). This meant that the organization of the 'forestes' survived throughout the Middle Ages and in the following centuries. In time, the word 'forestis' evolved in the Dutch language to 'forest', 'foreest', 'voorst' and 'vorst'; in German, to 'Forst' and 'vorst', and in French to 'forêt'.[13]

In England, a situation developed in the 11th century which was slightly different from that on the continent. After William the Conquerer had conquered England from Normandy in 1066, he introduced the so-called 'Forest Law' (Hart, 1966, p. 7; Darby, 1976b, p. 55; Cantor, 1982a,b; Tubbs, 1988, p. 67). The law served to protect the sovereign right of the king to all wild animals. This particularly concerned deer. An area which was covered by the Forest Law was known as a 'Forest'. The difference from the older 'ius forestis' on the continent is that the 'Forest Law' was also declared applicable to areas which did have a clear owner (Rackham, 1980, pp. 176–177, 184). This ownership had been established during the Anglo-Saxon period. As on the continent, Anglo-Saxon kings gave land (estates) in loan to the nobility in exchange for their loyalty and services (Aston, 1958; Page, 1972, pp. 45, 53–54). The nobility or lords owned a 'manor', the administrative unit into which medieval England was divided (Hart, 1966, p. 8; Page, 1972, pp. 88–90; Cantor 1982a,b; Hooke, 1998, p. 54). The 'commons', which have traditionally been used communally by local communities, were part of these. The users, or so-called 'commoners', had the right to collect firewood and timber, to graze animals ('rights of estovers, common and pannage'), which could be passed down (Hart, 1966, p. 8; Page, 1972, pp. 88–90; Cantor, 1982a,b). Therefore three parties were involved in the use of a 'Forest', the Crown, the lord or several lords, and the commoners.

Like a 'forestis', a 'Forest' applied to all sorts of terrain. In 1598, an English lawyer named Manwood defined a 'Forest' as: 'territory of woody ground and fruitful pastures, privileged for wild beasts and fowls of forest, chase and warren, to rest and abide there in safe protection of the King, for his delight and pleasure' (Tubbs, 1988, p. 67). Like a 'forestis', a 'Forest' was therefore the wilderness comprising trees and shrubs, grasslands, water, peat bogs, marshes and wild animals which were under the protection of the king. A 'Forest' had an organizational structure similar to that of a 'forestis', and 'forestarii' were responsible for carrying out the 'Forest Law'.

---

[13] Krause (1892), Kaspers (1957, p. 53), Ten Cate (1972, pp. 111, 189), Gilbert (1979, p. 10), Mantel (1980, pp. 65–68), Buis (1985, p. 1; 1993, pp. 26, 36, 37).

4.3 The Meanings of the Term 'Wald'

Areas for which the legal term 'forestis' was declared applicable during the Middle Ages were also described as 'Wald'.[14] In texts dating from the Middle Ages and later, the 'Wald' (also referred to as 'walte', 'waldes' or 'welde') was a place where livestock and birds (probably mainly tame geese) were grazed, and where food for bees was found (Remling, 1853, pp. 35, 67, 316; Mantel, 1980, pp. 225–226, 937),[15] as well as a place for collecting wood.[16] In the 'Wald', the 'blueme' (flowers), the 'blumenweide' (pasture with flowers), 'dess Plumenbesuchs' or the 'blumbesuch' (visits to the flowers) was carried out by livestock (Endres, 1888, p. 51; Hilf, 1938, p. 132; Mantel, 1968; 1980, pp. 194, 195, 972; 1990, p. 93). The 'Wald' was used for haymaking (Kaspers, 1957,

---

[14] For example, the 'Bayreuth Forstordnung' dating from 1493 states: 'Wo aber meinem gnädigen hern armleut [farmers], die der Welden oder forsten so nahent gesessen wern, daß die mit irem viech nit entpern [lack] oder meyden konnten, und darein gerechtigkeit zu treiben hetten' (Mantel, 1980, p. 969). See also Vanselow (1926, pp. 171–213), Kaspers (1957, pp. 18, 40–50, 143, 151–157, 168), Hesmer (1958, pp. 408–423), Streitz (1967, p. 39), Buis (1985, pp. 235–236, 304–307), Mantel (1980, pp. 925–996).

[15] See *inter alia* Remling (1853, p. 67), Habets (1891, pp. 351–360), Krause (1892b), Weimann (1911, pp. 4–6, 20–21), Kaspers (1957, pp. 40, 44, 123–125), Mantel (1980, pp. 925–995).

[16] A document issued in the Bishopric of Speyer in the west of Germany in 1404 states:

Item wir meynen auch, wann vnd zu welcher zyte daz ist, daz ecker in den obgenannten welden sint, daz dann die von Lachen mit ire sweine in die selben welde vnd eckern wol faren mogent, … Item wir meynen auch, daz die von Hambach mit yrem viehe zu weyde wol faren sollen vnd mogent in die von Lachen marcke vnd weyde (Remling, 1853, p. 35).

Another example is a document from the same bishopric, dating from 1408, in which King Rupert indicates that he has no rights to an area in the bishopric. With regard to the term 'Wald', the text states: 'doch also das die arme lute zu Ketsch vnd der der dasselbe ampt von des stiftes wegen innhat buwe vnd brenneholtz darinne hauwen mogen vnd auch der weyde desselben waldes mit yerem viehe geniessen' (Remling, 1853, p. 67). A document dating from 1462 states:

sin nachkomen vnd stiefft vnd die dorffer darumb gelegen vnd alle angehorigen des stieffts by aller ander des walds herlickeit, nutzunge, weidgang, holtzunge, eicheln, vogelweide, auch rehen, hasen, vnd fuchs zu fahen vngehindert vnd des stieffts arme lute mit atzunge oder frondiensten 'in keynerlei wise zu besweren (Remling, 1853, p. 316).

The 'Wald' by Soluthern in Switzerland in 1666 is described as: 'In anderen nächts umb die Statt ligenden Wälden, deren es zum Lust und Nothdurfft vil, hat die Burgerschaft neben dem Baw- und Brennholtz eine herrliche Viehe Weyd, auch das Acherumb für die Schwein' (Meyer, 1931, p. 392). In addition to the grazing of livestock, reference was also made to bees (seu apium pascuis) (Remling, 1852, p. 23; Hesmer, 1958, p. 393). A document dating from 1339, about the division of a 'Waldt' between the cities of Kleef and Cologne, states:

Vort mehr synt wy averdragen mit den anderen erffgenoten, dat wy mit oeren wille die anderen seven deele des vurs. waldes onder oen hebn laten slaen ende deilen tot hollantzschen rechten, in dess manier, dat sie ons ende onsen erven nae denselven hollantzschen rechten alingen tenden geven sullen, uit genaemen den thiende van den holte, van ekeren [to fatten the pigs on acorns] ende van weiden (Weimann, 1911, p. 57).

With regard to Westphalia, there is a record dating from 1569, stating: 'Es ist vast ein waeldig Land/ und darumb gechickt zum Vieh: Insonderheit hat es viel Eychwald/ darinnen die Schwein gemestet werden' (Ten Cate, 1972, p. 196). Also see: Schubart (1966, pp. 56–57); Kasper (1957, p. 152); Mantel (1980, pp. 462–465, 925–996; 1990, pp. 90–107 and the rest of that chapter).

p. 150).[17] The 'Wald' was therefore a place where livestock found fodder and where it was collected. In the Frankish language, the place where animals found food or where food was collected was described as a 'weide' (pasture) (De Vries, 1970, p. 249; Van Veen and van der Sijs, 1991, p. 817). Animals which were looking for food were engaged in 'weiden'. According to texts dating from the Middle Ages and after, the 'Wald' contained food, 'weide' (pasture) for livestock, birds and bees. The meaning of the term 'Wald' therefore also comprised the 'weide', the provision of the food for these animals and for wild animals. Grass and herbs which were, for example, referred to as 'waydt', served as food, as well as the foliage of trees and shrubs which was collected for fodder (Trier, 1963, pp. 1–38, 81; Rackham, 1980, p. 4; Mantel, 1980, pp. 941, 969, 984; Pott, 1983; Tack et al., 1993, pp. 80–81).[18] The fruits of trees also constituted 'weide' (pasture) for livestock. This applied in particular for the acorns of oak trees which were known as 'waid' for pigs.[19] Therefore in the Middle Ages, and for several centuries subsequently, the meaning of the term 'Wald' comprised the 'weide' with flowers for livestock and bees.

References to cutting leaf-fodder for livestock can be found in written sources as early as those dating from Roman times. The elm (*Ulmus*) was considered the best fodder, followed by rowan (*Sorbus aucuparia*), and ash (*Fraxinus excelsior*). In addition to these species, hazel, hawthorn and even conifers, such as juniper (*Juniperus communis*), and Scots pine (*Pinus sylvestris*), were cut as fodder throughout Central and Western Europe.[20] All sorts of deciduous trees and shrubs were used for cutting foliage for fodder because they have an enormous potential for regeneration.[21] Fodder was collected by cutting branches or twigs with foliage from the crown of the tree or shrub. Depending on the shape acquired by the tree by cutting the foliage, this was known as coppicing, pollarding or shredding the tree. Apart from cutting the foliage from trees, it was also possible to cut or strip the shoots sprouting from a tree stump or shrub (for example, see Rackham, 1980, p. 4; Pott, 1983; Tack et al., 1993, pp. 80–81).

[17] A description of the 'Wald' near the city of Tangermüne, a town in the east of Germany, approximately 40 km north of Maagdenburg, dating from 1651, reveals that specific hay fields formed part of the 'Wald', as well as grasslands. This description also states that a river could be part of the 'Wald'. It states:

> In der Nähe hat sie [the city] einen schönen lustigen Stadt-Wald oder Busch von fruchtbaren Eich-Bäumen, worin der Zeiten etliche Schock Schweine können fett gemacht werden, dieser Wald hat auf der einen Seite die Elbe, auf der anderen den Fluss die Tolarn, in sich viele nutzbare Wiesen und schöne Plätz (Krause, 1892b, p. 86).

[18] For example, the 'Kurfürstl. Oberpfälzische Forstordnung', dating from 1565, states in Article 28: 'Von betreybung whun/ und wayd der Wällde' (Mantel, 1980, p. 984). In a clause from 1561 it says: 'Da nun im andern fall ein Wald nit zum hauw, sonder allein zur Weyd verordnet' (Mantel, 1980, p. 96; also see Mantel, 1980, pp. 79, 82, 85, 95).

[19] A text dating from 1310 states: 'to pasture his pigs in the monastery's "welden"' (Ten Cate, 1972, p. 100). See also *inter alia* Habets (1891, p. 358), Weimann (1911, pp. 57, 66, 82), Kaspers (1957, p. 291), Mantel (1980, pp. 925–995) and later in this chapter.

[20] Trier (1952, pp. 96, 148; 1963, pp. 1, 19), Hart (1966, p. 42), Flower (1977, p. 109), Rackham (1980, pp. 4, 174, 223, 345), Pott (1983), Meiggs (1982, pp. 266–267; 1989); Austad (1990), Andersen (1990), Mantel (1990, pp. 102–104), Tack et al. (1993, pp. 80–81).

[21] Trier (1952, pp. 7–8, 11), Bühler (1922, p. 551), Rackham (1980, pp. 34–35), Koop (1987), Mantel (1990, p. 333), Mayer (1992, p. 429).

In the course of the Middle Ages, cutting or breaking the foliage in the 'Wald' was increasingly restricted and eventually even entirely prohibited because of the damage which was caused, particularly to trees.[22]

The above shows that the meaning of the term 'Wald' was broader in the Middle Ages and later than the more modern meaning of a wood. 'Wald' also meant 'weide' (pasture, i.e. grasses, herbs, nectar and foliage of trees and bushes) for animals, both for livestock and for wild animals. In view of this, the meaning of the term 'Wald' could therefore also have comprised grasslands as well as trees. The meaning of 'Wald' as a place with fodder also appears from the results of the research by the etymologist Trier (1963). On the basis of research into the meaning of 'Wald' in the early and later Middle Ages, Trier concluded that the term (and the Anglo-Saxon term 'weald') related to the foliage on a twig of a tree or a shrub and (in the second half of the 9th century) to the crown of the tree (Trier, 1963, pp. 39–42). In his opinion, this meaning survived up to the 20th century in Swabia in Germany, in Alsace in France, and in Switzerland, where all branches of a tree with foliage are collectively known as 'Wald' or 'Waldung'.

The 'Wald' was the crown of the tree. In these areas, foresters in the 18th and 19th centuries spoke disparagingly about a tree which had grown in light conditions and therefore had a large crown on a relatively short trunk as though it 'had a great deal of Wald and little trunk' (Trier, 1963, p. 40). According to Trier (1963), the term 'waldig' was used in the Alsace and in Switzerland up to the beginning of this century about trees formed in this way. A tree in which the crown died off as the result of a disease was known as 'Walddür' (with a dry crown), and cutting the branches and top off a tree once it had been cut down was known as 'Walden' or 'Auswalden'. Therefore, as regards their meaning, 'Wald' and 'Stamm' were opposites, like foliage and wood. Therefore, according to Trier (1963), a 'Wald' did not only contain trees, but a tree also contained 'Wald' (Trier, 1963, pp. 40–42). Trier thought that in the course of the Middle Ages the meaning of 'Wald' changed from foliage or a clump of foliage on a branch to all the foliage together, the crown of the tree, and subsequently all the crowns together, the 'Wald'. Thus the whole was given the name of just one part. According to Trier, the term 'Wald' became a *pars pro toto*.[23] Trier's conclusions regarding the term 'Wald' do not conflict with the finding that the 'Wald' was the wilderness where there was wood and food, 'weide' for wild animals and livestock, in the form of grass, herbs, foliage and acorns. On the contrary, this explains why it is plausible that 'Wald' (and related concepts such as 'welde', 'wold', 'woud' and 'weald') was used implicitly in medieval and later texts to mean the 'weide' for livestock and wild animals, and

[22] For example, the Bayreuth Forstordnung, dating from 1493, states: 'Item das schnayten von den paumen soll verbotten werden, bei 5 Kr., daß auch Nyemandt, dann nach anweysung thun soll, und die anweysung sol von den knechten gescheen, und die unartigen rauhen paumen, die zum zymern nit glich sein' (Mantel, 1980, p 969). Also see, *inter alia*, Endres (1888, p. 54), Pott (1983), Mantel (1980, pp. 103–104, 934, 941, 946, 969, 977, 984, 992).
[23] Referring to something by naming a part (Koenen and Drewes, 1987, p. 929).

sometimes a 'Wald' would be described as being extremely suitable for grazing livestock. It was also the place where 'weide' (= fodder) was available for livestock in the form of the foliage of trees and shrubs, as well as sprouting stumps.

From his research into the meaning of the term 'Wald' (also written as 'uuald' and 'uualt') in texts dating from the 8th to the 11th century, Borck concluded that at that time the term related to land which had not been cultivated. As such, it contrasted with land which had been cultivated, the so-called 'Feld' (Borck, 1954). In these medieval texts the term 'Wald' was also used in combination with the term 'wuostinna' (wilderness). According to Borck, that term served to specify the meaning of the term 'Wald' in more detail. For example, texts dating from the 8th to the 11th century refer to: 'wuastinna waldes', 'wastwald', 'wuastweldi' and 'in Waldes einote' (Borck, 1954). The Flemish word 'wastine' and the English word 'waste' are related to 'wuostinna'. These two terms also referred to land which was not cultivated (see Rackham, 1980, p. 401; Williams, 1982; Tack *et al.*, 1993, pp. 19, 20, 42–46). Rackham (1980) also found this contrast in Anglo-Saxon sources in the general topographical description 'wuda & feld'. He translated this as 'woods and fields'. In his opinion, the Anglo-Saxon terms 'wudu', 'weald', 'weldis' and 'wold' meant 'wood' in the modern sense of the word (Rackham, 1980, p. 128). The contrast mentioned by Rackham does not exclude the possibility of the contrast between land which had not been cultivated and that which had, in the terms 'wuda & feld', analogous to 'Wald' and 'Feld'.

According to Borck and Trier, 'Wald', as well as the term 'wold' and the Anglo-Saxon word 'weald', also referred to treeless areas (Borck, 1954; Trier, 1963, p. 45; see also Gove, 1986). The fact that this assumption could be justified is clear from data from the north of the Netherlands. In the present province of Groningen the terms 'wald' and 'wold' applied in the Middle Ages to expanses of treeless, raised bogs (see Ligtendag, 1995, p. 228). In the 11th century and subsequently, they were referred to as 'wolde(n)', 'wald', 'uualde', 'uualda', 'waldt' and 'walt'.[24] Once the land had been cultivated, these terms continued to be used in toponyms in the area (see Ligtendag, 1995, pp. 39, 41, 57, 65, 74, 77, 88, 176). These meanings of 'Wald' and related concepts support the abovementioned finding that in the Middle Ages, the term 'Wald' had a broader meaning than the modern word 'wood'. It was a wilderness which had not been cultivated and which also included open areas, such as peat bogs and grasslands.

Areas where the 'ius forestis' applied were at all events described in the German language as 'Waldgeleite', 'Waldtgleit' or 'Geleite von dem Walde' from the 14th to the 17th century. 'Waldgeleite' derives from 'Geleit' or 'Beleit', which meant 'boundary' (Weimann, 1911, p. 3; Kaspers, 1957, pp. 154–156,

---

[24] A protocol dating from 1513 states: 'Anno duisent vijffhondert dartijn hebbenn die schepperen vannden drie Delffzijlen, ennde eendrachtilicken beslooten in eene ghemeene werffdach upden Delffzijlle, bij poena tijn gollt gulldenn, dat neemandt nhae desen daege sall in dat Wollt turff graven ' (Ligtendag, 1995, p. 225).

166–168). Within the 'Waldgeleite', the 'Forstbann', the 'ius forestis', the 'Forstrecht' and the 'Waldrecht' applied. The 'Waldgeleites' was separated from the so-called 'Feldgeleites' ('Feldgeleite' or 'Veldtgleidt'), to which the 'ius forestis' did not apply. The reason that the 'ius forestis' did not apply there was that at the time that an area had been declared a 'forestis', it already contained settlements with crops and hay fields. Because it was cultivated, this land had a clear owner, according to the legal interpretation of the time, and could not therefore be declared as 'forestis'. The 'Feldgeleite' was an area where a plough or scythe had been used (Kaspers, 1957, pp. 155, 168, 238). Therefore the 'Feldgeleites' did not fall under the jurisdiction of the 'Waldgraf' and the 'vors-dinc'.[25] As remarked earlier, 'forestes' could contain settlements with cultivated land, namely, the traditional royal possessions such as farms ('curtes' or 'fisci'), or the new settlements in the 'forestis' established with royal permission, the so-called 'villae'. These were villages which were located within the borders of the 'Wald' ('villages, located within the perimeter of the "Wald"'). We also noted that the lord granted concessions to colonists to establish farms. In this case, these were farms which were part of the 'Wald' ('farms which belonged to the wood') (Kaspers, 1957, pp. 151, 154–155).

The above shows that uncultivated wilderness without a clear owner, for which the legal term 'forestis' had been declared applicable in the Middle Ages, consisted of 'Wald' as well as water, and that it cannot be concluded from these texts that the term 'Wald' (and related terms) was used to mean 'wood' in the modern sense of the word. The term also included open vegetation such as grasslands and peat bogs. The 'Wald' was the wilderness beyond the land which had been cultivated where there was wood and food, 'weide', for livestock in the form of grasses and herbs, as well as 'weide' in the shape of foliage and fruit from wild trees and shrubs.

Trier's conclusion (1963) that the 'Wald' was an area with a light cover of trees and shrubs which was used for trees and grazing livestock, outside the cultivated fields, is in line with this (Trier, 1963, p. 45). This view does not conflict either with the finding of Chapter 3, that prehistoric vegetation, the virgin wilderness, could have consisted of a grazed park-like landscape.

In a 'Wald' in the meaning of the park-like landscape where livestock grazed and foliage was cut, terms such as 'hochwald', 'hochgewelde', 'howaldt', 'howelde', 'hochgewälde', 'Hogewald', 'hohe Wäldte', 'hogen geweldts' (all meaning 'high wood') and 'Hinterwald' (behind wood) are used for trees, while 'niederwald', 'nederwald' (low wood) and 'Vorderwald' (before wood) are used for shrubs.[26] Both are places where foliage was cut. Trees were referred to as the 'hoge wald' because the foliage, the 'Wald', was high up, while bushes were 'nederwald' because the foliage was lower down. In this sort of park-like grazed

---

[25] An act dating from 1342 states that the waldgraaf had jurisdiction over everything 'was in dem walde, in dem wasser', or inside 'dem geleyde van dem walde … geviel of geschiede' (Kaspers, 1957, p. 166).
[26] See Hausrath (1928), Hilf (1921, p. 142), Trier (1952, p. 97), Hesmer (1958, p. 178), Mantel (1980, pp. 124, 330, 338, 955–957, 961, 977).

Fig. 4.1. A park-like landscape in the grazed Borkener Paradise, Germany. The long periphery of mantle and fringe vegetation of (flowering) blackthorn (*P. spinosa*) is the transitional stage of scrubland to grassland in wood-pastures. The trees are close together (a bosquet, bosket or bouquet of trees: a grove). The shrubs which form the mantle and fringe vegetation in this grazed area can be described as the 'nederwald', i.e. the foliage at the bottom (low situated 'Wald'). The trees which grow out above these have foliage at the top and can therefore be described as 'Hogewald' (high situated 'Wald'). The mantle vegetation of a grove can be seen in the foreground on the left. The 'nederwald' and the 'Hogewald' have also been referred to respectively as the 'Vorderwald', i.e. the foliage in front of the foliage high in the trees, and the 'Hinterwald', i.e. the foliage in the trees behind the 'Vorderwald' (see also Fig. 3.6). This mantle in front of the grove was also known as the 'vorholt', i.e. the holt (wood) in front of the trees. Looking at the front of the grove with mantle and fringe vegetation (see the right-hand side of the photograph), the 'vorholt' can also be seen as the 'onderholt' (or underwood), in relation to the trees. The trees in this picture are then the 'grote holt' or 'high wood', 'Oberholz' or timber (photograph, F.W.M. Vera).

landscape, the trees are behind the mantle and fringe vegetation, which explains why trees were described as 'Hinterwald', while the shrubs which formed a belt *in front of the trees* were known as the 'Vorderwald' (see Figs 3.6 and 4.1).

As noted earlier, the practice of cutting the foliage largely came to an end in the Middle Ages because it was increasingly prohibited. This also meant the end of some of the 'weide' in the 'Wald'. The 'weide' in the form of grass and herbs, the grassland, and the foliage eaten by the livestock themselves, continued to exist in the 'Wald' up to the 18th and 19th centuries. Then this form of 'Weide' came to an end in Western and Central Europe. Pasturing livestock was separated from the production of wood. This occurred partly on the insistence of foresters, who considered that the livestock eating the foliage and twigs of trees and shrubs, including coppices (see later in this chapter) had become unacceptable. This distinction between wood and 'weide' became possible because of changes in agriculture. Specially bred species of grasses and clover were introduced as fodder. In addition, the potato was introduced on a large scale not only as fodder for livestock, but also for human consumption.[27] 'Weide' was cultivated both in and outside the 'Wald', and what remained of the 'Wald' changed into a place without livestock fodder which was mainly used for cultivating trees to produce wood. The 'Wald' became the modern wood. As the 'Wald' had been the last uncultivated land, it is plausible to assume that the term 'Wald' developed into the modern meaning of a 'closed forest' in the 18th and 19th centuries.

4.4 The Meaning of the Terms 'Holt' and 'Bosch' in Relation to 'Wald'

In relation to the term 'Wald', the terms 'holt'[28] and 'bosch'[29] are found in historical texts. In the Dutch language, 'holt'/'hout' is older than 'bosch'. 'Holt'/'hout' is found in early medieval place names. The oldest mention of 'Bosch' is from a later date, from 1225 to 1230 (D.P. Blok, Nederhorst ten Bergh, 1997, personal communication). It is assumed that 'bosch' was taken from the old French terms, 'boscage' and 'bosquet' (bouquet) (Van Wijk, 1949, p. 86; D.P. Blok, Nederhorst ten Bergh, 1997, personal communication). From the 15th to the 17th century, the term 'bosch' was very common in the Dutch language area in descriptions of use in relation to 'holt' in the context of 'holt' that was taken from the 'bosch'. The word 'bos' was either a *neutral* or

---

[27] See Landolt (1866, pp. 49–52), Gayer (1886, p. 13), Hermann (1915), Bühler (1922, p. 611), Grossmann (1927, pp. 29, 33), Meyer (1931, p. 439; 1941, p. 103), Rodenwaldt (1951), Schubart (1960, pp. 69, 100), Streitz (1967, pp. 55, 69 e.v.), Musall (1969, pp. 176–181), Slicher van Bath (1987, p. 31), Mantel (1990, pp. 90–91, 182, 433–434), Bieleman (1992, pp. 104, 130).

[28] In Dutch: 'holt', 'hollt', 'holtz' and 'hout'; in German: 'Holtz', 'holtz', 'holz' en 'Holtzer'; in English: 'holt' (Habets, 1891, pp. 11, 17–18, 248–249, 357; Krause, 1898b; Sloet, 1911, pp. 28, 53, 66, 133, 327, 393; 1913, p. 64; Weimann, 1911, pp. 57, 92, 133; Hausrath, 1928; Trier, 1952, p. 44; Kaspers, 1957, p. 40; Hesmer, 1958, p. 96; Hesmer and Schroeder, 1963, pp. 133–150; Mantel, 1980, pp. 330, 338, 880–996; Buis, 1985, pp. 79, 104–105, 111, 345, 351, 431, 436; Elerie, 1993, pp. 86, 90–92; Rackham, 1993, p. 79).

[29] In Dutch: 'bos', 'boss', 'bosghe', 'bosch', 'bossch', 'busch', 'bussch', 'bus', 'buss', 'boechs', 'buysch', 'boisc', 'buisch' en 'boess'; in German: 'Busch', 'Büsch' en 'bussch'; in English: 'bush' (Habets, 1891, pp. 16–18, 49, 52, 297–299, 366; Krause, 1898b; Sloet, 1911, pp. 28–29, 32, 66, 119, 126, 133, 170, 393; 1913, pp. 39, 63, 101; Weimann, 1911, pp. 21, 28, 60, 135–137; Van Wijk, 1949, pp. 85–86; Trier, 1952, pp. 24–25; 1963, pp. 43, 164; Kaspers, 1957, pp. 124, 193; Hesmer, 1958, pp. 104–105, 453; Hesmer and Schoeder, 1963, pp. 133–140; Buis, 1985, pp. 111–112, 132, 246–247, 324, 431, 436; Elerie, 1993, pp. 90–92).

masculine word in medieval Dutch (Van Wijk, 1949, pp. 85–86).[30] The masculine article 'de' was used for the masculine word, i.e. '*de* bos bomen' (the bunch of trees) analogous with '*de* bos bloemen' (the bunch of flowers). At the time, the term 'bos' may have meant a collection, a 'bouquet' of trees. Regulations from Limburg dating from 1533 indicate that 'bos', at that time, had the meaning of bundling things together. For example, in times when the 'bosch' was left undisturbed, i.e. when no felling or grazing was permitted, no one was allowed to 'bosschen' (to bundle wood) or to transport 'busselen' (bundles of wood) through it (Habets, 1891, pp. 394–395).[31] This means that the word 'bos' could mean a collection of trees or shrubs in the sense of a group of trees or shrubs standing together in an area where they were not usually grouped together, or were even lacking altogether. This meaning of the term 'bos' fits in with the picture of a grazed, park-like landscape. The conclusion drawn by Trier on the basis of his research into the meaning of the terms 'Wald', 'Holz' and 'Busch', that the use of the terms, 'Busch' and 'Gebüsch', and related terms, refer to a park-like landscape where livestock fed on foliage, grasses and plants, supports this theory. According to Trier, foliage ('Wald') and wood ('Holtz') were taken from 'der Busch', while the open spaces between the 'Gebüsch' were available for livestock to feed on grasses and herbs (Trier, 1963, pp. 3–46, 81).

In a list of 361 decrees and regulations on the 'Wald', 'Forst' and 'Holz', dating from the 13th to the 16th century in the German-speaking part of Europe studied by Mantel, the introductions always refer to 'Waldordnung', 'Holzordnung', 'Wald- und Holzordnung', 'Holz- und Waldordnung', 'Forst- und Waldordnung', 'Forst- und Holzordnung', and 'Forstordnung'. The term 'Busch' is not found there. Kaspers referred to a 'Buschordnung' dating from 1587, and Hesmer to a 'Büschordnung' dating from 1692 (Kaspers, 1957, p. 124; Hesmer, 1958, p. 105). Mantel reproduced the complete texts of a number of decrees and regulations, in which the terms 'Wald' and 'Holz' were constantly used in combination with each other or in combination with the term

---

[30] For example, the provisions of the books of the Marks and documents state: 'hoir holt uuten bosch te voeren'. (1482) (Sloet, 1911, p. 53); 'uuyten bossche geen holt' be taken (1503) (Buis, 1985, p. 436); ' … des gemeynen buysch … ' (1539) (Habets, 1891, p. 297). In medieval Dutch, the word 'bos' was both masculine and neutral (Van Wijk, 1949, pp. 85–86). For example, a document dating from the middle of the 14th century read: 'der Bosch seer gehouwen ind vernicht is' (Habets, 1891, p. 359); a document in 1554 … '; a document dating from 1756 to 'de Speulder bosch', and the book of a mark dating from 1621 the words: 'van het bossch ofte gemeine malen' (Sloet, 1911, pp. 32, 428, 440–441). Also see: Habets (1891, pp. 16–17, 52, 75, 297–299, 392–394), Sloet (1911, pp. 28–29, 48–49, 63–66, 80, 119, 133, 166, 171, 175, 298, 327, 393, 439–440; 1913, pp. 39, 57, 63–64, 101).

[31] It also states:

Ende were het saeck dat emandt boschde in vorsch. Bosch, der niet en woont in eenig van den vorsch. XIIII Kirspelen, het were met waghen of sonder waghen, en dat dien emandt vonde, der woonde in den XIIII Kirspelen; die moghtem penden fut den hochsten keu, gelyck offem die forster gepandt hedden (Habets, 1891, p. 395).

'Forst'.[32] These texts reveal that the 'holz' is found in the 'Wald'.[33] In contrast with the terms 'Wald' and 'holz', the term 'busch' is rarely found in the texts of these regulations (see Mantel, 1980, pp. 880–996).[34] This indicates that in German-speaking Central Europe, the term 'Busch' was also a relative newcomer in relation to the terms 'Wald', 'Holz' and 'Forst'. Another indication of this is that from the end of the 12th and the beginning of the 13th century, 'Forst', 'Wald' and 'Holz' were used as names of the courts where the law of the 'ius forestis' was pronounced, i.e. the 'vorsthinc', 'holzgedinghe' and 'waltgedinghe' (Kaspers, 1957, p. 40; Buis, 1985, p. 226), while the 'geding' is not known in relation to the term 'bosch' or 'Busch'. A 'holzgedinghe' did refer to 'busschen.'[35]

The 'holt' ('Holtz') was taken as a branch, shoot or trunk from the 'Holt' ('Geholtz') or in the 'bosch'. 'Holt' (wood) was taken from shrubs and trees, which were referred to in the Middle Ages as 'bossch'. The 'holt' is the 'bossch' ('der Busch') where the wood was found. In the Middle Ages, the term 'holt'/'hout' acquired the meaning of 'wood', while in the Dutch language, the term 'bos' superseded the term 'holt'/'hout' as a word for wood (D.P. Blok, Nederhorst ten Bergh, 1997, personal communication). Decrees dating from the 15th to the 17th century in the Dutch-language area show that the 'Bosschen' were part of the 'Walt'. The 'bossch', as well as the 'Hout', were in the 'wald' or in the 'wolden' (for example, see Wartena, 1968; Buis, 1985, pp. 304–306, 345, 347; Ligtenberg, 1995, pp. 41, 88, 230–231). In (lower) German the 'Holtz' was taken from the 'Geholtz' which lay in the 'Wald' (Trier, 1952, pp. 44–45, 50; also see Hesmer and Schroeder, 1963, pp. 133–140; Mantel, 1980, pp. 877–996). In German, the term 'Wald' has replaced the terms 'Busch' and 'Holtz', to refer to woodland.

The term 'Wald' may have acquired the meaning of woodland, because, as we remarked earlier, it traditionally referred to uncultivated land with woods, and in the end, the woods were the only uncultivated land in the sense that it

---

[32] In the regulations quoted by Mantel, 'wald' and 'holz' are virtually always used together in the introduction or in the articles of 'Wald' and 'Forst' decrees, in the sense of 'in unnsen welden und höltzern'; 'unsern welden, und die armen leuten höltzer und welden'; 'den Wälden und gehöltzern'; 'Ordnung der weld und höltzer' ; 'alle weld und hölzer, dem wildpan'; 'unnd Kurfürsten welden höltzern'; 'da es den Wald und Hölzern am wenigsten schaden thut'; 'Der erst Tail / Redet von allerley nützlichen anstellungen / wie inn Wällden / vnnd höltzen'; 'so bibher Vorst oder Gehülz inn ihen verwaltung'; 'und mit den Wälden, und gehöltzern, mit dem gebrauch' (Mantel, 1980, pp. 932, 936, 968, 972; 973, 976, 982, 989, 994).

[33] For example, an undated document states 'Aber alle ander holze in den welden' (Mantel, 1980, p. 954). See also Mantel (1980, p. 957). In 1351, the following was written about the Forest of Hopedale in Wales: 'item que le bois en la foreste de Hopendale' (Linnard, 1980, p. 226).

[34] A text about the Waldrecht, dating from 1482, states: 'Item so we eichenholtz in desen welden dis lands Munster buschen aen orloff hauwet off voert ind ervulgcht wirt hauwen off varen, den mach der voerster penden vur v marc xx wyßpenninck deme heren' (Mantel, 1980, p. 963).

[35] For example, a document dating from 1474, from the area around Kleef, states: 'van holzgedinghe und verkoering upder Hese und anderen busschen im Lande van Moers. Das Holzgeding findet in Baerl stat' (Weimann, 1911, p. 21).

was not exploited. In English, the terms 'bush, 'weald', 'wold', 'wald' and 'welde' have been replaced by 'wood'. The original terms have survived only in toponyms, such as Buckholt, Andreasweald, Weald, Bruneswald (now Bromswold), and Weldis (now the Wilds) (Trier, 1952, p. 44; Hart, 1966, p. 257; Page, 1972, pp. 14–15, Rackham, 1980, pp. 124, 128; 1993, p. 79; Hooke, 1998, p. 140). 'Wood' became a *pars pro toto*. The 'wood' was both what was taken, and where it was taken. In the English language, the term 'bush' is still used in the sense of a collection of branches, shrubbery and groves (Ten Bruggecate, 1990, p. 131), a meaning it shared with 'Busch' and 'bossch'.

The picture of a park-like landscape with grazing means that terms related to 'holt' and 'bos', such as 'unterholz', 'underholt', 'onderholt', 'underbusch', 'onderbuss', 'underwood', 'brushwood', 'fürholz', 'vorholt' and 'vorholtz', as well as terms such as 'grote holt', 'Oberholz', and 'highwood', are derived from the place where the wood is found, as in the case of the term 'Wald'. In the Middle Ages, these terms were used to refer respectively to shrubs, shrubbery, groups of shrubs, the sprouting stumps of shrubs, coppices and trees where wood was taken for firewood and timber (see Fig. 4.1).[36] In the 16th century, the English term 'highwood', for example, was completely replaced by the term 'timber', and 'timber' was differentiated from 'underwood' (see Tubbs, 1964; Hart, 1966, p. xx; Flower, 1977, pp. 14–15, 21; Rackham, 1980, pp. 156, 174; 1993, pp. 62–63, 67). The term 'voorhout' ('fürholz', 'vorholt' and 'vorholtz') was taken from the mantle and fringe vegetation which always lie in front of the woods in a grazed, park-like landscape (see Figs 3.6 and 4.1). Standing in front of this sort of vegetation, the shrubs are under the trees, the highwood or timber (see Fig. 4.1). This could explain the use of the term 'underwood' ('Unterholz', 'underholt', 'onderholt', 'underbusch', 'onderbuss') for the shrubs.

In the Netherlands, 'bos' was defined in 1803 as a collection of trees on a sufficiently large area. In 1810, 'bos' was defined in the context of the exploitation and division of land act as: 'Land will not be defined as forest if there is no tall tree every twenty feet, or, in the case of coppice wood where there is not a shrub at most every four or five feet' (Buis, 1985, pp. 400, 410). In this definition, the term 'bos' can cover an infinite area. In my opinion, the meaning of a collection of trees covering a limited area in space, a 'bouquet' of trees, was lost as a result.

[36] For example, a document by the Archbishop of Maagdenburg, dating from 1368, states: 'Den fulteberch hat grote holt und dat onderholt dat reckene ick ut 5 marck geldes, ein jar helpe dem andern' (Hausrath, 1928, p. 347). See also in Justi (1744, cited by Bühler, 1922, p. 599), Hausrath (1928), Trier (1952, p. 97), Hart (1966, pp. 23–25), Rackham (1975, p. 24; 1980, p. 118), Flower (1977, p. 26), Buis (1985, pp. 110–111), Mantel (1980, pp. 330, 338; 977; 1990, p. 335), Dengler (1990, p. 265), Ellerie (1993, pp. 91–92), Best (1998).

4.5 What Was a 'Silva'?

Following the explanation of the terms 'Wald', 'Busch' and 'Holtz', the question arises how the term 'silva' relates to these. Nowadays, it is translated by the word 'forest', i.e. a limited and extensive area of trees. As we saw in Chapter 2, Gradmann (1901) and Cermak (1910) referred to the texts of the Romans, Caesar and Tacitus, to show that at that time, large, uninhabited, uncultivated areas of Central and Western Europe were covered by a great closed forest. They made this assumption on the basis of the fact that the Romans referred to these areas as 'silva'. In medieval texts, the term 'silva' was also translated as an extensive closed forest, which would suggest that the areas referred to in these texts as such, consisted of extensive forests (see, *inter alia*, Kaspers, 1957, pp. 30, 31, 89; Rubner, 1960, p. 37; Schubart, 1966, p. 18; Mantel, 1990, pp. 36, 335; Buis, 1993, pp. 36–37).

According to Caesar, a lightly equipped traveller could cross the 'Silva Hercynia' on foot in 9 days. On the basis of the fact that, according to Caesar, the Silva Hercynia extended in those days across the whole of southern Germany and along the Danube into Romania, the question arises whether it really was a closed forest. Caesar also wrote that there were aurochs (*B. primigenius*) in the Silva Hercynia, which the inhabitants of the Silva Hercynia trapped in pits (Caesar, Book 1, pp. 145–146). If there really were aurochs, this means that there must have been grasslands, because the aurochs, like its domesticated successor, the domesticated cow, was a typical grass eater. A more open landscape with grasslands is a much more probable habitat than a closed forest for this wild predecessor of the domestic cow.

In AD 98, Tacitus wrote about Germania: 'Terra, etsi aliquanto specie differt, in universum tamen aut silvis horrida aut paludibus foeda', which is generally translated in the literature as: 'In general, that land is terrible because of the forests and horrible marshes' (Blink, 1929, p. 174; Wehage, 1930; Mantel, 1990, p. 53). In fact, the word 'horrida' can also have the meaning 'thorny'. The word 'horrida' can mean 'frightening' in a metaphorical sense, but only when the normal meaning of 'thorny' is impossible. In combination with 'silva', 'thorny' is certainly possible. In connection with 'silva' and 'dumis', 'horrida' can mean a thorny shrub or thorny grove (Muller and Renkema, 1995, p. 408).

In classical Latin, 'silva' means wood, as the Romans had in their own highly developed environment. For an impenetrable forest of dark trees, with the emphasis on close vegetation, they used the term 'lucus'. After AD 1, the term 'silva' was also used for 'forest' in the sense of 'grove'. The Roman Lucanus used 'silva' for grove in the figurative sense for someone 'densam ferens inpectore silvam' (with a tight bunch of arrows in his chest). However broad this chest might have been, it was a spatially limited bundle, a bunch of arrows, not an amount spread out over an immense surface, a jungle of arrows. Consequently, 'silva horrida', as used by Tacitus, should not be translated as 'land, "horrida" because of its forests' (ablativus causae), but as 'land, "horrida" in its groves' (ablativus limitationis). This meaning corresponds with the land with 'horrida' forests. The characteristic of groves in that land is that they are 'horridae' (A.J. van Wolferen, Doorn, 1998, personal communication). Given the above, the sentence from Tacitus' book about Germania 'Terra, etsi aliquanto specie differt, in universum tamen aut silvis horrida aut paludibus foeda' can be translated as: 'The land, even though it is quite diverse, is generally either thorny in groves (ablativus limitationes) or swampy in marshes' (A.J. van Wolferen, Doorn, 1998, personal communication). Freely translated, it means: 'The land looks very different in many places, but in general, it is covered with thorny forests (either thorny trees or thorny groves) and unhealthy marshes' (see Figs 4.1, 4.2 and 4.3).

The meaning of 'silvis horida' is also illustrated in a well-known passage from Virgil's *Aeneid*, Book IX, line 382 *et seq.*: 'Silva fuit late dumis atque ilice nigra horrida quam densi complerant undique sentes', which means: 'The thorny forest grove with gorse and oak which was overgrown with dense, thorny scrub.' It is even possible that Tacitus used the description of 'silva horrida' to describe Germania precisely because it was known to the Romans, and therefore recognizable. So thorny forests or thorny groves could only have referred to thorny bushes, because there are no natural thorny trees in Europe, except for the wild pear (*P. pyraster*).[37]

The wild pear and the thorny shrubs indigenous in Europe cannot grow in a closed forest (Ellenberg, 1986, pp. 94–95; see Chapter 2). On the other hand, they are very common in grazed, park-like landscapes, where they flourished (also see below in this chapter). Moreover, Tacitus wrote about the land of the Chatti in Germania, that the Silva Hercynia there was less open than in other parts of Germania. In other words, Germania may not have been an area covered by dark primeval forests in Roman times. In those days, the term 'silva' may have referred to a landscape consisting of a mosaic of groves and grasslands with trees growing in thorny scrub. One can imagine that a great Roman army walked into an ambush in such a landscape and was destroyed, as was the case for the Roman Quintilius Varus in Germania in AD 9, during his campaign

---

[37] In the Dutch and German translations of the Latin text (see Cermak, 1910; Blink, 1929, p. 174; Wehage, 1930; Mantel, 1990, p. 53), this sentence was translated respectively as 'dat land [Germania] is in 't algemeen verschrikkelijk door wouden en akelige moerassen' (Blink, 1929, p. 174) and 'Die Beschaffenheit des Landes ist zwar sehr unterschiedlich, aber im allgemeinen ist es bedeckt mit schrecklichen Wäldern oder abscheulichen Sümpfen' (Mantel, 1990, p. 53). The word 'horrida' was translated by 'terrible'. In English translations, 'silvis horrida' is translated as 'bristling forests' (see Mattingly, 1986, p. 104). In the Dutch translation of 'Landscape and Memory' this phrase was also translated as 'bristling forests' (Schama, 1995, pp. 89, 620). 'Horrida' means bristling, thorny and terrifying (Muller and Renkema, 1995, p. 408). Both 'terrifying' and 'terrible' are 'horribilis' in Latin. Therefore both 'horrida' and 'horribilis' can mean terrifying. In the case of 'horrida', it is a matter of shaking with cold and getting gooseflesh (hair stands on end, i.e. as bristles). In connection with 'silva' and 'dumis', 'horrida' now means 'bristly' or 'thorny'. Therefore I believe that 'silva horrida' could also have referred to thorny groves. The meaning of bristly for 'horrida' is expressed in the plant species *Genista horrida*, which grows in southern France, amongst other places. It is a type of gorse which is characteristically extremely thorny (Bonnier and Layens, 1974, p. 70). Therefore the original translation of the text is a possible one, but not the only one. Thus the phrase used by Tacitus cannot be used to demonstrate that the virgin vegetation in Central and north-west Europe consisted of dense, closed forests or primeval forests.

Fig. 4.2. Mantle and fringe vegetation, shrubs, 'strubben', 'struiken', 'vorholt', 'onderboss' or 'underwood', with a young oak, a 'Waldrechter', growing in this on the left (photograph, F.W.M. Vera).

Fig. 4.3. A young oak tree growing in the middle of hawthorn scrub in the Borkener Paradise, Germany. The hawthorn protects the oak from being nibbled by large herbivores, such as, for example, cows and horses. The photograph illustrates the old English proverb that the thorn is mother to the oak (photograph, F.W.M. Vera).

against the Germanic tribes (see Grant, 1973, pp. 61–89; Schama, 1995, pp. 101–106).

Between the time that the term 'silva' was used by the Romans for the areas of Central and Western Europe, and the texts dating from the Middle Ages, which also contained this term, there are many centuries of which we know little or nothing regarding the meaning of the Latin terms used. In the Middle Ages, Latin was an artificial language, and everyday terms from living languages were translated into it (D.P. Blok, Nederhorst ten Bergh, 1997, personal communication). Medieval texts with sentences such as 'silvae vel forestes nostrae' (Kaspers, 1957, pp. 26, 30; Buis, 1993, p. 37) show that silva meant our 'forestes', i.e. 'silvae' = 'forestes nostrae' (D.P. Blok, Nederhorst ten Bergh, 1997, personal communication).

Areas described by the Romans as a 'silva' were declared 'forestes' by the Franks. In view of what was said earlier in this chapter about the nature of areas proclaimed as 'forestis', this indicates that what was described in the Middle Ages as 'silva' included grasslands. One example is the 'Silva Arduenna'. This extended in the area between the Rhine, the Meuse and the Moselle. It was declared a 'forestis' in the 7th century ('foreste nostra nuncupante Arduinna') (Kaspers, 1957, p. 93). In the period from 743 to 747, it was referred to in documents alternately as 'foresta nostra Ardinna' and 'silva nostra Arduenna' (Kaspers, 1957, p. 26). As we noted above, felling trees in a 'forestis' without the express permission of the king, was prohibited under the 'ius forestis'. Permission was granted to colonists to create a field only so that they could grow crops and meet their needs for timber (see Kaspers, 1957, pp. 93–96). Nevertheless, a 'forestis' was often used to provide the fodder needed for livestock, without any reference to cutting down trees to create pasture. This means that there must have been natural grasslands. Another clear indication of this is that at the time of the Romans, and when the Merovingians declared the 'Silva Arduenna' a 'forestis' in the 7th century, there were still aurochs in this area (Lebreton, 1990, pp. 29–37). In view of the fact that a 'silva' became a 'forestis', and a 'forestis' was also a 'Wald', amongst other things, and there are many indications which show that a 'Wald' consisted of a mosaic of grassland and groves, it is probable that a 'silva' was a wilderness consisting of this sort of landscape.

Further details about the structure of the vegetation and the way in which it developed in the wilderness, in particular with regard to the grazing of livestock, can be obtained from the regulations which were drawn up in the Middle Ages by the kings and lords on the basis of the 'ius forestis' for the use of the wilderness declared to be 'forestis', and by communities or so-called 'marken', 'gemeynten' or 'commons'. For the way in which the structure of this use evolved, reference is made to Appendix 1. Hereafter, these regulations on the use are examined to see whether this can provide information about the structure and development of the vegetation in relation to the grazing of live-stock.

4.6 Regulations on the Use of the Wilderness

The wilderness was used as pannage for pigs and to graze livestock, gather wood and find honey.[38] The pigs were 'outside', i.e. in the uncultivated wilder-ness, from a few weeks to about 4 months (Hesmer and Schroeder, 1963, p. 104; Ten Cate, 1972, pp. 130, 206). They were fattened on acorns and the fruit of wild pear, wild apple and wild cherry, berries of the whitebeam, sloe-berries, rosehips and hazelnuts. This fruit was known as the *mast*. The most important mast consisted of acorns.[39] These were also collected to feed the pigs when they were kept indoors (Hesmer, 1958, pp. 391, 412; Tendron, 1983, p. 23; Buis, 1985, pp. 181, 209). The pannage for pigs was known in Dutch as 'aecker', 'eycker', 'eckel', 'akeren', 'ekeren' or 'aten' (Habets, 1891, pp. 306, 358; Buis, 1985, pp. 47, 431; Elerie, 1993, p. 89; Tack *et al.*, 1993, p. 181); in German, 'Acker', 'Ecker(ich)', 'Geäcker', 'Äkeret', 'Acherum' (Hilf, 1938, p. 133); in English, 'pannage' (Rackham, 1980, p. 119), and in French, 'le panage' (Tendron, 1983, p. 22). In Dutch, an acorn was also called 'ecker', 'acker', 'aacker' or 'aker', from which is derived the word 'akker' (De Vries and Tollenaere, 1997, p. 56). According to De Vries and Tollenaere (1997), the word 'akker' is generally connected to the Latin Greek 'agô', which means 'I lead to fodder'. An 'acker' is the 'weide' (pasture), the fodder, to which the live-stock, the pigs were driven. For the pigs, the 'weide' is the 'acker', 'aacker' or acorn. The mast, or collection of acorns, was also known as 'acker' (Hilf, 1938, p. 134; Ten Cate, 1972, pp. 115, 129). In Anglo-Saxon, the word 'aecer' meant acorn (Rackham, 1993, p. 174). The Dutch word 'akker' (nowadays meaning field) is derived from the word 'acker'. In medieval texts, the 'acker' was a place where there were trees, i.e., oak trees and wild fruit trees, where the pigs were taken to be fattened on the mast. The term 'Acker', like the terms 'Wald' and 'Holt', has therefore become a *pars pro toto*, referring both to the

---

[38] Grossmann (1927, pp. 14–24), Hilf (1938, p. 136), Reed (1954, pp. 28–29), Kaspers (1957, p. 19), Hesmer and Schroeder (1963, p. 103), Schubart (1966, pp. 10, 143–144), Streitz (1967, pp. 36, 42), Buis (1985, p. 183), Mantel (1990, p. 151), Tack *et al.* (1993, p. 19).
[39] Endres (1888, p. 49), Hilf (1938, pp. 132–138), Nietsch (1939, pp. 111–112), Hesmer and Schroeder (1963, pp. 104, 128, 279), Schubart (1966, pp. 33, 111), Ten Cate (1972, pp. 9, 75), Duby (1968, p. 8), Slicher van Bath (1987, p. 52), Mantel (1990, p. 97), Tack *et al.* (1993, p. 174).

wilderness and trees, and to the mast produced by them.[40] That part of the wilderness was the 'Acker'. Thus, like the 'Holt' and the 'Bosch', the 'acker' was situated in the 'forestis' or 'Wald' (see Fig. 4.4).

In the Middle Ages, pork, and particularly bacon, was an essential source of energy for the winter, and therefore an important part of the daily winter diet (Reed, 1954, p. 32; Bogucki and Grygiel, 1983; Jahn, 1991, p. 395; Tack *et al.*, 1993, pp. 27, 175). Other livestock (sheep, cows and goats) provided meat, wool, hides and milk, as well as manure to put on the fields. The cattle were used to pull ploughs and carts. In the cultivated areas themselves, there was little room for grazing livestock. They could go into the fields only after the harvest, on the stubble and fallow ground, and after haymaking, on the wet hay fields by streams. Therefore for a large part of the year, the livestock had to graze on land away from the fields (Grossmann, 1927, p. 23; Hilf, 1938, p. 132; Schubart, 1966, pp. 12, 67; Buis, 1985, pp. 106, 209). Therefore the wilderness formed an essential, integral part of the system of farming at the time (see Appendix 2). Its use was regulated by the lord or by the local communities themselves, as for example, in the free commons (see Appendix 1).

The oldest recorded regulations on the use of uncultivated land date from the 6th and 7th centuries. These concern the use of 'forestes'. They are about the grazing of pigs and other livestock, cutting foliage, collecting honey and

[40] The description of a route in 1289 read: 'etliche Äcker, so daselbst vor Wobeck bey einem Boemlehren Busche belegen' (Schubart, 1966, p. 182) Here, the word 'Acker' is used to mean the place where pigs are put out to pannage, i.e. the place where oaks grow. This was the meaning of the word 'Acker' from the Middle Ages up to the 18th century, a period when pig mast played an important role. In addition, the term 'akker' also meant 'mast', i.e. the production of acorns by oak trees. Before the pigs went into the forest, 'den Acker' was examined to see whether 'derselbige busche follen acker habe' (Hilf, 1938, p. 134); i.e. whether the forest had a full mast, many acorns (Ten Cate, 1972, p. 129). Trees in the 'akker' had to be spared to provide mast for pigs. For example, the city of Goslar, in the Harz mountains in Germany, proclaimed in 1543: 'und sollen und wollen auf jedem Acker neben dem Bau und Nutzholz zwölf Hegereis über die, die beredt stehen, und hegen lassen' (Schubart, 1966, p. 191). A regulation dating from 1710, states that: 'laubholz stehen bleiben, auf dem Acker 32 stück' (Bühler, 1922, p. 301). A notice from the Quarter of Zutphen, dating from 1741, states that 'syne schaapen geen schade aan enig akkerhout komen brengen' (Buis, 1985, p. 353). In other words, 'Acker' refers to a place where there are oak trees and wild fruit trees which provide mast. Someone who had the right to use this mast was known as an 'Ackerman' (Schubart, 1966, p. 76; Ten Cate, 1972, p. 131; Janssen and Van de Westeringh, 1983, p. 39). The 'Acker' was also a measure for areas of trees (Cotta, 1865, pp. 136, 341; Wahrig, 1980, p. 112), where pigs were put out for pannage. It was also a measure for raised bog, where peat was collected (Ligtendag, 1995, pp. 224, 225, 227, 229, 230, 235, 237). The use of the measure 'acker' for peat can be explained when we assume that it was a measure which was used for wilderness that was exploited, but was not cultivated. As already mentioned, it was a measure applied to unexploited raised bogs, where peat was collected. In fact, this moorland consisted of a certain amount of 'akkers' of peat. I think it is highly probable that the English measure 'acre' has the same origin, and originally also had the same meaning. The 'acre' goes back to the Anglo-Saxon word 'aecer' (Rackham, 1993, p. 174; Flextner and Hauck, 1983, p. 17; Gove, 1986, p. 18), a term very similar to the German/Saxon 'Aecker', which means 'acorn' (Van Veen and Van der Sijs, 1990, p. 44). In that case, an 'aecer' or 'acker' could have been a unit with mast for one pig. This is indicated by the fact that in England, the density of pigs was often one pig per acre (0.4 ha) or less (Rackham, 1980, p. 120). The area of an 'acker' in Groningen was 1.6 ha (Ligtendag, 1995, p. 228), four times this area. On the European mainland, the density of pigs put out to pannage varied from one pig per 0.24 to one pig per 1.2 ha.

Fig. 4.4. November scene from a 16th-century French Book of Hours, showing pigs put out to pannage in the forest (photograph made available by Ten Cate, 1972, p. 126). One of the swineherds, a so-called Ackerman, is hitting the acorns out of the oak tree with a stick. When pannage took place in uncultivated land, this was called the 'acker', containing the fruitful trees which produced the mast, or acorns. The pigs peel the acorns (Ten Cate, 1972, pp. 267–268). The rooting of the pigs, in which they turned over the soil, was known as 'ackeren'. They were looking for worms, insects, slugs and other animal food, to supplement their diet of acorns.[41] Practice showed that without this 'Erdmast' (earth mast) or 'Wuhl', which was rich in protein, i.e. on a diet of acorns alone, the pigs fell ill.[42]

protecting trees, including those which produced food (mast) for pigs, such as oak, wild apple, wild pear and wild cherry.[43] They regulated what had to be paid to the lord for putting pigs out to pannage or livestock out to graze in a 'forestis' Kasper, 1957, p. 29; Rackham, 1980, pp. 124–125, 155). In addition, there were regulations about the use of trees and shrubs. Even in the earliest regulations, a distinction was made between those practices for which special permission had to be granted, and what could be used without special permission. In general, trees could not be freely used.[44] This applied in particular for trees

[41] Hobe (1805, pp. 178–182), Herrmann (1915), Meyer (1931, p. 286), Hilf (1938, pp. 133–134), Hesmer (1958, p. 391), Ten Cate (1972, p. XX), Mantel (1990, pp. 97–98).
[42] Hobe (1805, pp. 178–182), Ten Cate (1972, pp. 264–267). On this subject, Hobe wrote: 'Kein Sachkundiger wird aber leugnen können, daß es bey voller Mast den Schweinen äusserst nöthig wird, von Zeit zu Zeit aus der Holzung zum Wurmen getrieben zu werden, wenn sie nicht krepieren und kranken sollen' (Hobe, 1805, p. 181).
[43] Bühler (1922, pp. 65, 66, 381), Meyer (1931, p. 283), Kaspers (1957, p. 29), Trier (1963, p. 4), Ten Cate (1972, pp. 59–67), Mantel (1990, pp. 184–186).
[44] Endres (1888, pp. 36, 40–41, 43, 91, 125), Bühler (1922, pp. 66, 301), Vanselow (1926, pp. 21–22, 211), Hilf (1938, p. 158), Rodenwaldt (1951), Hesmer and Schroeder (1963, p. 49), Rackham (1980, p. 174), Mantel (1990, pp. 184–185, 329).

which produced the mast for pigs. They were described as 'fruitful trees' ('arbores fructiferae' or 'silva fructicans'), 'fruit trees', 'tragenden', 'tragbaren', 'beerenden' or 'bärenden Bäumen'.[45] The oldest regulations refer specifically to oak, beech, wild apple, wild pear, wild cherry and service trees (Bühler, 1922, pp. 65, 66, 381). Later, protected trees also included whitebeam, chestnut, walnut, hazelnut, wild cherry and alder buckthorn.[46] In all these regulations, the oak has a central place because of the importance of the acorns for pannage.[47] The express consent of the court, the 'Forst', 'Holz' or 'Waldding' (forest court) was needed to cut down these species of trees in the 'forestes', or damage them in any other way.[48] There were barbaric punishments for infringements of these regulations (Kasper, 1957, p. 56; Ten Cate, 1972, p. 109; Mantel, 1990, p. 184).[49]

The importance of these fruit trees to the lords can be related, in the first instance, to the high incomes from pannage, in comparison with, for example, wood. Up to the first half of the 18th century, this income was equal to 10–20 times, or even 100 times that from wood.[50] The importance of pannage is also

[45] See o.a.: Endres (1888, p. 49), Gradmann (1901), Herrmann (1915), Bühler (1922, pp. 65–66, 381), Vanselow (1926, p. 22), Grossmann (1927, p. 23), Meyer (1931, pp. 282–283, 298–299, 351, 410; 1941, pp. 115, 283), Hilf (1938, pp. 133–136), Nietsch (1938, pp. 111–112), Kaspers (1957, p. 29), Rubner (1960, p. 50), Hesmer and Schroeder (1963, pp. 133–141, 144, 148), Trier (1963, p. 4), Hart (1966, p. 127).

[46] Endres (1888, p. 82), Bühler (1922, p. 64), Vanselow (1926, pp. 22, 211), Grossmann (1927, p. 23), Meyer (1931, p. 286; 1941, p. 125), Hilf (1938, p. 133), Rodenwaldt (1951), Hesmer (1958, p. 71), Schubart (1966, pp. 16, 33, 48, 142), Streitz (1967, pp. 36–38, 53), Ten Cate (1972, p. 93), Flower (1977, p. 15), Dengler (1990, p. 290), Mantel (1990, pp. 325, 354).

[47] Hilf (1938, pp. 132–136), Hesmer (1958, p. 101), Hesmer and Schroeder (1963, p. 145), Schubart (1966, pp. 58, 68, 110–111), Streitz (1967, pp. 36 en 58), Ten Cate (1972, p. 9), Buis (1985, p. 53), Mantel (1990, p. 184).

[48] Endres (1888, pp. 41, 91), Hausrath (1898, p. 47), Bühler (1922, pp. 66, 301), Vanselow (1926, pp. 21–22, 211), Rodenwaldt (1951); Kasper (1957, p. 56), Hesmer and Schroeder (1963, p. 49), Ten Cate (1972, pp. 108–109), Musall (1969, p. 97), Mantel (1990, pp. 125, 329, 351).

[49] An article in a document from Eichelberg describes the punishment for illegally peeling the bark from the oak (the tanning acids in oak bark were used for tanning leather):

> wo der begriffen wirt, der einen stehenbaum schelett, dem were gnade nutzer dan recht. Und wan man den sollte recht thun, solle man ine by seinem nabel sein bauch uffschneiden, und ein darm daraus thun, denselben nageln an dem stame und mit der person herumber gehen so lang er ein darm im leibe hat. Darumb were ime gnade besser dan recht' (Mantel, 1990, p. 184).

[50] For example, in 1590, 8659 Maria guilders were received for driving 9039 pigs into the Lauensteiner Amtforst, compared with only 84 Maria guilders for wood. In 1594, 1110 Taler were received for the pannage of 2000 pigs on 6000 Morgen (2400 ha) in the Lauenförder Forst of the Solinger Wald, compared with 44 Taler for wood (Endres, 1888, p. 80; Meyer, 1931, p. 283; Mantel, 1990, p. 101). Iben (1993) refers to an income of 1100 Taler for the pannage of 2000 pigs, compared with 49 Taler for wood in 1595. Thus in these examples, the incomes from the mast are 10 to 20 or even 100 times higher than the income from wood. According to Endres (1888, p. 172), pannage filled the lords' coffers more than any other form of exploitation (Endres, 1888, quoted in Mantel, 1990, p. 101). For the importance of pig mast for the lords, also see: Endres (1888, p. 172), Herrmann (1915), Meyer (1931, pp. 283, 288, 294), Bertsch (1949, p. 105), Rodenwaldt (1951), Hesmer (1958, p. 390), Schubart (1966, pp. 66, 69), Streitz (1967, p. 68), Ten Cate (1972, p. 206), Mantel (1990, p. 100 e.v.), Iben (1993).

shown by the fact that in the medieval documents from the Netherlands, England and Germany, the size of an area was expressed in terms of the number of pigs that could be pannaged there (Herrmann, 1915; Ten Cate, 1972, p. 72; Rackham, 1980, p. 122; Stamper, 1988; Buis, 1993, pp. 30–33). To produce this mast, young oak trees were also coppiced at a height of a few metres. As a result, these oak trees formed a broad crown, low down on the trunk, and produced a relatively large number of acorns at a young age (Flörcke, 1967, p. 86; Pott, 1983b; Pott and Hüppe, 1991, pp. 30–31). The pannage was carried out by the community of holders of rights (Meyer, 1931, p. 294). All the inhabitants in a settlement kept pigs to meet their meat requirements, especially bacon; not only those who cultivated the fields, but also craftsmen, like the smith, and all the people who lived in towns.[51] Apart from the use of the trees and the pannage, there were a lot of regulations for the cultivation of the wilderness. Cultivation involved felling the trees and shrubs to create a field for growing crops. This was not permitted in a 'forestis' without the lord's express consent. In the commons (on the continent called 'marken' or 'gemeynten'), every commoner was initially free to cut down trees or shrubs to make a field for growing crops. Eventually this was no longer permitted without the express consent of a meeting of commoners (Kaspers, 1957, p. 236; Hesmer, 1958, p. 87; Streitz, 1967, p. 37; Mantel, 1990, pp. 61–62). A colonist in a 'forestis' had to pay a tax ('agrarium') to the lord after cultivating it. In the time of the Franks, the 'agrarium' was part of the harvest. As noted earlier, this was called a 'Medem' or 'Rottzehnt'. It was collected by the 'forestarii' as 'Holzkorn', 'Forstkorn' or 'Wildbannkorn' (Kaspers, 1957, p. 236).

Because the fruit trees were explicitly protected by the lord, they were also known in German-speaking parts of Europe as 'Herrenholz' or 'hovetbome' (trees of the court, the 'curtis', i.e. property of the lord) (Hilf, 1938, p. 168; Musall, 1969, p. 97, and see Sloet, 1913, p. 141; Hausrath, 1982, p. 347). This was in contrast to so-called 'herrenlose' wood (the wood that did not belong to the lord), also known as 'malae', 'unfruchtbarn holtz', unreal ('unecholt'; 'Unholtz'), useless ('unnützes'; 'unnützliches'), dead, dry or harmless wood ('douffholtz' or 'duisholt'). This referred to shrubs, trees which did not

[51] For example, in the 14th and 15th centuries, every house in the German town of Göttingen had a shed or a stable. Even in 1749, mast for the pigs was still very important for the people of that town (Schubart, 1966, pp. 68, 111). Cities in Switzerland such as Berne and Zurich issued regulations about pigsties in the city (Meyer, 1931, p. 441). A 15th-century regulation, which prohibited knocking acorns out of trees, shows that in the Dutch town of Utrecht there was even pannage in the churchyards (Ten Cate, 1972, p. 115). The abbot of the monastery in Deutz (near Cologne) in the area of the Lower Rhine in Germany had pigs come from Velp, and even from Rhenen in the province of Utrecht in the Netherlands (Weimann, 1911, pp. 91–92, quoted by Hesmer, 1958, p. 390). These Dutch towns were in the lands of this monastery (Gaasbeek et al., 1991, p. 28). The council of the German city of Dortmund complained in 1635 that the oak forest which had once fed 2000–3000 pigs was only able to provide pannage for at most 200 in that year (Hesmer, 1958, p. 97). Also see Endres (1888, p. 80), Meyer (1931, pp. 304, 393, 441), Hesmer (1958, p. 390), Schubart (1966, p. 111), Streitz (1967, p. 68), Ten Cate (1972, pp. 9, 115, 132).

bear fruit, and dead trees. These trees and shrubs could be used by the commoners to meet their own needs for firewood, without special permission from the court.[52]

Trees referred to as such in the records include hornbeam, birch, alder, lime, willow, dogwood, sycamore, field maple, elm, poplar, hazel, aspen, hawthorn and thorns in general (Endres, 1888, p. 99; Bühler, 1922, p. 416; Rubner, 1960, pp. 50–51; Musall, 1969, p. 97; Mantel, 1990, pp. 185, 326). In France, a distinction was made between 'bois vif' and 'mort-bois'. The 'mort-bois' could be collected to meet the people's needs. The 'bois vif' were trees which bore fruit, such as oak, but also species which did not bear fruit, such as hornbeam, aspen, sycamore and birch, which could not be freely used. The 'mort-bois' also included certain species of 'living wood', i.e. wood which did not belong to the lord, such as willow, thorns, dogwood, alder, juniper and elder (Rubner, 1960, p. 50). In England, a distinction was made from the 12th and 13th centuries between trees known as 'highwood' and the shrubs, or 'underwood'. The 'highwood' belonged to the lord and could be felled only with his express permission (Hart, 1966, p. 25; Flower, 1977, pp. 14–15; Tubbs, 1988, p. 67).[53] The underwood could usually be freely collected by the commoners for firewood (Rackham, 1980, pp. 134, 175). From the middle of the 15th century, a distinction was made between 'timber' and 'wood' or 'underwood'. Timber was used for trees which provided wood for carpentry. Examples of this were oak, ash and sweet chestnut. 'Wood' was used for shrubs which provided firewood and wood for fences and tools. These included hazel, field maple, holly and thorny shrubs (Hart, 1966, p. xx; Rackham, 1975; 1980, pp. 174, 181).

Although a commoner had the right to meet his own building needs from the 'forestis', he did have to show that this was necessary (Endres, 1888, p. 99; Reed, 1954, p. 33; Kaspers, 1957, pp. 95, 126, 185, 204–205, 214, 232). When this proof had been presented and permission granted to fell a tree, an official forester would show the commoner the tree, marking it with a special axe (Vanselow, 1926, p. 21; Hesmer and Schroeder, 1963, p. 145; Mantel, 1990, p. 329; Buis, 1993, p. 189).[54] In the free commons it was also prohibited

---

[52] A document dating from 1310 on this question, states: 'Alle ander holze [except for oak trees and beech trees] in den welden, welches daz ist, daz mag ein iglicher geniessen zu siner notdorfte one laube, wer aber buwen wolde in dem lande, der sal einem amptman das holze heischen, sine notdorfte unde nit me' (Endres, 1888, p. 41). See also: Endres (1888, p. 36), Sloet (1913, pp. 138, 144), Kasper (1957, pp. 108–109, 124, 239), Schubart (1966, p. 33), Mantel (1990, pp. 185, 325–326).

[53] In the 13th century, a distinction was once again made between trees which were and which were not suitable for timber. If a tree was suitable for construction purposes, it was called 'quercus' ('Quercus apta ad meremium'). If it was suitable only for firewood, it was called 'robur' ('robur ad focum'). Thus the names quercus and robur did not necessarily refer to oak trees (Hart, 1966, p. 21; Rackham, 1980, p. 182).

[54] A document dating from 1663 states that when someone wanted to build: 'sollen aus befehl des markemeisters [Waldgraf] der waldbereiter und förster dieselbige bäu besichtigen und erkennen, wasz und wieviel holz darzu vonnoeten, doch der mark unschädlich' (Endres, 1888, p. 43).

to fell trees which bore fruit without express permission. This permission was granted at a meeting of commoners. In the common, every commoner also had a right to as much wood for building as he/she needed (Endres, 1888, pp. 40–41; Hesmer and Schroeder, 1963, p. 145; Buis, 1985, pp. 249).[55] When the meeting of commoners had given a member permission to fell a fruit-bearing tree (almost always an oak), it would be pointed out to him by the 'holtrichter' (wood judge), who marked it with an axe (Endres, 1888, pp. 40–44; Hesmer and Schroeder, 1963, pp. 104, 145; Buis, 1985, pp. 46, 64, 249).

4.7 Regulations on Grazing Livestock

Regulations on grazing date back to the 6th century. Up to the 13th century, they were concerned only with the payments by commoners to the lord for the pannage of pigs in the 'forestes' and 'Forests'.[56] These sources rarely mention the regeneration of trees, and the regulations on grazing livestock were not concerned with this at all.[57] The absence of regulations about the regeneration of trees is because the regeneration took place without any action on the part of the users. After all, what happened naturally did not require regulations, and anything which did not require regulations was not laid down in practical rules.[58] This also applied for the mast. This was actually provided by God, and therefore did not need to be regulated in any way.[59] Only the number of pigs to be pannaged was regulated. As far as is known, the only thing that was regulated up to the 13th century regarding the grazing of livestock in Central Europe was the cutting of foliage for fodder. The oldest known regulation on this practice dates from the 7th century. This expressly stipulates that

---

[55] A document, dating from 1585, from the Grobholthauser Mark determined: 'So aber Jemantz etwas zu thunen oder timmerholt was bedurfftig hette, soll dem Holtrichter vnd Holtknechten angiben vund besichtigett werden, als dan nach befinding der nott, vnd gedrage seiner marcken rechten, mit der scharbeilen gewisset werden' (Hesmer and Schroeder, 1963, p. 147).

[56] Endres (1888, pp. 52, 82), Hermann (1915), Meyer (1931, pp. 115, 283), Kaspers (1957, pp. 29, 39, 149, 185), Hesmer and Schroeder (1963, pp. 104, 144, 150), Schubart (1966, p. 12), Ten Cate (1972, p. 59 e.v.), Rackham, 1980, p. 111 e.v.); Jansen and Van de Westerigh (1983, p. 41), Buis (1985, pp. 40, 66, 210), Mantel (1990, p. 99), Iben (1993).

[57] See: Bühler (1922, p. 258), Hesmer and Schroeder (1963, p. 150), Streitz (1967, p. 36), Mantel (1968), Rackham (1980, p. 16), Buis (1985, pp. 206, 210, 218–219, 416, 429; 1993, p. 77).

[58] Bühler (1922, pp. 258, 339), Vanselow (1926, pp. 23; 1957), Streitz (1967, p. 58), Rackham (1980, p. 16), Stamper (1988), Mantel (1990, p. 295).

[59] A document dating from 1310 from Selze in the south of Alsace, states: 'Dar nah wirt ein eckern und gerethe [stock and demand of food] von gottes gnaden uffe den welden, so sülent die burger zu demme closter gan unt sollent mittenander werden zu rate umbe daz eckern' (Ten Cate, 1972, pp. 99–101). In 1433, the lord of the castle asked the Markerichter of the Losser Marke: 'Item myn G.L.G. eyn ordell laeten fraegen, wanner de Almechtige Gott in der marcke eyn akern wolde geven, wo velle schweyne dan alsdan Syne Genaede als holtrichter in der marcke behoere te dryven' (Ten Cate, 1972, p. 115). In about 1575, the Raesfelder Markenrecht stated: 'Item wanner got ein ackern verleint' (Ten Cate, 1972, p. 122).

the 'fruitful' trees should be spared when foliage was cut. The cutting of foliage was probably regulated because it harmed the trees, including those which produced mast, such as wild fruit trees and oak.[60] One indication of this is that cutting foliage was prohibited in many places in Western and Central Europe from the 15th century (Endres, 1888, p. 54; Flower, 1980; Mantel, 1980, pp. 103–104, 934–992; 1990, pp. 103–104; Pott, 1983).

Initially, grazing livestock was free. The commoners were entitled to graze their livestock in the 'forestes' and 'Forests', as well as on the commons ('marken' or 'gemeynten'). This right had priority over the 'ius forestis' or the 'Forest Law'[61] in other words, the lord had to permit the grazing of livestock. According to medieval legal views, this was because everyone individually had to meet his own primary needs, i.e. in food and fuel. The economic system was based on autarchy (Kaspers, 1957, pp. 185, 205, 214). A commoner could graze as many animals on the common or in the 'forestis' as he needed to meet the household needs for food. They were allowed to graze livestock, as they were allowed to take firewood and wood for building, to meet their own needs.[62]

Equality with regard to meeting the people's needs was the starting point (Endres, 1888, p. 8). For the common, this meant that it had to remain closed. Nothing could be traded from the common, such as wood and livestock, but nothing could be brought in either, such as livestock or food or fodder for livestock. Any advantage to one member, e.g. if he sold products from the common, was seen as disadvantaging the other members of the community (Endres, 1888, p. 13). More or less from the time that the commons organizations were established, trading was forbidden, or subject to strict restrictions.[63] It was also prohibited to bring pigs or other livestock to the

[60] The 'Lex Visigothorum', dating from 654, of King Rekkesvind of the Visigoths (649–672) (Van Caenegem, 1967, p. 19) determined that the members of a local community were permitted to cut foliage. Travellers passing through could cut fodder for their oxen only for 2 days, provided they spared the 'fruitful' trees, i.e. the trees which produced the mast for pigs (Trier, 1963, p. 4; Pott, 1983).

[61] Grossmann (1927, pp. 28, 31, 62, 68), Kaspers (1957, pp. 126, 149, 185, 193, 204–205, 214), Hart (1966, p. 8), Page (1970, pp. 88–90), Rackham (1980, pp. 130, 134, 174–175, 183), Cantor (1982a,b), Buis (1985, p. 179).

[62] Endres (1888, pp. 7, 8, 36), Grossmann (1927, p. 61), Hilf (1938, pp. 125, 160), Rubner (1960, p. 52), Hesmer and Schroeder (1963, pp. 104, 145), Buis (1985, p. 150), Tack et al. (1993, pp. 32, 175).

[63] For example, a document dating from 1506, from Asselt (the Netherlands), states: 'Niemandt van den bueren van Astelt en sal siin gedeylde holt vercopen enighe buytenluyden' (Buis, 1985, p. 436). According to a regulation dating from 1458, in the book of the Ucheler Bos and Mark, it was not permitted to simply give wood away either. It states: 'Item so en sal men nyement gheen holt geven dan by oirloff der gemeynre mercken' (Sloet, 1911, p. 133). Also see Endres (1888, pp. 9 e.v, 116), Sloet (1911, p. 66), Weimann (1911, p. 89), Hilf (1938, p. 160), Hesmer and Schroeder (1963, pp. 145, 273), Buis (1985, pp. 155, 436).

common from outside (Endres, 1888, p. 52; Buis, 1985, p. 81).[64] If someone wanted to graze animals on the common, he had to breed them himself and feed them in winter on fodder collected in the common or in the 'forestis'.[65] The animals of commoners from a particular common were branded, so that it was possible to establish whether there were any 'foreign' animals in the common (Sloet, 1913, pp. 15, 357; Ten Cate, 1972, p. 121). As regards the grazing of livestock, it was therefore not possible to graze more animals than there was food produced by the common or 'forestis'. This automatically led to a sort of ceiling on the number of livestock that could be kept (Endres, 1888, pp. 32, 50), i.e. the number of animals which a common or 'forestis' could provide with fodder all year round. This number included the livestock consumed by the commoners.

Regulations on restricting the grazing of livestock were introduced only from the 13th century. Up to the 18th century, they virtually all stated that grazing was prohibited in coppices for a few years (left untouched) to prevent the livestock from eating the new shoots sprouting from the stumps.[66] In the regulations, this regrowth was known as 'spring'[67] (in old Dutch, 'opslage', and in old German, 'auffschlagh'). The 'spring' was distinguished from saplings grown from seed, which were known as 'seedlings' (in old Dutch,

---

[64] For example, the book of the mark of the Lierder and Spelder mark in Gelderland, the Netherlands, dating from 1515, states: 'Item tis verwilcoert dat nymant uutheymsche beesten op die merct vryen sal, op een peen van twee heren pundt, ind elcke beest op een heeren pundt' (Sloet, 1911, p. 118).

[65] On this subject, a document from the Selbolder mark dating from 1366 states: 'Auch wer swyne in der marcke hette, die er in synem huse ertzogen hatte, wie viel der ist, die mag er in die marck triben' (Ten Cate, 1972, p. 101). The members of the marks of Dorth, Oxe and Zuidlo determined that: 'Item nyemants en sall ander beeste dan hie dess winters op synen meess gehadt ind gefoerth hefft, ind den lantheren offte bouwman eygentlick toebehoren, annemmen, by ene pene van ene tonne bierss; thoe betaelen alz baeven' (Sloet, 1913, p. 169). Also see Hausrath (1898, p. 101), Endres (1888, pp. 52, 81, 112–113), Sloet (1911, pp. 113, 169), Grossmann (1927, p. 25), Kaspers (1957, pp. 149, 150, 185, 205), Ten Cate (1972, p. 101), Mantel (1980, p. 464), Jansen and Van de Westeringh (1983, p. 37), Buis (1985, p. 181).

[66] A bill issued by the Estates of Holland in 1555 expressed this as follows: 'de schapen, en beesten als koeyen, paarden, hockelingen … spruyten en looten boven afbijten soodat se daarna niet opschieten, noch tot perfect gewas komen mogen, dan naakte stobben blijven moeten' (Buis, 1985, p. 129). The general rule which applied in this respect was that the coppice 'ins Gehege zu schlagen bis sie dem Zahn des Weideviehs entwachsen waren' (Grossmann, 1927, p. 30).

Also see o.a.: Bühler (1922, p. 259), Vanselow (1926, p. 24), Grossmann (1927, p. 30), Meyer (1931, p. 406), Hess (1937), Hausrath (1982, pp. 207–208), Hesmer and Schroeder (1963, pp. 151–154), Hart (1966, pp. 29–30, 62–63, 79, 85, 95, 104), Schubart (1966, p. 37), Streitz (1967, p. 39), Rackham (1980, pp. 159–160), Buis (1985, pp. 50, 59, 67, 113, 128, 129, 179, 180, 197, 203, 209, 210, 274, 501), Mantel (1968; 1990, pp. 326–328), Putman (1996b, p. 25).

[67] In Dutch, the terms used were 'opslag' and 'opslage' (Buis, 1985, p. 81; 1993, p. 85), in German, 'Aufschlag', 'auffschlagh', 'vpslagh' and 'Jungholtz' (Hesmer and Schroeder, 1963, pp. 151–152; Schubart, 1996, p. 39), in English, 'spring' or 'sprynge' (Hart, 1966, pp. 62–63, 79, 85, 90, 95, 104, 299, 324; Rackham, 1980, pp. 159–160; Jones, 1998).

'Saielingen', in old German, 'erdkymen')[68] or 'waver' ('Wayver', 'weaver' 'weyverd'). For example, in 1462 it was decreed that 'sufficiaunt Wayvers after the custum of the contre' (sufficient young trees) must remain for the owner, and in 1657 that the lessee of the right to chop down firewood was instructed to make sure that 'all the said Springwoode [is] well and sufficiently weavered' (Jones, 1998, p. 62). The word 'waver' for young trees can in all probability be traced to the fact that after the thorny scrub, which protected them from being eaten, offered them shelter from the wind and supported them, was removed to be used as firewood, they gave an unstable, impression in the openness, they were 'wavery'.

4.8 Grazing Livestock in Relation to Coppices

It was discovered at an early stage that when the trunks of many species of trees are cut down when they are young, or when bushes are cut down, the stumps sprout, producing more wood after a few years (Hausrath, 1928; Mantel, 1990, p. 333). The use of the shoots of stumps which have sprouted certainly dates back 5000 years to the Neolithic era. This is clear from wooded trackways which have been dug up (Rackham, 1980, pp. 106–107; Caspare, 1985; Evans, 1992; Parker Pearson, 1993, p. 24). The potential for regeneration is greatest when the trunk of the tree is cut at a young age (Dengler, 1990, p. 263; Mayer, 1992, p. 429).

The stumps of virtually all the species of shrubs and deciduous trees found in the lowlands of Western and Central Europe have a great capacity for stooling (Bühler, 1922, p. 551; Rackham, 1980, pp. 34–35; Koop, 1987; Mayer, 1992, p. 429). With the exception of yew, conifers do not have this property. Moreover, a tree can reach a significantly greater age as a stool than as a tree (Rackham, 1980, p. 28). For example, an ash dies after 180–200 years if it is a tree, but as a stool, it can reach an age of about 300 years, and even ages of 500–1000 years are possible (Rackham, 1980, pp. 7, 29, 208). If they are not regularly cut down, hazel trees reach an age of 70–80 years (Savill, 1991). However, as coppiced stools, they easily grow to an age of 300 years (Rackham, 1980, p. 208). When a large small-leaved lime, 200–300 years old, is cut down, new shoots still sprout from the stool (Rackham, 1980, p. 243; Pigott, 1991). Even when an old small-leaved lime (*Tilia cordata*) falls over, the root clump will send out suckers (Pigott, 1991). Of all the deciduous trees, the beech is least able to stool.[69] However, on limestone soil, podsol soil and well-drained podsol, beech is not far behind other species (Hesmer, 1958, pp. 71–72; Koop, 1987). There

---

[68] In Dutch, seedlings were known as 'Saielingen' and 'grondelingen' (Buis, 1993, p. 85; Tack *et al.*, 1993, p. 100), in German 'Anflug' or 'erdkymmen' (Vanselow, 1926, p. 226; Hausrath, 1928; Meyer, 1941, p. 98; Trier, 1952, p. 13; Schubart, 1966, p. 112) and in English 'saplings' or 'wavers' ('weyverd, weavers') (Hart, 1966, p. 79; Jones, 1998).

[69] Bertsch (1949, p. 57), Hesmer (1958, pp. 71–72), Streitz (1967, pp. 54, 155), Rackham (1980, p. 7), Buis (1985, p. 740), Ellenberg (1986, pp. 50–51), Jahn (1991, p. 396), Mayer (1992, p. 429), Pott (1992).

is a good chance that the stool of the beech will die if the periods between coppicing are 30 years or more. Other species retain their capacity to regenerate longer; for hazel, the period is shortest: 40 years. For the oak, the period is approximately 60 years (Krahl-Urban, 1959, p. 136; Ellenberg, 1986, pp. 219–220; Dengler, 1990, p. 263; Peterken, 1992; Pott, 1992). Because the capacity of trees to regenerate is greatly reduced as they grow older, it is understandable that old trees were taken by being felled with the root-ball (see Tack *et al.*, 1993, pp. 109–110). The stool would not send out shoots anyway, while the roots provided a significant amount of wood. In contrast, removing stools from coppices was prohibited (see Sloet, 1911, pp. 70–71; Hilf, 1938, p. 132; Mantel, 1980, p. 87).[70] After all, in coppices, the primary consideration was the regrowth, the spring from the stump.

From the 13th century, there were regulated coppices, in the sense that the stools were cut down in lots according to an established rotation of the felling cycle.[71] The earliest references to this come from present-day Belgium and northern France (Alsace) and dates from the 12th century (Rubner, 1960, p. 37; Buis, 1985, p. 197).[72] In connection with the creation of the regulated coppices, there were the first regulations for grazing livestock. These entailed the temporary fencing off of newly coppiced wood from grazing livestock, to prevent the livestock from eating the young shoots or spring. These regulations show that it was the wish or the need to protect the shoots from the stools of trees and shrubs from being eaten by livestock, as well as increasing the

---

[70] A provision from the book of the mark of the Hoogsoerense bos in the Netherlands, dating from, 1608, reads:

> Alsoe men verleden jair bevonden heeft dat eenige erffgenamen hair toegedeylde holt in Hoghe Zoere bosch nyet geliicx ter eerden affgehouwen mair uuytgeroyt hebben, wairdoor die stammen nyet hebben connen uuytlopen, twelck zolde strecken tot onderganck van den bosch, als is verordonneert ende verwillecoert dat van nu voortaen geen erffgenamen ofte pechters holt sullen moegen uuytroyen mair gladt van die stamme affhouwen, by die pene op yder boem die uuytgeroyt wordt bevonden twee daler te betalen (Sloet, 1911, pp. 70–71).

[71] An early example of this development in German is a 'Forstordnung' of the bishopric of Magdenburg, dating from 1278. Schubart (1966, p. 37) gave the following translation of the original Latin text, which he included in the appendix: 'wurde auferlegt, den Wald bestandsweise auf den Stock zurückzusetzen und dabei zu veranlassen, wiederum auf den Stock zurückzusetzenden Schläge zu trennen'. Clearly, there were therefore also lots which had to be cut down consecutively. The Latin text states 'silvan dictam resecari', which shows that it refers to 'wood, twigs or bushes which are cut down' (see Schubart, 1966, p. 189, note 51).

[72] The regulations on cutting firewood are believed to have been a result of the increasing demand for firewood for households because of the increasing population and population density and a growing demand for charcoal and firewood for industrial purposes, i.e. the glass and metal furnaces. (Bühler, 1922, p. 259; Vanselow, 1926, p. 11; Endres, 1929; Streitz, 1967, p. 36; Schubart, 1966, p. 35, 40; Mantel, 1990, pp. 217, 219, 330–331; Perlin, 1991, p. 165 etc.; Buis, 1993, pp. 131, 173). Another indication of the increasing scarcity of firewood is that from the 14th to the 16th century, wood which was not 'fruitful', the 'non-fructiferae' or 'malae' were banned from being cut in the forests. This included birch, aspen, alder, ash, sycamore, field maple, hornbeam, holly, thorns and juniper (Rubner, 1960, pp. 50–51; Hesmer and Schroeder, 1963, pp. 145–146; Mantel, 1990, pp. 325–326). Therefore they could no longer be freely cut to meet the needs of the population.

production of wood, which was the most important reason for concentrating stools in lots (Bühler, 1922, p. 259; Mantel, 1990, pp. 326–329, 331).[73] In the commons, the coppicing was also regulated in lots for this reason.[74] It was at about this time that regulations were also issued in the lowlands of Western and Central Europe about saving a certain number of shoots when the trees were coppiced. These were then allowed to grow into trees.[75] The fact that coppices with surviving trees could provide the mast for pigs, as well as producing timber and firewood, may have contributed to the fact that this form of exploitation increased significantly in the course of the 15th, 16th and 17th centuries and reached a peak in the 18th century.[76]

In Latin, the coppice was referred to as 'subboscus', 'silva cedua' and 'silva caedera' (derived from 'caedere', to cut or fell), 'silva minuta' (shrub, small or low wood) and 'silva resecari' (cut wood or coppice); in German, as 'Nederwald', 'underholt', 'Geholz', 'Gehölze', 'Busch', 'Buschholtz', 'Schlagholz', 'Schlagwald', 'Strauch', 'Strauchwerk' and 'Berg'; in Dutch, as 'underbusch', 'holt', 'houw', 'bossch', 'struijk', 'rys', 'onderholt'; in English, as 'holt', 'under-

---

[73] In 1237, King Henry III of England issued an order for the Forest of Dean, stating: 'to take care that in the season when underwood should be cut, it should be so cut to grow again (revenire) and that no damage should befall the coppice (coepecia) … and places so assigned shall be well and sufficiently enclosed so that no beasts shall enter to browse there' (Hart, 1966, pp. 29–30).

The first Württemberg 'Landesordnung', dating from 1495, ordered that it was necessary to cut wood in areas (so not here and there), in order to protect the coppice wood against livestock by using fencing. Thus the regulation prescribed that 'eine Hege der Haue' must be introduced everywhere and, on penalty of a large fine and prohibited driving livestock into the protected 'Haue' (Bühler, 1922, p. 259; Mantel, 1990, p. 328). The Second Württemberg 'Landesordnung', dating from 1515, and the Third, from 1521, even ordered that coppices which were not protected against livestock and were therefore destroyed, had to be notified. The Fifth, dating from 1536, indicated that the poor condition of the coppices was caused by the unlawful cutting of wood and wandering of livestock (Bühler, 1922, p. 259). The first 'Forstordnung' issued for the Spessart in central Germany in 1666 is very explicit in this respect. It states: 'Nachdeme man auch befindet, daß das unordentlich plätzige hauen, so in den Wäldern hin und wider geschicht, Schaden bringet, dann solche Örter und Plätze zu keiner Heeg gebracht werden können, auch der Wind desto ehender einbrechen, und Schaden thun kan, derentwegen dann ordentliche Gehäw und Schläge angefangen werden müssen' (Vanselow, 1926, p. 24).

[74] For example, there is a report dating from 1568 on the Lavesumer and Lünzumer marks in the prince-bishopric of Münster, which states:

Item Dweil ider Holtrichter vnd Erfexen vernemen vnd spuren dat die Marcke Je lenger Je mer mit den dechlichen Houwen verwoestet, wollen Sie vor guet ansehen vnd raitesten achten, dat ein ort von der Marcken nemptlich die Westersidt den wegh benut von Sanct Annen Bergh langs den Berbbergh den wilgrims pat gnant na der Hulsemer Marcken thein iar mit houwen drifte vnd hoide gefriet, damit die vpslagh desto beter sinen wabdom muchte hebben' (Hesmer and Schroeder, 1963, p. 151).

[75] See, *inter alia*, Woolsey and Greeley (1920, pp. 489–490), Vanselow (1926, p. 27), Hess (1937), Reed (1954, p. 37), Rubner (1960, pp. 36–43), Tubbs (1964); Flower (1977, p. 24), Rackham (1980, p. 136), Buis (1985, pp. 127, 631–632), Mantel (1990, p. 336).

[76] Bühler (1922, p. 599), Vanselow (1926, pp. 23–24), Hesmer (1958, p. 390), Rubner (1960, pp. 40, 44, 263, 301 etc.), Hart (1966, pp. 100, 106), Streitz (1967, pp. 37, 39), Flower (1977, p. 24), Buis (1985, p. 636), Dengler (1990, p. 291), Mantel (1990, pp. 331, 336–338, 393).

wood', 'brushwood', and in French, as, *inter alia*, 'bois de renaissance'.[77] The names for the coppices were taken both from the way in which the wood grew and from the way in which it was taken. Thus, as for the terms 'Wald', 'Holtz' and 'Acker', the name seems to have been derived from the use, and they therefore have a utilitarian origin.

Initially the felling cycles of the coppices were short, i.e. 3–9 years.[78] Thus the capacity for regrowth was retained for all the species of deciduous trees (including beech) and shrubs, so that they could be cut down for centuries on end without much danger of the stools dying. There was no need to give any thought to the regeneration of the trees after each harvest of wood; it was merely a matter of preventing the livestock from eating the young shoots on the stools. This explains why the books of the commons and books of customary law included numerous regulations about leaving the coppice untouched, i.e. fencing it off from grazing by livestock, though hardly any regulations indicate additional or supplementary measures (see Buis, 1985, p. 113). Those that are mentioned are the replacement of stools which have died, and the custom of growing trees from seed (Hesmer, 1958, p. 326; Mantel, 1980, pp. 125–130, 345).[79] The dead stumps were replaced by planting young trees. There are reports of this practice in Flanders and England, dating from the 17th century (Flower, 1977, p. 28; 1980, pp. 159–160; Tack *et al.*, 1993, p. 103).

Apart from leaving them untouched, coppices were protected from livestock by digging ditches around them and creating earthen walls planted with dead or living thorny shrubs, such as hawthorn and blackthorn.[80] On the European

---

[77] Sloet (1911, pp. 28–29), Hausrath (1928), Trier (1952, pp. 24, 44–45, 51, 96, 148, 152), Kaspers (1957, p. 171), Hart (1966, pp. 23–25), Schubart (1966, p. 189), Rackham (1975, p. 24; 1980, p. 118), Flower (1977, p. 26), Buis (1985, pp. 40, 67, 110–111, 350), Meiggs (1989), Mantel (1990, p. 335), Tack *et al.* (1993, pp. 26, 96, 97), Best (1998).

[78] Tubbs (1964), Rackham (1980, p. 137), Buis (1985, p. 208), Mantel (1990, pp. 337, 393), Best (1998), Gulliver (1998).

[79] A problem in the interpretation of sources given by Mantel (1980) is the fact that Mantel does not make a distinction between the sowing of forests of shade-tolerant trees where areas have been felled, with species such as European silver fir, Norway spruce and beech, in the mountains, and coppices of deciduous trees in the lowlands or lower areas in mountainous regions. One can only be certain that he is referring to lowlands when there is a reference to a species which requires light, such as oak.

[80] The wall was known as a 'heijm' (= heim, hein, heining) or dyck. One report from Flanders, dating from 1483, described a wood as being surrounded by a dyke (Tack *et al.*, 1993, p. 211). Surrounding lots with dead or living (thorny) bushes was known as 'af-', 'uit-' or 'omtuinen' and 'af-' or 'beheijmen' (Buis, 1985, pp. 110, 619, 621, 623 and 625; Tack *et al.*, 1993, p. 143). A hedge was a living surround (Buis, 1985, pp. 621, 623). In Switzerland, the young wood had to be protected with a hedge and a ditch (Grossmann, 1927, p. 30; Buis, 1985, p. 623). In German, the words used were 'betuinen', 'beheinen', 'hegen', 'hainen', 'bezaunen' and 'begraben' (Hesmer and Schroeder, 1963, p. 207). In Dutch, closing these areas to livestock was known as 'gevriet' or 'in vrede leggen' (Buis, 1985, pp. 108, 167), and in German, 'befriedigung', 'befrechtigung' or 'Zuschlagung' (Hesmer and Schroeder, 1963, pp. 150–153, 207, 208). In English, it is described as 'fencing' and 'enclosing' (Tubbs, 1964; Hart, 1966, pp. 29, 30, 85, 90, 95; Rackham, 1980, pp. 159–160). Also see Grossmann (1927, p. 30), Hesmer and Schroeder (1963, pp. 152, 207), Hart (1966, pp. 85, 95), Rackham (1980, pp. 6, 158), Buis (1985, pp. 619–625), Tack *et al.* (1993, pp. 143, 211), Best (1998), Jones (1998).

mainland, it eventually also became compulsory for the livestock to be herded.[81] The lots where the livestock were herded were indicated with signs (Grossmann, 1927, p. 30; Mantel, 1990, p. 327). In addition, regulations were issued from the 13th century about planting young trees, usually oaks in wood-pastures. Commoners were often obliged to plant a single oak or a few oaks when they were allocated an oak to fell.[82] The young trees which were planted had to be protected from the livestock by planting them in thorny scrub, placing them in the same planting hole with thorny shrubs, or surrounding them with thorn bushes. These measures were adopted until the 18th century.[83] In the 15th and 16th centuries, regulations were issued for the protection of nurseries of young trees, which were established from that time.[84]

None of the regulations on grazing livestock which were issued in the lowlands of Western and Central Europe from the 13th to the 18th century were aimed at regulating the grazing of livestock in general. They even clearly state that the coppicing should be organized in such a way that it obstructed the rights to graze livestock as little as possible.[85] This continued to be the case until the 19th century.[86]

Grazing livestock was completely prohibited only in odd cases, as, for example, in the coppiced woodlands of the Swiss city of Zurich, which were completely closed to grazing in 1376 and 1477 (Grossmann, 1927, p. 20; Meyer, 1931, p. 304). In the Netherlands, grazing livestock, as well as felling trees and

---

[81] Endres (1888, p. 141), Sloet (1911, p. 462; 1913, p. 396), Grossmann (1927, p. 25), Reed (1954, p. 43), Hesmer (1958, p. 454), Mantel (1990, pp. 96, 327).

[82] Bühler (1922, pp. 258–263), Grossmann (1927, p. 114), Meyer (1941, p. 297), Hesmer (1958, pp. 81, 101–106, 333), Hesmer and Schroeder (1963, pp. 170, 201, 209), Schubart (1966, p. 76 etc.), Streitz (1967, p. 39), Buis (1985, pp. 13, 15, 89, 280 etc.), Mantel (1990, p. 341 etc.).

[83] See: Grossmann (1927, pp. 113–114), Meyer (1941, p. 125), Rodenwaldt (1951), Hesmer (1958, pp. 101, 104–105), Hesmer and Schroeder (1963, pp. 107, 158, 197, 207 e.v.), Schubart (1966, p. 76), Flörcke (1967, p. 45), Streitz (1967, p. 53), Koop (1981, p. 10), Buis (1985, p. 280 e.v.).

[84] See: Hesmer (1958, pp. 102, 333), Hesmer and Schroeder (1963, p. 197 e.v.), Buis (1985, pp. 282, 326–327; 1993, p. 109), Mantel (1990, p. 343).

[85] An example is the first 'Forstordnung' for the Spessart, dating from 1666, which stated: 'So sollen demnach unsere Forstbeambte über solcher Ordnung dergestalt halten, daß dieselbe Gehäwe [the coppice] also angestellt werden, damit es der Wildbahn und männiglich angebrachter Huet und Trifft, so viel müglich unschädlich seh' (Vanselow, 1926, p. 24). Also see Endres (1888, p. 91), Hausrath (1898, p. 101), Bühler (1922, p. 612), Vanselow (1926, pp. 24, 145), Hesmer and Schroeder (1963, p. 152), Mantel (1980, pp. 194, 195, 214). The commoners complained about the duration and extent of the restrictions on the grazing of livestock, and they took down the fencing. (Hausrath, 1982, pp. 208–209; Hart, 1966, pp. 125, 145, 186, 291–292; Flower, 1977, p. 110). In addition, heated legal proceedings sometimes continued for decades, as in the case of the town of Michelstadt in the Odenwaldt, which instituted proceedings against the counts of Erbach-Erbach from 1756 to 1814, about the right of its inhabitants to graze their livestock in the 'forestes' of that lord (Rodenwaldt, 1951).

[86] In his famous book, 'Anweisung zum Waldbau', a classical work in the literature on forestry, the first edition of which was published in 1816, and the last in 1865 (Dengler, 1990, p. 16), Cotta (1865) stated: 'Die Hutungen dürfen nicht ohne Noth erschwert oder gar durch die Schläge abgeschnitten werden' (Cotta, 1865, pp. 13–14).

taking humus, were prohibited for 40 and 60 years respectively in the Rheder forest and the Worthreder forest. This indicates that there cannot have been much of the coppice left (Buis, 1993, p. 108). Therefore it was almost certainly a last attempt to allow the coppice, which had been destroyed by over-exploitation, to recover.

To protect the coppice, one animal species became subject to regulations, or even total grazing prohibitions, virtually throughout Western and Central Europe over the years. This was the goat. The reason for this was that goats browsed particularly on buds, leaves and young shoots in the coppice.[87] After the goat, most restrictions were imposed on sheep (Endres, 1888, p. 113; Mantel, 1980, p. 134; Buis, 1985, pp. 130, 353). In many cases, sheep were treated in the same way as goats (Mantel, 1990, p. 97). The reason for this is that they destroyed the grass because they cropped it very short (Grossmann, 1927, p. 78). In many cases, the number of sheep that could be kept was deter-mined in the regulations (Mantel, 1990, p. 439). As a result of the emergence of a trading economy and the flourishing cloth industry in the 16th century, there was a great demand for sheep's wool, so that there was a great increase in the number of sheep despite the restrictions, and consequently in their effect on the vegetation (Mantel, 1990, p. 439; Bielemen, 1992, pp. 80, 84).

Many reports from the Netherlands, Germany and Switzerland show that, despite all the regulations, trees and shrubs were illegally cut down and felled in large numbers, and there was widespread illegal grazing, so that the trees and shrubs eventually disappeared. The cutting of firewood resulted in the greatest devastation of the commons and the forestes (see Appendix 3).[88] Reports from the lowlands of Central and Western Europe on damage by livestock are virtu-ally always related to the biting off of the shoots sprouting from the stools in cop-pices. The historical sources from these regions rarely mention the destruction of seedlings by livestock.[89] When there are references to this, they almost always

[87] Endres (1888, p. 97), Vanselow (1926, p. 74), Grossmann (1927, pp. 25–27, 77–78), Meyer (1931, pp. 25, 382; 1941, p. 106), Hausrath (1982, p. 209), Reed (1954, p. 43), Hesmer (1958, p. 454), Hesmer and Schroeder (1963, p. 150), Hart (1966, p. 115), Streitz (1967, p. 38), Addison (1981, p. 18), Mantel (1990, pp. 96–97).

[88] In the dukedom of Gelre in the Netherlands, a bill (decree) was issued against the 'hewing and stealing of wood' in 1663 (Buis, 1993, p. 85). An 'Edikt gegen Holzverwüstung' was referred to in the prince-bishopric of Münster (Germany) as early as 1560. This was followed by an edict in 1613, which applied to the marks as well as to the lord's 'Holz'. According to Hesmer and Schroeder (1963), the fact that such decrees were issued on average every ten years in the 17th and 18th centuries, shows the extent to which they were not observed (Hesmer and Schroeder, 1963, p. 117). The long list of historical records which they provide on the destruction of the marks and forestes in the lowlands of Lower Saxony, to the west of the Weser and in the so-called Münster Bocht, is very significant (see Hesmer and Schroeder, 1963, pp. 128–141 and Appendix 3). Also see Bühler (1922, p. 259), Vanselow (1926, p. 24), Meyer (1931, pp. 406–407), Hesmer and Schroeder (1963, pp. 117, 128–141), Hart (1966, p. 73), Streitz (1967, p. 37), Buis (1985, pp. 106, 128–129; 1993, pp. 85, 106), Mantel (1990, pp. 331, 337, 393, 424–425).

[89] See Bühler (1922, p. 259), Hesmer and Schroeder (1963, pp. 151–154), Hart (1966, p. 79), Rackham (1980, pp. 159–160), Buis (1985, pp. 128, 129), Mantel (1990, p. 326).

date from the 18th century and later.[90] An indirect form of damage caused by grazing was by cowherds who started fires and ringed trees to increase the area of grassland for the grazing livestock (Mantel, 1990, p. 95; Buis, 1993, p. 83; Tack *et al.*, 1993, p. 213).[91] Therefore grazing was *not* regulated in the lowlands of Western and Central Europe in the regulations dating from the 13th century (the earliest regulations) to the 18th century because it was a threat to the seedlings of trees in a regenerating forest, but because the livestock destroyed the young shoots on the stools in coppices. Thus the vegetative rather than the generative regeneration of wood was protected. The question is, whether the regenerative regeneration required protection from livestock, and if not, why not?

4.9 Seedlings of Trees in Scrub

From the 13th to the 17th century, the lords in the 'forestes' and 'Forests' sold a great deal of so-called underwood to iron and glass foundries as firewood to stoke their furnaces. The earliest agreements state the condition that the seedlings and young trees (in Dutch, 'heesters', in German, 'Heister', and in English 'lez Saplings') in the underwood could not be felled.[92] The earliest names for this underwood, dating from the 13th, 14th and 15th centuries, are in German, 'fürholze'[93] 'vorholt', 'vorholtz', 'strübchen', 'strauch', 'onderholt'

---

[90] See Cotta (1865, pp. 83–84), Landolt (1866, pp. 152, 152–155, 431–435), Gayer (1886, p. 13), Vanselow (1926, p. 226), Grossmann (1927, p. 39), Hesmer and Schroeder (1963, pp. 152–153).

[91] There were severe sentences of corporal punishment for shepherds who deliberately started fires. For example, the 'Waldordnung' of Waldeck, dating from 1516, stated: 'Item nachdem durch die koler hirten und ander von waid und reut wegen prändt geschehen, dadurch die wälde sehr erösigt und verderbt werden, soll dasselbig hinfüro bey Augen ausstechen verboten seyn, und wer darüber betretten oder erfaren würdt, on gnade die Herrschafft gestrafft' (Mantel, 1980, p. 220). The royal Prussian 'Forst-Jagd- und Grenzordnung', dating from 1738, stated:

> Da auch Unsern Privat- und gemeinen Holzungen sowohl, als auf der Leibeigenhörigen Höfen, Aeckern und Wiesen die Bäume abgeschälet oder geringelt, die Aeste abgehauen und sonst auf andere Art beschädet, desgleichen der Junge Aufschlag durch angelegtes Feuer ruiniret worden; als verordnen Wir hiemit alles Ernstes, daß die Schäfer und Hirten niemalen eine Axt, Beil oder Feuerzeug bei sich führen … sollen (Hesmer and Schroeder, 1963, p. 150).

[92] See: Hausrath (1982, pp. 28–29), Hart (1966, pp. 22–29, 51), Wartena (1968), Rackham (1975, p. 27), Flower (1977, p. 26), Buis (1985, pp. 304–305), Best (1998), Jones (1998). In Dutch, the underwood is termed 'onderbusch', 'onderholt' and 'strubben'; in German, 'underholz', 'underholt', 'Unterholz', 'Erdholtz', and in English, 'underwood', 'scrub', 'thicket' and 'brushwood' (Bühler, 1922, pp. 551, 599, 602; Hausrath, 1928; 1982, p. 347; Hart, 1966, pp. 29, 73; Streitz, 1967, p. 52; Musall, 1969, p. 186; Buis, 1985, pp. 110–111; Dengler, 1990, p. 265). The underwood was contrasted with the high wood. In German, this was referred to as the 'grote holt', 'Oberholz', 'Heisterholz' and 'hovetbome' (Hausrath, 1928, p. 347; Hilf, 1938, p. 168). The term 'hovet' is found in the book of the mark of Rekken (Borculo), dating from 1613, in the sense of 'gehoveden lande', meaning the land which belonged to the court or curtis (Sloet, 1913, p. 141). Therefore a 'hovetbome' referred to a tree belonging to the court of the king, the curtis. 'Fruitful' trees were referred to as such.

[93] For example, a licence issued in 1332 to someone for the material for establishing an enclosure around his property. The text reads: 'Wäre das N.N. an den hegen synes guets nit so vil funde, dass er syn guet gefaden [surrounded by an enclosure] möchte, so mag er [the wood needed for this] in dem fürholze und in den strübchen vor Riederholz suechen und nemen' (Trier, 1952, p. 116).

Fig. 4.5. Oak seedling in blackthorn scrub in the Borkener Paradise, Germany. When the 'thorns' were cut for firewood, the 'ius forestis' or 'Waldrecht' provided that a certain number of these young trees growing in the scrub had to be spared to grow into trees to produce mast. This tree was known as a 'Waldrechter' because it was protected under the 'Waldrecht' (photograph, F.W.M. Vera).

and 'busch'; in Dutch, 'strubben', 'boes', 'bosch', 'onderbusch', 'onderboss' and 'onderholt'; in English, 'bush', 'scrub', 'shrub', 'shrubbery', and 'underwood'; and in French, 'petit taille et bordure', 'bois', 'boccage' and 'buisson'.[94] Virtually all these names can also be read as descriptions of mantle and fringe vegetation, and are actually known as such (also see Trier, 1952, pp. 97, 115, 116). The clearest example of this is the name 'vorholt' (see Fig. 4.1).

A certain number of 'heesters' (young trees) had to be spared for every unit of area when coppicing took place. These were chosen and marked by a 'forestarius' or 'forester'.[95] These were the so-called 'noble' or 'fruitful' trees, or

[94] Sloet (1911, pp. 28–29, 133), Hausrath (1928), Trier (1952, pp. 24–25, 27, 87, 96, 97, 99, 115, 116; 1963, pp. 35, 166), Hart (1966, pp. 22–23, 26), Mantel (1968), Rackham (1980, p. 137; 1993, p. 62), Buis (1985, pp. 110–111).

[95] The accounts dating from the 15th century on the sale of in the Nederrijkswoud between Nijmegen and Kleef state that the purchaser of the underwood: 'honderd heisters dair men die ut den ganzen slach bi den tekenmeystern utnemen sal' (Wartena, 1968; Buis, 1985, p. 305). In 1595, for the sale of underwood in the New Forest, it was decided that 'great trees' and 'trees fit for timber' and 'saplinge of oak apt and fit to be timber' were reserved for the Crown (Tubbs, 1964, p. 96).

In the Forest of Dean, a contract of sale in 1615 determined that all the 'green and quick timber trees of oak' were reserved for the king and should be marked with a 'known fashionable sign or mark'. The purchasers of the underwood, iron foundries, had to mark six or at least five of the best oaks or other principal trees in every acre (0.4 ha), so that they would 'seed and replenish the ground' (Hart, 1966, p. 95).

trees which would produce mast and timber. Apart from the wild fruit trees, these above all included oaks (see Fig. 4.5). The trees were indicated on the basis of the 'ius forestis', 'Waldrecht' and the 'Forest Law'.[96] In German, these trees, especially oak, as fully grown trees were known through the centuries as 'Waldrechter' up to the 18th century, because they were spared on the grounds of the 'Waldrecht' (Dengler, 1935, p. 507; 1990, p. 291). The fact that the seedlings of trees in the mantle and fringe vegetation, the underwood or brushwood had to be protected when it was cut down, shows that there was regeneration of trees there.[97]

The earliest regulations on cutting down underwood or brushwood refer most to thorn bushes, hazel and holly.[98] Thorny bushes (blackthorn and hawthorn) were particularly popular for firewood (Rackham, 1980, pp. 352, 353; Tack *et al.*, 1993, p. 129). After the felling, the buyers of the firewood were obliged to protect the seedlings which had been spared and the stools of the shrubs which had been cut down, from being eaten by livestock. They had to dig a ditch around them, build an earthen wall and place a hedge on it, usually of thorny shrubs (Meyer, 1941, p. 115; Tubbs, 1964; Hart, 1966,

---

[96] A document dating from 1342 stipulated: 'dat die selve vorster ... geweist haven ain dem heister uff der statt, dair sey zo rechte unser beyder waltrecht wisen soelen' (Kaspers, 1957, p. 166). In the accounts of the sale of coppiced wood dating from 1417–1418, the waldgraaf of the Nederrijkswoud between Nijmegen and Kleef in the dukedom of Gelre, stated that: 'Men sal die alde heisteren laten stain ende van den besten jonge wail werder opmaken na waltrecht' 'die alde heisteren als die jonge heister sal ... laten stain ende van den besten jonge dair des neet wail werden upmaken na waltrecht.' In some cases, 'na waltrecht' was replaced by 'na alder gewoenten' (Wartena, 1968, pp. 36, 38; Buis, 1985, p. 304). The Forstordnung from Mainz dating from 1744 stipulated that: 'gesunde Eichen zu Wald-Recht stehen lassen ... von denen zu Waldrecht stehen gelassenen Eichen und Buchen' (Vanselow, 1926, p. 222; see Appendix 4, paragraphs 10, 12 and 16 of the Forstordnung of Mainz). In 1593, it was said that: 'it was underwood that was cut and timber trees were left standing' (Flower, 1977, p. 111). Also see Hausrath (1982, pp. 28–29), Tubbs (1964), Hart (1966, pp. 96, 100, 106), Flower (1977, p. 24), Mantel (1990, p. 336), Tack *et al.* (1993, p. 100).

[97] A route description dating from 1289 reads: 'etliche Äcker, so daselbst vor Wobeck bey einem Boemlehren Busche belegen' (Schubart, 1966, p. 182). As we read above, an 'Acker' indicated a place mainly with oak trees. Chapter 2 revealed that oaks regenerate particularly well in thorny scrub. Obviously, woods/scrub without trees were sufficiently remarkable to be used to indicate a route. In the 13th century, the king of England had many thousands of acres of underwood (1 acre = 0.4 hectare) cut down in the Forest of Dean to be sold as firewood in order to create a network of paths and roads through the forest. In addition, he had the existing routes widened by having the underwood removed on either side so as to increase the safety of travellers (see Hart, 1966, pp. 22–23). In the years following this, it was decided that the seedlings and young trees growing there should be spared (Hart, 1966, pp. 30, 79; Tubbs, 1964, p. 96).

[98] See Hausrath (1898, p. 44; 1928), Meyer (1941, p. 76), Tubbs (1964; 1988, p. 154), Hart (1966, pp. 29–30, 46–47, 128, 180–181, 308 etc.), Schubart (1966, pp. 21, 89–93), Streitz (1967, p. 52), Flower (1977, pp. 27, 63, 73).

pp. 62, 63, 79, 85, 90, 95, 104, 184; Flower, 1977, p. 28; Best, 1998; Jones, 1998).[99]

There are reports of the growth of seedlings of trees in thorny scrub in the historical sources from the whole of Central and Western Europe.[100] The oldest was referred to above, i.e. the reference by the Roman Tacitus in AD 98, that Germania looked very different in different places, but was generally covered with bristling woods (either bristling trees or bristling groves) and unhealthy marshlands. Medieval records indicate that thorny shrubs were widespread in the lowlands of Central and Western Europe at that time (Hilf, 1938, pp. 119–120; Hesmer and Schroeder, 1963, p. 50; Hadfield, 1974; Rackham, 1980, pp. 7, 118–119, 293, 352; Hausrath, 1982, p. 16).[101] According to Rackham (1980), thorny scrubs ('spineta') are mentioned in English medieval records as the most common species in the 'silvas' (Rackham, 1980, pp. 119, 352–353). From his study of the history of woodlands in England, Rackham (1980, p. 173) concluded that hawthorn and holly shrubs were a natural and essential phenomenon in wood-pastures, and that new oak trees grew in the middle of these, protected from being eaten by livestock. There is an old saying in the New Forest: 'The thorn is mother to the oak' (Penistan, 1974, p. 105; see Figs 4.2 and 4.3). Therefore in England, these thorny shrubs were sometimes described as the 'nursery crop' for trees (Addison, 1981, p. 95). Thorns and holly were actually considered so important for the regeneration of trees that a statute dating from 1768 laid down a punishment of 3 months of forced labour for damaging thorns and holly in the New Forest, starting every month with a number of lashes of the whip (Rackham, 1980, p. 173).

---

[99] 'and places so assigned shall be well and sufficiently enclosed so that no beasts shall enter to browse there' (Hart, 1966, p. 30). A text dating from 1572 about the Forest of Dean reads:

> The underwood together with the lopping and shredding of all those trees which heretofore have been used to be lopped and shred, growing in Maylescott bottom and Bucholemore coppice in the Forest of Dean, are meet to be sold this year to the oresmith in the same Forest. No timber-trees nor saplings of oak likely to prove to be timber, to be fallen by colour [marked] hereof. And the spring [regrowth] reserved (Hart, 1966, p. 79).

In 1595, the right to cut wood was granted for three areas, two of 40 acres and one of 30 acres, and the arrangement stipulated that 'great trees' and 'trees fit for timber', and all 'saplinge of oak apt and fit for timber' should be reserved for the Crown (Tubbs, 1964, p. 96).

[100] See Woolsey and Greeley (1920, p. 72), Meyer (1931, pp. 417–418; 1941, pp. 105, 115), Hesmer and Schroeder (1963, pp. 101–105), Tubbs (1964), Hart (1966, pp. 276, 298), Rackham (1975, p. 27; 1980, p. 173), Addison (1981, p. 95), Hausrath, 1982, p. 31).

[101] Following their study of the historical sources of the North German lowlands, Hesmer and Schroeder noted: 'Ferner treten in den Akten noch mehrfach Angaben über Sträucher auf, insbesondere über Dornsträucher ("Weißdorn", "Schwarzdorn", "Hagedorn", "Schlehe", meist jedoch einfach nur "Dorn"), ferner über "Faulbaum" und Brombeeren, die nicht weiter von Interesse sind' (Hesmer and Schroeder, 1963, p. 50). This remark also illustrates to what extent thorns were considered of subordinate or even of no importance in the historical study of the use of forests, and were therefore ignored in the studies of the archives (also see Landolt, 1866, p. 148).

Thorny shrubs and juniper spread in grazed grassland. Thorns are seen as irritating weeds which have to be destroyed. In many parts of Europe, blackthorn and juniper are still considered to be weeds (Grossmann, 1927, pp. 113–114; Ellenberg, 1986, p. 43; 1988, p. 20). They impeded grazing and took the place of more valuable sorts of wood (Hobe, 1805, p. 125; Landolt, 1886, p. 148; Vanselow, 1926, p. 223).[102] They were removed because they took up the area available for grazing.[103]

From the 14th century, young oaks were planted to encourage regeneration. To protect them from livestock and wild animals, they were surrounded by thorns or planted together with thorny shrubs in a single hole. Young trees were also planted in thorny scrub.[104] In fact, this imitated the process of regeneration in thorny scrub.[105] The fact that oaks were planted is very probably related

[102] In an argument for the enhancement of the marks and the improvement of grazing by promoting the growth of grass, Hobe (1805) raised the rhetorical question:

> denn wo ist eine Gemeinheit, die nicht mit alten bemooßten Maulwurfshaufen, Wacholderstauden, Dornen und dergl. angefüllt, und so beschaffen wäre, daß am Ende die Hude ganz darauf wegfallen oder immer elenderes Vieh darauf nur erhalten werden kann; die münsterischen, paderbornischen, märkischen, clevischen Gegenden, so wie auch die lünerburgischen, hannöverischen und andere Länder geben hiervor noch traurige Beweise (Hobe, 1805, p. 124).

[103] In the mark of Bronsbergen and Wichmond in Gelderland (The Netherlands), it was decided in 1753 that:

> De bouwlieden van den erven en landeryen aan de gemeinte liggende zullen gehouden ziin elk tegen ziin land de doornestruyken op de gemeinte uit te roeyen en te removeren, and wel sorge dragen, dat deselve niet wederom komen uit de spruyten and op te wassen, by poene dat diegene, welk sulx binnen een jaar tiids niet mogten gedaan hebben, na verloop van dien, en vervolgens zo duk en menigmaal enige doornestruyken op de gemeente sullen bevonden worden, vervallen zullen ziin in de boete van enen daalder, ten profyte van de markt te verbeuren (Sloet, 1913, p. 360).

A regulation dating from 1755 in Switzerland stated: 'Wie nun kein Wald, mit und neben welchem die Trift und Weyd besser bestehen kann, als eben mit dem Eichwald, bevorab wann ein solcher von allem Buchgewächs und Gestäud wohl gesaubert ist' (Meyer, 1931, p. 400). With the words 'denn, wo Dornen stehen, kann kein Vieh weiden', Hobe (1805) indicated that if those who were entitled to fell wood did not remove the thorns, the community ('der Communität') could do so on its own initiative (Hobe, 1805, p. 175). Also see: Hobe (1805, pp. 124–125), Gradmann (1901), Bernôtsky (1905), Grossmann (1927, pp. 113–114), Nietsch (1939, p. 156), Hausrath (1928; 1982, pp. 212–213), Hilf (1938, p. 10), Meyer (1931, p. 400; 1941, pp. 105–106), Hesmer (1958, pp. 104–105), Hesmer and Schroeder (1963, p. 207), Schubart (1966, p. 82), Musall (1969, pp. 138, 173).
[104] Grossmann (1927, pp. 113, 114), Meyer (1941, pp. 115, 125), Rodenwaldt (1951), Hesmer (1958, pp. 101, 104–105), Hesmer and Schroeder (1963, pp. 158, 197, 207 e.v.), Schubart (1966, p. 76), Flörcke (1967, p. 45), Streitz (1967, p. 53), Buis (1985, p. 280 e.v.), Pott and Hüppe (1991, pp. 85, 107).
[105] In 1609, Nordon, who made an inventory of the coppices in the New Forest, wrote in his report under the heading: 'To raise timber in open forests, parkes, chases and wastes without incoppicing' ['open' meaning free accessible for cattle and wild animals] 'Everye keeper in fforeste parke or chase, as also officers within his Majesty's Mannors upon wastes, are to be injoined to caste acornes and ashe keyes into the straglinge and dispersed bushes: which (as experience proveth) will grow up, sheltered by the bushes, unto suche perfection as shall yelde times to come, good supplier of "timber"' (Flower, 1980, p. 312). According to Nordon, the keepers could do this as they went round the forest (Flower, 1977, p. 28).
In 1553, it was decided in Münsterland (Germany) that thorns should be spared so that young oak trees could be planted there: 'Zum Vierten sollen die Dornen soviel muglich beschont werden, die Telgen dar Innen zu Potten vnd zu befridige' (Hesmer and Schroeder, 1963, p. 170). Also see Puster (1924), Grossmann (1927, pp. 113–114), Rodenwaldt (1951), Hesmer and Schroeder (1963, pp. 107, 170, 207), Flower (1977, p. 28; 1980, p. 312), Koop (1981, p. 19).

to the fact that the thorny scrub where regeneration took place spontaneously was considered to take up too much valuable grazing land. Nevertheless, it was considered essential for protecting the seedlings and young trees from livestock and wild animals (Mantel, 1980, p. 366).[106] In the 18th century, hawthorn and blackthorn were cultivated in 'nurseries' so that they could be used to protect young trees planted in wood-pasture (Schubart, 1966, p. 193; Pott and Hüppe, 1991, p. 73). In a number of areas in Germany, the young oak trees were planted at relatively large intervals so that they would grow into good mast oaks, i.e. oaks with large crowns, which therefore produced many acorns (Hesmer, 1958, p. 390; Pott, 1983).

There are historical reports, not only on oaks, but also on the growth of other species of trees in thorny scrub on grazed commons. Bühler (1922, p. 416) referred to Petrus de Crescentiis, who wrote in 1305 that apple trees, chestnut trees, pear trees and other trees that bore fruit, should be freed from thorns, and that when good trees suitable for timber were overgrown by thorns, these should be removed. According to a description dating from 1559/60, the marshy woodland along the River Rhine near Speyer contained 'Dornenhorst mit Obst- und Eichbäumen' (thorny scrub with fruit and oak trees) (Musall, 1969, p. 95). When trees were felled in Epping Forest in 1562, these included 618 crabtrees and hawthorns on approximately 35 ha (Rackham, 1980, p. 353). In the Forest of Dean, there were many crabtrees, next to thorny shrubs (Hart, 1966, pp. 80, 209, 276).[107] Elm and ash also grew in thorny scrub (see Tubbs, 1964; Flower, 1977, p. 24).

In addition to the regulations providing that seedlings and young trees should be left standing, there are also reports about thinning out scrub to provide light for young oaks, to allow them to grow.[108] In addition, not all young trees were nominated to be spared. There were even regulations issued that stated that not all trees should be spared. In a sense, these regulations promoted a certain thinning out of the trees to prevent there being too many. The reason

[106] A decree dating from 1590 states: 'desgleichen sollen auch Eichelns gesehet und gepflantzt und die Dorn in den Waidgängen nit gentzlich abgehauwen werden, damit die junge bäum desto baß [extraordinary, better]) uffwachsen … mögen [will grow]' (Mantel, 1980, p. 366).

[107] An inventory dating from 1633 states: 'The hazels, the crabtrees, the birches, the maples, hawthorns and hollies, growing there in great abundance' (Hart, 1966, p. 276).

[108] A notice dating from 1779 in Switzerland reads: 'Im Lentzhard in den Eychen Einschlag habe observiert, daß das vile Gestrüpp den jungen Eychen im wachsen hinderlich ist, deswegen … vorzustellen, ob Sie nicht ratsam finden, diesen Einschlag von dem Gestrüpp säubern zu lassen' (Meyer, 1931, pp. 417–418). According to Rackham (1980), there were regular reports in the 14th century on the care of oak trees in thorny scrub, as regards pruning and thinning out the trees. As an example, he refers to a text dating from 1362–1363, which reads: '60s [shilling] from shragg' & 'paryng young oaks, together with [sales of] thorns in various places on the woodbanks etc.' He considers that the term 'shraggyng' clearly meant pruning, while 'paryng' [in Latin, mandura (cleaning)] probably had this sort of meaning (Rackham, 1980, p. 159). 'Mudare' means cleaning (Muller and Renkema, 1995, p. 586) and probably meant the removal of unsuitable young trees, i.e. thinning out. Also see: Woolsey and Greeley (1920, p. 72), Bühler (1922, pp. 416–417), Meyer (1931, pp. 417–418; 1941, pp. 105, 115), Hesmer (1958, pp. 105, 455), Hesmer and Schroeder (1963, pp. 104–105), Schubart (1966, p. 112).

for this was that when too many trees grew in the scrub, the shrubs and coppices disappeared (Hart, 1966, p. 168; Rackham, 1980, p. 145; Simpson, 1998). I will return to this later in the chapter. Another reason for not saving all young trees was that if trees grow less closely together, the extra light they receive causes them to form a larger crown and therefore blossom more profusely. As a result, the oaks produced more acorns and therefore more mast for the pigs (Hesmer, 1958, p. 390). If there is no thinning out of young trees such as oak, the trees will grow more closely toegether. This could be the way trees, including oaks, with long trunks without branches occurred in grazed landscapes.

The regeneration of all the species of trees indigenous to the lowlands of Central and Western Europe still takes place in park-like landscapes, such as wood-pasture, for example, in England, Germany and France (see Figs 3.6, 4.1–4.3 and 4.5–4.7).[109] As we established in Chapter 2, Watt described this type of regeneration of trees in detail (1919; 1924; 1925; 1934a,b) to show how the original forest returned after the grazing of livestock came to an end. In this context, he also referred specifically to the New Forest.[110] On the basis of detailed research going back to the 13th century, it is possible to follow the relationship between grazing and the regeneration of trees, in particular, in the New Forest (also see Tubbs, 1964; 1988; Peterken and Tubbs, 1965; Flower, 1977; 1980). Therefore the New Forest is an ideal place to study the regeneration of trees through history and to shed light on the link between the generative regeneration of trees and the grazing of livestock.

4.10 The Regeneration of Trees and the Grazing of Livestock

Nowadays, the New Forest covers an area of approximately 37,500 ha. Of this, approximately 18,000 ha is open to livestock and deer. This part consists of approximately 14,500 ha of open grassland, heath and peat, and 4000 ha of

---

[109] Also see in: Watt (1919; 1923; 1924), Flower (1977, p. 112), Bürrichter *et al.* (1980), Rackham (1980, pp. 174, 188, 293, 324–325), Tubbs (1988, p. 157; 1987), Rodwell (1991, pp. 333–363), Pott and Hüppe (1991, pp. 25–26), Pott (1993, pp. 172–178), Oberdorfer (1992, pp. 81–106) and personal observation.

[110] Watt (1919) wrote about the New Forest:

Among such spiny plants as Ilex aquifolium, Prunus spinosa, Crataegus monogyna, etc., usually near the periphery of clumps of these, (oak) saplings of various heights were found growing up among the protecting branches. Their demands for light led them to incline their stems to the outside of this protection but rarely did I find any protruding twigs which would be liable to be nibbled by the cattle, ponies, sheep, deer, etc., which roam through the Forest, or if they do project they are promptly eaten back. Once these branches emerge from this protection sufficiently high up to escape the browsing animals, the future of the tree is assured. It is no uncommon thing to find a large oak standing in the centre of such a clump – an oak which has grown up with the thorny species, the latter affording it the necessary protection. This phenomenom has been recorded by numerous observers for trees in general and undoubtedly the protection thus afforded was the salvation of the oaks in question (Watt, 1919, pp. 196–197).

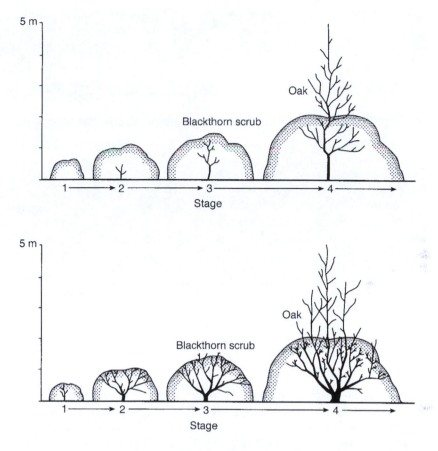

Fig. 4.6. Schematic representation of young trees including oaks, growing in blackthorn scrub, in the presence of large herbivores such as cows and horses. The young trees are nibbled up to as far as the thorns of the blackthorn reach. They grow along with the shrubs, and eventually grow above them (redrawn from Pott, 1993, p. 222).

deciduous woodland. This woodland is known as the 'ancient woods' or 'ornamental woods' (Flower, 1980). Figure 4.7 gives an illustration of these woods. Like the grasslands, heath and peat, they were traditionally grazed by commoners' livestock (Tubbs, 1988, pp. 67–68). In this woodland, there is little or no regeneration of trees, although there are some seedlings, for example, of beech and oak (Morgan, 1987a). With the exception of a few beech seedlings, these virtually all disappear (Peterken and Tubbs, 1965; Flower, 1977, p. 195; Morgan, 1987a; Putman *et al.*, 1989; Siebel and Bijlsma, 1998). The high densities of herbivores are seen as the reason for this (Peterken and Tubbs, 1965; Flower, 1977, p. 200; Harmer, 1995; Siebel and Bijlsma, 1998). Fenced-off

Fig. 4.7. A so-called 'ornamental wood'; the grazed grove or boscage of Mark Ashwood, in the New Forest, England (photograph, F.W.M. Vera).

areas in the forest are given as evidence for this. In these areas, there *is* prolific regeneration of trees (Peterken and Tubbs, 1965; Putman *et al.*, 1989).[111]

On the basis of measurements of the circumferences of trees and counts of the annual rings, Peterken and Tubbs (1965) concluded that in grazed forests which had never been fenced off from livestock and deer, three generations could be distinguished. One generation dated from the period 1648–1753; one dated from 1858–1915, and one from after 1938 (Peterken and Tubbs, 1965). In their opinion, these groups were correlated to periods of low densities of herbivores. Flower (1977; 1980) assembled all the circumferences of trees measured by Peterken and Tubbs (1965), and added new measurements. He determined the relationship between the circumference of the trunks and the age of the trees, on the basis of counting the annual rings and the thickness of the planted marked oaks. These oaks were planted respectively in 1700, 1756

---

[111] On this subject, Peterken and Tubbs (1965) wrote:

> In two Forestry Commission experimental sites in Mark Ash Wood, each less than 3 acre, beneath gaps in the canopy, and fenced against deer and stock, regeneration, mainly of beech, is prolific. This contrasts sharply with the complete failure of regeneration in similar but unfenced sites near by, and demonstrates strikingly how herbivores control natural regeneration (Peterken and Tubbs, 1965, p. 164).

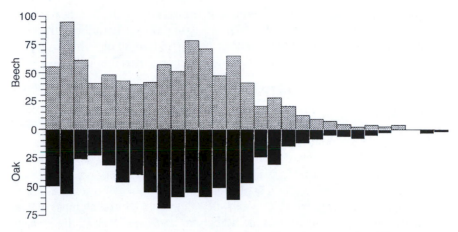

Fig. 4.8. The distribution of the number of oak and beech trees in the different categories of thicknesses in the 'ornamental woods' in the New Forest, England, as determined by Flower (redrawn from Flower, 1980, p. 327). The horizontal axis shows the numbers; the vertical axis shows the circumference of the trees in categories from 0.2 m per category, up to the largest category of 5.5–5.8 m. The oldest trees dated from the beginning of the 16th and the end of the 17th centuries. The figure illustrates that regeneration took place without interruption.

and 1775 (Flower, 1980). In this way he showed that there had been clearly separate generations through the centuries in the forests as a whole, and that regeneration took place all the time, in contrast with the findings of Peterken and Tubbs (Flower, 1977, p. 204, see Fig. 4.8). Periods for which there is sufficient information available on the density of livestock do show that there is a certain correlation between the density and regeneration, though regeneration occurred even with the highest known densities (Flower, 1977, p. 91; 1980). When Flower's data on the regeneration of trees (1980) are combined with those on the numbers of herbivores recorded by Putman (1986), this reveals that in the 18th and 19th centuries, there was regeneration with densities of one cow 4.5–5 ha^{-1}, one horse 9–15 ha^{-1}, and one deer 3–3.5 ha^{-1} (altogether 110–130 kg ha^{-1})[112] (Flower, 1980; Putman, 1986, pp. 11–12, 24–25). Peterken and Tubbs (1965) stated that with a grazing concentration of one so-called feeding unit ha^{-1}, there was virtually no regeneration.[113] Converting his

[112] The weights used to determine biomass are: one cow, 350 kg; one horse, 250 kg; one red deer, 80 kg; one fallow deer, 55 kg; one sheep, 40 kg and one roe deer, 20 kg (S.E. van Wieren, Wageningen, 1997, personal communication). For the unspecified term 'deer', the average was taken of the red deer and the fallow deer because the forest had been stocked with fallow deer as well as red deer. The biomass used for one deer is 67.5 kg.
[113] Peterken and Tubbs (1965) give as a conversion factor: 1 pony = 5 'feeding units'; 1 deer = 3 'feeding units' and 1 cow = 1 'feeding unit' per acre (0.4 hectare).

figures into the units used by Peterken and Tubbs (1965), the figures recorded by Flower (1980) and Putman (1986) showed that the regeneration of trees still occurred in the New Forest with densities of 1.4–1.5 feeding units ha^{-1}; i.e. densities at which Peterken and Tubbs considered that regeneration was completely out of the question.

In other forests in England there was also regeneration of trees, with relatively high densities of livestock and deer. According to Rackham (1980), there were 300 fallow deer, 200 sheep and more than 1000 cows, as well as several dozen red deer in Epping Forest in the second half of the 17th century, in an area of 2000 ha (together amounting to 187 kg ha^{-1}). This is a biomass which is in the order of that found in the game reserves in Africa (see Schröder, 1974; Drent and Prins, 1987). In view of the present number of trees, Rackham believes that there was prolific regeneration of trees, with the biomass of livestock and deer prevailing at the time (Rackham, 1980, p. 185). With regard to deer, it is stated that in natural forests the density amounts to 0.5–1 deer 100 ha^{-1} (= 0.4–0.8 kg ha^{-1}). In order to guarantee regeneration in forests, it is assumed by forestry experts that the density of red deer can be 0.5–3 animals, and of roe deer 4–5 animals 100 ha^{-1} (= 1.2–4.4 kg ha^{-1}) (Anonymous, 1988; Wolfe and Von Berg, 1988; Remmerts, 1991, p. 18). Densities of 10 red deer 100 ha^{-1} (= 8 kg ha^{-1}) are considered very high in forestry circles (Schröder, 1974).

The density of deer in the English forests was considerably higher, although regeneration did in fact take place there. Apart from livestock, it amounted to approximately 30 animals 100 ha^{-1} (= 20 kg ha^{-1}). The 1668 Enclosure Act for the Forest of Dean stated that the maximum number of deer was 800 (Hart, 1966, p. 293). In terms of the area at the time (9600 ha), this meant eight deer 100 ha^{-1} (= 6.4 kg ha^{-1}). This density was independent of the numbers of cows, sheep and goats which the commoners had the right to graze in the Forest. Yet an inventory dating from 1690 shows that there certainly was regeneration of trees. The inventory stated that there were 100,000 young trees in the Forest, some of which were 40 years old (Hart, 1966, p. 186).

Peterken and Tubbs (1965) assumed that the regeneration of the grazed forests ('ornamental woods') had to take place *in* the forests themselves. For this regeneration, they used Watt's gap phase model (1947). Thus they also looked for the cause of the failure of the trees to regenerate *in* the forest. They gave the destruction of seedlings by the excessive numbers of herbivores as a reason for this. Flower (1977) also assumed that the regeneration of the forests must take place in gaps in the canopy. As long as the gaps were large enough, he thought that the oak would also regenerate there (Flower, 1977, p. 195). However, Flower (1977; 1980) noted that the regeneration of trees had taken place without interruption throughout the history of the New Forest. Thus there were no waves of regeneration in the forest, as Peterken and Tubbs had assumed. Nevertheless, Flower did support their viewpoint by stating that there will always be regeneration in the New Forest, as long as the level of grazing does not exceed the critical level, as defined by Peterken and Tubbs (see Flower, 1977,

p. 204).[114] As we noted above, this limit was amply exceeded in periods when there was regeneration.

While there was no regeneration in the grazed areas of the New Forest, it did take place on the periphery of these forests in blackthorn scrub, especially the regeneration of oaks (Peterken and Tubbs, 1965; Flower, 1980; Siebel and Bijlsma, 1998). The forests spread in concentric circles in an expanding ring of mainly young, pedunculate and sessile oak, which emerged from the advancing blackthorn scrub by growing as a clone (Watt, 1919; Tubbs, 1988, pp. 153–157). Throughout the New Forest, this has resulted in a concentric expansion of forests in the form of successive generations of trees. The oaks also grew in unshaded holly scrub, as well as in blackthorn.[115] The woods originating from holly scrub are known as 'holmes' (islands) or 'hats' (Peterken and Tubbs, 1965; Tubbs, 1988, p. 154; Siebel and Bijlsma, 1998). These 'holmes' are regularly mentioned in inventories of the New Forest, dating from the 16th and 17th centuries (see Flower, 1977, pp. 70, 112, 139, 141–144). They developed because the holly colonizes the open heath where pedunculate and sessile oak, yew (*T. baccata*), rowan (*S. aucuparia*) and whitebeam (*S. aria*) then become established in the young scrub (Peterken and Tubbs, 1965). The young trees and holly then grow up together. The holly acts as a 'nurse', protecting the young trees from being eaten by horses, cows and deer. After about 40 years, the holly no longer grows up, but the trees continue to grow. Oak or other species of trees, which were the first to become established on the periphery of the still young holly scrub, thus eventually grow up above the scrub in the centre. This results in the shape of a hat. Many places in the New Forest have names incorporating 'hats' and 'holm' (Tubbs, 1988, p. 154). After 80–100 years, the trees in the holly scrub form a closed canopy (Peterken and Tubbs, 1965).

The trees have to become established in the holly in the first 20 years after the holly starts to grow, as colonization becomes impossible later because of the shade cast by the holly scrub, once it has closed over. The next opportunity for trees to become established does not arise until the holly scrub degenerates after 200–300 years (Tubbs, 1988, p. 154). The trees by no means always successfully become established in the holly scrub, as shown by the inventories which regularly refer to treeless or virtually treeless 'holms' (Hart, 1966, pp. 308–311; Flower, 1977, p. 141).

---

[114] On this subject, Flower (1977) wrote: 'At times it has been suggested that parts of the Forest are dying through lack of regeneration. The age profiles in part II show, to the contrary, that provided the grazing pressure does not exceed the critical level as defined by Peterken and Tubbs (1965), the Forest is quite capable of perpetuating itself' (Flower, 1977, p. 204).

[115] For example, an inventory dating from 1585 reads: 'set with thornes, holmes and oaks' (Flower, 1977, p. 143). A bill dating from 1597 reads: 'Godshill Bail: Sold by woodward in Sett thornes 59 acres of holms, thorn and hazell wood total, 44.3.4' (Flower, 1977, p. 112). An inventory from 1609 reads 'holly, white and black thorn, sheltering a vigorous growth of oak and ash saplings' (Tubbs, 1964, p. 98).

While the reason for the failure of the forests in the New Forest to regenerate was seen in the forest itself, there *was* regeneration *outside* the forest, i.e. in
the mantle and fringe vegetation of the 'ornamental woods' and in the open
fields, particularly of oak (Tubbs, 1964; Peterken and Tubbs, 1965; Tubbs,
1988, pp. 153–157; Siebel and Bijlsma, 1998). The width of the annual rings
and the shape of the oldest generation of trees in the grazed forest shows that
the regeneration goes back a long way, and that all these trees grew in relatively
open conditions, with a great deal of light (Peterken and Tubbs, 1965; Flower,
1980, pp. 88–89). In my opinion, this indicates that these forests developed in
blackthorn or holly scrub (see Fig. 4.9). The regeneration of trees in scrub
explains how this can take place, even with very high densities of livestock and
wild animals. In fact, seedlings were actually protected from deer and livestock
by the thorny scrub, which served as fences. The edge of the forest was able to
advance because the blackthorn spread into the grassland, sending its rootstock
underground. Blackthorn can advance at a rate of 0.3–1.0 m year^{-1}. A blackthorn seedling can in this way expand into a hurst of 0.1–0.5 ha in the space of
10 years, and in this way, a grove of a similar size can develop in the grassland
during the same period (Hard, 1975; 1976, p 186; Wolf, 1984; Wilmanns,

Fig. 4.9. The oldest generation of trees, approximately 300 years old, in Bradley
Wood in the New Forest. Note how many trees have thick branches low down on
the trunk. There is no shrub layer at all. As a result of the selective felling of oaks for
the shipbuilding industry, beech has become very dominant in this oldest
'ornamental wood' (Flower, 1980) (photograph, H. Koop).

1989; Schreiber, 1993). When the trees grow taller than the scrub, the black-thorn eventually disappears, as do the other thorny shrubs, because they require light (Puster, 1924; Watt, 1924; 1934b; Ekstam and Sjörgen, 1973; Ellenberg, 1986, p. 95; 1988; 1988, p. 157; Coops, 1988). On the basis of the size of the 'ornamental woods' in the New Forest, it is possible to get an idea of the area of the groves which develop in this way. In historical sources which describe the size of the groves with thorny shrubs where oaks grew, the size of the largest varies from 250–300 to almost 500 ha (Hart, 1966, p. 301; Flower, 1977, pp. 68–71, 86, 112).

After a while, the centre of the grove disintegrates because the oldest trees there die off first (see Fig. 4.10). Apart from the ageing process, this can also be the result of dry summers, as was found in the New Forest (see Flower, 1980; Koop, 1989, pp. 108, 110; Peterken, 1996, pp. 337, 359; personal observation). There is no regeneration in the gaps which develop in the canopy as a result, because of the presence of high densities of large herbivores, as revealed by the situation in the 'ornamental woods'. Grassland develops as a result of the browsing, particularly of horses and cows, and this increases in size as more trees die off in the centre of the grove. The gaps join together, and the area of grassland increases markedly (H. Koop, Wageningen, 1997, personal communication). Storms can

Fig. 4.10. Dead beech tree in Anses Wood in the New Forest. The tree appears to have grown in an open space. There are no young trees or shrubs at all (photograph, F.W.M. Vera).

Fig. 4.11. The degenerating centre of the Denny Wood grove in the New Forest. The trees are approximately 300 years old. Because of the (continuing) absence of thorny scrub, no seedlings have (yet) grown as a result of grazing. However, a grassland with a rich diversity of species has developed (Koop, 1989, p. 110; photograph, H. Koop).

also contribute to an increase in the area of grassland (see Fig. 4.11). If the light conditions are favourable, hawthorn, blackthorn or holly bushes may become re-established, and the process of regeneration of trees establishing themselves in thorny scrub starts all over again.

The regeneration of trees in thorny scrub can be traced back to the 16th and 17th centuries in the archives of the English forests. These often refer to thorny scrub and holly which protected young oaks, beech and other species of trees so that they could grow.[116] Apart from the trees which regenerate in black-

[116] An inventory of the New Forest dating from 1585 reads: 'set with oak, holmes [holly scrub] and thorne; set with ash, holmes and thorne; thornes, holmes and oaks' (Flower, 1977, p. 143). An inventory of the Forest of Dean, dating from 1680, reads: 'The other parts of the Forest as well inclosed as inclosed as uninclosed [i.e. for the grazing of livestock] consists of hills bare of wood and places called Lawnes in which nevertheless there are good store of bushes and cover convenient for the growth and shelter of young timber-trees, and where within the memory of man have been store of great oaks and as good timber as in other parts of the Forest' (Hart, 1966, p. 298). For example, in 1683, a commission considered that on the basis of an inventory, 8000 'cords', each valued at 8 s., could be felled in the underwood in the Forest of Dean every year: 'it being in many places so thick and high as to hinder the growth of young oaks' (Hart, 1966, p. 181). Similar reports appeared about the Forest of Dean in the 17th, 18th and 19th centuries (pp. 62–63, 73, 79, 85, 90, 95, 100, 104, 106, 178, 180–182, 184, 186, 192, 209, 273, 308–309, 311). For the forests in England, also see Tubbs (1964), Flower (1977, pp. 21, 28, 63–65), Rackham (1980, pp. 187, 293), Addison (1981, p. 85).

thorn and holly in the New Forest, the process also takes place in heather and gorse scrub (Fenton, 1948; Shaw, 1974; personal observation).

The sources referred to above also show that there was prolific regeneration of oak and other species of trees, while large numbers of trees were felled for the shipbuilding industry (Flower, 1980; Rackham, 1980, pp. 187, 293). Therefore the picture sometimes painted in the literature of all the forests being destroyed by felling trees for the shipbuilding industry is not correct as regards regeneration[117] (see Holmes, 1975; Perlin, 1991, pp. 176, 191 *et seq.*). Figure 4.12 shows how the different parts for building ships were taken from the oak. In those days, ships consisted of many rounded parts, produced by oaks which had grown in relatively open conditions. Thick branches which had grown in a curve were used for the ribs. These branches particularly grow on trees such as mast oaks, with a crown which starts low down on the trunk (Flower, 1977, p. 46). Therefore people were not interested only – or even particularly – in long, straight trunks. Both England and the Netherlands imported a great deal of this oak timber for the shipbuilding industry from Germany (Rodenwaldt, 1951; Hesmer, 1958, p. 84; Holmes, 1975; Buis, 1985, p. 513; Perlin, 1991, p. 160). The curved oak, as well as the straight trunks used for planks, and pine trunks which were used to make masts, were transported to the Netherlands down the Main and the Rhine on enormous wooden rafts flying the Dutch flag (Endres, 1888, p. 157; Buis, 1985, p. 879) (see Fig. 4.13). The rafts were 300 m long, 50 m wide and had a depth of 2.20 m. Using paddles, more than 550 people navigated the craft to the city of Dordrecht, where they were moored.[118] The timber was used to build ships there, or in Amsterdam or the Zaan region. In Dordrecht, some timber was sold to England, and was then transported there (Mantel, 1990, p. 272).

The regeneration of trees in thorny scrub and holly was a general phenomenon in the New Forest and the other Forests of England, and the oak was

---

[117] In 1633, 166,848 trees were counted in the Forest of Dean (Hart, 1966, p. 274). In 1667, only 200 of these were left. On 23 April 1680, i.e. 13 years later, a commission reported on the Forest of Dean, that of the 23,000 acres (9200 ha) comprising the forest, 'much the better half, dispersedly over the whole, is well covered with young wood of oak, beech, birch, hawthorn, hazel, and holly, in many places whereof are very hopeful young oaks and beeches of 40 years growth and upwards as well without the inclosures as within' (Hart, 1966, p. 178). Areas which had not been fenced in the previous years were also thickly covered with oak, beech and birch, which had grown 'past danger of cattle' (Hart, 1966, p. 298). In 1690, 23 years after only 200 large trees had been counted in the forest, an official from the Forest of Dean reported: 'But there are in the Forest near 100,000 young trees and saplings, some of them about 40 years' growth, which if carefully preserved may be of great use to the navy. Also valuable quantities of holly, hawthorn, crooked beech, birch, hazel and stroggal oak that may be cut away and converted into cordwood' (Hart, 1966, p. 186). Therefore these young trees already existed when the massive felling between 1630 and 1667 had reduced the number of trees to 200.

[118] Buis (1985, p. 506), Van Prooije (1990; 1992a,b,c), Mantel (1990, pp. 272, 279), Perlin (1991, p. 160), Van Wijk and Allebas (1995, pp. 37–59).

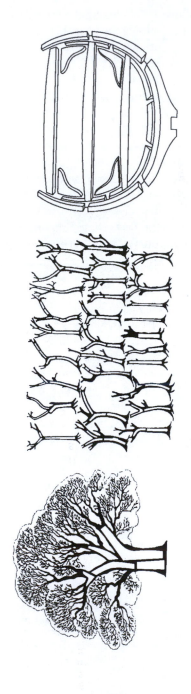

Fig. 4.12. Illustration of how the shipwright timber was obtained from oaks which grew in all sorts of shapes. The most important parts obtained, particularly from oaks grown in open spaces, as well as trees grown in open spaces, were the so-called 'ribs' of the ship (redrawn from Stagg, 1987, p. 24, and Miedema and Gevers, 1990, p. 57).

Fig. 4.13. *The descent of a timber raft on the Rhine, one hour above the city of Bonn* (engraving dating from 1783). Transporting timber in the form of a raft flying the Dutch flag, on the Rhine in Germany, in the 18th century. The men in the boats in the foreground are measuring the depth of the river. At the head of the raft, hundreds of people are paddling to keep the raft on course (photograph from the Dordrecht archives).

strongly dominant in this (see Watt, 1919; Rackham, 1980, pp. 289, 293, 294–297, 300; Tubbs, 1988, pp. 153–157). On the basis of his study of the history of wood-pasture, Rackham (1980) concluded that grazing did not prevent the regeneration of trees (Rackham, 1980, p. 187). The regeneration was restricted to the open parts of the wood-pasture. On the other hand, the regeneration of oak stopped in the wood-pasture as soon as grazing came to an end (Rackham, 1980, pp. 294, 295, 296).[119]

---

[119] In this respect, Rackham (1980) wrote with regard to Staverton Park: 'The decline of grazing was followed by a great increase of holly and birch, relatively sensitive trees. But oak, which had regenerated moderately freely under grazing, ceased to do so after grazing ended' (Rackham, 1980, p. 294). He also wrote: 'Oak regenerates embarrassingly well in the less-wooded parts of most wood-pastures' (Rackham, 1980, p. 295). On the absence of the regeneration of oak, he also wrote: 'Since about 1850 something has happened to prevent this turnover (regeneration), but without affecting the ability of oak to reproduce outside woodland. By the 1910s this was a widely recognised problem, the subject of a classic study by A.S. Watt [1919]' (Rackham, 1980, p. 296). The grazing in virtually all the 'commons' came to an end at about this time as a result of the abolition of grazing rights (see Rackham, 1980, p. 6).

4.11 Coppices-with-Standards (Standing Trees) Growing in Scrub

As we noted above, the concentrically spreading scrub and mantle vegetation in groves with seedlings and young trees were, in the first instances, cut for fire-wood. The seedlings and young trees were spared, and were protected from live-stock by being fenced off. The trees which were spared then grew up above the scrub, which was used for coppicing. This is how the coppices with standards developed (see Fig. 4.14). The shrubs were known as 'underholt', 'unterholz' or 'underwood', while the trees were referred to as the 'grote holt', 'Oberholz', or high wood (see Figs 4.1. and 4.2).

There are data to show that the number of young trees called 'heesters' which had to be retained, gradually declined. For example, the accounts of the Rijkswoud between Nijmegen (the Netherlands) and Kleef (Germany) show that the number of trees to be retained was reduced in the period 1417/18–1430 from 30–75 to 13 per morgen (a unit of land measurement). Later, the number increased again to 20–60 per morgen (see Wartena, 1968; Buis, 1985, pp. 305–306).

Fig. 4.14. Coppice with standing trees, in the Ardennes, France. The standing trees are mainly oak, the shrubs are mainly hazel. The original mantle vegetation of thorny scrub, hazel and young trees, has developed as a result of coppicing, and partly by generative reproduction, into a shrub layer known as 'underwood'. A limited number of standing trees are spared, as the coppice would otherwise decline significantly because of the shade (photograph, F.W.M. Vera).

In my opinion, these data reveal a trial and error approach. I believe that the explanation for this decrease is that when scrub was used for firewood, the young trees had to be spared, but not too much. Otherwise, the shrubs below the trees would eventually be negatively affected by the shade cast by the young trees which had grown into trees. This must have been the experience in wood-pasture, where it became clear that when the trees growing from the thorny scrub developed into groves with a closed canopy, the shrub level disappeared as a result of the shade (see Figs 4.9 and 4.10). According to Rackham (1980), the competition for light between timber (used for building) and underwood was generally recognized in England (Rackham, 1980, p. 145).[120] The people who had the right to cut underwood in the Forest of Dean tried to prevent the growth of 'timber', because this was at the expense of underwood (Hart, 1966, p. 168).

Reports are also known from German-speaking countries, which indicate that the number of standing trees was limited in favour of the underwood (Mantel, 1980, pp. 337–339; Hausrath, 1982, pp. 30–32).[121] From the point of view of the supply of firewood, it is clear why the strictly protected 'fruitful wood', such as oak trees, which produced the mast for pigs, was also described as 'harmful wood' ('schedlich Holz'), while the underwood was considered harmless ('unschedlich') (Endres, 1888, p. 36; Gradmann, 1901, p. 442; Hilf, 1938, p. 133). Therefore it is not surprising that the historical sources show that in the 16th and 17th centuries, when coppices with standing trees developed, most forests had few standing trees. This changed in the 18th century, as the following paragraph will show. As regards the young trees or shoots to be spared, there were no real regulations at first. A 'forestarius' individually marked each young tree to be spared when the underwood was to be cut down (see Kaspers, 1957, p. 166; Hart, 1966, p. 95; Wartena, 1968; Buis, 1985, p. 305). Not all young trees had to be spared. This means that some young trees were cut down as well as the shrubs. The stools of these trees produced shoots, and the next time, these were coppiced, together with the other shoots.

---

[120] A report from Northamptonshire dating from 1604 states: 'The tymber & decaying trees mentioned … may be taken when the coppices shalbe sold, for that they stand to the spoile of the vnderwoods therein.' The following report comes from Bradfield Woods, West Suffolk, and dates from 1604: 'A great part of the timber trees growing upon … Munsces Park were felled in or about 1656 and 1657 and on the fall of the same underwood of … Munces Park is become of a greater value by a third part than it was when the timber was standing.' The last quotation dates from the 18th century, and reads: 'The Timber, if not taken down, would howse your underwood and spoil its growth' (Rackham, 1980, p. 145).

[121] As early as 1531, a 'Waldordnung' was issued: 'Item nachdem in den Schlägen viel Hegereiser und etliche Bäum aufgezogen werden, die doch nicht mehr nutz sind, allein die Schläg dämpfen und das Junge Holz verderben'. It was subsequently determined that only 10 heesters (young trees) should be spared with each felling, and that the broad-crowned trees particularly should be felled, except when they were border trees. The Forest decree of Neuburg, dating from 1577, determined that, because the underwood was no longer growing well, as a result of the dense timber, only three to four large oaks and beeches, five to six medium trees for building wood and timber, and 6–8 'lassreitel' should be left standing per Jauchert. These sorts of regulations were also included in other Forest decrees (Hausrath, 1982, p. 31).

In this way, all the species of trees which were traditionally spared, such as oak, beech, birch, ash and lime, eventually also became part of the coppice.[122] After all, all these species of trees regenerated in the scrub which was coppiced, as we saw in this chapter and in Chapter 2. Subsequently, the regeneration of the standing trees could take place not only by sparing the shrubs, but also by leaving one shoot on the stool of an oak or other species.[123] This shoot was known in Dutch as a 'spaartelg', in German as a 'Laßreiser', and in English as a 'staddle'.[124] The regulations show that when the shoots to be spared were indicated, this was on the basis of the trees which were there (Mantel, 1980, pp. 396–401; Hausrath, 1982, p. 32). From the 17th century, young trees were also planted. This was known as 'to put in vacant places' (Flower, 1977, p. 28; Rackham, 1980, pp. 159–160; Tack *et al.*, 1993, p. 103). There was some spontaneous regeneration from seed in the coppices, although I found relatively little about this in the literature. For example, the regeneration of oak from seed is possible in a coppice, but only with a short rotation cycle (Rackham, 1980, p. 297; see Mantel, 1980, pp. 345, 389).[125] With a longer cycle, the shoots joining together on the stools cast too much shade for too long, so that the seedlings die off. For example, the crowns of hazel coppices join together after only 4 years, so that species of plants which require light have virtually disappeared by the 6th year. Young oak seedlings survive in the shade for a few years, but need to be put in the light again by the next cutting of shoots in the coppice, or they die off (Forbes, 1902; Rackham, 1980, pp. 80, 296–297). With short cycles of 3–9 years, such as those used in the Middle Ages (Rackham, 1980, p. 137; Buis, 1985, p. 208; Mantel, 1990, pp. 337, 339), the seedlings have light restored to them 'in time', so that they can continue to grow. In this way, they have a chance of growing taller than the shrubs in the coppice.

There may also have been regeneration from layering, as this is possible with all species of deciduous trees (Koop, 1987; Watkins, 1990, p. 92). I did not find any reports on this in historical sources in the publications consulted. Altogether there are few reports on measures for the regeneration of coppices, except for sparing the staddles and leaving the coppiced lots untouched (Buis, 1993, p. 113). This is not surprising, in view of the enormous capacity for regeneration of the stools of all the species of deciduous trees with relatively short felling cycles in the Middle Ages (3–9 years) (Rackham, 1980, p. 137;

---

[122] See, for example, Cotta (1865, pp. 120–127), Trier (1952, pp. 62, 81, 90), Schubart (1966, pp. 106, 169), Evans (1992), Mantel (1990, p. 333), Watkins (1990, p. 82).

[123] For example, a sale contract in 1484 prescribed: 'unde schal laten stan hovetbome und lathrise, als eine gemeine wonheit ist' (Hausrath, 1982, p. 29). The 'hovetbome' referred to trees which belonged to the court or curtis, or the trees which were protected by the 'Waltrecht'.

[124] Bühler (1922, p. 301 e.v.), Rubner (1960, p 40), Tubbs (1964), Flower (1977, p. 24), Rackham (1980, pp. 147–148), Hausrath (1982, p. 31), Buis (1985, pp. 127, 365), Mantel (1990, pp. 336, 338).

[125] For example, the 'Bemberg Forstordnung' states that the stool should not be cut taller than a shoe, so that the seedlings are not prevented from growing too much (Mantel, 1980, p. 345).

Buis, 1985, p. 208; Mantel, 1990, pp. 337, 339) and the extremely long period (centuries) during which this regenerative capacity was maintained.

My hypothesis is that, as explained above, the coppice with standing trees developed along empirical lines. In my opinion, the origin of the coppice with standing trees lies in the unshaded scrub/coppice where seedlings of trees could grow. Experience has shown that the canopy of the standing trees could not account for more than approximately 25%, as the underwood would otherwise be damaged too much by the shade of the trees (see Cotta, 1865, pp. 133–135, 137–138; Warren and Thomas, 1992, pp. 255–256). This cover would be approximately 50 large trees, 120–150 years old, per hectare (see Cotta, 1865, pp. 137–138).

This hypothesis is different from that of other authors.[126] They suggest that coppices with standing trees developed from the original closed forest. They believe that gaps appeared in the canopy when trees were felled, so that the forest gradually became more open and a shrub layer, the underwood, developed under the trees. This underwood was then coppiced. In my opinion, this hypothesis is incorrect, for the reasons given earlier. However, this does not mean that these woods did not become more open over the centuries because of the felling of trees in the groves in park-like landscapes, so that shrubs subsequently colonized the open spaces from the mantle and fringe vegetation of the groves. Then, these shrubs could also have been coppiced. In this way, groves could gradually have been transformed into coppices with standing trees.

The convex shape of the oldest walls around coppices, which served to keep the livestock out and protect young trees and sprouting stools from their browsing, is a strong indication of the development of coppices from thorny scrub which formed the mantle vegetation of the groves, and the subsequent transformation from groves to coppices. This form corresponds with the round shape of groves in grazed, park-like landscapes, which is in turn the result of the concentric expansion of groves as a result of the clonal advance of blackthorn by means of underground suckers (see Fig. 5.47). When the coppices developed, the outer perimeter of the blackthorn scrub which formed the mantle vegetation around the grove, formed the boundary of the brushwood which was used for coppicing in the first instance.[127]

4.12 The 'Hage' or 'Haye'

What has been said above about the origin of coppice and coppice-wth-standards throws a certain light on the meaning of the Anglo-Saxon word 'haga' or 'hege'. These terms were supposedly at the basis of the English deer park

[126] See Bühler (1922, p. 599), Hausrath (1928; 1982, p. 28), Rackham (1980, p. 300), Buis (1985, p. 632), Ellenberg (1986, pp. 49–53), Mantel (1990, pp. 333–338), Watkins 1990, p. 10).
[127] Rackham produces many historical maps of coppices which clearly show this concave shape of the oldest borders of coppices (see Rackham, 1975, pp. 16, 17; 1980, pp. 186, 189, 198, 278, 289, 325; 1993, pp. 15, 107, 127, 137, 277, 369). See also Peterken (1996, p. 359).

(Rackham, 1980, pp. 188–195; Hooke, 1998a,b, pp. 154–159). The terms 'haye' and 'hedge' are related to them (Rackham, 1980, p. 188; Cantor, 1982). The terms 'hage' and 'haye' often occur as toponyms in medieval England. In the early Middle Ages before William the Conqueror (1066), the term 'haga' in charters is related to catching deer (Rackham, 1980, pp. 188–191). In addition, it occurs in charters as a phenomenon related to boundaries; an impenetrable barrier and line of defence (Hooke, 1998a,b, pp. 155–157). The Anglo-Saxon term 'haga' was supposedly a screen intended for use in catching deer. In the 10th century, a 'haga' was also related to a forest with wild animals. These were permanent enclosures into which the wild animals were driven through a narrow opening. This technique was used for deer and also for wolves and wild boars, as evidenced by the identification of 'wulfhagan' and 'swinhagan' (Hooke, 1998a). In German-speaking areas, 'Jagen' and 'Hagen' are used in one single expression as 'Jagen und Hagen' (Hooke, 1998a). The term 'derhage' from the 11th century is supposed to indicate the relationship between 'haga' and deer. The 'Hage' would have been the origin of so-called deer parks. These are permanent enclosures, in which deer are kept permanently for hunting (Rackham, 1980, pp. 188–195; Hooke, 1998a,b, pp. 154–159). The screen of the deer parks developed through time into fencing made with poles (Rackham, 1980, pp. 191–192; Hooke, 1998a,b, p. 157). It can be concluded from the fact that the boundaries of the later deer parks were characterized by an earthen wall, that a 'haga' was also characterized by an earthen wall (Rackham, 1980, p. 193; Hooke, 1998a,b).

The term 'haga' also occurs in relationship with the 'haw' of hawthorn. The underwood, which was used as coppice wood, was also known as 'hag' (Rackham, 1993, p. 67; Gulliver, 1998; Hooke, 1998b, p. 154). The term would usually mean a forest with a hedge around it (Rackham, 1980, p. 188). This indicates that a 'haga' was a barrier consisting of thorns. In the previous paragraphs, the terms 'hagen' and 'hegen' have been seen in the meaning of protecting the felled thorny scrub, either in combination with an earthen wall and ditch or not. In German, this was called laying coppice wood in 'Hag und Graben'.

The common denominator of all these terms is the aspect of impenetrability for animals. This is a feature also shown by mantle and fringe vegetations around a grove (in Old English: graf) in the uncultivated wilderness. The fact that the underwood used as coppice wood was called a 'hag' leads us to suspect that there is a relationship between the coppice wood and the 'haga'. The 'hag', 'haga' or 'haya' would be the mantle and fringe vegetation surrounding a grove. Just as this 'hag' could keep animals outside in the sense that it formed an impenetrable barrier within which young trees could grow protected from being eaten, it could also keep the animals inside a grove once they were there. A 'haga' would have been the mantle and fringe vegetation that acted as a screen to keep animals driven into the grove where they were (see Figs 4.1 and 4.2). The hunt would then take place as follows. In a park-like landscape, the wild animals would be rounded up and finally driven through an opening in the mantle and fringe vegetation in a grove, which had been selected for size and because it was completely surrounded by a closed mantle and fringe vegetation, the 'hage'. The animals would be inclined to flee into the grove through an opening, because a grove offers more protection from predators than open terrain. The grove itself was good for hunting because of the openness of the growth there. As indicated earlier, a storey of young trees and bushes is missing there (see Fig. 4.9). There was good visibility there and you could move around in it on horseback.

Subsequently, a hunting lodge could have been built near the 'haga' (Rackham, 1980, p. 197). This is probably the origin of the oldest building, the Ridderhof – the former hunting lodge of the Dukes of Holland – in the Hague (Den Haag in Dutch, La Haye in French). The original name of the town is Die Haghe (Don, 1985, p. 146). In the Netherlands, the city is also named 's Gravenhage (in full: des Graven Hage, i.e. the 'hage' of the duke). The hunting lodge was located in 's Graven Wildernissen (the wildernesses of the dukes) (Buis, 1985, p. 14). In addition to these indications of the meaning of 'hage' as mantle and fringe vegetation of a grove, there is also another. In texts in Medieval Dutch from the 13th to the 15th centuries, the combination of 'bosch' and 'hage' occurs very frequently as 'bosch ende haghe' (grove and 'hage') (De Haan, 1999; M.J.M. de Haan, Roelofarendsveen, 2000, personal communication).

The mantle and fringe vegetation of a 'haga' selected for hunting would have to remain closed. Through the years, openings would have appeared in the 'hage', which would have to be filled. Planting dead or living thorny bushes could have done this. In time, poles could have been used instead, because it would be easier and because they offered more security in terms of exclusion. The poles could also have been used to keep the screen in place. In this way, 'haga' could have come to have its current meaning, i.e. a piece of land surrounded by a palisade (see Rackham, 1980, p. 191; Hooke, 1998a).

Another argument for relating the phenomenon 'haga' with groves encircled by a mantle and fringe vegetation in a park landscape is the fact that the toponym 'haga' occurs repeatedly within a distance of a few kilometres. This is also the case for groves in a park landscape (see Fig. 5.47). The shape and size of deer parks originating from the 'haga' also indicates a relationship with groves. Analogous to the oldest boundaries of coppice wood, they all have a concave shape, were located in a Forest and were generally not so large, namely 40–80 ha, although there were also very large ones (1600 ha) and very small ones (6 ha) (see Rackham, 1975, pp. 16, 17; 1980, pp. 186, 189, 191, 198, 278, 289, 325; 1993, pp. 15, 107, 127, 128, 137, 146, 194, 196, 277, 369; Cantor, 1982; Hooke, 1998a,b, p. 154). The earthen walls that surrounded the deer parks originating from 'haga' could have arisen in three ways. Either it was adopted from the practice of keeping the coppice wood around the 'haga' as a protective screen, or the 'haga' was also used for coppice wood, or they date from before the time the coppice wood was used. In that case, the 'haga' could have served as an example for an efficient protection of storing the 'haga' used as coppice wood.

As a result of the fact that people started to keep deer permanently inside the 'haga', the grove would have become more and more open, because there was no regeneration in the grove as a result of the presence of large ungulates. It would have become more open from the centre out, because the oldest trees were located in the centre of a grove. Regeneration only took place on the edges of a mantle vegetation. In this way, the park would have become synonymous with a certain type of landscape, namely semi-open, with large single standing old trees without an undergrowth of bushes or young trees (see Fig. 4.11).

4.13 The Vegetative and Generative Regeneration of Trees in Relation to the Grazing of Livestock

At the beginning of the 18th century, there was a change in the demand of household and metal furnaces for firewood in the German states. Not only did the demand increase as such, but people no longer wished to have the wood delivered in bundles of twigs or sticks (faggots), but in blocks. A particular measurement was used for this; in Dutch, it was called 'vademhout', in French, 'bois de corde', and in German, 'Malterholz' (Schubart, 1966, p. 98; Tack *et al.*, 1993, p. 127). Delivering blocks meant there was a need for wood from the trunk which split easily. To obtain this thickness, the usual coppice cycle was doubled, or even tripled, to 30–50 years. Later, it was extended even further to 60–80 years (Vanselow, 1926, p. 153; Schubart, 1966, pp. 98–99, 108, 126–127; Mantel, 1990, p. 366). As a result of this, the coppices changed from a shrub layer under standing trees, into a so-called pole-forest, known in German as 'Stangenholz' or 'Heisterwald' (Vanselow, 1926, p. 153; Schubart 1966, pp. 98, 108; Mantel, 1990, p. 366). These forests of deciduous trees first appeared between 1700 and 1730 in Hessen (Hausrath, 1982, pp. 63–69; Mantel, 1990, p. 336). The pole-forest comprised the traditional 'fruitful' trees (almost always oaks) as standing trees which grew above the coppice, which had grown into a pole-forest. They were kept because of the importance of the mast for pigs. Therefore there was actually still a coppice-wth-standards, i.e. a lower level for firewood and an upper level for the mast and timber. The only change was that the lower level had acquired the appearance of a low tree level because of the longer cycles (Schubart, 1966, p. 100). As beech produced the best charcoal, as well as excellent household firewood, the development of this form of exploitation was accompanied by a marked increase in the area of beech trees. This was achieved particularly by planting young beech trees.[128]

[128] Vanselow (1926, pp. 8–9, 153), Endres (1929), Hesmer and Schroeder (1963, pp. 165–196, 209, 274), Schubart (1966, pp. 98–99, 102, 126, 127, 162), Mantel (1990, p. 441), Tack *et al.* (1993, p. 120).

The longer coppice cycle must have led to problems for the regeneration of beech, as the stool of this species sprouts little or not at all, with cycles over 40 years long.[129] Therefore new young beech trees had to be planted after every felling. For this purpose, increasing numbers of beeches were grown from seed in nurseries ('kampen' or 'Kämpe') in the 18th century (see Hesmer and Schroeder, 1963, pp. 174–196, 209, 274). It became common to increase the cycles of the shoots, saplings or trunks to 60 or 80 years. When the coppice cycle was increased, there was some spontaneous growth from seed on the forest floor (e.g. see Hobe, 1805, p. 68).[130] Shoots of beech coppices actually develop flowers and seed after only 20–30 years (Ellenberg, 1986, pp. 219–220). This resulted in a pole-forest with beech seedlings, in many cases with oaks growing as standards because they had been spared (see Vanselow, 1926, pp. 26, 35; Hesmer and Schroeder, 1963, pp. 151, 152).

Beech seedlings can survive for years under a virtually closed canopy in certain soils (Kraft, 1894; Bühler, 1918, p. 443; Vanselow, 1949, p. 31; Dengler, 1990, pp. 64–67; Korpel, 1995, p. 9). Both recently germinated seedlings, and dormant seedlings which have been under the canopy for years, will grow without any problem when they receive more light when old beech trees above them are removed (Bühler, 1918, p. 443; Woolsey and Greeley, 1920, p. 76; Vanselow, 1949, p. 77; Leibundgut, 1984b, p. 18; Mayer, 1992, pp. 317–318). This happened when trunks were removed from the pole-forest. In about 1740, this process was systematically introduced in Hessen (Bühler, 1922, pp. 302–303; Schubart, 1966, pp. 100–103, 125–127; Mantel, 1990, p. 357). The canopy was thinned out by felling part of the poles in the wood in a regular pattern, so that the beech seedlings had an opportunity to grow. As the young beech trees continued to grow, more and more old beech trees were successively felled until all the poles were finally cleared and the whole area was covered by a new generation of beech (Vanselow, 1926, p. 76; Mantel, 1990, p. 357). This method of regeneration by harvesting wood is described in detail in the 'Hanau-Münzenberger Forstordnung', dating from 1736 (Bühler, 1922, p. 303; Hausrath, 1982, p. 64; Mantel, 1990, p. 357). This was adopted to the letter in the 'Mainzer Forstordnung', issued in 1744 (see Vanselow, 1926, p. 222 *et seq.*; Appendix 4). This regulation still refers to the oaks which were spared on the grounds of the 'Waldrecht' as 'fruitful' trees, while the beech trees in the stool forest under the oaks are described as seed trees.

---

[129] See Cotta (1865, p. 30), Landolt (1866, p. 314), Hausrath (1982, p 37), Ellenberg (1986, pp. 219–220), Mantel (1990, p. 366), Pott (1992).

[130] For example, a report dating from 1710 about the Bokeler Berg in Northwest Germany states: 'Bokeler Berg ist ein Berg Büchen Holzes [beech coppice] und Voll Junges aufschlages so in gute stande, davon Järlich auf die Ledige plätze verpflanzet werden' (Hesmer and Schroeder, 1963, p. 163) and in 1721, the 'Amtrentmeister' from Ahaus wrote that 'einige Tausent aufgeschlagene junge Büechen zum vorschein kommen' (Hesmer and Schroeder, 1963, p. 151).

Thus the 'Mainzer Forstordnung' actually formed a continuation of a method of exploitation (coppicing) which had been used for centuries. This is clear from the fact that livestock were still grazed in the pole-forest in accordance with the traditional rule which applied to coppices, that a regenerated lot had to be opened up to grazing when the 'growth' (i.e. in this case, from seed) had grown up above the reach of the animals (see Appendix 4). In 1774, the 'Forstordnung' was followed by the 'General-Verordnung', which determined that when the regeneration from seed was the main aim, no livestock could graze in the forest at all.[131] In the lowlands of Central and Western Europe, this was a break with the past. This regulation also determined that the cycle of the forest should be extended from 80 to 140 years. This sort of cycle results in a forest now known as a productive standing forest. In my opinion, this type of forest, in which the regeneration of the trees must take place *in* the forest, therefore developed in the lowlands of Central and Western Europe only in the second half of the 18th century (Schubart, 1966, pp. 100–101).

The regeneration of trees from seed then became a problem in relation to the grazing of livestock in the lowlands of Central and Western Europe, as formulated in the prevailing theory. Because the livestock trampled and ate the seedlings *in* the forest, and the aim was to regenerate trees *in* the forest, the livestock prevented the regeneration of trees and therefore of the forest. This was not the case when the regeneration of trees took place *outside* the forest in thorny scrub in the grasslands and in the mantle and fringe vegetation of groves in wood-pasture, as was traditionally the case.

At the beginning of the 19th century, there were coppices and tree forests side by side; the coppices regenerated vegetatively, while the tree forests regenerated generatively. For the coppices, the traditional rule of thumb was still used, that a regenerating lot could be opened to grazing when the shoots had grown above the reach of the livestock. In lots regenerated from seed, there was to be no grazing at all. The reason for this was that shoots growing from seed are not safe, even when they are higher up than where the livestock can reach them. Shoots growing from seed are still so thin that the animals can easily knock them over to get to the tips of the shoots and bite them off. Thus from the middle of the 19th century, there was an increasing insistence in forestry circles that grazing livestock in forests made regeneration impossible and should there-

---

[131] Article 4 of the 'General-Verordnung' of the Spessart, dating from 1774, states: 'Wäre sogleich nach völlig ausgehauen und gesäuberten Schlägen auf die Wiederaufbringung eines ordentlichen Anflugs ein Hauptaugenmerck zu richten, die genaueste Heege anzulegen, alles Eintreiben des Viehes, Grasen Mähen, und überhaupt was einem Schlage nur schädlich fallen könne, schärfstens zu untersagen' (Vanselow, 1926, p. 22 *et seq.* and Appendix 4).

fore be stopped altogether.[132] Grazing was then seen as the greatest enemy of forests (Landolt, 1866, p. 152).[133]

The strong rejection by foresters of allowing livestock to graze in the forest can be explained in two ways. The first is the difference in growth of seedlings compared with the shoots on a stool of a particular species of tree (see Cotta, 1865, pp. 84–85). The second is that the seedlings are no longer protected in the forest by thorny scrub, as was the case in the past.

A shoot on a stool grows to a much greater height and thickness in the first years of growth than the stem of a seedling. Seedlings of pedunculate and sessile oak reach a height in the first year of respectively 20 cm and 16 cm, while a shoot on the stool of an oak grows at least 2 m, i.e. ten times more and reaches a thickness of 2.5 cm. As Fig. 4.15 shows, a seedling grows to this height only after 6 or 7 years (Turbang, 1954; Trier, 1963, p. 179; Watkins, 1990, p. 89; Rackham, 1993, p. 65). Therefore it takes much longer for seedlings to grow

[132] With regard to the length of time that the coppice had to be enclosed, Cotta used the general rule, as he wrote himself: 'das Holz muß dem Maule des Viehes entwachsen sein' (Cotta, 1865, p. 84).

On the other hand, for regeneration from seed, he considered that the rules in forestry science should the determining factor, rather than traditional practice. On this subject, he wrote: 'Die Viehutungen bestehen in den meisten Waldungen gesetzlich oder vertragsmäßig, Zeit und Art der Schonungen sind also gewöhnlich schon dadurch bestimmt; hier ist aber nicht die Rede von dem, was Gesetze und Verträge bestimmen, sondern von der Schonungszeit, welche durch die Grundsätze der Forstwissenschaft geboten wird' (Cotta, 1865, p. 84). He appended this with

die allgemeine Regel: 'das Holz muß dem Maule des Viehes entwachsen sein' ... ist unzulänglich. Versteht man darunter eine Höhe, die größer ist, als daß das Vieh mit dem Kopfe dahinlangen kann, so ist das zu wenig; den das Vieh überreitet viel größeres Holz und beschädigt die Spitze daran. Versteht man aber eine Größ, wo dieses nicht mehr möglich ist, so müßte die Schonungszeit so hoch gefeßt werden, daß bis dahin überhaupt keine Weide wegen Mangel an Gras mehr geübt werden kann, weil das größere Vieh ziemliche Stangen überreitet (Cotta, 1865, p. 85).

In high woods (for which Cotta used the term 'Hochwaldungen') of beech, silver fir, hornbeam and oak, Cotta considered that areas where trees have been seeded should not be opened to livestock in less than an average of 20 years, and in woods of elm, ash and sycamore, not after less than 15 years. He did not think that it would be possible to achieve such a long period of enclosure, because the rights of the commoners to meet their own grazing needs covered a much shorter period (Cotta, 1865, pp. 84–85). Also see Landolt (1866, pp. 49–52, 152–155, 199, 314, 390, 431–435), Gayer (1886, pp. 12–13), Bühler (1922, p. 611), Vanselow (1926, p. 145), Grossmann (1927, pp. 30–31, 33, 35, 38–40, 108), Meyer (1941, pp. 102–103), Schubart (1966, p. 96), Mantel (1990, pp. 91, 182).

[133] The Swiss forester Landolt (1866) thought that grazing rights made any improvements in forestry impossible (Landolt, 1866, pp. 49–52). Landolt went further than Cotta (1865) by considering all domestic animals as being harmful to the forest (Landolt, 1866, pp. 152–155, 431–435). Therefore he argued for the abolition of grazing rights (Landolt, 1866, pp. 49–52). He thought that it was possible to replace the grazing rights of livestock by separating areas of land suitable for making pasture from the rest of the 'Wald' and lay these to grass (Landolt, 1866, p. 51). Thus he suggested separating the functions and dividing them between different areas of land.

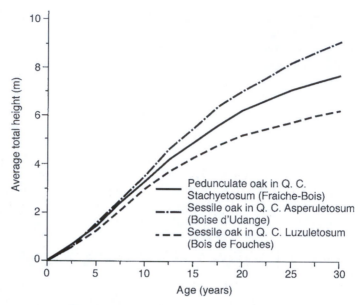

Fig. 4.15. Height of pedunculate and sessile oak seedings in different types of forest in France (see Turbang, 1954, p. 104).

tall enough to be out of reach of the livestock than the shoots on a stool. An English manual about cultivating oaks, dating from 1609, said that a plot of land sown with acorns should be closed to grazing livestock for at least 20 years so that the seedlings could grow without any risk from the livestock, while coppices had to be fenced off for only 7–9 years (Flower, 1977, p. 28).[134] Figure 4.15 shows that oak seedlings are between 5 and 7 m tall after 20 years. Figures 4.15 and 4.16 show the difference in growth between seedlings and shoots on a stool. As Fig. 4.16 shows, the growth of a shoot on a stool is much greater than for a seedling for the first few years. Because it grows much taller (and thicker), a shoot on a stool grows at a much faster rate than a seedling. This explains why Cotta (1865) stated that for regeneration from seed, the traditional rule that shoots had to grow tall enough to be out of reach of the livestock was no longer appropriate.[135] In that period, the seedlings could not grow tall

[134] The experience that a much longer period of closing land to grazing livestock was needed for oaks to grow from seed led in the second half of the 17th century to the promulgation of a series of laws in England for the individual forests, the so-called Enclosure Acts, which determined that parts of the forests should be permanently enclosed from grazing livestock to allow oaks to grow from seed there (Hart, 1966, p. 169; Darby, 1970; Rackham, 1980, p. 185; Putman, 1986, p. 18; Tubbs, 1988, pp. 71, 76; Perlin, 1991, p. 217). It was not possible to carry out these laws in the end, because of the protests of the commoners who destroyed their fences and sent the livestock into the seeded parts of the forests (Hart, 1966, pp. 125, 145, 199, 209; Tubbs, 1988, pp. 74–75).

[135] On this subject, Cotta wrote: 'Die Servitute. In einem Walde, dessen Schläge die Hutungsberechtigen im sechsten Jahre des Holzalters mit dem Viehe behüten dürfen, ist in vielen Fällen der Niederwald räthlicher als der Hochwald, es sei denn, daß man die Schläge mit großen Stämmen bepflanzen wollte' (Cotta, 1865, p. 113, paragraph 111).

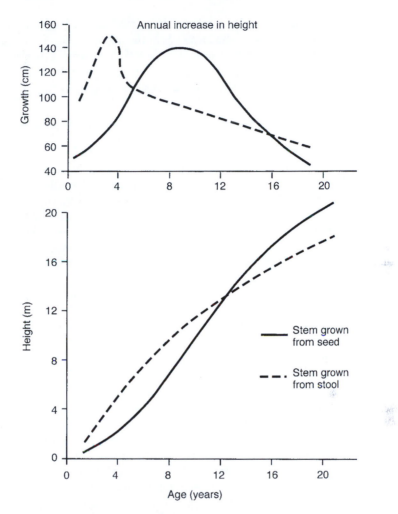

Fig. 4.16. Growth curves of grey alder (*Alnus glutinosa*) in a good location for growth. It is clear that the spring from a stool, i.e. vegetative regeneration, grows much more rapidly in the first few years than growth from seed, i.e. generative regeneration (see Mayer, 1992, p. 198).

enough to be out of the reach of livestock, while they were not protected by thorny scrub. This also explains why coppices were closed to livestock for only 3–5 years under this traditional rule in the 16th and preceding centuries (see Streitz, 1967, p. 39; Mantel, 1980, p. 135; Hausrath, 1982, p. 207; Buis, 1985, p. 50; 1993, p. 108).

In a relatively short time the shoots on a stool grew tall and thick enough to no longer be at risk from livestock. Therefore the damage to seedlings by grazing livestock only really applied in Central and Western Europe after the tree forest had developed from coppices as a way of producing wood. According to written sources, the damage to seedlings by livestock *in* a regenerating tree forest only became a problem in the lowlands in the 18th century for the generative regeneration of the forest after this method of production had been generally introduced. Virtually all the preceding regulations about grazing livestock in relation to the regeneration of the forest relate only to vegetative regeneration in coppices.

4.14 The Development of 'Natural' Regeneration

In the 19th century, the above-mentioned regeneration technique for the tree forest led to the so-called shelterwood system (see Appendix 5). Described briefly, this technique meant that increasingly large gaps were made in the forest canopy by means of thinning out trees at intervals from several years to a decade. As the canopy becomes thinner, the seedlings of the standing trees grow taller. The term 'shelterwood' system is based on the fact that after every felling, the remaining trees are so spread out that they form a screen that shelters the young trees. Finally, after the last felling or clearing, there is only an open area left with a new generation of growing trees.

This technique was eminently suitable for the regeneration of beech in a beech forest. However, it was not successful for regenerating oak in an oak forest, or in a forest consisting of oak and beech. It was only following a modification in the sense that the canopy of the oak forest was thinned out much more than was usual for the regeneration of beech, that the oak was also successfully regenerated with this technique.[136] This still always required a great deal of human intervention, such as removing other sorts of trees, such as lime, hornbeam, elm and beech, which would otherwise win out in the competition with the oak.[137] Confusingly, this technique, like the sprouting of stools in coppices, is described as 'natural' regeneration.[138] Since the 19th century, coppicing is no longer or barely used as a form of exploitation in forestry, and 'natural' regeneration is now defined as: *regeneration with seedlings which grow from seed dispersed by the trees forming the canopy* (Bühler, 1922, p. 257; Dengler, 1990, p. 47). Measures such as working the soil or destroying unwanted plants, shrubs and trees are part of this 'natural' regeneration. Cotta (1865) made a distinction between the 'artificial' and 'natural' regeneration achieved by human intervention, on the one hand, and the regeneration which takes place in the wilderness without any human intervention, on the other

---

[136] See Bühler (1922, pp. 310, 312, 331), Vanselow (1926, pp. 63, 87–88), Hausrath (1982, p. 76), Krahl-Urban (1959, p. 146 and Appendix 5).
[137] See Bühler (1922, pp. 218, 295 e.v.), Vanselow (1926, p. 27), Tangermann (1932), Hess (1937), Turbang (1954), Hesmer (1958, p. 261), Krahl-Urban (1959, p. 214), Nüßlein (1978), Klepac (1981), Evans (1982), Tendron (1983, pp. 57–63), Raus (1986), Dengler (1990, p. 274), Lüpke and Hauskeller-Bullerjahn (1999).
[138] See Cotta (1865, p. 2), Landolt (1866, p. 197), Gayer (1866, pp. 32, 43, 68), Bühler (1922, p. 257), Vanselow (1949, p. 17), Dengler (1990, p. 47).

hand.[139] He called the latter 'Holzwildwuchse'. After Cotta (1865), this distinction between 'natural' regeneration and the regeneration 'which occurs naturally in the wilderness' was no longer made. Both types of regeneration were described as 'natural regeneration'.[140] This is extremely confusing, because it suggests that 'natural regeneration' is a process which takes place in untouched nature.[141] In fact, this is not the case.

The reason that 'natural' regeneration of the oak using the technique of the shelterwood system initially failed, is the greater amount of light required by the

[139] On this subject, Cotta wrote:

Bei dem Waldbau ist es nicht nothwendig, wie bei dem Feldbau, daß man allezeit vorher säen oder pflanzen muß, um zu ernten; sondern es läßt sich die Ernte auch so betrieben, daß der Wiederwuchs des Holzes eine natürliche Folge davon wird, indem man durch richtige Bewirtschaftung die an vorhandenem Holze in Thätigkeit schon begriffen Naturkräfte nach seinen Zwecken so leitet und durch Hinwegräumung der Hindernisse so unterstützt, daß der Wiederwuchs durch die freie Wirkung der Natur erfolgt. Diese Art der Holzerziehung nannte man früher die natürliche Holzzucht. Hier stellte man die kunstliche zur Seite und verstand darunter den Holzanbau durch Ausstreuung des Samens von Menschenhänden und durch Pflanzung sowohl mit Wurzeln als ohne Wurzeln (durch Stecklinge) und durch Ableger. Die naturliche und die kunstliche Holzzucht standen sonach den Holzwildwuchse gegenüber wo Holz ohne alles menschliches Zuthun wächst, mithin auch solches, das unseren Zwecken oder unserem Nutzen nicht entspricht (Cotta, 1865, p. 2).

In this respect, it is striking that Cotta stated that the way in which trees in the wilderness regenerated and grew did not comply with the requirements regarding timber. However, 'natural' regeneration does. Therefore it is a technique which developed purely on a pragmatic basis.

[140] In his book, 'Theorie und Praxis der natürlichen Verjüngung im Wirtschaftswald', Vanselow (1949) wrote on the regeneration of forests: 'Ist diese Lebensgemeinschaft naturgemäß und befindet sich in ungestörter Entwicklung, im biologischen Gleichgewicht, so ist die natürliche Verjüngung eines solches Waldes eine Selbstverständlichkeit, genau so, wie der Wald sich 'Jahrtausende lang als Urwald erhalten und verjüngt hat' (Vanselow, 1949, p. 17). Also see in Landolt (1866, p. 197), Gayer (1886, pp. 32, 43, 45, 68), Forbes (1902), Morosow (1928, p. 71), Vanselow (1949, p. 17), Dengler (1990, p. 47).

[141] Forbes (1902) described natural regeneration as: 'Simply the germination of the seeds which fall from mature trees, and their development into saplings and trees' (Forbes, 1902, p. 239). Later in his article, he wrote: 'Where natural generation is entirely left to Nature, as in the case of large forests or waste land ungrazed by domestic animals' (Forbes, 1902, p. 243). On the same page, he also wrote: 'South America has many large tracts covered by virgin coniferae forests holding undisputed possession of the soil. All these, and many others existing in various parts of the world, are entirely the result of natural regeneration' (Forbes, 1902, p. 243). Later still, in the same publication, he wrote: 'and that millions of seedlings should be allowed to perish, year after year, in our woods, for want of a little assistance. To leave everything to natural regeneration would undoubtedly be a mistake' (Forbes, 1902, p. 245). In the first quotation, Forbes described natural regeneration as it is used in the strict meaning of forestry in the sense of Bühler (1922, p. 257) and Dengler (1990, p. 47). In the second quotation, he clearly makes a distinction between natural regeneration in nature, when he also indicates that this situation arises by keeping out livestock, and natural regeneration which does not take place in nature. In the third quotation, natural regeneration refers to regeneration in virgin nature, while in the fourth quotation, natural regeneration means spontaneous regeneration, i.e. regeneration without human intervention. All in all, in this article, the term 'natural regeneration' has three different meanings which are all used at different times, which suggests that natural regeneration in the sense used in forestry, i.e. as used by Bühler (1922, p. 257) and Dengler (1990, p. 47), is the equivalent of regeneration in nature, which is actually not the case. Harmer (1994a) also noted that the use of the term 'natural regeneration' does not really means what it suggests.

oak, compared, for example, with beech (Vanselow, 1926, pp. 63, 87–88; Krahl-Urban, 1959, p. 146). Under a canopy of thriving beech trees, young oaks have too little light to regenerate. A detailed description of this technique of 'natural' regeneration and all the problems which this entailed for the regeneration of oak, is given in Appendix 5.

The light required by oak is also the reason that it cannot be regenerated by means of the so-called selection system in an oak forest or in a forest where oak and beech both grow, in contrast with beech, which is more tolerant of shade (Bühler, 1922, p. 566; Boden, 1931; Seeger, 1938; Dengler, 1990, p. 294). This selection system is another form of 'natural' regeneration of forest which was developed in the 19th century. It consisted of felling one or more trees here and there, creating gaps in the canopy where young trees would then spontaneously grow from the seed of species surrounding the gap. This technique is used for the regeneration of forests consisting of species which are tolerant of shade, such as beech, Norway spruce (*Picea abies*) and silver fir (*Abies alba*). Appendix 6 gives a more detailed explanation of this technique.

A combination of the shelterwood system and the selection system resulted in the group selection system. This technique is also suitable for the regeneration of beech forests without any problems, though it is not suitable for oak forests or forests of oak and, for example, beech. With this technique, the oaks also need a great deal of extra help, because the seedlings cannot survive in the presence of species which tolerate shade better, such as beech. Appendix 7 contains a detailed explanation of this technique.

The development of 'natural' regeneration shows that the oak needs a great deal of human intervention to be able to regenerate in forests. Without this help, it is impossible for the oak. Paradoxically enough, natural regeneration creates problems for the oak in forests without clear felling (Von Lüpke and Hauskeller-Bullerjahn, 1999). The development of the tree forest from coppices via pole-forests also showed that the hazel cannot survive in a tree forest because of the shade cast by the closed canopy (Schubart, 1966, p. 168).[142] The hazel, which had been prolific for centuries in the shrub level of coppices with standards, or as a coppice, disappeared, while according to the prevailing theory, this species formed a shrub level in the original natural forest in Central and Western Europe.

4.15 Grazing Livestock and the Destruction of the Forest

On the basis of the prevailing theory, 20th-century authors believe that the original vegetation of the lowlands of Central and Western Europe was a closed forest. For this reason they submit that all the measures taken from the 13th

---

[142] In 1759, it was said, regarding the forest of the city of Göttingen (Germany): 'Häselnholtz findet sich auf dem Walde genug, das Haseholtz vergeht aber von selbst, wenn der Hey 16–18 Jahre alt wird und Büche und anders hartes Holtz die Haselbüsche überwächst, wodurch denn kommt, daß z. Zt., wenn ein Strich Holtz haubar ist, gar kein Haselholtz mehr darauf zu befinden ist' (Schubart, 1966, p. 168).

century onwards to regulate the grazing of livestock are in line with those issued in the 19th century to protect seedlings *in* a tree forest. They were *all* aimed at protecting seedlings *in* the forest.[143] In this way, they extrapolate the generative regeneration of closed forests by seedlings, back to the Middle Ages. This also applies to the tree forest itself, with the conclusion that the remaining virgin areas of Europe, which were proclaimed as 'forestes', and where regulations on the grazing of livestock were passed over the years, were originally closed forests. For example, Bühler (1922) stated: 'The destruction of the forests was prohibited in many places in the Middle Ages. As livestock grazing destroyed the forest, grazing in forests was regulated' (Bühler, 1922, p. 300). An additional argument which supports this hypothesis is the measures taken for the protection of fruitful trees, which go back to the early Middle Ages. They are actually seen as a measure for regeneration in the form of the shelterwood system. According to this view, the measures relating to the grazing of livestock were additional measures, and served to protect seedlings in the closed forest.[144]

As we established in this chapter, the provisions in the lowlands of Central and Western Europe dating from before 1700, on which the theories are based, *all* relate to protecting recently coppiced areas, as well as the protection of nurseries for young trees, the so-called 'kampen'. It is only the provisions on the grazing of livestock dating from *after* 1700 that relate to the regeneration of trees by means of seedlings *in* a forest.

The arbitrary cutting of wood, which was prohibited by regulations in the course of the 15th century, is also considered as a form of exploitation analogous to the selection system, which therefore automatically resulted in 'spring', or the regeneration of trees (see Vanselow, 1926, p. 58). As we noted earlier, the 'spring' referred to the growth of the shoots from the stools, and not to the growth of seedlings. Cutting wood here and there entailed cutting the shoots on stools, spread throughout the 'Wald'. In my opinion, the cutting of wood at random, as referred to in the earliest regulations on the grazing of livestock, concerned the haphazard cutting of shoots of stools in the wilderness. The sprouting stools spread throughout the 'forestis' or 'Wald' were an easy target for the livestock, which initially wandered freely around the 'Wald'. Therefore the spring could not be effectively protected against livestock, which ate the young shoots. As we saw in the earliest regulations on the grazing of livestock, the effective protection of sprouting stools against livestock was actually one of the most important reasons for regulating the cutting of shoots on stools, in the

[143] See Bühler (1922, pp. 258, 300, 301, 339), Vanselow (1926, pp. 7, 26, 59, 64; 1949, pp. 17, 79), Grossmann (1927, pp. 108, 116), Meyer (1931, pp. 345, 386, 442), Hausrath (1982, pp. 33, 206–209), Hesmer (1958, pp. 90, 454), Streitz (1967, pp. 53–54, 68–69), Hesmer and Schroeder (1963, pp. 151–153), Mantel (1980, pp. 91–92, 116), Buis (1985, pp. 40, 51, 105, 113, 179, 273).

[144] See: Bühler (1922, pp. 301, 309), Vanselow (1926, pp. 26, 59, 79), Meyer (1931, pp. 345, 386, 442), Hess (1937), Hausrath (1982, pp. 33, 206–209), Hesmer (1958, pp. 90, 260, 454), Rubner (1960, p. 38), Hesmer and Schroeder (1963, pp. 152–153, 173), Schubart (1966, p. 15), Streitz (1967, pp. 53–54, 68–69, 155), Buis (1985, pp. 40, 51, 105, 113, 179), Dengler (1990, p. 291).

form of parcelling coppices and periodically fencing them off from livestock with fences and earthen walls. This was the only way in which the young shoots could be effectively protected against livestock. As we saw above, this resulted in the 'modern' coppice.

As we noted earlier, the grazing of livestock was seen in the 19th and 20th centuries as the factor which destroyed the forests. This is understandable, when one considers the stories which prevailed in those days about the grazing of livestock (Mantel, 1990, p. 425). The increase in the population and an increase in the demand for firewood and pasture resulted in an enormous pressure on coppices and wood-pasture. According to the written sources, brushwood, grasslands, heathland, gorse, juniper, thorny scrub and shifting sands had replaced the original forests throughout the lowlands of Western and Central Europe.[145] Fig. 4.17 show this destruction. The fact that the grazing of livestock was considered to be responsible for the destruction is not surprising when one examines the density of livestock and the biomass recorded in the literature. For example, in 1784 in Prussia, there were 19 horses, 53 head of cattle and 215 sheep on 100 ha of forest (319 kg ha^{-1}) (Mantel, 1990, p. 425).[146] A document dating from 1664 states that the commoners in the Forêt de Fontainebleau from 2154 households in 17 communities had a total of 10,381 cows and 6367 pigs in the 'Forêt' with an area of approximately 14,000 ha (259 kg ha^{-1}) (Tendron, 1983, p. 23). In addition, the commoners also had the

---

[145] One of these reports dates from 1839, and concerns the County Mark in Germany. The report comes from Pfeil, and reads:

Durch ungeregelte Benutzung, ganz vorzüglich aber durch Mangel an Schonung, da die Markegenossen überall gemeinschaftlich hüteten und keine Einschonungen duldeten, waren schon in der Zeit, als Friedrich d. Gr. die Regierung antrat, die Gegend von Holz entblöbt und die kahlen Berge brachten wenig mehr als einige verkrüppelte Sträucher, Heidekraut und Ginster. Ein starkes Streurechen und Plaggenhauen, um den sehr vernächlässigten Ackerbau zu erhalten, brachte den noch bestandenen Boden sehr herunter. Nicht weniger waren auch die Communalforsten verwüstet, da auch hier dieselben Uebel stattfanden und besonders nordwärts der Ruhr, wo die weideberechtigungen größtentheils nicht dem Grundbesitzer, sondern Fremden zugehörten, gar keine Schonung statt fand (Hesmer, 1958, pp. 93–94).

Statistics dating from 1843 reveal that at that time there were 11,000 ha of shifting sands on the Veluwe in the Netherlands (Van der Woud, 1987, p. 214). Also see Hobe (1805, pp. 14, 28–29, 74–112, 124), Landolt (1886, pp. 152, 431–435), Gayer (1886, p. 13), Bühler (1922, pp. 264, 300, 610), Vanselow (1926, p. 23), Grossmann (1927, p. 83), Hausrath (1982, pp. 208–209, 257–259, 291), Rodenwaldt (1951), Reed (1954, pp. 81, 92, 94), Hesmer (1958, pp. 56, 57, 77, 91 etc.), Hesmer and Schroeder (1963, pp. 102, 130–133), Streitz (1967, pp. 38, 75, 156), Holmes (1975), Tendron (1983, pp. 9, 26), Buis (1985, pp. 63–64, 173, 286, 369 e.v., 521), Mantel (1990, pp. 424–425, 439).

[146] The numbers of animals are converted into kilograms of biomass/hectare on the basis of the following weights: one sheep 40 kg; one cow 350 kg; one horse 250 kg; one pig 70 kg (S.E. van Wieren, Wageningen, 1997, personal communication). The pigs have been left out of consideration in the comparison, as they were put out to pannage in the woods for only a few weeks to 4 months. Initially the other livestock grazed there throughout the year (Hesmer and Schroeder, 1963, p. 104; Ten Cate, 1972, pp. 130, 206).

Fig. 4.17. Desertification in Friesland, the Netherlands, as a result of tree felling, grazing and cutting turves. Nowadays, these sandy areas, which developed as a result of the non-sustainable exploitation of nature, are described and preserved as natural areas (photograph F.W.M. Vera).

right to collect acorns, beech nuts and other fruits, and to take litter. When these rights came to an end in the Bramwald in Germany in 1870, there were 1700 head of cattle, 3880 pigs and 17,500 sheep on an area of 1800 ha (719 kg ha^{-1}) (Krahl-Urban, 1959, p. 21). In one particular part of Hessen in Germany, with an area of 2409 ha, there were 15,100 sheep in the 19th century (250 kg ha^{-1}) (Gothe, 1949), while in one area in the west of Switzerland, 135 cows and 155 horses grazed on 250 ha (344 kg ha^{-1}) (Meyer, 1941, p. 125). Moreover, grazing occurred even when there was no right to graze, i.e. there was illegal grazing. Therefore the actual densities were higher than those suggested by the official figures (Hesmer, 1958, p. 391; Peters, 1992). But even the figures given here are extremely high for a 'forest' situation.

As a result of developments in agriculture, such as the cultivation of clover, the breeding and propagation of certain species of grass, and the large-scale introduction of the potato, the grazing of livestock in uncultivated land came to an end in the 18th and 19th centuries. This would not have happened without this breakthrough in agriculture, and the pressure of all those who felt responsible for supplying wood, because there was no time limit on the right to graze livestock. As a result of the changes in agriculture, a larger number of livestock could be kept on a smaller area.[147] The cultivation of animal fodder in the form of grassland and potatoes meant that more livestock could be fed indoors than in the past. More people could be fed on the potatoes grown in a particular area than on the pigs which were fed on acorns in that area. Moreover, the pigs could also be fed with potatoes. The returns were further increased with the introduction of artificial fertilizers. An alternative to the wood-pasture developed as a result of all these developments, so that it was possible to terminate grazing rights in the wood-pastures, coppices and high forests in favour of timber production.[148]

This new method of intensive agriculture was propagated in the 18th century by the physiocrats, who believed that land was the primary production factor, agriculture the only form of production, and farmers the only productive population group (Buis, 1985, pp. 520–521, 590; Van der Woud, 1987, p. 536). They endeavoured to remove the restrictions which had been traditionally imposed on agriculture by lords (the government) on the basis of the 'ius forestis', such as the prohibition on felling trees. Their motto was 'laissez faire et laissez passer'. They argued for the abolition of the commons and the distribution of the common land among the commoners. This would benefit the agricultural productivity, because every individual farmer would make the greatest possible effort to get as much as he could from the land. After all, he would reap the benefits of his own efforts himself, rather than benefiting the other commoners with rights, which applies in the case of common ownership.

The abolition of the commons and the distribution of the common land took place during the course of the 18th and 19th centuries (Hobe, 1805, p. 113; Grossmann, 1927, p. 29; Buis, 1985, pp. 389, 520–521, 590; Mantel, 1990, pp. 179, 182). This also led to a division and separation between pasture and the cultivation of wood, which still applies today. Certain areas of land were designated as pasture; others for the production of wood. Foresters had been

---

[147] Hobe (1805) outlines the advantages of the cultivation of clover for livestock in great detail. With regard to those who still grazed livestock in the old-fashioned way, he remarked:

Bey allen Vortheilen, welche ich aus der Landwirtschaft zog, und wenn ich auf dem Gute geblieben wäre, noch mehr hätte ziehen können, jemehr durch ordentliche Eintheilung und herumdüngen die Ländereyen in Stand gekommen wären, stutzen doch nur meine Nachbarn, und ausser einigen Wenigen blieben sie den ihrer alten schlechten Wirtschaft, kehrten sich an keinen Kleebau, trieben ihr Vieh in's Holz, auf magere ausgehungerte Felder. Denn wenn das Land 5 Jahr genutzt worden, dann bleibt es wieder liegen, und treibt oft nur mageres Gras, die sogenannte Hudeblume (Hobe, 1805, p. 134).

[148] See: Hobe (1805, pp. 87, 106–107, 118–119, 126–130, 134–137), Hermann (1915), Grossmann (1927, p. 33), Meyer (1931, pp. 349, 439), Rodenwaldt (1951), Schubart (1966, pp. 69, 100), Streitz (1967, pp. 55, 69 e.v.), Musall (1969, pp. 176–181), Buis (1985, p. 775; 1993, pp. 151, 181), Slicher van Bath (1987, p. 31), Mantel (1990, pp. 68, 90, 182, 433), Bieleman (1992, p. 130).

Fig. 4.18. Removing the 'mast trees' from the 'acker', after the oaks and other 'fruitful' trees were no longer important for producing mast, as a result of the introduction of the potato. Later, potatoes and root crops were grown on the 'acker' (illustration from Andrews, 1853, taken from Darby, 1970, p. 201).

insisting on this division for some time.[149] In this way, wood-pasture and coppices were on the one hand, turned into grassland and open fields for the cultivation of root crops and cereals, and on the other hand, into closed forests. Where pigs had previously fed on the mast of trees in the 'acker', the mast trees, the oaks, were removed (see Fig. 4.18), and then potatoes and root crops were cultivated for pigs on the 'acker'. In this way, the 'acker' changed from an area of uncultivated wilderness with 'fruitful' trees, into the present open fields of root crops.[150] The 'forestis', 'forest', 'forêt', 'Forts', 'Wald' or 'woud' changed into the uncultivated area that was only intended for timber production (see Fig. 4.19).

---

[149] See Grossmann (1927, pp. 31, 33, 35, 40, 101, 108), Vanselow (1926, p. 145), Meyer (1941, p. 103), Hesmer (1958, pp. 85, 135, 159–160, 391), Schubart (1966, p. 96), Buis (1985, pp. 391–392, 404, 409, 413, 418, 603), Mantel (1990, pp. 65, 91).

[150] On the arrival of the potato, Meyer exclaimed (1931): 'tatsächlich ist der Eichwald der Erdäpfeln gewichen! Der einheimische Baum einem exotischen Kraut!' (Meyer, 1931, p. 439). Also see Hobe (1805, pp. 100–107), Vanselow, (1926, p. 40, 109, 148), Meyer (1931, pp. 362, 439), Hesmer (1958, p. 327), Schubart (1966, pp. 100), Streitz (1967, pp. 30, 48), Buis (1985, pp. 391, 404, 604; 1993, p. 150), Dengler (1990, p. 299), Mantel (1990, pp. 91, 368, 378, 434–435 etc.), Jahn (1991, p. 386).

Fig. 4.19. An oak that has grown in a wood-pasture, surrounded by beeches in a closed forest near Höxter, Germany. The forest arose after grazing of the wood-pasture by specialized grazers like horses and cattle ended. The oak has the characteristic shape of open-grown trees, i.e. its crown starts low on the trunk. The beeches are typical forest-grown trees, i.e. they have a tall, straight trunk without branches and long, narrow crowns (photograph F.W.M. Vera).

4.16 'Having but Little Wood, and that Oke like Stands left in our Pastures in England'

In 1605, the Englishman Rosier described an area along the St George's River in the current state of New York in the United States as

> good ground, pleasant and fertile, fit for pasture, for the space of some three miles, having but little wood, and that Oke like stands left in our pastures in England ... And surely it did all resemble a stately Park, wherein, appear some old trees with high withered tops, and other flourishing with living greene boughs ... the wood in most places very thinne, chiefly oke and some small young birch.

In 1607, Captain Gilbert described the trees in the coastal area of Main in roughly the same landscape context 'the most p^t of them ocke and wallnutt growinge o greatt space assoonder on from the other as our parks in Ingland and no thickett growing under them' (Day, 1953, pp. 335–336). In 1634, the landscape of New England was described by William Wood as

> And whereas it is generally conceived that the woods grow so thicke, that there is
> no more cleare ground than is hewed out of labour of man, it is nothing so, in
> many places divers acres being cleare, so that one may ride a hunting in most
> places of the land, if he will venture himselfe for being lost; there is no underwood
> saving in swamps and low grounds.

In 1654, Edward Johnson wrote that the forests were relatively 'thin of Timber in many places, like our Parks in England' (Bromley, 1935, p. 64; Whitney and Davis, 1986, p. 74). Descriptions from the 17th century of trees from Nieuw Nederland (the current New Jersey) include stately oaks with broad grown-out crowns, so-called 'pasture' oaks (Russell, 1983; Whitney and Davis, 1986). Such trees must have grown up in an open landscape (Rackham, 1998). This tree shape will have appeared familiar to the English colonists from the wood-pastures they knew in England. In addition, trees were described with a straight branchless trunk of up to 20 m (Van der Donck, 1655, p. 15). These trees must have grown up much closer to each other. Both growth shapes are also known from white oaks (*Quercus alba*) elsewhere in the east of the United States from the pre-colonial time (see Gordon, 1969; Whitney and Davis, 1986; Ruffner and Abrams, 1998). Both growth shapes can occur in park-like landscapes, as discussed in section 4.9.

In the east of the United States, trees were also described as so-called witness or corner trees (Gordon, 1969; Whitney and Davis, 1986; Abrams and Downs, 1990; Nowacki and Abrams, 1992; Mikan *et al.*, 1994). Witness or corner trees were trees that served as the corner points of pieces of land given to the colonists. The species and appearance of these trees were described. Often they were remarkable trees, because they served as a point of recognition in the field. For example, a boundary marker described a piece of land in Concord in Massachusetts in 1700 as 'Beginning at a great white oake marked by ye upland on ye easterly corner of sd Hardys meadow land the line runs partly westerly to another white oake marked by ye meadow land and then starting on a line westerly to a markt maple and then to a great white pine in ye swampe that is markt' (Whitney and Davis, 1986, p. 74). The white oak is named particularly often as a witness tree in the description of the so-called presettlement forests (see Gordon, 1969; Abrams and Downs, 1990; Cho and Boerner, 1991; Abrams and Ruffner, 1995; Abrams and McCay, 1996). There were some with short trunks and wide spread-out crowns, as well as some with long trunks without branches. In certain parts of the east of the United States, black oak (*Q. velutina*) and chestnut oak (*Q. prinus*) form a large share of the witness trees (Gordon, 1969; Mikan *et al.*, 1994).

These and other historical descriptions of the east of the United States all indicate the presence of open, park-like landscapes with open forests and areas where forests alternated regularly with grasslands.[151] Until very recently, there

---

[151] Friederici (1930), Bromley (1935), Day (1953), Gordon (1969), Thompson and Smith (1970), Russell (1981; 1983), Pyne (1982), Cronon (1983), Whitney and Davis (1986), Covington and Moore (1994).

were still open growth or 'pasture' oaks in what are now closed forests (Marks, 1942; Cottam, 1949; Gordon, 1969; Ehrenfeld, 1980; Leitner and Jackson, 1981). The growth shape of the trees is not the result of forest with a closed canopy becoming more open, for example due to grazing by European livestock introduced by the colonists. In that case, the trees would have to have long trunks and a small crown, as a result of competition with other trees in the forest (Rackham, 1998). Trees develop an open grown appearance if they grow in openness directly after they are established, such as in park-like or savanna-like landscapes. This and other data mentioned indicate that at the time of the colonization of the east of the United States, there were open conditions in places where a forest with a closed canopy is now present or assumed as climax. Nothing is known on the regeneration of those oaks, because there are no written sources from the time before the arrival of the European colonists.

4.17 European Mental Models in North America

Open, park-like landscapes in the east of the United States are sometimes disregarded as misunderstood by the colonists or as propaganda, because such descriptions would have served to attract the colonists from Europe to America. They should therefore be viewed sceptically (Raup, 1937; Russel, 1981). The first colonists would not have had the idea that these landscapes were mainly caused by fire by Native Americans (Raup, 1937). As Cronon (1983, p. 22) points out, the Europeans would have viewed the landscape they found in terms of their own cultural concepts, their mental models. It has been indicated in previous sections that the park-like landscape was one of these concepts. This is shown by comparisons made by English colonists for example with English pasture oaks and the [deer] parks, among others. The land of origin will have played a role in other ways as well. From the cultural concept of the Dutch colonists, the large amount of wood that could be used as firewood was amazing (see Van der Donck, 1655, p. 15). In the 17th century already, there was almost no firewood left in the Netherlands. There, peat was used as fuel. The amount of peat burned around 1650 in the Netherlands was equivalent to approximately 800,000 ha of coppice wood (Buis, 1985, pp. 488–489). The oaks in the east of North America that were suitable for shipbuilding were also noticed. As mentioned in section 4.10, the Dutch collected timber for shipbuilding from Germany and the Baltic states at that time, because there were almost no suitable oaks left in the Netherlands.

What also fitted in the European mental model of the 16th and 17th century was the growth of young trees in thorny bushes or other bushes unsuitable for livestock in the grasslands grazed by livestock. That Rosier went through 'lowe Thicks', consisting of young bushes and young trees (saplings) in 1605, should be seen in this context (Cronon, 1983, p. 27). The use of the term 'underwood' by Morton (1632) for example in his description of burning by Native Americans, should also be seen in this context. Morton wrote 'The

reason that mooves them to doe so, is because it would other wise be so over-
growne with underweedes, that it would be all coppice wood, and the people
would not be able any wise to passe through the country out of a beaten path'
(Day, 1953, p. 335). In Europe, scrub that was used as coppice wood grew in
grazed park-like landscapes. Coppice would then be meant as scrub growing in
the open or a mantle and fringe vegetation bordering on a grove and not a bush
level in a closed forest as is now generally assumed. The underwood in the statement by Wood (1634) on fire that 'consumes all the underwood and rubbish
which otherwise would overgrown the country, making it unpassable and spoil
their affected hunting' could have had this meaning (Cronon, 1983, p. 49). The
fact that the Native Americans burned the vegetation would have amazed the
Europeans from the point of view of their mental model, because this was at a
cost to the young trees. At that time, the young trees within the bushes were
strictly protected in feudal Europe by the 'ius forestis'. There was illegal burning of vegetation in Europe by shepherds who increased the surface of the pasture lands for livestock in this way. There was severe corporal punishment for
this, such as gouging out of both eyes (see Mantel, 1990, p. 83; Tack *et al.*,
1993, p. 213 and see footnote 91).

Given the fact that trees regenerated in grazed grassland in the east of the
United States as well (see Scot, 1915; Bromley, 1935; Marks, 1942; Stover and
Mark, 1998), the introduction of livestock by the colonists would, in principle,
not have caused any problems for the regeneration of trees, analogous to the situation in wood-pastures in Europe. It is known that in the east of the United
States, bushes and trees were removed from the pastures using scythes and grub
hoes (Bromley, 1935). Analogous to the European situation, the use of trees and
shrubs as coppice wood will not have caused any problems for regeneration of
the wood, if the species of tree had the ability to grow from stools. In the east of
the United States, this is the case for oak and other deciduous trees (Ward,
1961; McGee, 1981).[152] There are even indications that genera in common
with Europe had this ability to a late age (see McGee, 1981). For example, the
American beech has a greater ability to grow from stools than the European
beech (Ward, 1961). The rotation cycles used in the east of the United States
varied from 15–40 up to 60 years (Raup, 1964; Whitney, 1987; Mikan *et al.*,
1994; Abrams, 1996; Ruffner and Abrams, 1998) and evidently did not cause
any problems in terms of vegetative regeneration. In the scientific literature, this
coppice culture is presented as clear felling (Loftis, 1983). Like in Europe, there
were only problems when there were attempts to regenerate oaks in a forest climate based on forestry techniques for timber production developed in Europe

---

[152] For example, Dwight (1821) wrote

when a field of wood is, in the language of farmers, cut clean, i.e. when every tree is cut
down, so far as any progress is made, vigorous shoots, sprout from every stump, and having
their nourishment supplied by the roots of the former tree, grow with a thrift and rapidity
never seen in stems derived from seed. Good grounds will thus yield a growth, amply
sufficient for fuel, once in every fourteen years (Bromley, 1935, p. 64).

in the 18th and 19th centuries, such as using the shelterwood system. Before then, this was done using total felling. In the shelterwood system, oaks can only survive if other, more shade-tolerant species of tree are resisted.[153] These data indicate that, like in Europe, oaks, which require light, are not species of the closed forest as climax vegetation.

Analogous to the wood-pastures in Europe, there was no regeneration of trees and bushes in the east of the United States *in* the forest due to grazing by a specialized grazer like cattle (see Bromley, 1935; DenUyl *et al.*, 1938; Marks, 1942; DenUyl, 1945; Steinbrenner, 1951). Modern forestry techniques, whereby regeneration of trees takes place within the forest, arrived from Europe in the 18th and 19th centuries. The mental model of European foresters, that livestock grazing prevented the regeneration of trees, came with them.

In addition to the mental models there is also language, as expression of the mental model. As described earlier in this chapter, terms like 'forest' in English, 'forêt' in French, 'Forst' and 'Wald' in German and 'woud' in Dutch, did not mean the same in the 16th and 17th centuries as they do now, namely a vast closed forest. In the 15th and 16th centuries, English, French, German and Dutch colonists took the European meaning of these terms at that time with them in their minds when they went to the east of North America. Descriptions of the landscape with these terms should therefore be read in this context. This is also true, for example, for the term 'wood' in English. In the 16th and 17th centuries, Dutch and German had two terms for the English word 'wood' (see Section 4.4). There was wood as material for building houses and ships. In Dutch this was 'hout' ('Holz' in German). For example, Van der Donck wrote in 1655 'In nieuw Nederlandt zijn de Landen doorgaans houtachtigh / dat is: over al bezet met Bomen / Struycken en Strugelen / en de schaapen daer door en omtrent wayende / verliesen veel van haer Wol daer aen /' (Van der Donck, 1655, p. 34). (In Nieuw Nederland the lands are generally woody, that means: trees, bushes and low, densely growing shrubs occur everywhere and the sheep that graze around them and among them lose a lot of their wool.) On the other hand, there was a word in the meaning of bunches of trees together ('bossch' in Old Dutch and 'Bush' in German), where pigs were fed acorns for example. Van der Donck (1655, p. 33) wrote: 'by eenighe Jaren zijnder soo veel Aacker inde Bosschen / dat de Verckens daer een handt dick Speck by konnen zetten' (in certain years there are so many acorns in the forests that the pigs can make a handbreadth of bacon). In the original texts, the word 'wood' was therefore not the same as the modern 'wood'. A non-forest tree-land, such as an open landscape with standing trees spread out, also contained wood. The word 'woud', as Van der Donck used it in his description of 'het Wilde Vee in Nieuw Nederlandt' (the wild livestock of Nieuw Nederland) in the sentence 'noch ander slagh van grof vee ... Wout-ossen, die sy in 't Latijn Boves Silvestris noemden' (Van der Donck, 1655, p. 37) (another type of rough livestock ... wood oxes that they call

---

[153] Loftis (1983), Beck and Hooper (1986), Abrams (1996), Cook *et al.* (1998), Brose (1999), Brose *et al.* (1999).

Boves Silvestris in Latin) must be interpreted in the European meaning of that time for the term 'woud'. As discussed earlier in section 4.3, 'woud' ('Wald' in German) was not cultivated land, that could be both a park-like and a treeless landscape. So it did not mean that these animals lived in a forest in the modern meaning of the word at all.[154]

The statement by Cronon (1983, p. 22) on the tendency of the Europeans to see the landscape in terms of their own cultural concepts is now also true in the sense that the landscape from the past is seen from the point of view of current scientific concepts. The current scientific concept is that the original vegetation in the lowlands of Europe and the east of the United States was a closed forest in the past. This forms the context from which we view the past in the east of the United States. The open landscapes that appear in historical descriptions would be the result of human intervention, namely the burning of the vegetation by Native Americans. As indicated in Chapter 2 and in this chapter, grazing by large grazers such as cattle is seen in the concept as a process whereby regeneration of trees does not occur. This premise is partly based on forestry techniques developed in the 18th and 19th centuries in Europe, whereby regeneration of trees takes place in the forest, such as the shelterwood system.

The development of closed forests took place in the east of the United States, analogous to Europe, without the presence of large grazers. Initially, the indigenous bison was driven off and secondly the cattle introduced by the Europeans were removed, because this species did not belong in the original vegetation. However at the time of the colonization by the Europeans, there were large numbers of bison in the east of the United States (Branch, 1962, pp.52–65; McHugh, 1972, pp. 59, 270, etc.; Joke and Sawtelle, 1985, cited in Crow *et al.*, 1994; and see Fig. 2.3.). For example, around 1770 more than 10,000 bison were present in North Pennsylvania (McHugh, 1972, p. 270 *et seq.*). The influence of these animals on the vegetation could have been great, analogous to specialized grazers like cattle and horses in Europe. The fact that they left traces in the landscape is shown by the fact that the first railroads from the east in a westerly direction through the Appalachians were laid on bison trails (Branch, 1962, p.56; McHugh, 1972, p. 59). The bison disappeared soon after colonization of the east of the United States. Of the more than 10,000 bison in the north of Pennsylvania in 1770, there were none left 70 years later in 1840 (McHugh, 1972, p 270, etc.; Day, 1989). In 1825, the bison was extinct in West Virginia (Day, 1989).

The bison is also placed in the prevailing scientific concept, i.e. in what is assumed to have been the climax: forest. These eastern bison are therefore identified as 'wood buffalo'. As far as we know, these bison did not differ from the 'plain bison' (Geist, 1991; 1992), i.e. they were specialized grazers. The bison

---

[154] The original Dutch text shows that they were bison (*B. bison*), although the English translation (O'Donnell, 1968, p. 46) translates 'woudossen' as moose or elk (*Alces alces*). The text assigned to bison (*B. bison*) in this translation can be found in the original Dutch text under the heading 'eelanden' (moose or elk (*A. alces*)). See Van der Donck, 1655, p. 36; O'Donnell, 1968, p. 45.

would have extended from the prairie in an easterly direction, after the Native Americans had made the forest open using fire, or had even made it disappear (Rostlund, 1960; Branch, 1962, p. 53; McDonald, 1981, pp. 102–104; Pyne, 1982). This hypothesis is also based on the idea that the original growth was a closed forest. The question is how likely that is if, in the east of the United States, there was originally, i.e. without human intervention, not a closed forest, but a more open landscape with grasslands. In that case, the biotope of the bison as specialized grazer was originally present. The presence of the bison could in turn explain the permanent presence of open, park-like landscapes from the historical descriptions, if the effect of this grazer was analogous to that of cattle in the lowlands of Europe. In turn, that could explain the permanent presence of light-demanding oaks in prehistoric times without the presence of fire, such as is shown in pollen analyses (see Clark, 1997 and Chapters 2 and 3).

It is the case that grazing reduces the chance of fire as a result of the reduction of the amount of flammable material by grazing. In this way, grazing facilitates the establishment of woody species (Sinclair, 1979a). This has been determined in North America (Archer, 1989; Abrams, 1992; Covington and Moore, 1994; Fuhlendorf and Smeins, 1997; Chambers et al., 1999). Now fire is assigned the role of keeping the landscape open and creating conditions for regeneration of oaks. Within the framework of the scientific concept of the closed forest as climax, we cannot imagine how else an open landscape could have arisen and how the oaks, which need light, could have survived.[155] From the point of view of that concept, burning by Native Americans has also appeared as an explanation for the presence of open landscapes and the regeneration of oaks (see Abrams and Seichab, 1997). However, there seems to be no historical evidence for large-scale burning of forests by Native Americans (Day, 1953; Russell, 1983). Russell (1983) states the openness of the landscape in itself as such is used as the evidence of large-scale burning. Because there were open forests everywhere, there must have been burning everywhere, because otherwise the phenomenon cannot be explained within the current scientific concept. According to Russell, the Native Americans would only have made fires at camps and settlements, whereby the fire would have incidentally escaped. This fire did add to fire caused by lightning (Russell, 1981). If we assume that the burning by Native Americans was not responsible for the general openness described in historical sources, then this does not mean that this burning did not have any effect on the growth patterns. The landscape will certainly have become more open as a result of burning bushes and scrub with young trees. In that case, the landscape created by the Native Americans is a derivative of the original open, park-like landscape.

---

[155] See Day (1953), Monk (1961b), Raup (1964), Curtis (1970), Thompson and Smith (1970), Russell (1983), Whitney and Davis (1986), Mikan et al. (1994), Adams and Andersson (1980), Abrams et al. (1995), Abrams (1996), Abrams and Seichab (1997), Dodge (1997), Ruffner and Abrams (1998).

4.18 Conclusions and Synthesis

This chapter examines the use of the wilderness since the early Middle Ages, the changes which took place as a result in the wilderness, and the terms used for this. We looked at how the use has changed, and how terms have consequently developed a new meaning. The most important findings and some explanatory notes are presented below.

> • The meaning of the words 'silva', 'forest', 'forêt', 'Forst', 'Wald', 'wold', 'weald', 'woud', 'wald', 'bos', 'Busch', 'wood' in the context in which these terms were used in the Middle Ages, shows that they did *not* entail the meaning of the wilderness being a closed forest.

All these terms have in common the fact that they relate to the use of the wilderness. The place was the 'forestis' (German: 'Forst'; French: 'forêt'; English: 'forest', and Dutch: 'foreest', 'voorst' and 'vorst') and the 'Wald'. The 'forestis' was a legal term to indicate that the king had rights relating to the use of the 'Wald'. The 'Wald' was uncultivated land, declared to be 'forestis'. The 'forestis' comprised water as well as 'Wald'. The 'Wald' was the uncultivated land which comprised treeless moors and grasslands, as well as forests. The 'Wald' was therefore not forest in the modern sense of the word. The forest (German: 'der Busch'; English: 'bush'), the wood (Saxon/German: 'holt', 'Holt', 'Holtz'; French: 'bois') and the animal fodder (Saxon/German: 'waydt') formed part of the Wald, which was a park-like grazed landscape with groves, grassland and peat. The wood came from the shrubs and trees and the 'fodder' (the foliage, the acorns, the grass and the plants) came from the grassland scrub and the groves.

For terms such as 'holt' and 'acker', the use determined the name. The place became synonymous with what was taken from it. For example, 'wood' was taken from 'the wood', and pigs went out to pannage on the 'ackers' (i.e. on acorns, in Anglo-Saxon, aecern) in the 'acker', i.e. the place where the acorn (acker, aecer) grew. The pannage took place in the 'forestis' or 'Wald'.

> • All these terms related to the uncultivated wilderness which consisted of a mosaic of grasslands, scrub, trees and groves, as shown by the range of uses.

The wilderness ('forestis', 'Wald') was used for obtaining firewood and timber, for the production of peat and metals, for grazing livestock, cutting foliage for fodder, collecting honey, and the pannage of pigs. This use was covered by the 'ius forestis', also known as 'forest law' or 'waldrecht', laid down in all sorts of regulations. The origin of the term 'forestis' is a legal one. It served to determine the rights of the king. Terms related to the concept of the 'forestis' also acquired a legal meaning, because they also related to the wilderness, which had been declared by the king to be his 'forestis'. Certain types of use of the wilderness were declared to be the king's privilege because the wilderness was the forestis. For example, this applied to hunting particular species of animals, which became royal game, such as red deer, bison and aurochs.

- The oldest regulations about the grazing of livestock in the wilderness, dating from before 1300, concerned the granting of permission for the use of the wilderness and the regulations on payment. They did not contain any rules about maintaining the wilderness.

The regulations dating from before 1300 related only to the pannage of pigs. There were not yet any regulations on the grazing of livestock, horses, goats and sheep. These regulations do not contain any indication about the possible occurrence of retrogressive succession from the original forest to grassland and heathland.

The regulations about the grazing of livestock after 1300 up to about 1800 are all concerned with the protection of the vegetative reproduction, the spring on the stools of felled trees and shrubs, and sparing young trees growing *outside* the forest in thorny scrub.

The regulations about livestock, for example, concerned thorny shrubs in which seedlings developed of trees, which were then used for coppicing. In the oldest regulations, these shrubs are described with terms such as 'underwood', 'voorhout', 'Onderhout', 'underbusch' and 'nederwalt'. These terms indicate that they concerned the mantle vegetation around groves. Initially, the young trees, which have grown through the coppice, are still referred to as such.

- The texts of the regulations show that the regeneration of trees takes place *with* grazing, and that the regulations were *not* for the generative regeneration, but only for the vegetative regeneration.

The historical data show that in the last wildernesses in the lowlands of Central and Western Europe, the 'forestes', the process of the generative regeneration of trees took place in thorny scrub. Moreover, written sources indicate that this method of generative regeneration took place throughout history, even in the presence of very high densities of large herbivores. The seedlings of trees and shrubs were naturally well protected by thorny shrubs. They could grow even with a biomass of herbivores per hectare up to 40 times higher than that which is considered necessary nowadays for the successful regeneration of trees, according to the prevailing theory about forests.

- There are strong indications that the English deer park originates from park-like landscapes from a grove selected for hunting, surrounded by a mantle and fringe vegetation, the 'haga' or 'haye'.

The Anglo-Saxon toponyms 'haga' and, after the Norman Conquest, 'haye', occur in relationship to mantle and fringe vegetations. These mantle and fringe vegetations can be derived from park-like landscapes. From groves surrounded by mantle and fringe vegetation in park-like landscapes, we can derive in turn the deer park as a certain grove selected for hunting surrounded by mantle and fringe vegetations, een 'hage', into which the game was driven and then killed there. Later, the game was kept there permanently. The mantle and fringe vegetation was in time replaced by a wooden palisade.

- Livestock became a threat only when man exposed the seedlings to large browsing herbivores by cutting the thorny scrub to collect firewood.

The disappearance of trees and shrubs was the result of cutting down thorny shrubs as coppices, and the inadequate protection of the young trees which were exposed. In combination with the exorbitantly high densities of livestock in the 18th and 19th centuries, this form of exploitation led to the disappearance of trees and shrubs from large areas. Therefore the destruction of the original vegetation was, in the first instance, the result of collecting firewood and over-exploitation by man.

- In the 18th century, the changing demand for wood led to the development of the tree forest where the regeneration of trees was attempted in the forest itself. Thus large areas of continuous tree forest developed only after 1700.

The changing methods of wood production were a reaction to a change in the demand for wood. There was a demand for thicker pieces of wood, which required thicker, and therefore older, trunks. This changing demand then initiated longer cycles for the trunks. This resulted in the pole-forest, and subsequently in the modern tree forest. Thus both these types actually developed from the coppicing cultivation. Because vegetative regeneration from the stool was certainly no longer possible with beech when longer cycles were used, this resulted in the regeneration of the forests by planting young trees. The tree forest that developed where the regeneration of trees takes place in the forest, is not analogous to the natural vegetation, and is therefore incorrectly used as a reference in the prevailing theory for the original vegetation of the lowlands of Central and Western Europe.

> • The grazing of livestock proved to be harmful for the tree forest, because the livestock destroyed the seedlings of the trees in the forest, and therefore made regeneration impossible in the forest. As the tree forest was considered as a reference for the natural vegetation, it was thought that the grazing of livestock was harmful for the survival of the original forest.

The tree forest was vegetation which in principle developed where there were no large herbivores. The presumed analogy of this vegetation with the original natural vegetation led to the conclusion that in the original vegetation, large herbivores had not played a significant role either. This circular argument actually formed the basis of the circular arguments in the theories about succession and the natural vegetation.

> • The regulations on the grazing of livestock from before and after 1800, respectively, reveal the differences in the methods used for the regeneration of forests. Before 1800, the protection was focused particularly on vegetative regeneration, while after 1800, it was focused on generative regeneration *in* the forest.

The provisions dating from before 1800 show that grazing was prohibited for 3–6 years to allow trees to regenerate, while this prohibition lasted 15–20 years after 1800, which was not considered unrealistic in terms of the traditional rights relating to grazing livestock. After the technique of 'natural regeneration' was developed in the 18th century, foresters increasingly insisted on the division of woodland and pasture. This could be achieved after 1800 only as a result of changes in agriculture which led to an alternative to wood-pastures. These changes were the introduction of clover and potatoes as livestock fodder.

> • The terms used in Europe for park-like landscapes, such as 'forest', 'forêt', 'Forst', 'Wald' and 'woud' and the related images, travelled as mental models in the minds of the European colonists to the east of the United States, where they found their way into historical descriptions of the landscapes the colonists found there.

There is no reason to doubt the images of the open, park-like landscapes that appear in the historical descriptions of the east of the United States, because those images would not have been familiar to the European colonists. On the contrary, they were very familiar with them. Much of what is described in the east of the United States fits the image of the wood-pastures in Europe. This includes images of the regeneration of trees in grasslands grazed by specialized grazers such as cattle.

> • There is no reason based on historical texts to assume that in the east of the United States a naturally present closed forest degraded to open grassland as a result of livestock pastures, as a result of there being no regeneration of trees.

The image of degradation of forest to grassland as a result of the cattle preventing the regeneration of trees is an image derived from forestry techniques from Europe from the 18th and 19th centuries. It assumes a closed forest as the naturally present vegetation. As evidenced by the regeneration of trees in grazed grasslands, this image is demonstrably incorrect for both the lowlands of Europe and the east of the United States. It is only true for regeneration in the forest, not for regeneration of trees in open grassland.

> • Fire as an explanation for the openness of the original growth in the east of the United States is a result of the prevailing scientific concept of the closed forest as the climax. Incorrectly, grazing by indigenous bison as a possible explanation is not considered.

Fire is given as an explanation for the openness of the original landscape, because no other explanations have been found within the framework of the prevailing theory of the closed forest as the climax in the east of North America. Because the landscape was open, fire must have been the cause of this. A possible role of large grazers like the bison is not considered at all in this. If this role is included, a number of phenomena can possibly be explained in mutual context, such as regeneration of oaks and open landscapes.

To summarize, it may be said that the arguments used by plant geographers and ecologists, such as Moss, Tansley and Watt, as well as the palynologist Iversen, from forestry and historical texts are incorrect, namely, that these show that the grazing of livestock was traditionally seen as a threat to the original forest. This interpretation of the written sources is purely based on an extrapolation of the results of grazing livestock on a type of forest and a technique for replacing that type of forest which was only developed in the 18th century. Neither this technique nor the type of forest that was regenerated were analogous to the vegetation and the regeneration of trees which were present at the time that the first regulations on the grazing of livestock were issued. The picture which they present on the basis of these sources, i.e. that there were forests with grazing which were to be preserved, does not accord either with the historical data, and is therefore incorrect. Nor do the historical data demonstrate that the forest disappears as a result of the grazing of livestock. They do show that groves did develop as a result of the grazing of livestock. Therefore these data cannot be used as arguments for the theory of retrogressive succession; nor do they provide any arguments to support the theory that the original vegetation was a closed forest.

Spontaneous Succession in Forest Reserves in the Lowlands of Western and Central Europe

5.1 Introduction

In this chapter, I would like to answer the question: can pedunculate and sessile oak and hazel survive in a spontaneously developing closed forest in the absence of large herbivores such as cows and horse? To answer this question, I will look at the secondary succession in forest reserves in Western and Central Europe which were grazed by livestock, but where this and other forms of human intervention have come to an end. As the last chapter showed, there were wood-pastures throughout Central and Western Europe. Before they were proclaimed reserves, the areas described in this chapter were the last remnants of these. They are spread throughout Europe on all sorts of different types of soil (see Appendix 11). Therefore I believe they can be seen as being representative of the former wood-pastures in the area of this study. In accordance with the theories on succession, the forests in these reserves would develop into natural forests via secondary succession.[1] If these theories are correct, all the indigenous species of trees which are found in the forest reserves should regenerate spontaneously. This applies to the species which pollen studies have shown became established in the European lowlands after the last Ice Age without any human intervention, and survived for thousands of years. Examples of these are: pedunculate and sessile oak (*Quercus robur* and *Q. petraea*), broad-leaved lime and small-leaved lime (*Tilia platyphyllos* and *T. cordata*), the different species of elm (*Ulmus* spp.), sycamore (*Acer pseudoplatanus*), ash (*Fraxinus excelsior*),

---

[1] See *inter alia* Doing-Kraft (1958), Dietrich *et al.* (1970, p. 8), Lödl *et al.* (1977), Genssler (1980), Lemée (1985; 1987), Koop (1982), Falinski (1986, p. vii; 1988), Broekmeyer and Vos (1990), Broekmeyer *et al.* (1993).

189

beech (*Fagus sylvatica*), hornbeam (*Carpinus betulus*), birch (*Betula* spp.), poplar (*Populus* spp.), willow (*Salix* spp.), rowan (*Sorbus aucuparia*) and hazel (*Corylus avellana*).[2]

The situation towards which these forests developed is known as the potential natural vegetation (PNV) (Tüxen, 1956). PNV means the vegetation which develops under the prevailing abiotic conditions as the final stage of the succession, after human influence has come to an end. This vegetation is the climax vegetation as defined by Tansley (1935), Watt (1947), Dengler (1935) and Leibundgut (1959; 1978). Watt based his gap phase model for the regeneration of the climax on this sort of development. The PNV is the same as the vegetation which is believed to have been present before humans intervened, where human intervention did not lead to irreversible changes (Tüxen, 1956, pp. 5, 15). This vegetation is known as the real natural vegetation (RNV) (Tüxen, 1956, p. 5). The PNV is the same as the RNV, if there have been no irreversible changes in the soil as a result of human activity, and there have been no changes in the climate, or new species have appeared on the scene as a result of human intervention. Human intervention can lead to irreversible changes so that a different climax vegetation develops. Therefore the PNV is not necessarily the same as the RNV. However, it is assumed that wherever trees can grow on the basis of the prevailing climatological and soil conditions, the PNV is forest. In that case, the structure of the vegetation is determined, though there is some uncertainty about the composition of the vegetation that can be expected. In the following sections, I will describe the spontaneous developments in different forest reserves. The location and abiotic characteristics of these reserves are briefly outlined in Appendix 11. In the study of these spontaneous developments I am looking for facts about the way in which different species of trees and shrubs behave. In view of the area of this study, I am particularly interested in pedunculate and sessile oak and hazel.

In so far as there are data available on the history of the reserves, I will include these in my study. However, as the data available on the reserves varies, this part of the chapter is slightly unbalanced. In one case, I present the data on the reserves in a cluster because this is also how they were presented by the researcher(s) in the publication. In that case, it is not possible to divide up the data.

At the end of this chapter, I will discuss the spontaneous development of the vegetation in forest reservations in the east of the United States. I do this because the forests and the processes that take place there are considered an analogy of those in the lowlands of Europe (see Whitmore, 1982; Westhoff, 1983; Peterken, 1991; 1996, p. 230; Holeska, 1993; Pontailler *et al.*, 1997).

---

[2] See *inter alia* Lüdi (1934), Firbas (1949, pp. 50–51), Meusel (1951/52), Iversen (1960; 1973, pp. 65–70, 105, 108), Glavac (1968), Trautmann (1969), Janssen (1974, p. 57), Grime (1979, p. 152), Zimmermann (1982), Ellenberg (1986, pp. 75–76), Jahn (1991, p. 411, etc.).

5.2 La Tillaie and Le Gros-Fouteau in the Forêt de Fontainebleau, France

5.2.1 A brief description and history of the reserves

La Tillaie and Le Gros-Fouteau are biological reserves in the Forêt de Fontainebleau near Paris in France. They were established between 1853 and 1861. La Tillaie has an area of 36 ha, and Le Gros-Fouteau has an area of 21 ha. They are both reserves in a strict sense. Since their establishment, there has been no further human intervention in the vegetation (Tendron, 1983, p. 37). They are amongst the oldest nature reserves in Europe (Koop, 1989, p. 77). The oaks in the reserves are sessile oaks.

Both reserves are part of the Forêt de Fontainebleau, a former 'forestis' (Tendron, 1983, p. 21). The local population kept pigs there until the 19th century, grazed their livestock, collected acorns and took the litter. According to one document dating from 1664, there were about 6367 pigs and 10,381 cows on a surface area of approximately 14,000 ha (Tendron, 1983, p. 22). An inventory dating from 1716 indicates that the area was in a deplorable condition at that time. Half of it consisted of heathland with juniper bushes, birch trees here and there, and young trees which had been planted, but were not flourishing. There was only 847 ha of standing forest left with trees in poor condition. The reserves which were later established were part of this (see Tendron, 1983, pp. 6, 27–29). From 1720, trees were systematically planted in the Forêt de Fontainebleau. In 1786, the Scots pine (*Pinus sylvestris*) was introduced. Up to 1830, between 5000 and 6000 ha were planted. Between 1831 and 1847, old trees were cleared and another 6000 ha were planted with Scots pine (Tendron, 1983, pp. 28–29). Romantic painters and tourists opposed this measure at the time. This resulted in the establishment of so-called 'reserves artistiques' in 1853 and 1861. The present reserves of La Tillaie and Le Gros-Fouteau are two of these (Tendron, 1983, pp. 32–33, 37).[3]

Historical texts contain indications that even before La Tillaie was pronounced a 'reserve artistique' there had been no further interventions in the forest complex where La Tillaie and Le Gros-Fouteau were located. In fact, this complex was described as a 'reserve' in a text dating from 1750 (Domet, 1873, cited by Lemée, 1987). This concerned an area of 600–750 ha, virtually the whole area still covered by standing trees, according to the inventory dating

---

[3] Old oaks with dead tops in the Forest of Fontainebleau, surrounded by heathland, juniper berries and shifting sands, with occasional enormous rocks, inspired the Romanticist and painter, Jean-Jacques Rousseau and other romantic painters, in what is known as the school of Barbizon (Reed, 1954, p. 101; Tendron, 1983, p. 30). They endeavoured to conserve the beauty of the landscape in the forest. Therefore they opposed measures which were taken at the time to regenerate the forest, such as felling the old oaks which the foresters considered to be finished, and planting young trees, usually Scots pine (Tendron, 1983, p. 28). The artists' efforts resulted in 624 ha being proclaimed as so-called 'reserves artistiques' in 1853. In 1861, a decree increased this area to 1097 ha (Tendron, 1983, p. 35).

from 1716. 'Reserves' were established by law in the 16th and 17th centuries so that there would be timber for building ships for the war fleet, in case it was needed (Woolsey and Greeley, 1920, pp. 223, 488–489).[4] The last reference to the felling of trees in La Tillaie for timber dates from 1372–1373 (Grand-Mesnil, 1982, cited by Lemée, 1987). Texts dating from 1664 and 1716 show that in La Tillaie and Le Gros-Fouteau, the trees were mainly beech and to a lesser extent oak, hornbeam and lime. The trees were between 150–250 and 200–350 years old. Therefore in the 17th century the area contained trees which were suitable for being appointed as a 'reserve'. According to the sources, there were young beech trees, hornbeam and lime under these trees, which indicates that these species were regenerating (Van Baren and Hilgen, 1984, Appendix 5; Lemée, 1987, p. 333; Koop, 1989, pp. 75–76). Another indication for keeping the trees as a 'reserve' in La Tillaie and Le Gros-Fouteau is that the annual rings of trees in Le Gros-Fouteau and the adjacent forest show that the sessile oaks in both reserves are extremely old. The oldest dates from 1520 (see Table 5.1). The establishment of the oldest beeches goes back to the beginning of the 17th century (Lemée, 1987; Koop, 1989, p. 77). The fact that there had been no human intervention before the establishment of the reserves can also be deduced from the regeneration which was taking place.

When the reserves were established in 1853, there were many hornbeam, beech and oak seedlings germinating in La Tillaie, varying in age from 1 to 30 years. In addition, there were young trees 100 years old and older, as well as single standing oak and beech trees which were deteriorating, and places where there were no trees left standing. In the area, there were also still a few hornbeams between 100 and 300 years old (Van Baren and Hilgen, 1984, Appendix 5). In view of the large numbers of livestock in the 17th century, and the fact that beech, hornbeam and lime regenerated in La Tillaie and Le Gros-Fouteau, it is very probable that the two reserves were no longer grazed in the period that they functioned as a 'reserve'.

As I mentioned above, the study of annual rings showed that the oaks in the reserves La Tillaie and Le Gros-Fouteau and in the direct vicinity dated from about 1500. Table 5.1 shows that the establishment of the oak trees goes back to the period 1520–1696, with the exception of two trees. One oak which was

---

[4] In 1594, a decree in France stated that on common land, the trees must be spared on one-third of the area of coppices-with-standards, so that there would be timber available for building in case of need, i.e. as a strategic reserve (Woolsey and Greeley, 1920, pp. 223, 488–489). Thus, on this land, there may eventually have been a sort of forest of trees with hardly any or no coppice left as a result of the shade of the standards. In 1561, it was decided that 'reserves' must also be planted in all the royal forests. The area was reduced from one-third to one-quarter. After this regulation was withdrawn in 1580 by King Henry III, it was incorporated again in the ordinance promulgated by Louis XIV in 1669. This stated that on land used as common land, or owned by the Church, trees must be kept in reserve. The reserve had to be a separate area so that an inspector could easily determine whether the legal provision had been met (Woolsey and Greeley, 1920, pp. 223, 488–489; Reed, 1954, p. 41).

Table 5.1. The age and the year of establishment of large oaks in the reserve Le Gros-Fouteau and the adjacent forest. The dates were determined based on the tree rings (redrawn from Lemée, 1987, Table 1).

Year of felling	Age in years at the base (approx. height 30 cm)	Year of establishment
1869	*c.* 470	*c.*1400
1896	463	1520
1896	452	1520
1896	452	1520
1896	446	1526
1967	425	1542
1971–72	402	1570
1971–72	390	1582
1980	388	1594
1971–72	386	1586
1971–72	386	1586
1952–53	378	1575
1967	377	1590
1971–72	377	1595
1971–72	375	1597
1967	370	1597
1971–72	357	1615
1971–72	354	1618
1971–72	352	1620
1971–72	342	1630
1967	342	1625
1971–72	340	1631
1971–72	334	1638
1971–72	305	1667
1971–72	304	1668
1967	300	1667
1971–72	296	1676
1967	276	1696
1971–72	230	1742

not mentioned dated from 1399. This died in 1869 (Domet, 1873, cited by Lemée, 1987). Research by Koop into the annual rings of beech trees in two core areas shows that, as Fig. 5.1 illustrates, the beech regenerated with one clear interruption. The oldest generation in one core area dates from the period 1651–1693. In the following period no trees were established. This period lasted until 1851, then trees were established almost continuously until 1950. In the other core area, there were two waves of regeneration. The first lasted from 1609 to 1723, after which there was no further establishment of trees until 1869. After 1869, a generation was established almost continuously until 1963. For the periods in both key areas, after 1950 and 1963 respectively, there

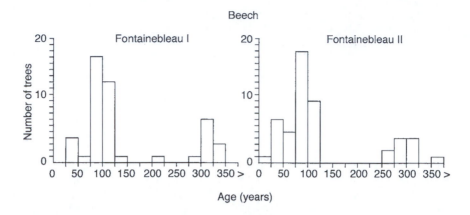

Fig. 5.1. Distribution of the trees per age category, in two sampling sites (I and II), in La Tillaie in the Forêt de Fontainebleau, France. The age of the trees is determined on the basis of annual rings observed by drilling (redrawn from Koop, 1989, p. 78).

are no more trees shown. The trees from that period were thinner than 10 cm, the diameter which was used as the lower limit for the samples (Koop, 1989, p. 78).

According to Koop (1989, pp. 90–91), the interruptions in the regeneration took place because there was a wave of regeneration after the tree felling in 1374/1375, which, in the first instance, led to the establishment of trees of all more or less the same age. He thought that the trees which became established after the tree felling, all reached the optimal stage at about the same time (in the same way as beech primeval forests are regenerated) (see Korpel, 1982). Figure 5.2 shows the similarity between succession in beech primeval forests and that in La Tillaie after the tree fellings in 1374/1375. As the figure shows, there was absolutely no regeneration under the closed canopy in La Tillaie during the stage when the trees from the wave of regeneration had become full grown, as in the case of primeval beech forests, because of the deep shade cast by the closed canopy. Because the trees were all of the same age, they also all reached their physiological end at more or less the same time so that the forest collapsed on a fairly massive scale rather like the primeval beech forests. In the 17th century, this resulted in the second wave of regeneration which was restricted to a period of 50 years. This produced the oldest generation of beech trees in the present reserve. The same process can be seen in the second sample (Koop, 1989, pp. 90–91). These data confirm what the historical sources say and what the results of the research into the annual rings of the oak trees shows. In view of the age of the oak and beech trees, it seems that young beech trees grew under the oak; i.e. that beech regenerated under the oak. In fact, as Table 5.1 shows, most oaks date from before 1609 and 1651, the dates of birth of the oldest beech trees in Koop's samples (1989). The oaks that were still pre-

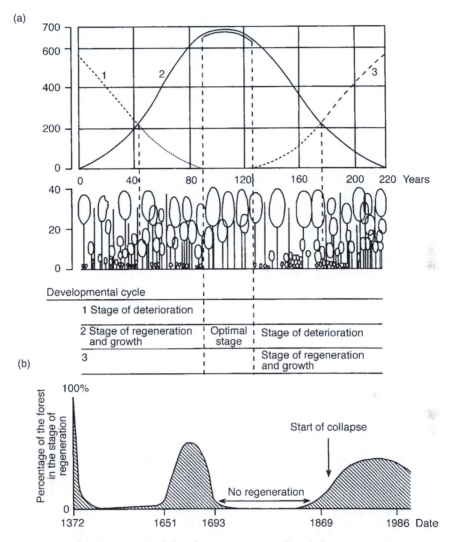

Fig. 5.2. Model of the cyclical development in virgin beech forests in Slovakia (a), according to Korpel (1982; 1987). The developmental cycles (1–3) are repeated and overlap. They are demonstrated in the changing structure of the forest. (b) The model as modified by Koop (1986) for the spontaneous succession in the forest reserve of La Tillaie in the Forêt de Fontainebleau, France (redrawn from Koop, 1989, p. 91).

sent in the reserves are all sited on poorer soil, in the *Fago–Quercetum*, where the beeches are up to 3 m lower than the oaks due to poorer growth. In the parts of the reserves with richer soil, in the *Melico–Fagetum*, the beeches through time have caused the death of all the oaks because they grew taller (Koop and Hilgen, 1987; Ponge and Ferdy, 1997 and see Fig. 5.5).

5.2.2 The present situation

Nowadays, both La Tillaie and Le Gros-Fouteau consist virtually entirely of pure beech forest. After these, the hornbeam is the most important species (Koop and Hilgen, 1987; Pontailler *et al.*, 1997). In both reserves, beech trees are found in virtually every category of diameter, as shown in Figs 5.3 and 5.4. On the other hand, the sessile oak is represented almost exclusively in the larger categories. Sessile oaks grow only in localized groups. Apart from this the trees in the reserves are ash, hornbeam, birch and rowan (Lemée, 1978; 1985; 1987). The birch and rowan are restricted to the pioneer stage in gaps in the canopy and do not grow into mature trees (H. Koop, Wageningen, 1997, personal communication). In La Tillaie, there are (still) two large large-leaved limes left (H. Koop, Wageningen, 1997, personal communication).

In both reserves, the number of old oak trees is steadily declining. In La Tillaie, there were still 3.13 oaks per hectare in 1902, compared with 1.38 per hectare in 1986. This is an average annual mortality of 0.97%. In Le Gros-Fouteau, the numbers were respectively 23.8 oaks per hectare in 1902,

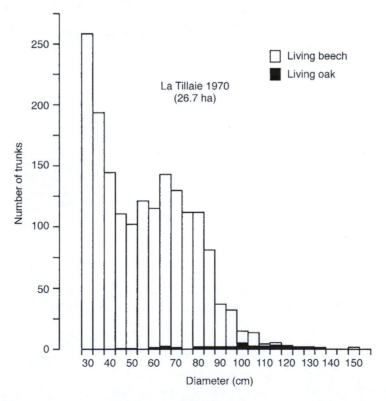

Fig. 5.3. Distribution of sessile oak and beech in diameter categories of 5 cm in the whole of the reserve of La Tillaie (redrawn from Lemée, 1978, p. 86).

compared with 7.9 oaks per hectare in 1986. This represents an average annual mortality of 1.31% (Lemée, 1987). One of the causes of the disappearance of oak is that as stated earlier the beech trees grow taller (Koop and Hilgen, 1987; Ponge and Ferdy, 1997; and see Fig. 5.5). When they grow past the oak trees, the latter die (see Fig. 5.6).

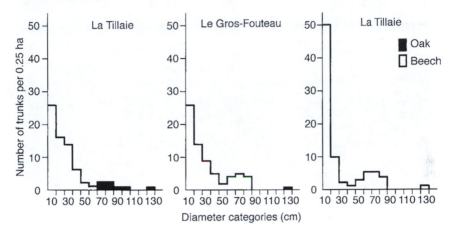

Fig. 5.4. Distribution of sessile oak and beech in diameter categories of 10 cm, in a sampling area of 0.25 ha, in forest with a closed canopy, in La Tillaie and Le Gros-Fouteau (redrawn from Lemée, 1978, p. 85).

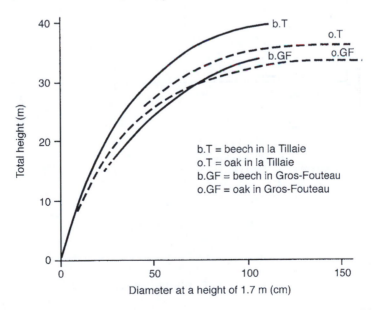

Fig. 5.5. The relationship between the height and diameter of sessile oak and beech in the forest reserves of La Tillaie and Le Gros-Fouteau (redrawn from Lemée, 1966, p. 307).

Fig. 5.6. Dying sessile oak in La Tillaie, and prolific regeneration, mainly of beech, which fills the gap in the canopy (photograph, H. Koop).

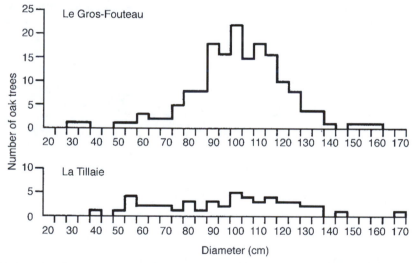

Fig. 5.7. The distribution in 20 cm diameter categories of sessile oak in La Tillaie and Le Gros-Fouteau (redrawn from Lemée, 1987, p. 338).

Another cause, as indicated by Fig. 5.7, is that there is virtually no new generation of oak trees (Lemée, 1978, pp. 85, 86; 1987; Koop, 1989, p. 78). Young oak trees are only found sporadically, and then virtually only as isolated specimens. The distribution of oak diameter classes shows a bell curve/normal curve distribution that is typical for a population that is becoming extinct (Koop and Hilgen, 1987, see Figs 5.3 and 5.7). According to Lemée (1987), this is not due to a lack of acorns. In good mast years he maintains that the predation of acorns was almost negligible, although there have been great differences through the years. However, the mortality of seedlings was very high. The seedlings survived longest in these gaps in the canopy, but according to the data in Table 5.2, they did not survive for more than 5 years. However, Lemée's research showed that in a number of cases there were young oaks of a greater age in the gaps in the canopy. In La Tillaie, he found approximately 150 young oak trees in 43 gaps, and in Le Gros-Fouteau, there were 31 young oak trees in 13 gaps. In virtually every case these were isolated specimens badly affected by browsing deer. Instead of the usual 5 m, they were only 80 cm tall. In five former gaps in the canopy, there were even several larger oak trees grouped together.

Lemée deduced from the architecture of the trees in the gaps that these gaps had originally had an area of 500–700 m² (0.05–0.07 ha) (Lemée, 1985; 1987). For the three gaps in Le Gros-Fouteau, for which the architectural records are shown in Figs 5.8, 5.9 and 5.10, Lemée (1987) reconstructed the process of regeneration. This reconstruction shows that birch trees were the first to become established, together with the odd oak; 16 years later, new oaks appeared for a period of 6 years, together with beech and hornbeam (Lemée, 1985; 1987).

Table 5.2. The development of sessile oak seedlings, following the mast of 1982, in numbers per m², under three different conditions (redrawn from Lemée, 1987, Table 4).

	Time of observation				
Place of observation	17-6-1983	9-9-1983	6-6-1984	8-6-1985	19-9-1986
Situation 1					
No. of seedlings	98.00	80.45	0.00		
No. of seedlings which have disappeared	17.0	17.55	98.00		
Survival cf. starting situation (%)	100	82	0		
Situation 2					
No. of seedlings	11.35	18.15	2.15	1.20	
No. of seedlings which have disappeared	0	3.2	9.2	10.15	11.35
Survivals cf. starting situation (%)	100	92	15.7	10.8	0
Situation 3					
No. of seedlings	17.75	17.25	7.00	6.75	2.25
No. of seedlings which have disappeared	0.50	0.50	10.75	11.00	15.50
Survivals cf. starting situation (%)	100	97	39	13	6

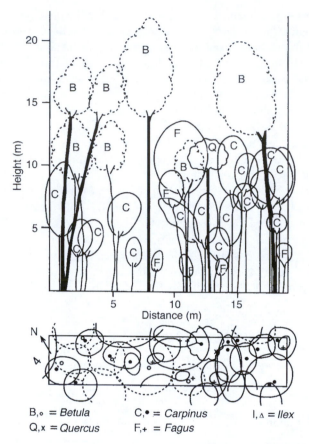

Fig. 5.8. The architecture of the regeneration of trees in a gap in the canopy in Le Gros-Fouteau (redrawn from Lemée, 1985, p. 6).

Another 14 years later, there were 57 oaks, 52 hornbeam and 23 beech trees left in the gaps, per 0.01 ha. Of the oaks, 7.3 specimens remained per 0.01 ha, on average, after 19 years; a decline of 80% (Lemée, 1987). The mortality was greatest amongst the smallest specimens. Where oaks had survived, the beech trees had grown to just below or to the level of the crowns of the tallest oaks, as illustrated in Figs 5.8, 5.9, and 5.10. The rate of growth of the beech is greater than that of the oak (Lemée, 1985). As Fig. 5.5 indicates, the beech trees will grow above the oaks in both reserves so that the young beech trees will eventually win the competition against the oaks in the large gaps (Lemée, 1987). Figures 5.8 and 5.9 also show that the hornbeam grows to the height of the oaks.

As regards the size of the gaps in the canopy, a frequency distribution of the area shows that there are very few gaps with a size for which Lemée (1985) found that oaks had become established, i.e. 500–700 m² (0.05–0.07 ha) (see Fig. 5.11). Of the 320 gaps which were found, only eight were larger than 0.03 ha (Lemée, 1985). Storms in 1967 and 1990 did result in an increase in the area of

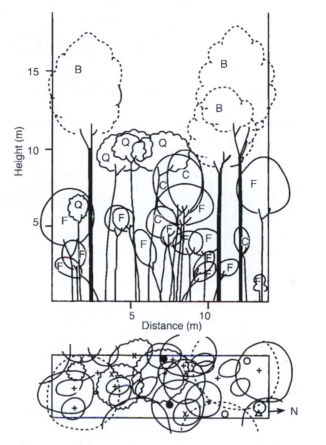

Fig. 5.9. The architecture of the regeneration of trees in a gap in the canopy in Le Gros-Fouteau (redrawn from Lemée, 1985, p. 7).

the gaps in the canopy (see Fig. 5.12). After the storm of 1990 in Le Gros-Fouteau – the heaviest storm – 20.6% of the surface area of the reserve consisted of open terrain. A storm surface of almost 1.5 ha (12,805 m^2) even appeared as a result of 155 trees being blown over (Pontailler *et al.*, 1997). However, all this open terrain did not lead to successful establishment of the oak. There were oak seedlings which grew in the gaps created by the storms, but these were soon suppressed by the rapidly growing young beech trees which had already germinated before the storm and grown into seedlings below the still intact canopy (Lemée *et al.*, 1992; Peltier, 1997; Pontailler *et al.*, 1997; H. Koop, Wageningen, 1997, personal communication). On the basis of these data, we must conclude that there are few or no opportunities for oak to become established. This explains the steady decline in oak trees, which was found in both reserves, as indicated above.

The progressive suppression of oak by beech in La Tillaie and Le Gros-Fouteau has occurred for only a few centuries, as shown by pollen frequencies in a podsol (see Fig. 5.13) (Guillet and Robin, 1972, cited by Lemée, 1987).

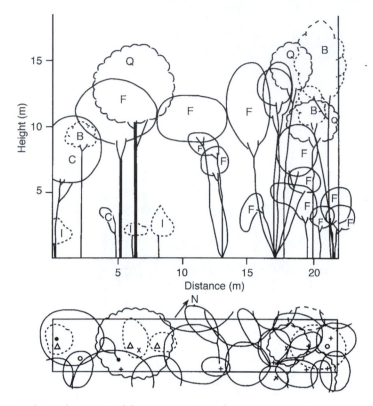

Fig. 5.10. The architecture of the regeneration of trees in a gap in the canopy in Le Gros-Fouteau (redrawn from Lemée, 1985, p. 8).

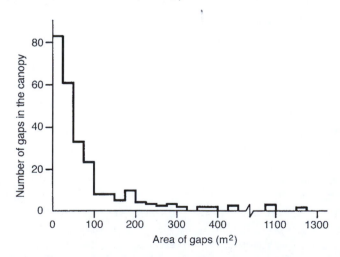

Fig. 5.11. Distribution of the areas of the gaps which formed in the canopy in the reserve of La Tillaie during the period 1980–1981 (gaps larger than 5 m²) (redrawn from Lemée *et al.*, 1986, p. 172).

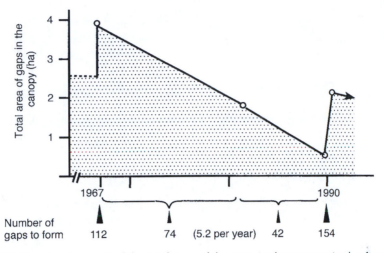

Fig. 5.12. The development of the total area of the gaps in the canopy in the forest of La Tillaie (redrawn from Lemée *et al.*, 1992, p. 988).

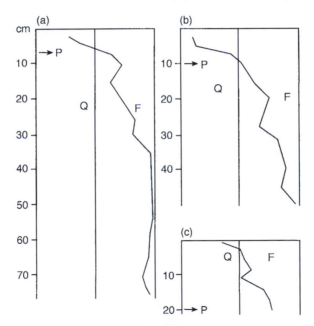

Fig. 5.13. The development of the relative frequencies of oak pollen (Q) and beech pollen (F), compared in three soil profiles in the reserves of La Tillaie and Le Gros-Fouteau. The total pollen sum of the two species is set at 100%. The vertical line down the figure is the 50% level. (a) 'podzol humique forestier' in La Tillaie (from Guillet and Robin); (b) 'podzol humo-ferrugineux' from a site bordering on Le Gros-Fouteau; (c) 'sol podzolique' (upper part) in Le Gros-Fouteau; P→ the point at which pollen of the Scots pine (*Pinus sylvestris*) appeared, which dates the introduction of this species to the early 19th century (redrawn from Lemée, 1987, p. 334).

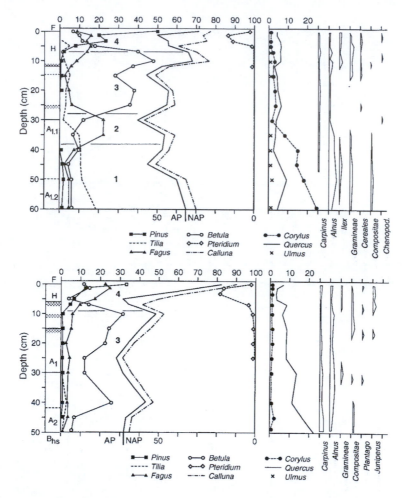

Fig. 5.14. Pollen diagrams with arboreal pollen (AP) and non-arboreal pollen of grasses, plants and heather (NAP), in heathland (upper figure), and young podzol soil in an open area where trees have been felled (lower figure). The relative percentage of pollen is shown on the horizontal axis on the top of the figure. For the arboreal pollen, the scale is at the top, and goes from left (0%) to right (100%). The bottom scale division shows the percentage of non-arboreal pollen, going from right (0%) to left (100%). The lines AP and NAP show the total pollen sum of arboreal pollen and non-arboreal pollen respectively. The right of the figure shows several species of trees and shrubs, such as hazel and juniper, and divides non-arboreal pollen into grasses (*Gramineae*) and grains (*Cereales*), etc. The left of the figure, on the vertical axis, shows the depth at which the samples were taken from the soil. The numbering 1, 2, 3 and 4 corresponds to the different periods in prehistoric times from which the samples date, and which are shown in Fig. 1.4. These periods are: 1. Sub-Boreal period (starting at 4500 BP; see Fig. 1.4); 2. Sub-Atlantic period, up to the transition from the Bronze Age to the Iron Age; 3. Celtic period (up to the beginning of the 18th century), when the Scots pine was introduced (*Pinus*); and 4. The modern period.

Because the modern pollen rain in both reserves shows that the oak is over-represented in the pollen rain there (Lemée, 1987), the stronger decline in the relative frequency of oak pollen in Fig. 5.13 reveals a stronger decline in the oak in comparison with beech, than suggested by the relative pollen frequencies. As regards the change in the species diversity in the 'forestis' as a whole, pollen diagrams of different places in the Forêt de Fontainebleau show that in the Sub-Boreal period (5000–2500 BP), lime and hazel were present, as well as the oak, beech and hornbeam described above (see Figs 5.14 and 5.15). In some sampling places, the percentage of lime pollen was even higher than that of oak pollen. Lime and hazel are now almost entirely lacking. The lime disappears from the pollen diagram in the Sub-Atlantic period (2500 BP–today). At the moment, there are still two large-leaved lime trees in La Tillaie (H. Koop, Wageningen, 1997, personal communication).

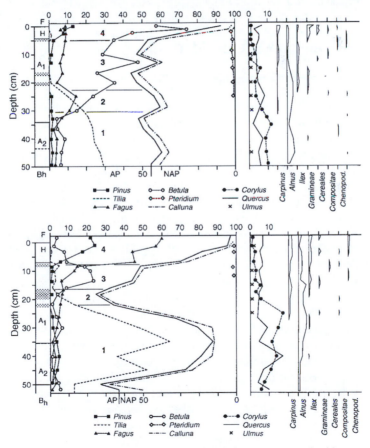

Fig. 5.15. Pollen diagrams of respectively, a birch forest on a 'podzol humo-ferrugineux' (upper figure), and a beech forest (lower figure), in the vicinity of the birch forest, on the same soil. For the keys, see Fig. 5.14 (redrawn from Lemée, 1981, pp. 194–196).

It may be concluded that La Tillaie and Le Gros-Fouteau are reserves with oak trees dating from the 16th and 17th centuries. Since then, there has been no further regeneration of oak. Everything points to the fact that these two reserves have been managed as reserves since the 16th century, and have not been grazed since then. This means that even during the period before La Tillaie and Le Gros-Fouteau were proclaimed reserves in the middle of the 19th century, no trees were felled. In that case, the 'no intervention' management goes back much further than the middle of the 19th century. Historical sources and studies of annual rings show that beech, hornbeam and lime have regenerated since the 16th century. Gradually, beech has become increasingly dominant, and has ultimately almost completely suppressed the oak. This process is still going on today. Even in the largest gaps made in the forest canopy by storms, there is no regeneration of (sessile) oak. In the end, spontaneous developments will therefore cause the (sessile) oak to disappear from the forest altogether in a relatively short time.

5.3 The Neuenburger Urwald, Germany

5.3.1 A brief description and history of the reserve

The Neuenburger Urwald lies in the north-west of Germany in the federal state of Lower Saxony. It is a former wood-pasture. It was pronounced a reserve in 1870 by Count Nikolaus Friedrich Peter von Oldenburg, the lord. His reason for creating this reserve was the beauty of the enormous oak trees which grew in a park-like landscape. The count was influenced by the ideas of the romantic artists in Germany, united in the so-called 'Hainbunde'. The 'Hain' was the name of the holy oak forest of the ancient Germanic people (Koop, 1981, p. 11). The Count designated the Hasbrucher Urwald, which is described in the next section as a reserve at the same time as the Neuenburger Urwald, and for the same reason. These two reserves are the oldest of their kind in Germany. For a detailed description, refer to Koop (1981). The following data are taken from his publication unless indicated otherwise.

The Neuenburger Urwald is situated south of Wilhelmshafen. It has an area of approximately 25 ha. Nowadays, it is part of a natural forest reserve of approximately 100 ha. In its turn, it is part of the Neuenburger Holz, which covers 627 ha. The Neuenburger Holz is a former common where cattle and pigs grazed and wood was cut for firewood and timber. As in many commons in Central and Western Europe, a commoner who cut down a tree was obliged to plant a few young oak trees and surround them with thorns to protect them from being eaten by livestock (Koop, 1981, p. 5). In the 17th century, the Holz consisted mainly of oak trees. In 1667 and in 1690, the sovereign lord decided that trees had to be planted.[5] A description dating from 1779 (cited and translated by Koop, 1981, p. 103)

---

[5] In 1667, the Danish King Christian V, the lord at the time, passed an order that 'Hesterkämpe angelegt werden, aus denen heranwachsenden Hester dann in den Wald verpflanzet und mit Dornen geschützet werden sollen' (Pott and Hüppe, 1991, p. 84).

indicates that in most of the Neuenburger Holz there were old, dying oak trees, many of which were hollow and rotting away. The undergrowth comprised many thorns, holly bushes and alder. The description also shows that in the past there were a large number of 'Weidekämpen' (enclosures with meadows). These were also called 'Deelen'. They were assigned to the commoners (Koop, 1981, p. 103). These open areas were surrounded by an earthen wall and a fence. These 'Weidekämpen' must have been hay fields, fenced off against the livestock. These hay fields were probably granted to save the rest of the area from being used for collecting litter, a custom that was known to be extremely harmful for the forest.[6] A description dating from 1780 (cited and translated by Koop, 1981, p. 103) indicates that the 'grosse Schaar', of which the present reserve is part, contained oak trees of many different sizes, and that amongst these there were many young planted oak trees between 12 and 18 years old. In addition, there were some pollarded beech trees and hornbeam, as well as thorns, holly and hazel.

At the end of the 18th century, commoners from three bordering villages in the Neuenburger Holz grazed a total of 234 horses, 961 head of cattle, 660 pigs and 1282 geese (Otto, 1780, cited by Pott and Hüppe, 1991, p. 85). This large number of animals would have destroyed the undergrowth (Nitzschke, 1932, p. 12). In view of the area of the Neuenburger Holz at the time, approximately 630 ha (see Koop, 1981, pp. 2 and 8), this represented an enormous pressure of grazing. When the 'Weidekämpen' were also grazed after haymaking, the biomass of horses and cows amounted to 627 kg ha^{-1}.

From 1780, attempts were made to limit grazing rights by introducing a grazing period and deciding that cattle had to be looked after by a cowherd. The cattle could be taken into the wood-pasture only from sunrise to sunset (Koop, 1981, p. 184). On the other hand, until 1883, horses were allowed to stay in the Neuenburger Holz day and night. After this, they also had to be accompanied (Koop, 1981, p. 104). The areas where grazing was not allowed were marked with bundles of straw. In 1852, 330 head of cattle, 50 horses and 20 geese grazed in the area. After 1883, only cattle grazed there. Between 1883 and 1893, their numbers fluctuated between 104 and 145 (Koop, 1981, p. 105). Table 5.3 shows the development in the number of cattle.

In 1870, the 'grosse Schaar' was proclaimed a reserve. After that date the commoners could no longer cut down trees, but only collect dead wood and graze cattle (Koop, 1981, p. 11). At that time, it was a park-like landscape with groups of impressive gnarled oaks and beeches, alternating with pasture with thorny scrub (Focke, 1871, cited by Koop, 1981, p. 11).[7] The right to graze

---

[6] Endres (1888, p. 53), Vanselow (1926, pp. 19, 80, 145), Endres (1929a,b), Gothe (1949), Rodenwaldt (1951), Hesmer (1958, p. 85), Streitz (1967, p. 69), Mantel (1990, pp. 102, 104).
[7] The botanist Focke (1871) from Bremen, described the Neuenburger Urwald as follows:

Mächtige, knurrige Eichen und dicht belaubte Buchen, die bald in Gruppen zusammengedrängt sind, bald auch grössere grüne rasenflächen freilassen. Das weidende Vieh lässt nur stacheliges Gebüsch aufkommen, … Der erste Eindruck, den man beim Eintritt in dem Urwald empfängt ist keineswegs der einer grossen Wildheit, vielmehr wird man zuerst an schöne Parkszenen erinnert … Vorherrschend ist im allgemeinen die Eiche an einzelnen Stellen aber auch die Rotbuche (Focke, 1871, p. 313, cited by Koop, 1981, p. 11).

Table 5.3. Development in the number of livestock in the Neuenburger Holz in the 18th century, and from 1852 to 1893 (redrawn from Koop, 1981, p. 104).

Year	Cows	Horses	Geese
18th century	961	234	1282
1852	330	50	20
1862	210	50	20
1872	197	50	0
1882	169	0	0
1883	119	0	0
1884	111	0	0
1885	110	0	0
1886	129	0	0
1887	125	0	0
1888	117	0	0
1889	120	0	0
1890	104	0	0
1891	131	0	0
1892	145	0	0
1893	131	0	0

cattle still exists, but according to Koop, it has not been exercised since 1930 (Koop, 1981, p. 6). Before that time, Nitzschke claimed that cattle still occasionally grazed in the area (Nitzschke, 1932, p. 13).

Following the strong decline in grazing which finally stopped altogether, the main trees to grow were young hornbeam and beech. There was no young oak at all (Nietsch, 1927; Nitzschke, 1932). Initially, holly (*Ilex aquifolium*), blackthorn (*Prunus spinosa*) and hawthorn (*Crataegus* spp.) were still widespread, and there were thick copses of hazel (Wehage, 1930; Nitzschke, 1932, pp. 18–19). According to Nitzschke, these species are characteristic in areas where oak trees grow. He found no oak trees in the age category of 10–50 years. Some oak seedlings did appear, but they disappeared after 3 or 4 years. The pasture in the reserve became covered with beech (Nitzschke, 1927; 1939, p. 97). On the other hand, according to Nitzschke, beech was represented in every age category. In the end, beech supplanted the large oaks which died. In his opinion, the so-called primeval forest in this way turned into a pure beech forest (Nitzschke, 1932, pp. 14, 16, 19–20, 24, 28–30).

In 1933, beech trees were felled in the Neuenburger Urwald in order to provide space for the oaks. This was done to safeguard the character of the wood-pasture. In the winter of 1949–1950, mostly beech trees were felled for firewood in more than half of the area of the reserve at the time (Koop, 1981, p. 12; Pott and Hüppe, 1991, p. 85). The part of the 'grosse Schaar' where this took place is no longer part of the reserve today (Koop, 1981, p. 12). In 1968–1970, another 50 large beech trees were felled in the reserve to create space for the old oaks, with the motto: 'The oak murderers must be removed' (in the words of the forester at the time, Meinrenken, cited by Koop, 1981, p. 12). There was also an aim to preserve the original wood-pasture as a cultural monument. With this aim in mind, young oak trees were planted.

5.3.2 The present situation

Nowadays, the Neuenburger Urwald consists of a closed forest with oak–beech forest (*Fago–Quercetum*), millet grass–beech forest (*Milio–Fagetum*) and oak–hornbeam forest (*Stellario–Carpinetum*). (Koop, 1981, pp. 17–31, 39; Pott and Hüppe, 1991, pp. 80, 90–92). The oaks which grow there are pedunculate oak.

A large part of the whole reserve has a thick undergrowth of holly. In some places, there are striking pollarded trees and trees which have been trimmed into the shape of candelabras. The latter are mainly hornbeams. The remaining old oaks are distinguished by their enormous size and shape. They are 25–30 m tall, and have a broad crown which starts low down on the trunk. This indicates that they grew in open conditions (Pott and Hüppe, 1991, pp. 80, 86, 89, 90). The beech trees are predominant in the reserve and the few oak trees are being overgrown by beech. They have partly died away (Koop, 1981, pp. 17, 23). Underneath the trees, there is a shrub layer consisting of holly (Koop, 1981, pp. 17–21, 23). In part of the forest, there is a thin tree layer of oak, with a second tree layer of hornbeam and beech below this. The beech, holly, rowan and hornbeam regenerate in the gaps in the canopy. In the oak–hornbeam forest, the holly and beech are least widespread. In these areas, the main trees to regenerate are rowan and hornbeam. However, where beech seedlings are numerous, they eventually replace the rowan and hornbeam.

More and more young beech trees are appearing throughout the reserve. It is as though the beech is successively taking over the whole forest. In the oak–hornbeam forest, this process is taking place more slowly than in the other two types of forest. In the oak–beech forest, this development has already resulted in a pure beech forest, while beech forms the shrub layer everywhere in the millet grass–beech forest (Koop, 1981, pp. 18–21, 23, 30–31; H. Koop, Wageningen, 1997, personal communication).

There is virtually no regeneration of oak at all. In his transects, Koop (1981) found a single young oak about 1 m tall in an open gap. He did not find any hazel or blackthorn at all (Koop, 1981, p. 31; H. Koop, Wageningen, 1997, personal communication). From groups of young trees, historical sources and a reconstruction of the regeneration which took place in the last century, Koop (1981) deduced the decline per hectare in the number of oak trees. The results are presented in Table 5.4. This table shows that the share of the oak trees in 1780 declined by almost 75% by the 1980s, and that alder disappeared altogether (Koop, 1981, pp. 31, 52; 1982). The annual mortality of oaks is of an order similar to that in the Forêt de Fontainebleau. The few young oak trees which do grow in the Neuenburger Urwald are all badly affected by being eaten

Table 5.4. Reconstruction of the increase or decrease in the number of trees per hectare in the Neuenburger Urwald, based on the historical data and dead trees in regeneration units. For every column, the number is also expressed as a percentage of the total (redrawn from Koop, 1981, p. 53).

Species of tree	Otto data (1780)	Reconstruction of the situation 100 years ago	Present number of trunks	Annual increase or decrease (%)
Pedunculate oak	122 (91%)	100 (60%)	20 (26%)	−0.9
Hornbeam	9 (7%)	66 (40%)	31 (39%)	+0.6
Beech	+	+	27 (35%)	+1.3
Alder	3 (2%)	−		−0.5
Total	134 (100%)	166 (100%)	78 (100%)	−0.3

and trampled by wild animals. On the other hand, many of the young beech trees have managed to survive under the closed canopy. They responded to the creation of a gap in the canopy by rapidly growing upwards (Koop, 1981, pp. 18–19, 34–37). The share of beech in the tree layer has increased. This trend is continuing, as is revealed by new recordings of the transects in 1980, 1985, 1990 and 1995 (Koop, 1981, p. 51; H. Koop, Wageningen, 1997, personal communication).

A pollen diagram from the Lengener Moor, west of the Neuenburger Holz (see Fig. 5.16) shows that since the beginning of the diagram, the Bronze Age, oak (*Quercus*) has been represented fairly constantly in the regional pollen rain with a percentage of 20–30%. In fact, the oak even achieves the highest percentage of all the species of trees. In addition, ash pollen is represented with relatively high levels. Beech pollen appears only in the Iron Age. The relative percentage never reaches the level of oak pollen (Pott and Hüppe, 1991, p. 83). The relatively high percentage of hazel is striking. It is highest in the Bronze Age. As we saw in Chapter 3, this indicates open conditions. After the Middle Ages, the percentage of grasses increases significantly. This may have been the result of the landscape becoming ever more open (O'Connell, 1986, cited by Pott and Hüppe, 1991, p. 82), but may also have arisen due to an increase in the surface area of hayland (see Hjelle, 1998 and Chapter 3).

To summarize, it may be concluded that oak and hazel do not regenerate in the closed forest of the Neuenburger Urwald, and that oak, hazel, hawthorn and blackthorn have lost or are losing the competition with beech and hornbeam. Nevertheless, the data relating to the prehistorical and historical situation reveals that the area where the Neuenburger Urwald is situated was traditionally populated by oak in the company of other shade-tolerant tree species, such as elm, lime, ash, beech and hornbeam. In historical times, when it was a wood-pasture, oak was widespread. In those days, the presence of oak may have been due to human activities, because young oaks were planted. Nevertheless, it is striking that once the grazing of cattle stopped, there was no further regeneration of oak at all, while there are all sorts of indications that this regeneration did occur in the period when there was grazing.

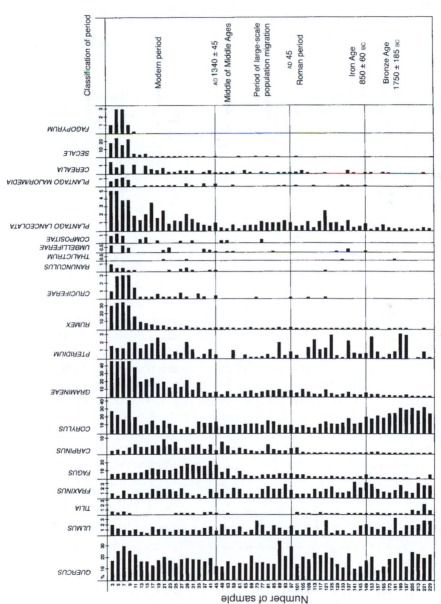

Fig. 5.16. Pollen diagram of the Lengener Moor at Spolsen, Germany (amended by Pott and Hüppe, 1991, from O'Connell, 1968; redrawn from Pott and Hüppe, 1991, p. 82).

5.4 The Hasbrucher Urwald, Germany

5.4.1 A brief description and history of the reserve

The Hasbrucher Urwald lies halfway between the towns of Bremen and Oldenburg. It consists of two parts, the 'Heu' with an area of 6.1 ha and the 'Gruppenbührer Seite' with an area of 9.4 ha. Nowadays, the 'Urwald' is part of a much larger nature reserve with an area of approximately 150 ha. This in turn is part of an even larger forest complex, the Hasbruch, with an area of 650 ha (Koop, 1981, pp. 1–2; H. Koop, Wageningen, 1997, personal communication). The oak trees in the Hasbrucher Urwald are pedunculate oak.

In 1578, there was pannage for several thousands of pigs. This points to the presence of a large number of oak trees.[8] In 1667, a report was sent to the lord that the whole of the Hasbruch was open and that therefore many hundreds of thousands of trees could be planted. This report was repeated in 1705 (Pott and Hüppe, 1991, p. 73). In 1767, approximately 100 ha of the Hasbruch was cultivated. Its size diminished to 750 ha (cf. Koop, 1981, pp. 3, 9; Pott and Hüppe, 1991, p. 73). In 1779, there were 1312 cows, 397 horses, 240 sheep and 502 pigs in the remaining wood-pasture (Pott and Hüppe, 1991, p. 73). Thus, as in the Neuenburger Urwald, there was a relatively high density of livestock. Grazing animals (horses and cattle) accounted for 745 kg ha^{-1}. In 1780, the main trees in the Heu and the Gruppenbührer Seite were oak trees which had all been planted (Otto, 1780, cited by Koop, 1981, pp. 103–104). Planting young oak trees with thorny scrub was very probably a common practice in the Hasbruch, as revealed by a report from 1794, which states that there were exclosures ('Kämpe') for cultivating hawthorn and blackthorn. A forestry plan dating from 1889 shows that the commoners were always obliged to fence off the parts of the Hasbruch where grazing cattle was not permitted. The foresters provided the wood and thorns needed for this free of charge (Koop, 1981, p. 105).

In 1870, the Heu and the Gruppenbührer Seite were proclaimed reserves (Pott and Hüppe, 1991, pp. 73–74). Grazing last took place in 1882 (Koop, 1981, pp. 6, 11, 105). The vegetation in the Heu and the Gruppenbührer Seite was described as containing single oaks or groups of oaks older than 150 years, as well as 40–50-year-old alder and pollarded hornbeam, as well as one American oak. Since then, virtually no trees have been felled or planted (Koop, 1981, p. 6). When grazing came to an end, only hornbeam and beech grew in the Hasbrucher Urwald, just as in the Neuenburger Urwald. There was no more regeneration of oak at all. The pastures became covered with beech (Nietsch, 1927; 1939, p. 97). As a result, the area gradually changed from open wood-pasture to a closed forest.

---

[8] In 1578, the following comment was made about the Hasbruch: 'Das Aste-Bruch [=Hasbruch] … Vor zweien Jahren ist die gemeine Sage gewesen, das etliche Tausend Schweine uff dem aft-Bruche vett geworden' (Pott and Hüppe, 1991, p. 73).

5.4.2 The present situation

In the present Hasbrucher Urwald, there are no longer any grasslands. It consists of millet grass–beech forest (*Milio–Fagetum*) and oak–hornbeam forest (*Stellario–Carpinetum*), with a high tree layer of pedunculate oak. Underneath this, there is a second layer of hornbeam, under which more and more beech trees are appearing (Koop, 1981, p. 33, and H. Koop, Wageningen, 1997, personal communication). Under this second layer there is another, third layer of beech trees. There is holly in the shrub layer almost everywhere. In the wettest areas, there are some oak trees with alder and hornbeam growing underneath them. In these areas, beech is also gaining more ground. In parts, there are many hawthorn shrubs (*Crataegus laevigata*) under the hornbeam. Apart from the shrubs and beech trees, there is no regeneration of other species of trees there. The growing beech trees, particularly, prevent this (Koop, 1981, pp. 33, 34, 38, and H. Koop, Wageningen, 1997, personal communication).

The enormous size of the old oak trees in the Hasbrucher Urwald shows that they developed in an open landscape. They are 25–30 m tall and have broad crowns (Pott and Hüppe, 1991, p. 90). Since 1978, Koop has not found a single young oak tree in his transects of the forest, and only one small tree 1.5 m tall outside it, which died in 1995 as a result of being trampled by wild animals (H. Koop, Wageningen, 1997, personal communication). In contrast, young beech trees managed to survive under the closed canopy despite being trampled by wild animals. This advance regeneration responds quickly to gaps in the canopy with a strong growth spurt (Koop, 1981, pp. 18–19, 34–37).

To summarize, the share of oak is also falling significantly in terms of density, while the share of beech continues to increase (Koop, 1981, pp. 31, 51–52; 1982). This trend continued until 1995, when the last records were drawn up in the permanent transects (H. Koop, Wageningen, 1997, personal communication).

5.4.3 Gaps in the canopy in the Hasbrucher Urwald and Neuenburger Urwald

On the basis of dead trees and the diameter of the regeneration units, Koop (1981) determined the distribution and size of gaps which had developed in the canopy of the Neuenburger Urwald and the Hasbrucher Urwald in the past 100 years for every type of forest. The measurement he used for the size of the gaps was the ratio between the diameter of the open spaces and the maximum height of the trees standing on the periphery of the gaps. The height of these trees was 30 m (see Koop, 1981, p. 52). Figure 5.17 shows the frequency of gaps of different sizes. Regeneration areas larger than 2–3 times the maximum height of the tree (approximately 0.2–0.3 ha) were found only in low numbers (Koop, 1981, pp. 50–52; 1982). The average ratio was smaller than 0.5 times the height of the tree (= 175 m^2 = 0.018 ha). Separate gaps can join together to

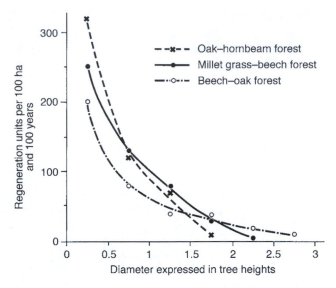

Fig. 5.17. Frequency distribution of the size of regeneration units, expressed as the product of the tree height of neighbouring trees on 100 ha, over a period of 100 years in the Neuenburger and Hasbrucher Urwald (redrawn from Koop, 1981, p. 52).

form larger ones. Thus several gaps created by one tree develop into regeneration units with an elongated shape (Koop, 1981, p. 52). In these larger gaps, pioneers such as birch, rowan as well as pedunculate oak become established. The latter are overtaken in height by the other two species in the course of time (Koop, 1982). In the oak–hornbeam forest, the open spaces reached a maximum diameter of 1–1.2 times the maximum height of a tree (= 30–45 m = an area of 700–1600 m^2 = 0.07–0.16 ha); in the millet grass–beech forest, 1.2–2 times the height of a tree (= 45–60 m = an area of 1600–2800 m^2 = 0.16–0.28 ha); and in the beech–oak forest, 2–2.2 times the height of a tree (= 60–75 m = an area of 2800–4400 m^2 = 0.28–0.44 ha) (Koop, 1981, pp. 51–52; 1982). The size of these gaps and the frequency distribution corresponds to that in La Tillaie in the Forêt de Fontainebleau, as shown in Fig. 5.11 (see Lemée *et al.*, 1992). The size of these gaps also corresponds to that found in deciduous forests in the temperate zone in the east of the United States and Japan (see Bormann and Likens, 1979a, p. 132; Barden, 1981; Runkle, 1982; Cho and Boerner, 1991; Clinton *et al.*, 1993; Abe *et al.*, 1995; Tanouchi and Yamamoto, 1995). There, the maximum size is between 0.1 and 0.15 ha (Runkle, 1982); an area which was found in La Tillaie after the storm of 1967 (Lemée *et al.*, 1992).

It can be concluded from the figures presented above that the gaps which develop in the canopy in European forests are representative of those which develop in forests in temperate conditions in the northern hemisphere. On the basis of the developments both in La Tillaie and in the Neuenburger Urwald and the Hasbrucher Urwald, it must be concluded that in these 'normal' gaps in the canopy, some seedlings of sessile and pedunculate oak do germinate and can even grow into young trees, but ultimately they are not able to become permanently established.

5.5 Sababurg in the Reinhardswald, Germany

The reserve of Sababurg in Germany was established in 1907. It is part of the Reinhardswald. This lies to the north of the city of Kassel and has an area of 27,000 ha. The reserve itself covers 92 ha (Swart, 1953; Flörcke, 1967, pp. 3, 47). The oaks found there are sessile oak and turkey oak (*Quercus cerris*). The area is named after the castle of Sababurg, which was built by the lord near the Reinhardswald. By way of protection, the castle was surrounded in the 16th century by thick hedges of thorny scrub, known in German at the time as 'Hecken'. Apparently, it was this thorny scrub which inspired the Brothers Grimm to write the fairy tale *Sleeping Beauty* (Flörcke, 1967, pp. 38–39).

The inhabitants of the settlements on the edge of the Reinhardswald traditionally had the right to collect firewood and timber in the 'Wald' and to send pigs out to pannage and cows, horses and sheep to graze (Flörcke, 1967, p. 35). According to a regulation dating from 1748, they used to drive 6000 head of cattle, 3000 horses, 20,000 sheep and 6000 pigs into the 'Wald' every year in the 16th and 17th centuries (Flörcke, 1967, pp. 42–43; Anonymous, 1978). This results in a grazing density of 107 kg ha^{-1}.

In the adjacent town of Hofgeismar, as many as 20,000 pigs were sent out to pannage in the Reinhardswald in good mast years (Flörcke, 1967, pp. 42–43). The regulation dating from 1748 also states that one-third of the area had to be fenced off from grazing livestock (Flörcke, 1967, p. 43). The so-called 'Blumenhute' was established for large livestock at that time. This meant that grazing was restricted in the 'Wald' to the period from 1 May to the end of August; the period when flowers bloom, hence the name 'Blumenhute'. The cattle had to graze in the presence of a cowherd. Sheep and pigs could be sent into the Reinhardswald in autumn and winter (Anonymous, 1978). As in many places, young oaks were cultivated in nurseries and when commoners were allocated a tree for timber, they were obliged to plant a certain number of young oak trees after taking their tree, and to protect them against livestock by surrounding them with thorns (Flörcke, 1967, p. 45; Anonymous, 1978).

The sessile oaks in the 'Urwald Sababurg' look like oaks which grew in an open area, i.e. they have large crowns starting low down on the trunk (Flörcke, 1967, pp. 63, 85, 91). Figures 5.18 and 5.19 show some oak trees. Flörcke (1967) found only three young, spindly oak trees, approximately 20–30 cm tall, in the whole reserve. One of these was freestanding while the other two grew underneath young beech trees. The seedlings were crooked and did not have a central stem (Flörcke, 1967, p. 96). The number of large oak trees declined between 1875 and 1978 by 31%, i.e. from 172 to 119 (Anonymous, 1978).

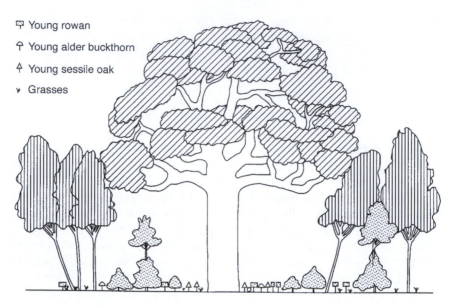

Fig. 5.18. Under an old mast-producing sessile oak (crown with diagonal shading) in the Sababurg Reserve in Germany, young birch trees (vertically shaded crowns) and young beech trees (dotted crowns) have grown spontaneously on a partly grassy forest floor. In addition, there are rowan, alder, buckthorn and sessile oak seedlings under the old oak tree (from Flörcke, 1967, p. 107).

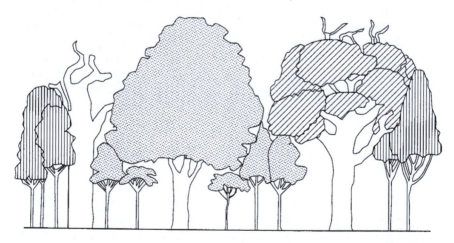

Fig. 5.19. Mixed oak–beech forest (the crown of the oak has diagonal shading, the beech crown is dotted, the birch crown has vertical shading). One of the oaks has died: to the left, between the birch and the beech. The young birch and beech trees are between 12 and 15 years old. There are no seedlings at all. The forest floor is covered with a thick layer of leaf mould (from Flörcke, 1967, p. 56).

This is an annual mortality of 0.36%; a mortality rate lower than that in La Tillaie, Le Gros-Fouteau and the Neuenburger Urwald. Other species of trees and shrubs also disappeared from the reserve. In the 1930s, wild apple trees (*Malus sylvestris*) was still widespread and hawthorn was very common (Bock, 1932/33). In the 1970s, there were only four to eight wild apple trees left of which only three or four blossomed (Flörcke, 1967, p. 68). There was hardly any hawthorn or blackthorn left (Flörcke, 1967, pp. 67, 69, 99). On the other hand, the beech trees did regenerate. They grew under the oak trees and replaced them (see Figs 5.18 and 5.19). Therefore the prognosis is that the 'Urwald Sababurg' will also develop into a pure beech forest (Flörcke, 1967, p. 104; Anonymous, 1978).

Flörcke (1967) attributed the lack of regeneration in the former wood-pasture to browsing by wild animals. He explained this by referring to fenced-off parts of the Reinhardswald, where there were many young oak trees as well as the large-scale regeneration of other species (Flörcke, 1967, p. 96). The density of wild animals in these fenced-off parts amounted respectively to 1.8 and 3.3 deer 100 ha^{-1} (Flörcke, 1967, p. 58). Flörcke (1967) does not indicate what he means by the regeneration of oak in these fenced-off areas. Therefore it is not possible to determine whether this involves oak seedlings or developing young trees. Nor is it clear whether this is a case of 'natural regeneration' in the sense used in forestry, i.e. regeneration which is accompanied by widespread and intensive human intervention on behalf of the oak, as we see in Appendix 5. Therefore it is not possible to draw any conclusions from Flörcke's comparison (1967). However, the failure of oak to regenerate in this reserve once again coincides with the end of grazing.

5.6 Rohrberg in the Spessart, Germany

5.6.1 A brief description and history of the reserve

The Spessart is a forest area in Central Germany which covers more than 100,000 ha of forest nowadays. It includes two forest reserves, namely Rohrberg, with an area of 10 ha, and a reserve in the forest area of Rotenbuch, with an area of 8 ha (Endres, 1929e; Lödl *et al.*, 1977). In Rohrberg, the spontaneous developments in the forest were studied (see Lödl *et al.*, 1977). It lies in the centre of the Spessart, halfway between the towns of Aschaffenburg and Würzburg. The oaks found there are sessile oaks.

The Spessart was one of the last wildernesses in Central Europe to be cultivated (Schubart, 1966, pp. 39, 145). It was given in loan to the Archbishop of Mainz by Duke Otto I, with the consent of Emperor Otto II, as a 'Bannwald' or 'forestis' in 982 (Vanselow, 1926, p. 1). As lords, the Prince-Bishops of Mainz gave colonists the right to settle in these 'wildbanne' in the 14th century (see Vanselow, 1926, p. 171 *et seq.*). In the following centuries, cattle were sent to graze, pigs were put out to pannage, wood and litter were collected, and a great deal of charcoal was burnt for glass and iron furnaces. Above all, burning charcoal in combination with grazing cattle, resulted in a partial deforestation and vegetation consisting of heather (Vanselow, 1926, pp. 7, 17–19, 144–146, 158). In the middle of the 16th century, oak and beech were referred to as the most important species of trees, as well as birch, aspen (*Populus tremula*), sallow (*Salix caprea*), hazel, wild cherry (*Prunus avium*), wild apple and wild pear (*Pyrus pyraster*) (Vanselow, 1926, pp. 6, 190). The protection of 'fruitful' trees, such as oaks, in accordance with the 'Waldrecht' continued until the 18th century (Vanselow, 1926, p. 23, also see Appendix 4). Pigs were sent out to pannage in the Spessart, as late as 1914 (Herrmann, 1915). The commoners' right to graze livestock and take foliage lasted until 1952.

As we saw in Chapter 4, the Spessart had an important place in the development of the shelterwood technique. The 'Mainzer Forstordnung', issued by the Prince-Bishop of Mainz in 1744, and the subsequent 'General-Verordnung', dating from 1774 (see Vanselow, 1926, pp. 37, 222–227), played an important role in this. These sources, as well as other sources dating back to the 15th century, do not mention the regeneration of oaks or measures aimed at this. Vanselow (1926) concluded from this that there was no regeneration as a result of the grazing of livestock (Vanselow, 1926, pp. 45–46, 165). According to the regulations dating from the 18th century, such as the 'Mainzer Forstordnung' and the 'General-Verordnung', beech trees did regenerate during this period (see Vanselow, 1926, pp. 65, 222–227). As we saw in Chapter 4, references to the regeneration of beech refer to the pole forests developing at the time.

The two reserves were established in the Spessart in 1928 and the last vestiges of the wood-pasture were to be preserved there. These were the remains of approximately 5000 ha of wood-pasture which still existed in the forest area of the Spessart in 1837 (Vanselow, 1926, p. 75; Endres, 1929b; 1929e). The oaks in both reserves are sessile oaks. They are between 200–300 and 500–600 years old, and possibly 800 years old. Since the creation of the reserves, there has been no human intervention, except that valuable pieces of the trunks of old dead oak trees were removed (Endres, 1929e; Lödl *et al.*, 1977). Livestock no longer graze in these former wood-pastures. I do not know when grazing ended in the Rohrberg reserve described below.

5.6.2 The present situation

In the Rohrberg reserve, research has been carried out into the spontaneous developments in the vegetation. The stimulus for this was the finding that the original open oak forest in the natural reserve was gradually changing into a closed beech forest (Lödl *et al.*, 1977). According to Lödl *et al.*, the question was how the regeneration of pedunculate and sessile oak, which require light, had taken place in the original primeval forest in Europe in the presence of the shade-loving beech trees, when the oak is not only disappearing in the reserves in the Spessart, but also in other forest reserves. In his opinion, this question had not been satisfactorily answered. The data given below have been taken from the results of the study carried out by Lödl *et al.* (1977). They linked this study to their activities in the oak reserve, Johannserkogel, in the 'Lainzer Tiergarten' in the Wienerwald in Austria. The results of this study are examined below.

The trees in the Rohrberg reserve consist of 11.2% sessile oak, 87.8% beech and 1.0% conifers. Judging by the shape of their crown and trunk, the sessile oaks grew in an open situation. They are trees with a large crown. Only one young oak was found in the reserve. This had a trunk with a diameter of 2 cm and a crown with a diameter of 2 m. This young oak tree was not considered viable, because of the danger of being eaten or trampled by wild animals, and because it was growing under a canopy. Therefore there is no new generation of oak trees. This contrasts with beech trees, which are represented in very large numbers in the younger categories with trees with smaller diameters (see Fig. 5.20). Beech grows up so vigorously that the shortest large oak trees in the reserve are being surpassed by beech trees and are dying. Lödl *et al.* (1977) expect that only the tallest and most vital oaks will survive.

Historical sources indicate that 70% of the forest consisted of oak and 30% of beech. The ratio of 11% oak and 88% beech found by Lödl *et al.* (1977) shows that there was a high mortality among the oaks. Gaps created by the death of oak trees are immediately filled with young beech trees. Because no young oaks grow into trees, the oak population increasingly consists of tall old oak trees which are gradually being replaced by new generations of beech trees. Figure 5.21 illustrates this process in terms of the height of the trees. In an adjacent oak forest under forestry management, the Eichhall reserve, in which the tree layer is dominated by oaks, 350 years old, and beech is widespread in the lower layer, the oak population is being maintained because the beech trees are removed. This illustrates the fact that oaks can survive in areas where beech are

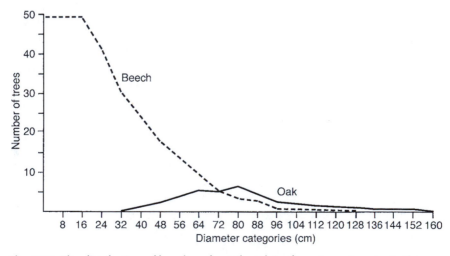

Fig. 5.20. The distribution of beech and sessile oak in diameter categories in the Rohrberg reserve in the Spessart, Germany (redrawn from Lödl *et al.*, 1977, p. 299).

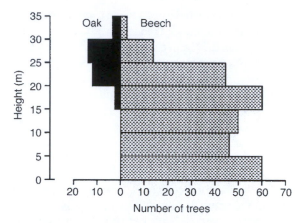

Fig. 5.21. The distribution of sessile oak and beech in the Rohrberg reserve in terms of height (redrawn from Lödl *et al.*, 1977, p. 301).

also found only if the beech trees are regulated. This is also the experience of other productive forests where oak and beech grow side by side.[9]

The final conclusion of the study is that in the Spessart, the open wood-pasture dominated by oak, will change into a closed beech forest within 100 years as a result of spontaneous development. Although it is not clear when grazing came to an end, the results of the study demonstrate that there is no regeneration of oak in this reserve, and that it will be replaced by beech if there is no grazing and no further intervention by man.

5.7 Priorteich in the Southern Harz, Germany

5.7.1 A brief description and history of the reserve

The forest reserve of Priorteich lies in the southern Harz mountains near Walkenried in Germany. It has an area of 7.8 ha, and consists of pedunculate oak, approximately 300 years old, beech with an average age of 170 years, hornbeam 115 years old, ash (*Fraxinus excelsior*) and sycamore. In addition there are wild cherry, fluttering elm (*Ulmus laevis*), sallow and rowan. The forest is classified as a millet grass–beech forest (*Milio–Fagetum*) (Raben, 1980, p. 6). The area of the study was fenced off. For more detailed information, refer to Raben (1980), and Jahn and Raben (1982). The data given below have been taken from Jahn and Raben (1982), unless indicated otherwise.

For centuries, the forests, of which the Priorteich is part, were coppice-with-standards to meet the needs for firewood and charcoal for metalwork furnaces. The

---

[9] See: Schwappach (1916), Bühler (1922, pp. 328, 358, 433, 566), Vanselow (1926, pp. 92–93, 110–115), Endres (1929b), Boden (1931), Meyer (1931, pp. 356–357), Wiedemann (1931), Seeger (1930; 1938), Hesmer (1958, p. 57), Krahl-Urban (1959, pp. 165, 188, 212), Dengler (1990, p. 224).

ore was mined in the Harz mountains. Timber (oak for half-timbered houses), grazing, hay-making and collecting litter were less important. In 1755, it was described as a coppice consisting of birch, aspen and hornbeam, with a few solitary oaks and the odd solitary beech. In view of the light in the coppice, it must have been a very open forest. Grazing livestock was allowed in the freshly cut coppice only when the young shoots on the stools of trees had grown out of the reach of browsing livestock. In 1896, the coppice-with-standards was transformed into a standing forest. It was described as a forest of oak, beech, hornbeam and a few birch and aspen trees, between 50 and 140 years old, as well as ash, sycamore and field maple in the wetter areas of the forest. At this time, the coppice was 22–24 years old and consisted of hornbeam, birch, aspen, smooth-leaved elm (*Ulmus procera*) and hazel. The transformation involved allowing the shoots of the coppiced wood to grow into trees. In 1936, the forest was proclaimed a reserve ('Landschaftsschutzgebiet'). In the context of the regeneration measures, trees were then felled on the periphery. In another part, the large old oaks were felled in 1960.

5.7.2 The present situation

As Fig. 5.22 indicates, beech, hornbeam and sycamore form a young stand of trees, which can be identified as the new developing generation of these species. The figure also shows that there is no regeneration of pedunculate oak in the

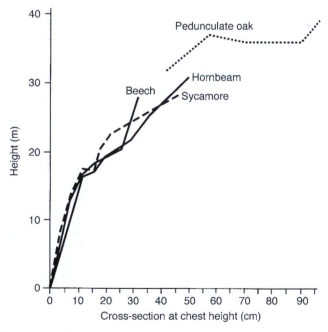

Fig. 5.22. The growth of pedunculate oak, hornbeam, beech and sycamore (height) in relation to the diameter of the trees at chest height in categories of 4 cm (redrawn from Jahn and Raben, 1982, p. 724).

Table 5.5. Number of seedlings and young trees up to a height of 1.3 m ha^{-1}, in three types of forest in the Priorteich reserve, Germany (redrawn from Jahn and Raben, 1982, Table 2).

Species	Type 1	Type 2	Type 3
Pedunculate oak	1,000	900	—
Beech	1,000	900	600
Hornbeam	5,000	200	500
Sycamore	28,600	30,200	33,100
Birch	—	100	—
Wild cherry	1,100	500	500
Ash	300	200	—
Wych elm	—	100	100
Rowan	700	200	100
Alder buckthorn	100	—	—
Total number of seedlings	37,800	33,300	34,900

reserve. There are seedlings, in some areas between 900 and 1000 ha^{-1} (see Table 5.5). However, they do not grow into the categories with thicker trunks. They all appear to die. There is also a high mortality rate of sycamore seedlings. Of the average 30,000 seedlings per hectare, only a small percentage survives, though the species does grow into the category with thicker trunks and therefore does eventually regenerate.

The species referred to by Jahn and Raben (1982) as light-demanding species, wild cherry and sallow, are suppressed. They are in the minority compared with the shade-tolerant species, beech, hornbeam and sycamore (see Fig. 5.23). Only a few wild cherry trees with thicker trunks survive. Aspen, which were widespread in the past, are completely lacking. In addition, the study shows that the pedunculate oak has virtually disappeared from the scene, and that the wild cherry is only rarely found as a tree. Jahn and Raben (1982) expect that in the future there will be no oak, birch, rowan and sallow, and that there will be very few wild cherry trees left. According to Jahn and Raben (1982), the oak has no chance of regeneration in the future, because when gaps do develop in the canopy, there is already prolific regeneration of other species, which have already germinated under the canopy. They then rapidly grow into trees in the gap (Jahn and Raben, 1982).

The same development took place in two other forest reserves that were coppice-with-standards. The first, the Bechtaler Wald is located in South Baden in Germany and is 12.5 ha in size. The data that follow come from Reif *et al.* (1998). It would originally have been a high forest with mainly oaks. Around 1750, it would have been transformed to a coppice-with-standards. The current crown consists of 80% of hornbeam and pedunculate oak. In addition, the current vegetation consists of 3–5% beech. Additionally, wild cherry, common maple, sycamore, Norway maple and a number of exotic trees such as walnut (*Juglans* spp.), locust (*Robinia* spp.) and red oak (*Quercus rubra*) occurred. Until 25 years ago, the crown was practically completely closed. In general, small

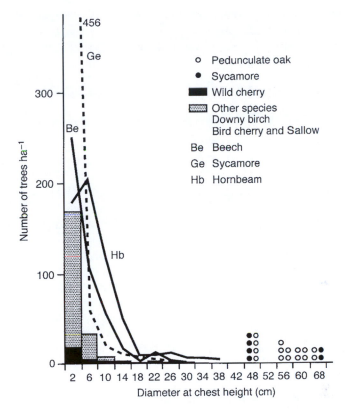

Fig. 5.23. The distribution of the diameters of pedunculate oak, sycamore, hornbeam, beech and wild cherry (*P. avium*) and other species, including silver birch (*B. pendula*), bird cherry (*P. padus*) and sallow (*S. caprea*) per hectare in categories of 4 cm (redrawn from Jahn and Raben, 1982, p. 725).

gaps have arisen in the canopy of >100 m^2. As a result of a hurricane in 1991, three additional large gaps (>300 m^2) have arisen.

Of the Norway maple, up to 30,000 plants ha^{-1} in the class >20 cm were found in the forest below 13 separate old and large gaps in the canopy, of the sycamore up to 16,000 ha^{-1} and of the hornbeam up to 4000 ha^{-1}. The oak, together with beech, common maple, and bird cherry in the same class, was one of the rarest species. Of these species, the beech was the most numerous, the oak the least common. Of the oak, up to a number of thousands of seedlings in the class > 20 cm were found. However, it did not appear at all in the classes above 130 cm. Sycamore and Norway maple were mainly represented this class. In particular the advance regeneration of the Norway maple and hornbeam made immediate use of the increased amount of light on the ground, which resulted from the gaps that had arisen in the canopy, to quickly grow in height. The young oaks occurred more in larger gaps than in smaller ones. Here the more shade-tolerant common maple and Norway maple were represented in much greater numbers. Nevertheless, the light-demanding oak does not succeed in penetrating the higher size classes (>130 cm). More shade-tolerant species such as sycamore, Norway maple and hornbeam do succeed in doing this. Therefore, everything indicates that the pedunculate oak does not get a chance in the regeneration of the forest and disappears. The forest will change into a vegetation dominated by sycamore, Norway maple and hornbeam. Because the beech can survive under a canopy of the Norway maple and sycamore, it is expected that they will grow above these two species in the end and suppress them. Therefore, in the long-term the beech would become the most important species in the forest (Reif *et al.*, 1998).

The second forest is called Wormley Wood and is located in Hertfordshire, UK, and is 140 ha in size. It is previous coppice-with-standards of sessile oak and hornbeam. The following data come from Le Duc and Havill (1998), unless stated otherwise.

The structure of the diameter classes of the hornbeam create an inverse J-shaped curve, which indicates that there is much regeneration, which also penetrates the higher diameter classes. The structure of the oak is a bell curve/normal curve distribution, like in the reserves La Tillaie and Le Gros-Fouteau in France of an older population that is dying out (Koop and Hilgen, 1987; Le Duc and Havill, 1998). Even if the structure in diameter is not such a good measure for age, then these data still show that there were little or no young oaks present, because only the young trees are thin. The sessile oak is progressively being replaced in this reserve by the hornbeam (Le Duc and Havill, 1998).

5.8 Forest Reserves in North Rhine-Westphalia, Germany

5.8.1 General characteristics

The survival of seedlings of pedunculate and sessile oak, small-leaved lime, beech and hornbeam has been studied in four forest reserves in North Rhine-Westphalia: in the Kottenforst and the Chorbusch, near Bonn, and the Geldenberg and the Rehsol in the Reichswald, near Kleef (Wolf, 1982; 1988). The reserves vary in size from 17 to 22 ha. The study was carried out in permanent quadrats in parts of the forest and outside, which had been fenced off from wild animals. All the reserves were exploited as coppices with standards until approximately 1850–1870. They are no longer grazed. For a detailed description of the area of this study, refer to Wolf (1982; 1988). Unless specified otherwise, the data presented below were taken from these publications.

5.8.2 The present situation

Except in the Rehsol, all the test areas have a virtually closed canopy. The crown cover varies from 80 to 100%. The cover in the Rehsol amounts to 67%. In the Kottenforst, the forest consists of an upper tree layer (height 21–30 m) of pedunculate oak, beech, hornbeam and small-leaved lime; a middle layer (height 17–18 m) of pedunculate oak, hornbeam and small-leaved lime (i.e. no beech); and a lower layer (height 5–12 m) of beech, hornbeam and small-leaved lime (i.e. no pedunculate oak). In the Chorbusch, there is no beech. The upper tree layer (height 24 m) is formed by pedunculate and sessile oak, and both the middle layer (height 13–15 m) and the lower layer (height 6–7 m) consist of hornbeam and small-leaved lime. In the Geldenberg, the upper tree layer (height 24 m) consists of sessile oak and beech, and the middle layer (height 17 m) consists of beech. There is no lower tree layer. In the Rehsol, the upper tree layer (height 19–24 m) consists of pedunculate and sessile oak, beech and hornbeam. The middle layer (height 14–15 m) consists almost solely of beech. Hornbeam occurs only sporadically. There is no lower tree layer here either. In the Geldenberg and the Rehsol, there is no small-leaved lime. In terms of the sociology of plants, the forests of the Kottenforst and the Chorbusch are classified as oak–hornbeam forest (*Stellario–Carpinetum*) and in the Geldenberg and Rehsol as oak–beech forest (*Fago–Quercetum*). In the test area, the experimental sites were partly fenced to keep out roe deer (*Capreolus capreolus*), fallow deer (*Dama dama*), wild boar (*Sus scrofa*) and rabbit (*Oryctolagus cuniculus*).

The numbers of seedlings in the experimental sites which were not fenced (A), and which were fenced (B), are shown in Table 5.6. In 1977, the seedlings

Table 5.6. Numbers of seedlings per m² in a pedunculate oak–hornbeam forest and an oak–beech forest, outside (A) and inside (B) fencing in the test sites (redrawn from Wolf, 1982, p. 480).

| Species | Year/ sample | Pedunculate oak–hornbeam forest | | | | Oak–beech forest | | | |
| | | Kottenforst | | Chorbusch | | Geldenberg | | Rehsol | |
		A	B	A	B	A	B	A	B
Pedunculate oak[a]	1977/9	2297[a]	2312[a]	432[a,b]	132[a,b]	9[b]	4727[b]	31[b]	5471[b]
Sessile oak[b]	1978/5	4	0	2	0	4	0	4	4
	1979/5	0	0	0	0	0	1	3	0
	1980/6	0	5	0	0	0	1	0	0
Beech	1977/5,7	56	52	1	0	14	227	1	7
	1978/5	37	20	0	0	2	0	0	5
	1979/5	0	2	0	0	23	0	16	7
	1980/6	13	2	0	0	0	2	46	44
Hornbeam	1977/5,7	1910	30	3685	3607	0	0	0	0
	1978/5	5303	267	2155	665	0	0	172	90
	1979/5	275	30	180	180	0	0	1	17
	1980/6	92	2	115	105	0	0	0	2
Small-leaved lime	1977/5,7	2620	652	612	160	0	0	0	0
	1978/5	442	70	335	57	0	0	0	0
	1979/6	617	107	295	227	0	0	0	0
	1980/6	85	60	775	542	0	0	0	0

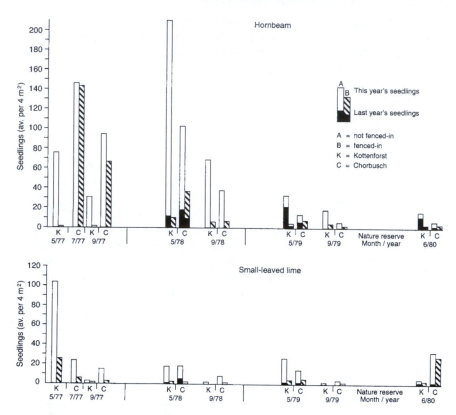

Fig. 5.24. The numbers of hornbeam and small-leaved lime which have germinated in an oak–hornbeam forest, and the extent to which they survive for the first year, outside (A) and inside (B) the fencing in the test sites (redrawn from Wolf, 1988, p. 169).

grew from a full mast (= up to 200 acorns m^{-2} (Dengler, 1990, p. 63)) in 1976. During the period 1970–1980, virtually no acorns were produced, so that there were no new seedlings. In fact, the period 1977–1980 shows the development of the cohort of oak seedlings following the full mast in 1976. Table 5.6 shows a rapid decline in the number of seedlings. The highest numbers were achieved in 1977, followed by a strong decline in numbers in the subsequent years.

In contrast with the oak, the hornbeam and small-leaved lime produced fruit every year, resulting in seedlings every year, as shown in Fig. 5.24. The difference between experimental sites which were fenced and those which were not fenced is particularly great in the oak–beech forest. In the part which was not fenced off, nearly all the acorns were eaten before they germinated. In the oak–hornbeam forest, there were some oak seedlings, but the hornbeam with 30,000–50,000 seedlings per hectare was by far the most prolific, as shown in Fig. 5.25.

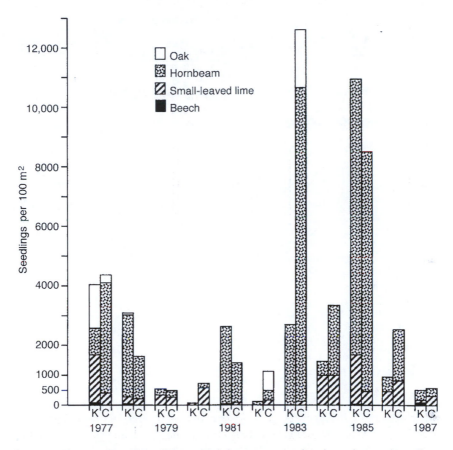

Fig. 5.25. The number of seedlings which have grown, of pedunculate and sessile oak, small-leaved lime, beech and hornbeam per 100 m², in the forests of Kottenforst (K) and Chorbusch (C), in a period of 11 years. The values are the averages of test sites which were and were not fenced in (*n*=40) (redrawn from Wolf, 1988, p. 169).

A much higher percentage of pedunculate oak seedlings survived in the oak–hornbeam forest in the Kottenforst within the fenced-off area than outside it (see Fig. 5.26). After 4 years, there were still 23% of the original number of seedlings left in the fenced-off area, while there were only 3% left outside. In the oak–hornbeam forest in Chorbusch, there was a much more rapid decline, as indicated by Fig. 5.26. Observations in the oak–hornbeam forest in the Kottenforst and the Chorbusch over a period of 11 years, reproduced in Fig. 5.25, clearly show that the presence of oak seedlings is related to the masts of 1976 and 1983. This figure shows that when there is a full oak mast in the Chorbusch, there is a significantly larger number of hornbeam seedlings, and in every year, the hornbeam has been represented by a significant number of seedlings both in the Kottenforst and in the Chorbusch. Compared with the oak,

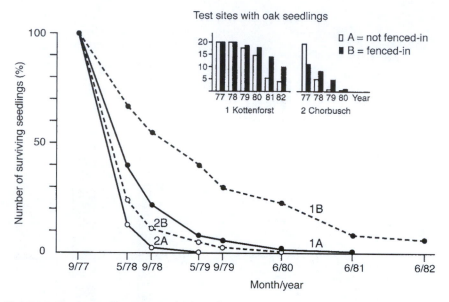

Fig. 5.26. The percentage survival of seedlings of pedunculate and sessile oak which germinated after the mast of 1976 in test sites which had not been fenced in (A), and sites which had been fenced in (B), in the forests of Kottenforst (1) and Chorbusch (2). The number of test sites to which the observations related in that year is shown top right (redrawn from Wolf, 1988, p. 170).

the small-leaved lime is also strongly represented. Furthermore, seedlings of these species grow up every year, as indicated in Fig. 5.24, so that there are seedlings of these two species in years when there are no oak seedlings; the hornbeam seedlings even occur in relatively large numbers.

However, in both reserves in the oak–hornbeam forest, there were no young oaks >50 cm. On the other hand, hornbeam and small-leaved lime were represented in the lower tree layer (height in the Kottenforst, 5–12 m, and in the Chorbusch, 6–7 m), the hornbeam by 11 and 17 trunks ha^{-1}, and the small-leaved lime by 63 and 52 trunks ha^{-1}, respectively, in the Kottenforst and in the Chorbusch. This indicates that the seedlings of these two species develop into trees, which is not the case for pedunculate and sessile oak. Thus all the seedlings of the pedunculate and sessile oak eventually disappear. This suggests that the pedunculate and sessile oak are suppressed by hornbeam and small-leaved lime. Figure 5.25 gives a clear indication that the oak has no chance of survival because of the predominance of small-leaved lime and hornbeam, and the success of these species in growing into trees. In view of this, this even applies when there is a full mast, as was the case in 1976. Predation by wild animals also has a negative effect on oak, as fewer oak seedlings survive in the areas which are not fenced off than in the fenced-off areas, according to Fig. 5.26, although more small-leaved lime and hornbeam seedlings survive there, according to Fig. 5.24. Fencing off areas of forest against wild animals therefore benefits oak, and the predation by wild animals in the forest has a negative effect on oak. In view of the number of oak seedlings in the fenced-off areas of the experimental sites, it is not probable that this practice can in itself lead to the successful establishment of oak, as it still has to compete with small-leaved lime and hornbeam in the fenced-off areas.

5.9 The Oak Reserve Johannser Kogel in the Wienerwald, Austria

5.9.1 A brief outline and history of the reserve

The oak reserve Johannser Kogel in the Lainzer game reserve contains pedunculate oak, 200–300 years old. In addition to this species, there are also turkey oaks. The reserve was faced with the problem of the regeneration of sessile oak. The forest had reached a stage where there should have been regeneration of sessile oak according to Leibundgut's regeneration model (1959; 1978) for primeval forests. As we saw in Chapter 2 this stage is also known as the ageing phase ('Alterphase') or sometimes the breakdown or dieback phase ('Terminalephase'). The regeneration should theoretically take place at the start of this stage. The young trees have an opportunity to grow because the canopy of the forest has become thinner as a result of the loss of branches and death of trees. In the breakdown or dieback phase ('Terminalephase'), the seedlings grow into young trees, which evolve into the future forest. However, the theoretically expected regeneration of sessile oak did not take place in the Johannser Kogel reserve (Mayer and Tichy, 1979). According to Mayer and Tichy (1979), this problem also arose in the oak reserve of Rohrbrunn in the Spessart.

The area of study lies in the north-western part of the Lainzer game reserve. Records dating from 1457 refer to the establishment of a game park in this region. Between 1711 and 1740, the imperial family acquired a large part of the game park which existed at the time. Empress Maria Theresa had a wall 22.6 km long built around the game park between 1772 and 1781. This enclosed area became the exclusive hunting domain of the imperial court. The oaks in the reserve date from before the time the wall was built. It is assumed that after the completion of the wall, the density of game increased significantly, to 50 head 100 ha^{-1}. In 1921, the game park was opened to the public, and in 1941, the oak reserve was established (Tichy, 1978, p. 6; Mayer and Tichy, 1979). The data presented below are taken from Tichy (1978), Mayer and Tichy (1979) unless indicated otherwise.

5.9.2 The present situation

The area of this study consists of oak–hornbeam forest (*Galio–Carpinetum*) and beech forest (*Fagetum*). The oak–hornbeam forest covers 62% of the area and the beech forest the remaining 38% (Mayer and Tichy, 1979). The broad crowns of the sessile oaks show that they grew in very light conditions (Mayer and Tichy, 1979). The distribution of the trees, which determines the appearance of the forest, in different categories of thicknesses is shown in Table 5.7.

Table 5.7. Distribution in categories of thicknesses, of different species of trees per hectare (see Tichy, 1978, Table 4).

Forest type	Species	Thickness categories (cm)									Total sum of all thickness categories
		2–6	8–14	16–22	24–34	36–50	52–70	72–90	92–110	112+	
1	S.o.	2	1	0	0	0	0	0	0	2	5
	T.o.	3	2	1	0	0	1	4	1	0	12
	Hb	1539	278	5	15	38	10	1	0	0	1886
	F.m.	50	17	0	0	2	0	0	0	0	69
	Bee	232	23	0	0	0	0	0	0	0	255
	Ash	6	14	1	0	0	0	0	0	0	21
	Bir	47	22	0	1	1	1	0	0	0	72
	Total	1879	357	7	16	41	12	5	1	2	2320
2	S.o.	4	0	0	0	0	12	66	44	13	139
	T.o.	1	0	0	0	7	26	10	3	1	48
	Hb	1044	78	4	28	39	11	1	0	0	1205
	F.m.	0	0	0	0	0	1	0	0	0	1
	Bee	11	3	0	1	1	1	1	1	0	19
	Ash	20	4	1	1	1	0	0	0	0	27
	Total	1080	85	5	30	48	51	78	48	14	1439
3	S.o.	0	0	0	0	0	1	4	6	4	15
	T.o.	10	130	1	0	0	2	3	1	0	147
	Hb	6800	2470	1	3	6	5	1	0	0	9286
	F.m.	30	170	0	0	0	0	0	0	0	200
	Bee	0	230	0	0	0	0	0	0	0	230
	Total	6840	3000	2	3	6	8	8	7	4	9878
4	S.o.	0	1	0	0	1	0	2	5	2	11
	T.o.	0	1	0	0	1	2	2	1	0	7
	Hb	285	53	8	2	15	10	1	1	0	375
	F.m.	60	18	7	2	9	12	4	0	0	112
	Ash	8	2	60	24	4	0	0	0	0	98
	Total	353	75	75	28	30	24	9	7	2	603
5	S.o.	0	0	0	0	0	0	0	0	1	1
	T.o.	0	0	0	0	0	0	0	1	0	1
	Hb	468	64	2	8	14	3	1	0	0	560
	F.m.	39	9	5	12	3	0	0	0	0	68
	Bee	92	3	0	2	5	10	12	8	4	136
	Ash	2	9	8	2	0	0	0	0	0	21
	Bir	5	7	0	0	0	0	0	0	0	12
	Total	606	92	15	24	22	13	13	9	5	799
6	Bee	0	0	0	0	0	136	34	0	0	170

S.o., sessile oak; T.o., turkey oak; Hb, hornbeam; F.m., field maple; Bee, beech; Ash, ash; Bir, birch. Forest types 1–4 belong to the oak–hornbeam forest (*Galio–Carpinetum*), and 5 and 6 to the beech forest (*Fagetum*). Type 1, *Galio–Carpinetum circaeetosum*; type 2, *Galio–Carpinetum luzuletosum*; type 3, *Galio–Carpinetum typicum*; type 4, *Galio–Carpinetum aceretosum capestri*; type 5, *Asperulo–Fagetum circaeetosum*; type 6, *Luzula–Fagetum typicum*.

This shows that the sessile oak is not – or is hardly – represented in the category with the lowest thickness, which includes seedlings and young trees, and is completely absent in the following categories. In so far as seedlings grew in the reserve after it was fenced off, as, for example, after the masts of 1972, 1975 and 1976, they subsequently disappeared. Hornbeam is found in large numbers in the categories of lowest thicknesses. Beech and field maple are also well represented in these categories. Beech seedlings are most common in the beech forest.

Figure 5.27 shows the distribution of heights in a number of categories of the sessile oak and turkey oak, which require light, on the one hand, and hornbeam and beech, which tolerate shade, on the other hand. In this figure, Mayer and Tichy show the change of generations in the reserve by means of chronosequences. The process of regeneration, which they call the transformation stage, is shown from top left to bottom right. It starts with what Mayer and Tichy call the climax forest, where the oak is still well represented. According to Leibendgut's model, this is followed by the replacement of old trees, including oaks by a new regeneration of trees. Going from top left to bottom right, it is clear that in this stage, species which tolerate shade such as hornbeam, beech and field maple, grow on a large scale, moving into the categories of greater thicknesses. Sessile oaks are almost entirely absent in the bottom categories, with the exception of a single specimen. The oaks found there are virtually all turkey oaks. According to Mayer and Tichy (1979), the single sessile oak which occurred in the bottom layer of vegetation vegetated and had absolutely no chance of penetrating into the upper layer of the vegetation.

Turkey oak, which is found in the categories of lower thicknesses in the transformation stage, tolerates shade better than sessile oak. However, the turkey oaks which were found in the advanced transformation stage were confronted by a much larger number of young trees of species which tolerate shade even better than they do themselves. The last figures, shown bottom right, indicate the end of the transformation stage, i.e. the conclusion of the regeneration. They show that the species which tolerate shade have penetrated into the categories of the greatest thicknesses and that the number of oaks has dramatically reduced. The cycle represented by the figure illustrates that as the forest passes through this regeneration cycle more often, the number of oaks is further and further reduced. This is also apparent from the distribution of the trees in categories of thicknesses at the end of the terminal stage; the stage in which the regeneration penetrates into the tree stage. In the stage in which, according to Leibundgut's regeneration model, the regeneration of the reserve of the trees which form the canopy is concluded, the oak plays no role at all. This stage is completely dominated by beech and hornbeam. The final stage of the successive regenerations will be the complete replacement of sessile oak by species which tolerate shade such as beech, hornbeam and field maple. The final result will be a climax consisting solely of shade-tolerating species, such as the beech climax.

Almost everywhere in the reserve, the most common seedlings of shade-tolerant species, which are most numerous in the bottom layer of the vegetation,

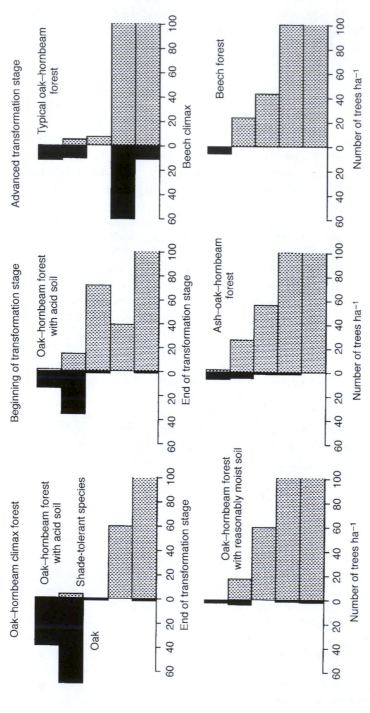

Fig. 5.27. From left to right, the figure shows the process of regeneration in the Johannser Kogel reserve, as constructed on the basis of chronosequences. The forest goes through the stages in accordance with Leibundgut's regeneration model (1959; 1978). The regeneration is called the transformation stage because the forest changes from a structure consisting of old trees into one consisting of increasing numbers of young trees. Sessile oak does not regenerate in this. The young oaks in the so-called typical oak–hornbeam forest are all turkey oaks. Only shade-tolerating species such as hornbeam, field maple and beech regenerate. Eventually a climax consisting purely of shade-tolerating species develops, such as the beech climax, bottom right (redrawn from Mayer and Tichy, 1979, p. 207).

are those of hornbeam (up to almost 10,000 specimens per hectare) (see Fig. 5.28). Throughout the reserve, the hornbeam also dominates the lowest layer of young trees in the forest (see Fig. 5.29). In a number of areas of the reserve, the hornbeam even represents 80–99.5% of the total number of trees in this layer. In some of the reserve, this species accounts for 80% of the trees, in the middle layer, and in one part, even for 80% of the highest tree layer.

These facts show that in the oak–hornbeam forest, the hornbeam is replacing the sessile oak and turkey oak. In the beech forest, sessile oak and turkey oak have already virtually disappeared. Only beech and hornbeam regenerate in the gaps which develop in the canopy of this type of forest, and hornbeam is dominant in the larger gaps. In conclusion, it may be said that everything indicates that in general, sessile oak is disappearing from the reserve, while the closed forest will be dominated by beech and hornbeam.

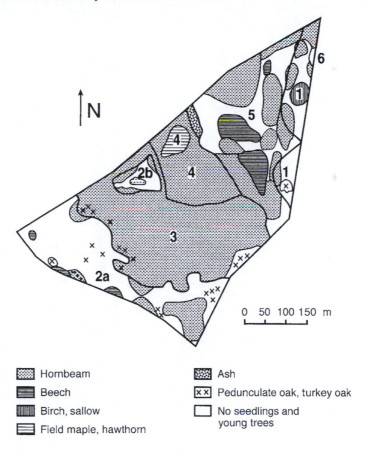

Hornbeam

Beech

Birch, sallow

Field maple, hawthorn

Ash

Pedunculate oak, turkey oak

No seedlings and young trees

Fig. 5.28. The transformation in the Johannser Kogel reserve. The figure shows the cover of young trees with a diameter up to 17 cm at chest height. The numbers indicate the different types of forest shown in Table 5.7. The density of young trees varies in the reserve (redrawn from Mayer and Tichy, 1979, p. 216).

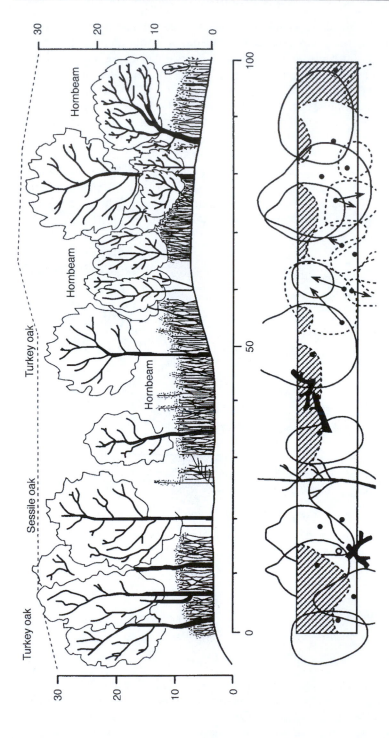

Fig. 5.29. A cross-section of an old stand of oak (sessile oak and turkey oak) in an oak–hornbeam forest, with prolific regeneration of hornbeam. The bird's eye view shows this regeneration with shading. The measurements are in m (redrawn from Mayer and Tichy, 1979, p. 217).

5.10 The So-called Primeval Forest Reserve of Krakovo, Slovenia

5.10.1 General characteristics

Mayer and Tichy (1979) gave the so-called primeval forest reserve in Krakovo in eastern Slovenia as an example of a forest where the oak survives in the presence of the shade-tolerant hornbeam. The reserve has an area of 40.5 ha and is close to the Croatian border. The oaks which grow there are pedunculate oaks. The forest canopy is usually thin and consists mainly of pedunculate oak. Below this there is a middle layer which is alternately dense and thin, and consists mainly of hornbeam. In addition, the trees in this layer are field maple, fluttering elm (*Ulmus laevis*) and native alder (*A. glutinosa*) (see Fig. 5.30). The vegetation is characterized as a type of oak–hornbeam forest. The forest has been used as wood-pasture (Peterken, 1996, pp. 41–42). Part of this reserve is flooded regularly because of the high water levels. In this part there is virtually no hornbeam. For this reason, Mayer and Tichy distinguish this wet part from the rest and call it a *Pseudostellario–Quercetum* (referred to as below as *Quercetum*). They describe the rest of the reserve as *Pseudostellario–Carpinetum* (referred to below as *Carpinetum*).

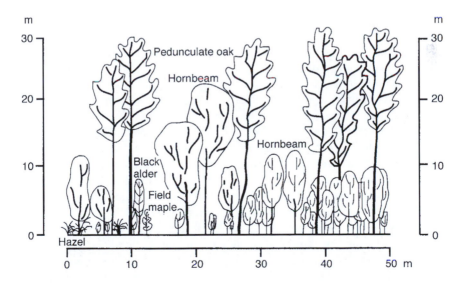

Fig. 5.30. A transect in the oak–hornbeam forest in the Krakovo reserve, Slovenia. There are no young pedunculate oaks growing at all. On the other hand, there is regeneration of hornbeam (redrawn from Mayer and Tichy, 1979, p. 222).

5.10.2 The present situation

In the Krakovo reserve, enormous numbers of seedlings of pedunculate oak and hornbeam have become established. In the *Carpinetum*, the number of seedlings of pedunculate oak amounted to 20,000–155,000 per hectare, compared with 8000–156,000 of hornbeam seedlings per hectare, while in the *Quercetum*, there were 20,000–342,000 pedunculate oak seedlings per hectare and between 0 and 18,000 hornbeam seedlings per hectare. The age of the large majority of seedlings (85%) varied from 1 to 4 years, while the remainder (15%) fell in the age categories of 5–10 years.

According to Mayer and Tichy, the regeneration of the forest in the reserve starts after the start of the terminal stage. This begins with the decay of one or more trees. As indicated in Section 5.9.2 on the Wienerwald, this is the start of the transformation during which the regeneration of the forest takes place according to Leibundgut's model of regeneration (1959; 1978), and the old generation of trees is ultimately replaced by a new generation. Because there is extra light in this stage, Mayer and Tichy believe that the *Carpinetum*, where the number of pedunculate oak and hornbeam seedlings are more or less in balance, will consist of a 'Dickung' consisting of both species. (A 'Dickung' is a layer of young trees up to 15 cm tall, the lower branches of which have not yet died as a result of a lack of light (Dengler, 1990, p. 169).) As the forest declines, they believe there will be competition between pedunculate oak and hornbeam. According to Mayer and Tichy, pedunculate oak has the advantage that when a gap develops in the canopy, the seedlings can respond to the increase in the amount of light by growing upwards very powerfully. On the other hand, they believe that hornbeam has the advantage that it can survive underneath the canopy because of its higher tolerance to shade. This species can wait underneath the canopy until a gap develops. In the *Quercetum*, which is regularly flooded, Mayer and Tichy believe that a 'Dickung' will develop consisting solely of pedunculate oak because of the enormous number of pedunculate oak seedlings in comparison with those of the hornbeam (ratio 45 : 1). The competition will then take place only between the young pedunculate oak themselves, so that the regeneration eventually consists of only pedunculate oak at the start of the terminal stage.

Therefore, Mayer and Tichy (1979) expect that when the forest in the reserve enters the terminal stage of decay, and the canopy becomes thinner because of the successive death of trees, the oak will eventually penetrate into the top layer. They did not think that this stage had been reached in the whole reserve, or even at all when the vegetation was recorded. They indicated that in the *Carpinetum*, there were no young oak trees at all underneath the closed canopy. On the other hand, there was a middle layer consisting of hornbeam, field maple and fluttering elm (see Fig. 5.30). Therefore, there was some advance regeneration of these shade-tolerant species, but not of pedunculate oak.

Mayer and Tichy made a prediction about the regeneration of oak which I believe to be questionable. In the first place, this is because of what they saw in the Wienerwald themselves, i.e. that hornbeam completely excludes oak, and in the second place, because of what they found in this reserve, i.e. that hornbeam does penetrate the middle layer, but oak does not. Mayer and Tichy based their prediction to a large extent on a publication by Acetto (1975) on the Krakovo reserve. They based the figures on the dominance of pedunculate oak and the fact that 85% of the young oak trees were in the age categories 1–4 years, while 15% fell in the category 5–10 years. It is not clear whether the oaks grew on into even older age categories. Neither Acetto's publication, nor that of Mayer and Tichy mentions this. The oak is the only tree in the reserve present in the canopy. Therefore oak dominates the plant layer as a seedling, and the crown layer as a tree. The above-mentioned recordings of the vegetation of the primeval forest by Mayer and Tichy (Fig. 5.30) reveal, as indicated above, that in the *Carpinetum*, only hornbeam, field maple and fluttering elm penetrate to the middle layer, while oak does not. On the basis of the information provided by Acetto and Mayer and Tichy, I believe that it must be concluded that as far as the regeneration of oak in the reserve is concerned, only seedlings and young trees up to 10 years old are found in the presence of a shade-tolerant species such as the hornbeam, which is also represented by many seedlings and which penetrates the higher tree layers. This situation is also found in all the other reserves described up to now, without this resulting in the successful regeneration of oak. Therefore, in my opinion, the expectation that oak will survive in the Krakovo reserve next to hornbeam is unrealistic. The reserve can therefore not be given as an example of an area where oak survives as a species which requires light, next to the shade-tolerant hornbeam, in a spontaneously developing oak–hornbeam forest, even though Mayer and Tichy describe it as such. There is clear evidence to the contrary: the advance regeneration of hornbeam, field maple and fluttering elm, so that it is more likely that pedunculate oak will eventually be ousted by these shade-tolerant species.

5.11 The Unterhölzer at Donaueschingen in the Black Forest, Germany

The Unterhölzer at Donaueschingen is the second reserve which Mayer and Tichy (1979) gave as an example of an area where oak survives in the presence of shade-tolerant species; in this case, beech. Other authors (Reinhold, 1949; Kwasnitschka, 1965) who carried out research in this forest, also shared this view. The oaks in this reserve are pedunculate oak. The reserve is situated in the eastern part of the Black Forest. It is surrounded by forests of silver fir (*Abies alba*) and Norway spruce (*Picea abies*). The forest itself is an oak–hornbeam forest (*Galio-Carpinetum*). According to Mayer and Tichy (1979), no tree had ever been felled in this forest up to the middle of the 18th century. In 1939, it was proclaimed a reserve (Kwasnitschka, 1965).

There is an evaluation of this forest area in the archives dating from 1787 (Reinhold, 1949). The forest to which this evaluation referred had an area of

380 ha. This source states that it was a very thin forest (20–40 trunks ha^{-1}), dominated by a layer of trees of different ages, with pedunculate oak 300–500 or 600 years old, a height of 24–26 m and a trunk diameter at chest height of 180 cm. In addition, there were 250-year-old beech trees with a trunk diameter at chest height of 120 cm. According to the evaluation, the oak and the beech were accompanied by many wild fruit trees such as crab apple, wild pear and wild cherry. In addition, there were lime trees (it does not specify which species), as well as sycamore and hornbeam, some extremely fast-growing, long crowned Norway spruce, some silver fir, as well as Scots pine and sallow. In addition, the evaluation stated that there was a great deal of hazel, hawthorn and juniper (*Juniperus communis*) (Reinhold, 1949; Mayer and Tichy, 1979). Reinhold (1949) described the forest as a natural oak–hornbeam forest on the basis of this evaluation, and therefore did not recognize that oak was being replaced by beech at all. According to Mayer and Tichy, oak survived, despite the fact that there were a great deal of beech trees in 1787. They thought that the reason for this was that the soil consists of clay, a soil which is ideal for pedunculate oak, and that there are late frosts, which beech does not tolerate very well. In addition, they thought that there was a particular type of pedunculate oak adapted to the area concerned.

In 1782, one fence was placed in the Unterhölzer Wald for pigs, and one for deer. During the Napoleonic Wars, inhabitants from the area brought their domestic cattle to the area for safety (approximately 6000 head of cattle). The publications do not indicate clearly in what area this occurred. Subsequently, Reinhold (1949) indicated that grazing rights continued to apply for 600 cows. In his view, this meant that the regenerated oak trees were destroyed in a short while, while the regenerated beech were able to survive locally. The number of old oak trunks was reduced to 5 ha^{-1}, and in some places to 40 ha^{-1}, where trees were exploited for timber. Hazel, alder, sallow and thorns gained the upper hand at the same time. Subsequently, the species were completely chopped down in 1830 when Norway spruce, Scots pine and larch (*Larix* spp.) were sown. Only a small area of 270 ha with 600-year-old oak and beech up to 250 years old, as well as a few wild fruit trees, survived. This was protected as a park for fallow deer and red deer (*Cervus elaphus*). In the 1980s there were 155 ha left, with 20–40 oak trees per hectare and a very grassy forest floor.

On the basis of the evaluation in 1787, Reinhold (1949) reconstructed the composition of the population of oak, beech and wild fruit trees, as well as the heights of the oak trees at that time, using the data on the diameters of the trees (see Fig. 5.31). According to Reinhold, the spread of ages of the oak trees which are up to 660 years old, is very even. From this, Reinhold concluded that oak survives in the presence of beech, and that the character of the remaining forest is close to that of a natural forest. As a reference for the natural forest he uses a selection system forest, i.e. a forest with trees of different ages where there is constant regeneration, as described in Chapter 4, in gaps in the canopy which form when one or a few trees are removed. According to Kwasnitschka (1965), the share of the oak in the species diversity has remained unchanged since 1787, i.e. 74.5%. The oak trees were up to 600 years old and had an average

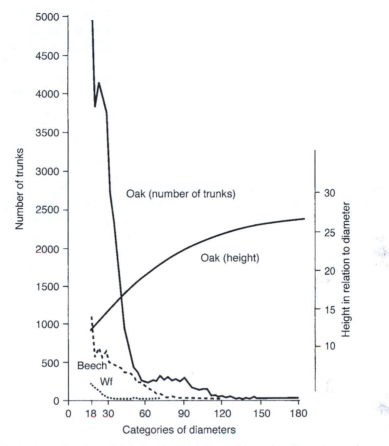

Fig. 5.31. A reconstruction of the population composition of pedunculate oak, beech and wild fruit trees, such as wild apple, wild pear and wild cherry, based on an evaluation dating from 1787. Wf = wild fruit trees (redrawn from Reinhold, 1949, p. 694).

age of approximately 300 years. According to Kwasnitschka (1965), the oak successfully regenerated to an age of 40, and on average to an age of 20, over a large area. The regeneration usually started in rings under the outer edge of the crowns, and then developed concentrically. According to Kwasnitschka, the whole of the forest floor will be eventually be covered with seedlings in this way.

The fact that at the time of the evaluation in 1787, there were species such as hawthorn, blackthorn, juniper and wild fruit trees, indicates that there was a wood-pasture at that time. This also applies for the population structure of oak. As we saw in Chapters 2 and 4, there are proportionally many young oaks on the outer edge of thorny scrub. The pattern of the concentric regeneration of oaks described by Kwasnitschka, is very similar to the regeneration of oaks in clonally expanding thorny scrub of blackthorn in wood-pastures; a process which was described in detail in Chapter 4 as a phenomenon in the New Forest. This also explains the large number of young and very young oaks described in the evaluation. As we showed in Chapters 2 and 4, there are many oak trees on the fringes of thorny scrub. As a result of the clonal expansion of blackthorn with the help of rootstocks, they advance concentrically into the grassland with trees following in their wake. According to the historical sources there was some grazing in the Unterhölzer, so the process described above was one of the possibilities. The data available do not clearly show when this grazing came to an end, but in view of the age of the young oaks given by Kwasnitschka, this cannot have been much longer than 50 years ago, unless they were young oaks which had been planted. The thorny scrub, which was clearly absent at the time of Kwasnitschka's observations (1965), but which was widely present at the time of the evaluation in 1787, may have disappeared because of the shade of the young growing oaks; a process which, as we noted in Chapters 2 and 4, was described in the 'commons' in the Chilterns in England by Watt (1934b), among others. It may also have been cut down; a measure which was widely used in forestry in the last century, because thorns were considered to be weeds that had to be eliminated, as we saw in Chapter 4 and in Appendix 4. This had already happened once before in the Unterhölzer in 1830. I believe that at the time of the evaluation in 1787, contrary to the views of Reinhold, Kwasnitschka, Mayer and Tichy, there was no closed forest in the reserve where oak survived in the presence of beech. The situation suggested by the evaluation has all the characteristics of a wood-pasture where the regeneration of oak takes place in thorny scrub, in the presence of grazing, as we saw in Chapters 2 and 4.

5.12 Dalby Söderskog, Sweden

5.12.1 A brief description and history

Dalby Söderskog (hereafter referred to as Dalby) is a National Park in the extreme south of Sweden in the province of Scania. It has an area of 36 ha, and was proclaimed a National Park in 1918. Since then, there have been no interventions in the vegetation (Malmer *et al.*, 1978). Biogeographically, southern Sweden includes the distribution area of pedunculate oak, sessile oak, small-leaved lime, wych elm (*Ulmus glabra*), beech, hazel, one-styled and two-styled hawthorn (*Crataegus monogyna* and *C. laevigata*) and blackthorn (see, *inter alia*, Mayer, 1992, p. 96 *et seq.*). The region where Dalby is situated is just outside the range of large-leaved lime and sycamore, but just within that of sessile oak and hornbeam (see Mayer, 1992, pp. 99, 106, 111 and Fig. 1.5). The border of the range of pedunculate oak is much further north (see Mayer, 1992, p. 111). Southern Sweden is also part of the original range of the large herbivorous species such as aurochs (*Bos primigenius*), European bison (*Bos bonasus*), red deer, roe deer, elk (*Alces alces*) and European wild horse (*Equus przewalski gmelini*) (see Heptner *et al.*, 1966, pp. 479, 487, 499, 863).

  Dalby was an unfree common of a lay sovereign (see Appendix 1). As common ground, it was grazed by domestic cattle, and wood was taken for firewood and building purposes (see Andersson, 1970, pp. 33–36). The area next to it was used above all for grazing horses from a studfarm which owned Dalby (Lindquist, 1938, pp. 260–262). The grazing ended at the end of the 19th century. At that time, it was mainly horses which grazed there. In 1914–1916, many trees were felled in the area, including large, heavy oak trees. In 1918 Dalby was proclaimed a National Park with the aim of retaining the vegetation at the time (Malmer *et al.*, 1978). The data given below have been taken from Malmer *et al.* (1978) unless indicated otherwise.

5.12.2 The present situation

There are floristic data available for the area dating from 1925, 1935 and 1969, as well as other supplementary data from 1909 (see Lindquist, 1938; Persson, 1980). The oak in this reserve is pedunculate oak. An evaluation from 1900 shows that this did not regenerate at the time, possibly because of competition for light with elm, beech, ash and hazel (Lindquist, 1938, p. 263). Because the hazel formed a closed layer of shrubs in many places, casting a deep shade, Lindquist (1938) expected that the hazel would prevent trees from germinating and growing. Therefore he expected Dalby to develop as a pure hazel wood (Lindquist, 1938, p. 266).

Measurements of the diameter of more than 1000 pedunculate oak in the reserve in the 1970s by Malmer *et al.* (1978) show that the larger diameters predominated at the time (see Fig. 5.32). Only 2.5% of the trees had a diameter smaller than 25 cm, therefore there were hardly any young oaks. Since 1935, the number

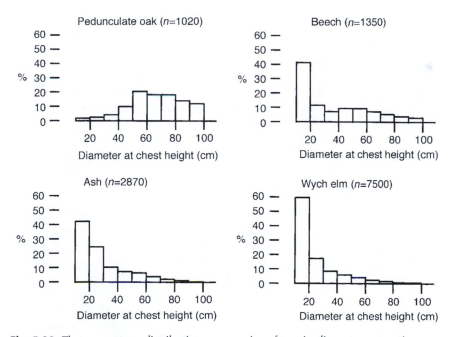

Fig. 5.32. The percentage distribution per species of tree in diameter categories per species of pedunculate oak, beech, ash and wych elm in Dalby Söderskog, Sweden. Only trees with a trunk diameter of >10 cm at chest height are included (redrawn from Malmer *et al.*, 1978, p. 20).

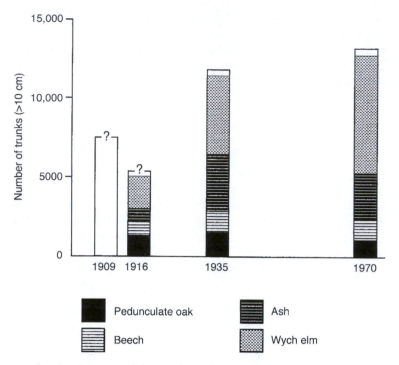

Fig. 5.33. The development of the number of trunks with a diameter of >10 cm at chest height, of the most important species of trees in Dalby Söderskog in the period 1909–1970 (redrawn from Malmer et al., 1978, p. 21). The level for 1909 is an approximate level, without further specification of the species of trees. Malmer et al. do not give any further details about these figures. As their publication gives both the volume of wood and the number of trunks, and the figures date from just before the extensive felling which took place during the years 1914–1918, it is probably an inventory of the supply of timber.

of young oaks has, moreover, declined by 26% (see Fig. 5.33) while, according to Malmer et al. (1978), at the same time, the total number of trees, particularly of wych elm as well as beech, has increased significantly, though to a much lesser extent. Although there were also relatively large numbers of young ash trees, this species has declined in absolute numbers since 1935, by 40%, as the figure indicates. Figure 5.32 shows that beech, ash and wych elm are very strongly represented by the smallest diameters, although the young trees of these species grow on into the larger diameter categories, so that there is successful regeneration.

According to Malmer et al. (1978) the wych elm dominated the highest shrub layer and the lower and higher tree layers. Between 1969 and 1975, the wych elm doubled in the lowest shrub layer, and the proportion of this species continued to increase in the highest layer, as indicated above. Therefore the wych elm was regenerating successfully. Pedunculate oak was completely absent in the highest shrub layer, and beech appeared only sporadically. In the highest shrub layer, there was hazel as well as wych elm. However, this cover had halved since 1935. The two-styled hawthorn, the hawthorn which is more tolerant of shade than one-styled hawthorn seemed to have disappeared from the higher shrub layer, though it had survived in the lower layer. Other species of shrubs and trees which require more light, such as the dogwood (*Cornus sanguinea*), spindle tree (*Euonymus europaeus*), blackthorn, guelder rose (*Viburnum opulus*) and crab apple, were only found on the periphery of the forest and on the edge of open gaps in the forest (see Fig. 5.34).

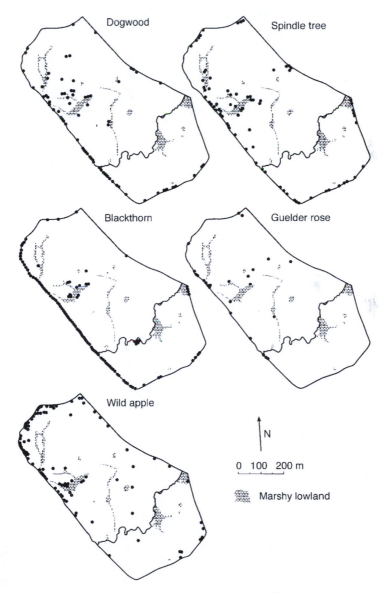

Fig. 5.34. The occurrence of dogwood, spindle tree, blackthorn, guelder rose and wild apple in Dalby Söderskog in the period 1934–1936 (redrawn from Lindquist, 1938, p. 121).

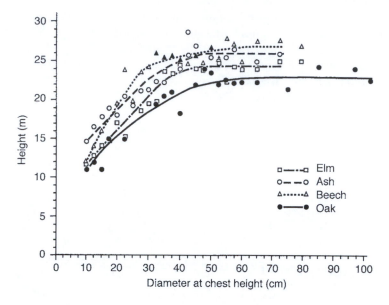

Fig. 5.35. The relationship between the average height and the trunk diameter at chest height of the most important species of trees in Dalby Söderskog from records from 1970 (redrawn from Malmer *et al.*, 1978, p. 22).

According to Lindquist (1938), many oak trees were about 250 years old, and had therefore grown in the first half of the 18th century, when cattle still grazed in Dalby. He believed that most of the large beech trees, ash and wych elm also dated from that time. According to Lindquist, grazing was less intensive then, although he did not provide any further concrete data to substantiate this.

In general, Dalby has become a denser forest since 1916. The number of trees increased, and the crown cover of trees increased from 76 to 93%. In contrast with Lindquist's expectations, the cover of hazel declined. In 1935, it was dominant, while the cover in 1970 had declined by approximately 50%. In addition, the cover of pure oak stands declined from 23 to 16%. The cause of this is probably that more and more other species reached the highest tree layer. Wych elm, ash and beech grew taller than the oak, as indicated in Fig. 5.35. As a result, the oak trees died (Lindquist, 1938, p. 263; Malmer *et al.*, 1978). In contrast with the first count, the total number of vascular plants species also declined, by 50%; it halved in only 45 years. The species of plants which dated from the time that cattle grazed, increasingly disappeared. The structure of the plant layer changed, in the sense that apart from the wettest areas, the whole forest floor became covered with perennial dog's mercury (*Mercurialis perennis*).

From the above, it can be concluded that the succession towards the potential natural vegetation (PNV) in Dalby, means for the vegetation as a whole, that it becomes less and less diverse. Oak and species of grasses and plants which are

Table 5.8. The number of trunks per hectare in the various height categories in the Vardsätra nature reserve, Sweden (from Hytteborn, 1986, p. 25).

Tree species	Height categories (m)			
	<0.5	0.5–1.3	1.3–4	>4
Pedunculate oak	0	0	0	4
Ash	88,000	2,000	860	247
Wych elm	34,000	770	696	374
Sycamore	5,000	240	54	14
Bird cherry	1,100	940	680	65
Aspen	160	60	12	0
Silver birch	0	0	0	34
Rowan	0	4	29	56

characteristic of grazed land will eventually no longer form part of this vegetation. The fact that species which are characteristic of wood-pasture such as blackthorn, hawthorn and crab apple (see Chapter 4) grow only on the periphery of the forest, could be the result of this development. In this case, these species were forced out to the edges of the forest.

The developments observed in Dalby can also be seen in the Vardsätra reserve in southern Sweden. This was also a former wood-pasture. It was proclaimed a reserve in 1912, when cattle ceased to graze there. The dense canopy of trees which then developed led to a strong decline in hazel. Records for which the results are shown in Table 5.8, show that the oak does not regenerate, while wych elm spreads and will eventually dominate the former wood-pasture (Hytteborn, 1986).

5.13 The Forest of Białowieza, Poland

5.13.1 A brief description of the forest of Białowieza

The forest of Białowieza and its two reserves, the National Park of Białowieza and the reserve for the thermophile oak forest at Czerlonka are extremely important for the subject matter of this study. I will start with a brief description of the characteristics of the forest of Białowieza. This also applies to the two reserves, because they were an integral part of the forest until the beginning of this century. The characteristics which distinguish the two reserves from the forest as a whole will be described in the paragraphs where I deal with each individually. This particularly concerns the regeneration of species of trees such as oak, small-leaved lime, hornbeam, wych elm and Norway maple (*Acer platanoides*) in the spontaneous developments which have been taking place in these reserves in the past few decades.

The forest of Białowieza lies to the east of Warsaw, on either side of the border between Poland and Belorussia, and covers an area of 125,000 ha. Of this,

58,000 ha lie in Poland and 67,000 ha in Belorussia (Falinski, 1986, p. 8). For a more detailed description of the soil and other abiotic characteristics, refer to Falinski (1986, pp. 15–35) and to Appendix 11. The dominant trees in the forest of Białowieza are Norway spruce, Scots pine, alder, (hairy) birch (*Betula pubescens*), aspen, pedunculate oak, hornbeam, small-leaved lime, ash, white willow (*Salix alba*) and crack willow (*S. fragilis*). In addition, there are bird cherry, wych elm, fluttering elm (*Ulmus laevis*), smooth-leaved elm (*U. minor*), Norway maple (*Acer platanoides*), rowan and silver fir (see Falinski, 1986, p. 43). In the forest of Białowieza the different species of trees reach exceptionally great heights. The silver fir grows to a height of 55 m, Scots pine, pedunculate oak, small-leaved lime and ash reach a height of 40–42 m, and the hornbeam a maximum of 30 m. The trees have tall, slender, branchless trunks and a narrow crown (Falinski, 1986, p. 52). There is virtually no layer with real shrubs, such as blackthorn, hazel or one-styled or two-styled hawthorn in Białowieza (Lautenschlager, 1917, p. 65; Falinski, 1986, p. 401). In the National Park, there is some hazel, as in an area devastated by storms in 1984 (H. Koop, Wageningen, 1997, personal communication). In addition, there is no beech in the forest, though this has been an important tree in the forest reserves described up to now. The reason for this is that the forest of Białowieza lies to the east of the distribution area of the beech. Other species apart from oak which were described in the other six reserves, such as small-leaved lime, hornbeam and wych elm, which are also known as species which tolerate shade, like beech, are found in the forest of Białowieza. The sessile oak is rare, although increasing numbers of oak trees are identified as sessile oak. Up to now, these two species have not been distinguished very well (H. Koop, Wageningen, 1997, personal communication). The forest area forms the eastern boundary of the distribution area of the sessile oak (Falinski, 1986, pp. 18, 20).

The forest of Białowieza is very important for the question posed by this study. As I will explain in the next section, it is one of the last areas of wilderness in Central and Western Europe to be colonized by farmers, and all the fauna of large herbivores and carnivores from the whole of Europe survived longest in the region where the forest was located. Aurochs, European bison, elk, red deer, roe deer, wild boar, European wild horse, beaver (*Castor fiber*), wolf (*Canis lupus*), lynx (*Lynx lynx*) and brown bear (*Ursus arctos*) were found in Poland up to the 17th century.[10] This fauna had been indigenous in Western and Central Europe since the Allerød (11,800 BP).[11] In addition, pollen diagrams from the central part of the forest where the National Park is situated now, show that the vegetation of this area had consistently included oak, lime and elm since the Boreal period (9000 BP), as well as hornbeam since the beginning of the Atlantic period (8000 BP) (Dabrowski, 1959; Huntley and Birks, 1983, pp. 142–160; Mitchell, 1998, Mitchell and Cole, 1998; see Fig. 5.36). Hazel was present from the

[10] Eichwald (1830, p. 249), Genthe (1918), Hedemann (1939, p. 310), Pruski (1963), Szafer (1968), Heptner *et al.* (1966, pp. 480, 491–499, 865), Volf (1979), Söffner (1982).
[11] Degerbøl (1964), Söffner (1982), Graham (1986), Aaris-Sørensen *et al.* (1990), Current (1991), Housley (1991), Stuart (1991, p. 477), Coard and Chamberlain (1999).

Boreal period, in comparison with oak, lime and hornbeam with high relative percentages of pollen. Moreover, the forest of Białowieza is extremely important for this study because changes in the area, which took place in Central and Western Europe in the Middle Ages, did not take place there until the 16th century. The first colonist-farmers settled there at that time and introduced cattle. Furthermore, the area is important because it is one of the few lowlands of Europe where small-leaved lime and hornbeam are traditionally found as well as oak. These species were not first coppiced and then returned as trees in the forest when the coppice evolved in the 18th and 19th centuries as in the rest of the lowlands of Central and north-west Europe.

5.13.2 The history of the forest of Białowieza

From the 15th century, the respective lords used the forest of Białowieza for hunting (Derkman and Koop, 1977, p. 8; Falinski, 1986, p. 26; Schama, 1995, pp. 48–65). It was one of the last wildernesses in Europe to be colonized. In the 16th century, the Duke of Lithuania gave colonists the right to settle, cultivate and exploit what is now the forest. This was done in accordance with the rules which applied for the 'forestes' in Central Europe (Hedemann, 1939, p. 301; De Monté Verloren and Spruit, 1982, p. 192).[12] The settlement was restricted to the periphery of the wilderness (Hedemann, 1939, p. 305; Jedrzejewska *et al.*, 1997). Apart from being allowed to create fields to cultivate cereals, the colonists also had the right to graze their cattle in the 'forestes' and to make hay and take litter and wood. In the 16th century grazing cattle and making hay were the most common uses of the wilderness (Eichwald, 1830, p. 251).

In accordance with the applicable legislation, the colonists were in servitude to the sovereign lord. This meant, among other things, that they had to make hay to feed the bison and had to work as beaters for the royal hunt for 14 days a year. In the 'forestis' itself, settlements were first established in the 17th century. The village of Białowieza in the centre of the present forest dates from 1696 (Eichwald, 1830, p. 249; Derkman and Koop, 1977, p. 9; Falinski, 1986, pp. 370, 372). As a result of this and following settlements and cultivation, the area of uncultivated wilderness significantly declined in the course of the centuries, by 45% from 1664 km^2 in 1639 to 1026 km^2 nowadays (Derkman and Koop, 1977, p. 9; Falinski, 1986, p. 28). Figure 5.37 shows this decline.

Lithuania and the former East Prussia were the last refuge of the aurochs in Europe in about 1400 (see Heptner *et al.*, 1966, p. 480; Szafer, 1968). According to Hedemann (1939, p. 310), the aurochs lived in the 'forestis' of Białowieza itself until the 15th century. The last aurochs to survive lived in the

---

[12] In Poland, a translation of the Saxony Spiegel appeared in both Polish and Latin. This is a description of a mainly Eastphalian right which was issued between 1215 and 1235 by the knight, Eike von Repgau. It reflected the land rights of East Saxony (Eastphalia) (De Monté Verloren and Spruit, 1982, p. 192).

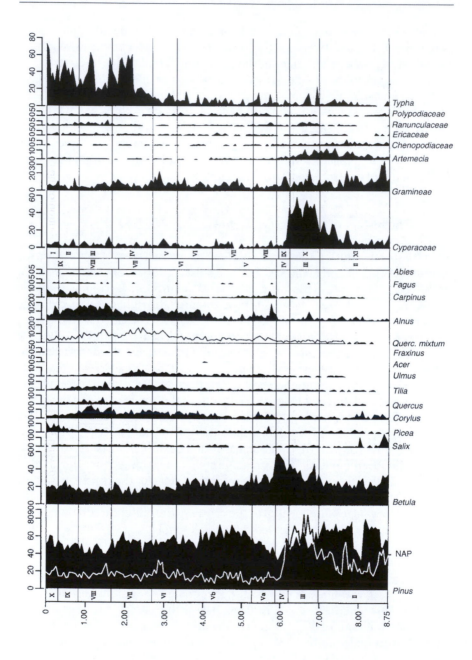

Typha
Polypodiaceae
Ranunculaceae
Ericaceae
Chenopodiaceae
Artemecia
Gramineae
Cyperaceae

Abies
Fagus
Carpinus
Alnus
Querc. mixtum
Fraxinus
Acer
Ulmus
Tilia
Quercus
Corylus
Picea
Salix
Betula
NAP
Pinus

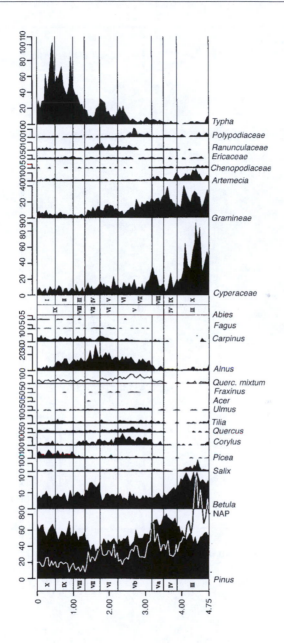

Fig. 5.36. Pollen diagrams from two peat bogs in the centre of the National Park of Białowieza (redrawn from Dabrowski, 1959, pp. 237, 238). The classification into pollen zones is taken from Firbas (see Table 1.1). The top diagram covers the period from the Allerod period (12,000 BP) to modern times in the Sub-Atlantic period. The bottom diagram starts with the Young Dryas (10,800 BP) and also ends with the modern period. Note that the distribution of the pollen frequency of grasses and plants (non-arboreal pollen, (NAP)) is shown the other way round from the pollen of trees and hazel shrubs. The sum of non-arboreal pollen is shown in the far left column in comparison with the pollen of trees and hazel.

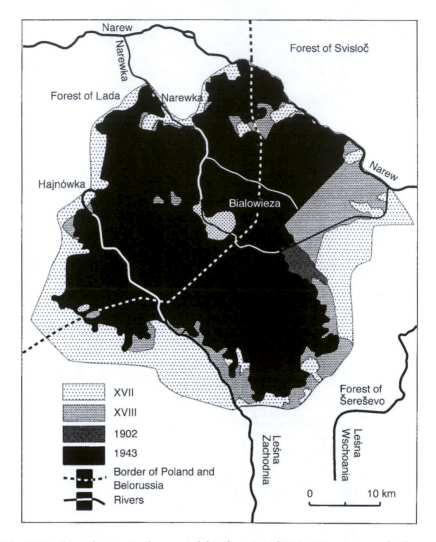

Fig. 5.37. The reduction in the area of the 'forestis' of Białowieza as a result of cultivation from the 17th to the 20th century. During this period an area of settlements appeared between the oldest known border and the present edge of the forest (redrawn from Falinski, 1986, p. 28).

16th and 17th centuries in the wilderness of Jaktorowska, an area of 20,000 ha, 60 km south-west of Warsaw. The last specimen, a cow, died there in 1627 (Szafer, 1968). An important cause for the extinction of the aurochs may have been the competition for food resulting from the pannage of pigs and grazing of horses and cattle (Szafer, 1968). When the aurochs was on the point of extinction in the 16th century, the European bison was still found in a relatively large area. This species died out in the first half of the 17th century in the steppes of

Russia, in 1762 in Romania and in 1793 in Saxony (Heptner *et al.*, 1966, pp. 491–499). Finally, Poland was also the last refuge of this species. The last wild specimen died in the forest of Białowieza in 1919. At that time there were only 54 left in zoos. In 1992, the number of European bison had increased to 3600 worldwide. More than 600 of these lived in Poland. Apart from the forest of Białowieza, where European bison were released in 1952, there are four other reserves in Poland where European bison roam free (Krasinski, 1993).

In the 17th century, the lay sovereign imposed stricter rules on the use of the wilderness of Białowieza because of the decline of the 'forestis' (Hedemann, 1939, p. 305). However, in the 18th and 19th centuries, the respective rulers, the Polish king and the Russian tsar, sold many trees to seafaring nations, such as the Netherlands, England and France, where they were used to build ships (Lautenschlager, 1917, p. 58; Wiecko, 1963, pp. 347–348). In addition, hunting continued to be important and had a great influence on the fauna. In about 1800, so many animals were killed, including many elk and European bison (Derkman and Koop, 1977, p. 9), that at that time there were virtually no large wild animals left in the area (Lautenschlager, 1917, p. 77). In the middle of the 18th century, there were only 50–60 red deer left. Eventually, this species was wiped out before 1800. In about 1800, the beaver, and in 1861, the brown bear were wiped out (Wiecko, 1963, pp. 347–348; Krasinski, 1978; Jedrzejewska *et al.*, 1997).

Jedrzejewska *et al.* (1997) indirectly came to an estimate of about 830 European bison in 1752, compared with a count of 700 by Brincken (1828) at the end of the 18th century. That number would have dropped to slightly more than 200 in 1802 (Hartman, 1939, cited by Jedrzejewska *et al.*, 1997). According to Wiecko there were 722 European bison left in 1822, and 898 in 1857 (Wiecko, 1963, p. 347). On the other hand, Jedrzejewska *et al.* (1997) mention a number of 1898 for the year 1858. In 1890, the red deer was reintroduced and the fallow deer was introduced (Jedrzejewska *et al.*, 1997). In the period from 1880 to 1915, the number of wild ungulates increased greatly. This period is well known as a time when the forest was actually a game park, where the Russian tsar could go hunting (Falinski, 1988). Because large numbers of animals were released into the forest, the density of game increased in that period (Falinski, 1986, p. 28). Between 1870 and 1890, the wolf was wiped out and until 1915 all immigrating wolves were shot (Jedrzejewska *et al.*, 1997). Figure 5.38 shows the development in the numbers of animals as reflected in the records. Fallow deer were introduced in 1892 (Falinski, 1986, p. 163). The records on the development of the number of European bison in that period are not in agreement. Wiecko (1963) states that in 1889 there were 80 living in the forest. On the other hand, Karcev (1903) states that in 1890 there were still 400 (Karcev, 1903, cited by Falinski, 1986, p. 163; Wiecko, 1963, p. 163). However, by the turn of the century it is clear that there were still approximately 700 bison in the forest of Białowieza (Falinski, 1986, p. 163). In 1914, there were 737 European bison, 59 elk, 6778 red deer, 1488 fallow deer, 4966 roe deer and 2225 wild boar (Escherich, 1917, cited by Peters 1992, p. 26). These

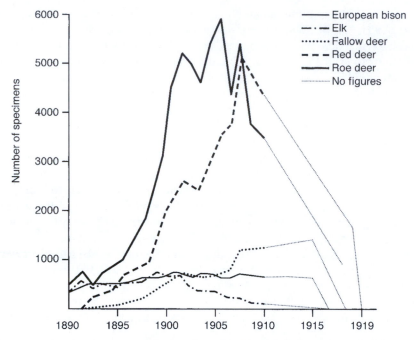

Fig. 5.38. The development in the number of wild animals from the end of the 19th to the beginning of the 20th century (redrawn from Krasinski, 1978, p. 37). In the period after 1910 the wild animals were not counted, because of the hostilities of the First World War and the chaotic situation which followed.

numbers resulted in a biomass of 20 kg ha^{-1}.[13] By way of comparison, by the end of the 1980s/beginning of the 1990s, the biomass in the exploited part of the forest and in the National Park was respectively 11 kg ha^{-1} and 16 kg ha^{-1}. In 1919, the last European bison in the wild was killed in the forest of Białowieza in the aftermath of the First World War; this meant a decline of 700 specimens in a period of 5 years.

At the turn of the century there were also large numbers of cattle grazing in the forest of Białowieza, which belonged to the inhabitants of settlements on the edge of the forest and in the forest, who had the grazing rights. No data is known from before 1780 regarding the number of livestock grazed. The numbers that are known are related to ten different years, starting with 1780, when the number was 5000–7000. The numbers are rough estimates. In 1795 livestock grazing was forbidden. Data on the subsequent period show evidence that cattle were still grazed in the area. The animals would initially have been kept

[13] The weights used for the large ungulates are: European bison, 400 kg; elk, 200 kg (Jedrzejewska *et al.*, 1994); red deer, 80 kg; fallow deer, 55 kg; roe deer, 20 kg; wild boar, 70 kg (S.E. van Wieren, 1997, personal communication, Wageningen).

on fenced pastures. After 1840, the animals would have grazed in the forest again (Jedrzejewska *et al.*, 1997). On the basis of the tsar's income from the commoners for grazing their cattle, Wiecko (1963), calculated that in 1888 there were 6348 head of cattle in the forest of Białowieza (Peters, 1992, p. 26). Jedrzejewska *et al.* (1997) reach a number of officially registered cattle between 1886 and 1909 of 6000 to 8300. They believe that the actual figures were probably much higher. About 40% of the forest was pasture at that time. Moreover, at the end of the 19th and the beginning of the 20th century, 3000–4000 head of cattle and 1000 sheep were driven by Russian merchants through the forest as the shortest route to the markets of a number of surrounding towns (Wroblewski, 1927, cited by Jedrzejewska *et al.*, 1997). According to Wiecko, pigs and sheep could enter the area freely, i.e. without payment. The actual number of livestock was therefore higher. For the period 1900–1902, Wiecko estimated the number of cattle at 4500–6000 in a grazed area of a total of 28,644 ha (Peters, 1992, p. 26).

Figure 5.39 shows the grazed areas in 1902 and 1964. The grazing was officially forbidden in the Polish part of the forest in 1973. At that time, there were still 1000–1500 head of cattle. Between 1870 and 1971 cattle formed 15–80% of the total number and 37–80% of the total biomass of ungulates (Jedrzejewska *et al.*, 1997). That means that at the time of high densities of wild ungulates around the turn of the century, the wild ungulates formed a minority at certain moments. In the Belarussian part there is still grazing (J.W.G. van de Vlasakker, Amerongen, 1999, personal communication). In the beginning of the 1990s, the number was 600 on 69 km^2. Grazing was mainly on open grassland in the forest and river valleys. Cattle formed 5% of the total number and 10% of the biomass of the total number of ungulates (Jedrzejewska *et al.*, 1997). Falinski (1966) referred to Karcev (1903), who wrote that by the end of the 19th century, there were 10,000 registered cattle grazing in the forest (Peters, 1992, p. 26). On the basis of this number and the grazed area indicated in Fig. 5.39 in 1902, this results in a biomass of cattle of 280 kg ha^{-1}; a density corresponding to that in the Forêt de Fontainebleau in the 17th century. As we saw in Chapter 4, this was 259 kg ha^{-1}. Compared with the biomass in the Neuenburger Urwald (627 kg ha^{-1}) and the Hasbrucher Urwald (745 kg ha^{-1}), this is not extremely high, but it is not so low either that it can be assumed that the cattle had no effect on the vegetation. Moreover, the density of livestock was almost certainly higher, because sheep and pigs could be sent into the area without payment, and Karcev (1903) and Jedrzejewska *et al.* (1997) also said that there was illegal grazing. The following observation illustrates that there might have been a great difference between the number of officially registered cattle and the actual number present in the forest. In 1964, 1400 cattle still had the right to graze there. On the basis of his own observations, Falinski stated that the actual number was still at least 5000 at that time; a difference by at least a factor of three (Peters, 1992, p. 26). Converted into the grazed area, this is a biomass of cattle of 665 kg ha^{-1} (= 1.9 cow ha^{-1}); a very high density for a 'forest'. By way of comparison, the biomass of wild animals (red deer, European bison, roe deer and wild boar) amounted to 29 kg ha^{-1} at that time

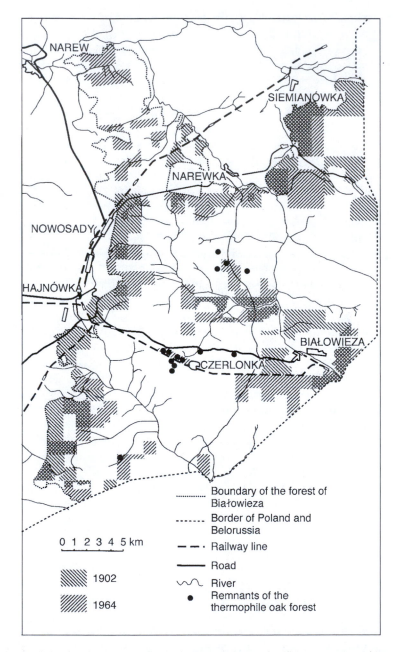

Fig. 5.39. The grazed area in 1902 and 1964 respectively, shown in units of Russian square miles (= 1.14 km² = 114 ha (Lautenschlager, 1917, p. 59)). The last remnants of the so-called thermophile oak forest (*Potentillo albae–Quercetum*) are also shown. This type of forest was believed to have developed as the result of grazing by cattle. In the 20th century, grazing was limited to the areas surrounding settlements (which are not all shown on this map) (from Falinski, 1986, p. 373).

(Jedrzejewska *et al.*, 1994). As mentioned earlier, grazing cattle was forbidden in the whole forest in 1973.

As Fig. 5.39 indicates, grazing in the 20th century is restricted to areas around the forest and the periphery of the forest. This concentration may be the result of the efforts which rulers have made since the 17th century to restrict grazing. For example, in the 17th century it was decided that cattle could venture only a quarter of a mile into the forest (Hedemann, 1939, cited by Peters, 1992, p. 25). (At that time a mile was 7.15 km (Hedemann, 1939, cited by Jedrzejewska *et al.*, 1997).) As noted above, grazing was officially stopped in the whole of the forest of Białowieza in 1964. Nevertheless, Derkman and Koop observed a flock of sheep in the thermophile oak forest near Czerlonka in the summer of 1977 (Derkman and Koop, 1977, p. 12).

In addition to grazing, there were always oaks in the forest in historic times. As mentioned above, in the 18th and 19th centuries, these were sold to countries such as the Netherlands, France and England, where they were used for shipbuilding. Descriptions dating from the 19th century indicate that at that time there were numerous oaks throughout the forest of Białowieza. Apart from the tallest trees, the middle-aged categories were also represented (Falinski, 1986, p. 228). These oaks, like hornbeam and broad-leaved lime, are represented by trees of 200, or even 300 and 400 years old (Falinski, 1988). Nowadays, the pedunculate oak is found throughout the forest as a solitary tree or in groups (Lautenschlager, 1917, p. 78; Pigott, 1975). Together with small-leaved lime and Norway spruce, these pedunculate oaks form a canopy above the hornbeam, elm, birch and willow (Falinski, 1986, p. 61). There are a few more or less pure groups of pedunculate oak in the categories of 60, 120 and 200 years old. They cover a surface area of approximately 350 ha. Foresters attribute the origin of these oaks to seedlings sown by humans (Lautenschlager, 1917, p. 78). The sessile oak is found in concentrated groups in only a few places, and never as a solitary tree. It should be noted with respect to these figures that, as remarked earlier, sessile oak is probably more common than indicated by these figures, because the species were not always accurately identified. There are no indications that oaks were planted throughout the forest. Apart from the few groups of oaks which are 200 years old and which were grown from seed, the regeneration of oaks clearly took place when there was grazing in the forest, and before, when the original wild fauna of aurochs, European bison, elk, red deer, roe deer, wild horses and wild boar were still found in the 'forestis'. This was confirmed by pollen research in the National Park (Mitchell, 1998; Mitchell and Cole, 1998).

5.13.3 The present situation in the National Park of Białowieza

The National Park of Białowieza lies in the central part of the present forest complex. This reserve has a very special place in the forest reserves of Central and Western Europe, because it is considered by many authors[14] as one of the most

[14] Pigott (1975), Rackham (1980, p. 253), Peterken (1981, pp. 4, 305), Falinski (1977; 1986, p. vii), Leibundgut (1993, p. 19), Walter and Breckle (1994, p. 76), Peterken (1996, p. 73), Bernadzki *et al.* (1998), Abs *et al.* (1999).

untouched, and therefore most natural, forests in the lowlands of Europe (see Fig. 1.1). The National Park was established with an area of 4700 ha in 1923 (Pigott, 1975; Falinski, 1986, p. 28). Every form of exploitation came to an end at that time, including the grazing of cattle (Falinski, 1986, p. 371; 1988; Peters, 1992, p. 26). Since then, the National Park has been strictly managed as a forest reserve (Pigott, 1975; Falinski, 1977; 1986, p. 8; 1988). In 1996, the National Park was increased to 10,500 ha (Bernadzki *et al.*, 1998; Bobiec, 1998).

The vegetation in the reserve is extremely varied (see Falinski, 1986, p. 119; Abs *et al.*, 1999). Almost 45% of the area consists of a continental form of the lime–hornbeam forest (*Tilia–Carpinetum*) (Pigott, 1975). The most important species in this forest are hornbeam, small-leaved lime, Norway spruce, pedunculate oak, Norway maple, aspen and wych elm (Falinski, 1986, pp. 79–81, 119). For large areas the small-leaved lime accounts for 20–30% of the canopy, and in a restricted area, even for 50%. Hornbeam forms a virtually continuous canopy underneath small-leaved lime, pedunculate oak and Norway spruce (Pigott, 1975). Other types of forest which account for a significant proportion of the reserve are the *Pino–Quercetum*, which is a transition between a deciduous and coniferous forest; the *Querceto–Picetum*, which has the character of a mixed deciduous and coniferous forest with Norway spruce, as well as pedunculate oak, aspen, birch and silver birch (*B. pubescens* and *B. pendula*); and the *Peucedano–Pinetum*, which is found only in sandy habitats. This last type of forest is dominated by Scots pine and Norway spruce; the former becomes strongly established after fire, and the latter when there was a fire long ago or never any fire at all (Falinski, 1986, pp. 89–94).

The deciduous trees in the National Park have massive, unbranching, cylindrical trunks. The forest floor is usually in deep shade with the exception of localized gaps in the canopy. There are virtually no real shrubs at all in the National Park. There are only sporadic examples of spindle tree and hazel. It is only in the southern part which was most intensively grazed until 1923, that a great deal of hazel was established in a space created by storms in 1984. According to Koop (H. Koop, Wageningen, 1997, personal communication), this evolution could have had a mainly vegetative origin, i.e. it could have grown from root systems which were already present. In that case these hazel trees could date from the period when there was grazing. There are forbs growing everywhere under the canopy, while there may be a thick covering of nettle (*Urtica dioica*) in places where more light penetrates to the forest floor (Pigott, 1975).

Pigott studied the regeneration of small-leaved lime in the National Park. He found that there were groups of seedlings and poles mainly of hornbeam, small-leaved lime, wych elm and Norway maple, and to a lesser extent, Norway spruce, everywhere under the closed canopy in associations of pedunculate oak, small-leaved lime and hornbeam. The wych elm seedlings were often correlated with thick vegetation of nettle (*Urtica dioica*), both under the closed canopy and under gaps in the canopy. In his opinion, there were virtually no seedlings of pedunculate oak, and there were no young trees at all. The seedlings and young trees of Norway maple were solitary trees, while those of other species grew in groups. These groups almost always consisted of one species. There were no seedlings of small-leaved lime or young trees of this species under old or young lime trees. They were found under the crowns of hornbeam, which, in their turn, also grew in groups. According to Pigott, this structural relation of groups of small-leaved lime under groups of hornbeam is a widespread phenomenon in the forest. Annual rings in a regeneration group of small-leaved lime indicate that if there is a gap in the canopy, some examples from a regeneration group can rapidly grow taller and fill the gap (Pigott, 1975). This is shown in Fig. 5.40.

Pigott concluded that the establishment of small-leaved lime, hornbeam, sycamore and wych elm is therefore not dependent on the development of gaps in the canopy. Admittedly they may be established there, but these gaps are also favourable for the growth of nettle, which then excludes the establishment of virtually all species of trees except for wych elm. The highest density of seedlings were usually found in the gaps in the canopy where there was no nettle. In contrast with what was found under the canopy, there were several different species growing together in that case. Pigott assumed that the unequal distribution and the occurrence of seedlings in groups under the intact canopy is related to the rooting of wild boar. They root up the mineral soil, leaving a bare floor, which then serves as a seed bed. Pigott found evidence of this in the irregular occurrence of places where pigs rooted and the irregular appearance of groups of seedlings which were, in fact, the same size as these places where the wild boar rooted.

With regard to small-leaved lime, Pigott concluded that the tolerance to shade of this tree when it is young, its great height and the enormous age which it can reach, make small-leaved lime potentially dominant in the forest. Hornbeam can

Fig. 5.40. The architecture of a transect 2 m wide, through a regeneration group of wych elm and small-leaved lime. The symbols are T = *T. cordata* (small-leaved lime); C = *C. betulus* (hornbeam); U = *U. glabra* (wych elm) and A = *A. platanoides* (Norway maple). On the left of the picture, there is a group of young limes which have grown where a gap appeared in the canopy. The stump of the tree that has disappeared is in the centre of the group (redrawn from Pigott, 1975, p. 162).

then form a sub-crown layer and be less dominant than it was in his observations. According to Pigott, the place of pedunculate oak is difficult to gauge in the whole picture, because there were only a few isolated old trees, hardly any young trees, and only sporadic seedlings. According to Pigott, little has changed as regards pedunculate oak since Paczoski (1930) studied the situation of pedunculate oak in the National Park. At that time it was also rare and there were no young trees. The density of pedunculate oak was then between 0 and 5 trees per hectare (see Table 5.9). According to Paczoski (1930, pp. 565, 570, 571), oak is disappearing in better soil as a result of competition from hornbeam, and therefore oak is surviving only in poorer soil (Paczoski, 1930, p. 571).

Derkman and Koop (1977) studied the structure and regeneration of the forest in the National Park. The results of the different recordings are shown in Figs 5.41, 5.42, 5.43 and 5.44 for each of the test hectares which were sampled. The figures show that hornbeam, small-leaved lime, Norway maple and Norway spruce were most numerous in the smaller diameter categories, i.e. in the lowest categories of the young tree stage, in the lowest categories of the tree stage and in all the categories of the pole stage. This shows that they are regenerating. This also applies for Norway maple, wych elm, silver fir and aspen, which were also represented in the lowest diameter categories. Like Pigott, Derkman and Koop found many regeneration groups of wych elm in patches of nettle (Derkman and Koop, 1977, pp. 24, 27; Koop, 1989, p. 93). There was virtually no pedunculate oak in the lowest categories in the different recordings. Derkman and Koop (1977) found three pedunculate oaks in only one recording of a pole stage. According to Koop (1989, p. 93), this was in a larger gap in the canopy. The pedunculate oak was thought to have become successfully established there, but did not have any chance against the more numerous hornbeam and small-leaved lime in this stage (Koop, 1989, p. 95). Despite this prognosis, these oaks still looked healthy in 1992 and had not been overgrown (H. Koop, Wageningen, 1997, personal communication). Derkman and Koop (1977) did find seedlings of pedunculate oak in all the test hectares (Derkman and Koop, 1977, p. 27 and appendices). Apparently, these

Table 5.9. The distribution of tree trunks in different thickness categories at chest height in a test hectare in 1930 in the National Park of Białowieza in the *Carpinetum– Pinetosum* (from Paczoski, 1928; 1930, from Walter and Breckle, 1994, p. 163).

Tree species	Diameter categories (cm)							
	10–20	20–30	30–40	40–50	50–60	60–70	70–80	80–90
Pedunculate oak	1	3	5	3	3	1	0	1
Hornbeam	14	24	66	50	6	1	0	0
Norway spruce	81	29	15	15	13	15	8	1
Scots pine	0	0	0	1	0	5	1	1
Aspen	0	0	0	0	4	0	2	0
Norway maple	1	2	1	0	0	0	0	0
Silver birch	0	1	0	0	2	1	0	0

NB There were also 35 hazel shrubs.

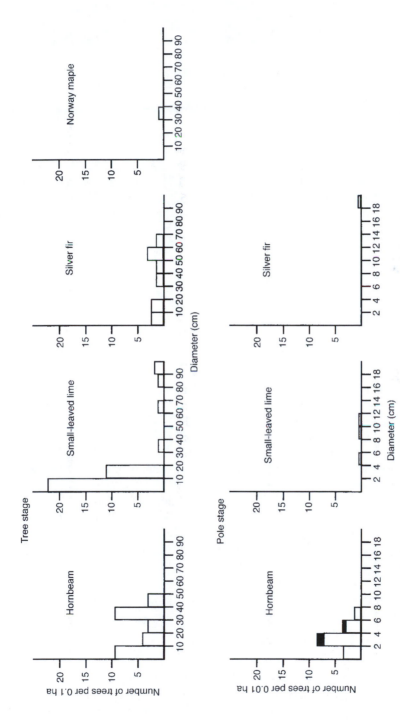

Fig. 5.41. The distribution in thickness categories in cm of the most important species of trees in the National Park of Białowieza in sector 342. For the tree stage, this is a representative test hectare for all the structures in the forest, and includes a transect of 100 m × 2.5 m. The results are shown in numbers of trees per 0.1 ha. The young trees stage and the pole stage can be described as regeneration. They were sampled in test circles of 0.002 ha, which included all the trees >2 m. The results of these recordings are shown in numbers of trees per 0.1 ha. The black areas represent dead trees. This explanation also applies for Figs 5.42–5.44 (redrawn from Derkman and Koop, 1977, figures I4, I5, II4, II5, II6, III4, III5, III6, III7, IV4, IV5 and IV6).

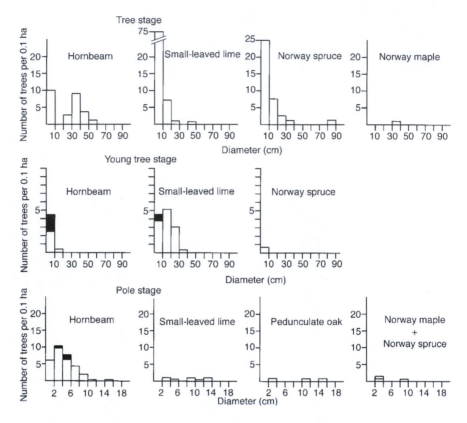

Fig. 5.42. The distribution in thickness categories in cm of the most important species of trees in the National Park of Białowieza in sector 369.

do not penetrate into the larger categories. Aspen and silver birch (*Betula verrucosa* = *pendula*) were found in an area of 2–3 ha with a young forest, where trees had been felled 200 years ago, next to hornbeam and small-leaved lime (Derkman and Koop, 1977, pp. 26–27). According to Koop (1989), the first vegetation to become established in this area where trees had been felled was a pioneer vegetation of aspen, silver birch and pedunculate oak, followed by hornbeam and small-leaved lime. He claimed that one oak tree remained from this pioneer stage (Koop, 1989, p. 93).

The results of the study confirm Pigott's finding (1975) regarding the regeneration of small-leaved lime, hornbeam, Norway maple, wych elm (in nettle) and pedunculate oak (see Derkman and Koop, 1977, pp. 23–33). The only difference is the presence of hazel in a pole stage in one of the test hectares. These hazel bushes were in a part of the National Park where there was grazing until 1923, so that there is a possibility that their presence is a remnant of the grazing. Other research (Abs *et al.*, 1999) also shows that the oak is represented as a seedling in the National Park, as is the hazel, with average values in

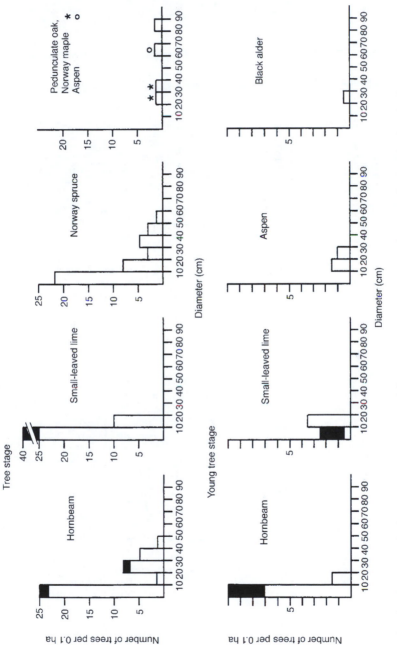

(Continued overleaf)

Fig. 5.43. The distribution in thickness categories in cm of the most important species of trees in the National Park of Białowieża in sector 256.

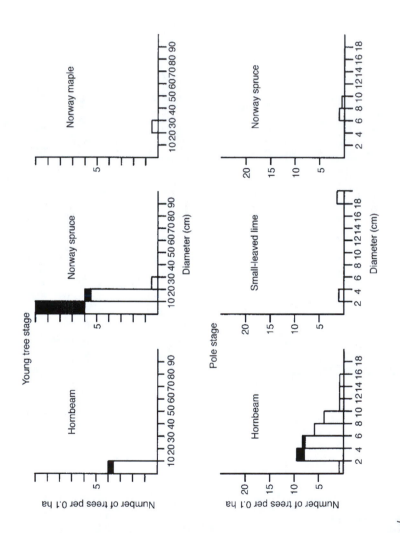

Fig. 5.43. *Continued*

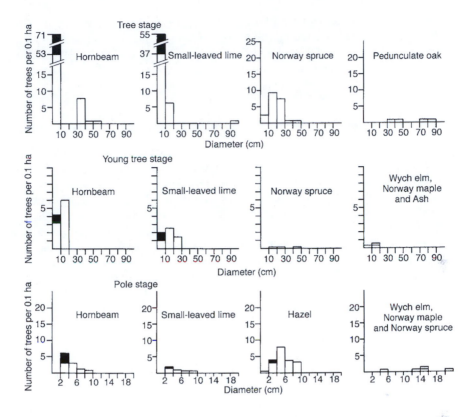

Fig. 5.44. The distribution in thickness categories in cm of the most important species of trees in the National Park of Białowieza in sector 371.

test areas of 1400 and 1950 plants respectively ha^{-1}. On the other hand there are values of hornbeam of 10,000 ha^{-1} and lime of 2250 ha^{-1}. Only these shade-tolerant species penetrate into the year class of 0.2–1 m; the oak does not. In contrast to other research, the hazel plays a role worth mentioning according to Abs *et al.* (1999). It is not clear here whether the seedlings penetrate into the bush level, or that there is a receding population of hazel that dates from the period before grazing. Furthermore, the hazel also establishes itself in open grassland after grazing has been stopped (Jakubowska-Gabara, 1996; Bodziarczyk *et al.*, 1999) The hazel could also date from then.

Samples of annual rings taken by Koop (1989) from eight pedunculate oaks in the National Park show that they were established in the 17th, 18th and 19th centuries. This finding corroborates Falinski's conclusions (1986, p. 288) from historical sources, namely that the regeneration of oak took place until very recently, because in the 19th century, the medium-aged categories were also widely represented in the forest. Pollen analyses confirm the permanent presence of the oak in the centuries mentioned (Mitchell, 1998; Mitchell and

Cole, 1998). There was a clear hiatus in the regeneration of hornbeam, and in particular of small-leaved lime, which coincides with the high densities of wild animals in the period 1870–1923 (see Koop, 1989, pp. 92, 95, 103). After 1923, the small-leaved lime regenerated almost continuously (Koop, 1989, p. 95). Pigott (1975) also failed to find a single small-leaved lime dating from the period 1870–1923. The cause for the absence of regeneration could be a combination of exorbitant densities of cattle and wild animals (H. Koop, Wageningen, 1997, personal communication).

Falinski (1986) gives conflicting reports about the regeneration of oak in the forest of Białowieza as a whole. On the one hand, he wrote that the sessile oak, in contrast with the pedunculate oak, regenerated well and was widespread in the undergrowth in certain reserves (Falinski, 1986, p. 61), though it is not clear whether he was referring only to seedlings or to the successful growth of seedlings into young trees. On the other hand, he said that in these reserves, oak does not regenerate (Falinski, 1986, p. 88). He also wrote that the regeneration of oak and Scots pine was completely absent or extremely restricted (Falinski, 1986, p. 228; 1988). In his opinion, this could be relatively recently, as both species were represented in large numbers in the middle and higher age categories (Falinski, 1986, p. 228). He did say that there were seedlings in the plant layer (Falinski, 1986, p. 231). Forestry inventories dating from the late 1970s and late 1980s, the results of which are shown in Fig. 5.45, show that in the National Park of Białowieza, oaks are found in the age categories of over 60 years only, and mainly in those of over 120 years (Jedrzejewska *et al.*, 1994). Hornbeam, and as we noted above, small-leaved lime, are also found in these categories, though they are represented with much larger numbers than oak in the categories over 60 years. These species clearly profited from the declining density of wild animals and cattle after 1923. The lime, of which the leaves are preferred by large ungulates, showed a particular increase (Edlin, 1964). The oak, on the other hand, benefited little or not at all from the reduction.

According to Fig. 5.45, the younger age categories of oak were found only in the part of the forest of Białowieza that has been exploited by forestry. Their presence there is due to forestry measures taken for the regeneration of oak; especially the planting of young trees (Jedrzejewska *et al.*, 1994; Bobiec, 1998; H. Koop, Wageningen, 1997, personal communication). This figure also shows that the young oaks in the forest that are assumed to have a natural origin, are

Fig. 5.45. *Opposite* The composition of ages, which a particular species of tree takes up in a particular age category, in a particular stand, in the forest of Białowieza. The left-hand column refers to the 'primeval forest' in the National Park; the right-hand column to the productive forests. The productive forests are in turn divided by the vertical dotted line into: the stands believed to have originated naturally (right) and the stands which were virtually always created by planting trees (left). The differences in the structure of the composition of ages are shown as ** $P<0.01$ and ***$P<0.001$. The composition of the ages of all the stands of trees in the exploited forest differs significantly from that in the primeval forest (redrawn from Jedrzejewska *et al.*, 1994, p. 660).

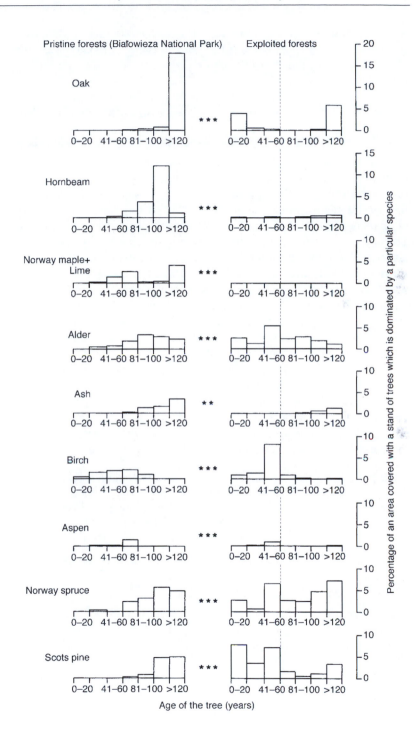

Age of the tree (years)

Percentage of an area covered with a stand of trees which is dominated by a particular species

hardly represented in the higher age classes, if at all. As a result of there being no upcoming year classes of oak, the number of oaks has decreased in all types of forest in the National Park; most strongly in the mixed deciduous forest and relatively little in the deciduous forest. Table 5.10 shows that the total number of oaks in all types of forest with a diameter at chest height of at least 5 cm almost halved in 56 years while the number of hornbeam almost tripled and the number of lime even increased by a factor of 25. Permanent sample transects from 1936 show that there were many oak in smaller diameter classes then, i.e. shortly after the end of grazing in 1923, and it showed the inverse J curve characteristic of natural forests, i.e. a very high number in the higher classes (Bernadzki *et al.*, 1998). Until then regeneration seems to have been successful. In 1992, the same samples showed that the oak had the bell curve distribution of an older population that is dying out, as in the other forest reserves covered. The lime, ash and hornbeam, on the other hand, showed the inverse J curve, i.e. a very high number of plants in the smallest diameter classes and a much lower number in the higher classes. The hornbeam, in particular, seems to be an upcoming species in this (see Bernadzki *et al.*, 1998). The lime also came up strongly. The increase of hornbeam and lime are both phenomena that occur when grazing is stopped in a forest.[15] Hazel also increases at first when grazing is stopped, namely by establishing itself in the open grassland (Jakubowska-Gabara, 1996; Bodziarczyk *et al.*, 1999).

These figures and the previous figures all indicate that what has been written recently about the regeneration of oak refers only either to the regeneration resulting from forestry measures or to the spontaneous growth of seedlings which then disappear after a few years. This finding is in stark contrast with the continuous presence of oak since the Boreal period (9000 BP) up until now, as indicated by the pollen diagrams of the National Park (see Dabrowski, 1959;

Table 5.10. Changes in the summarized number of trees in five transects (width 4–60 m; length 200–1380 m) in five types of forest in the National Park of Białowieza. 1. Coniferous forest; 2. Mixed coniferous forest; 3. Mixed deciduous forest; 4. Deciduous forest; 5. Flood plain forest. Each tree with a DCH (diameter at chest height) of at least 5 cm was identified. The total sampled area was 14.9 ha (from Bernadzki *et al.*, 1998).

Tree species	Number in 1936	Number in 1992	Annual increase or decrease (%)
Pedunculate oak	563	295	−1.5
Hornbeam	1175	3203	+2.3
Lime	101	2525	+7.3
Elm	45	74	+1.1
Norway spruce	6349	2325	+2.3
Scots pine	471	261	−1.3

[15] See Matuszkiewicz (1977), Falinski (1986, p. 201 *et seq.*), Kwiatkowska and Wyszomirski (1988), Kwiatkowska *et al.* (1997), Bradshaw *et al.* (1994), Jakubowska-Gabara (1996).

Mitchell, 1998; Mitchell and Cole, 1998; see Fig. 5.36). Therefore the oak must have been able to regenerate. In view of the above, it must be concluded that what is happening in the National Park is not a modern analogy of the natural development of vegetation in the lowlands of Europe, because two important components of the prehistoric vegetation, oak and hazel, seem to be disappearing. Clearly the factor or factors which these species require to survive are missing.

5.13.4 Regeneration of oak in the thermophile oak forest (*Potentillo albae–Quercetum*)

Apart from the National Park, the reserves with the so-called thermophile oak forest (*Potentillo albae–Quercetum*) are important for the hypothesis of this study because oak is widespread there, and these forests were grazed until the 1960s. Until recently, there were thermophile oak forests in several places in the forest of Białowieza (see Fig. 5.46) but they have now almost completely disappeared. In the reserve at Czerlonka, there has been a study into the spontaneous development of this type of forest after grazing came to an end. This showed that when the cattle disappeared from the forest, the forest was colonized by hornbeam, and all sorts of characteristic species of plants disappeared (Matuszkiewicz, 1977; Falinski, 1986, p. 201 *et seq.*; Kwiatkowska and Wyszomirski, 1988; Jakubowska-Gabara, 1996; Kwiatkowska *et al.*, 1997; H. Koop, Wageningen, 1997, personal communication).

Thermophile oak forests are thin oak forests found mainly in Central, Eastern and south-east Europe (Matuszkiewicz, 1977; Ellenberg, 1986, p. 249; Jakubowska-Gabara, 1996; Kwiatkowska *et al.*, 1997). They contain a wealth of vegetation consisting of forbs and grasses, with many species which thrive on light and heat and which also grow in fields (Matuszkiewicz, 1977; Ellenberg, 1986, p. 773; Falinski, 1986, pp. 87, 212; Jakubowska-Gabara, 1996; Kwiatkowska *et al.*, 1997). It is striking that species are present together in the thermophile forest that occur separately in forest or open terrain outside the thermophile forest (Kwiatkowska *et al.*, 1997). Some of the characteristic species are listed in Appendix 10. In the thermophile oak forest in the forest of Białowieza there are more species of the pea family (*Leguminosae*) than anywhere in the forest (Derkman and Koop, 1977, p. 12). Certain species of forbs, such as *Carlina aucalis* and *Dracocephalum ruyschiana*, are not found anywhere outside the thermophile oak forest of Białowieza (Falinski, 1986, p. 87). In general, the thermophile oak forest is the richest plant community in Poland from a floristic point of view (Kwiatkowska and Wyszomirski, 1990; Jakubowska-Gabara, 1996; Kwiatkowski *et al.*, 1997). One hundred and forty species of vascular plants were found on 512 m^2 (Matuszkiewicz, 1977). The main species of plant is *Potentilla albae* (Falinski, 1986, p. 87). The main trees are oaks, predominantly pedunculate oak. In general, only a few hundred square metres to a maximum of 1 or 2 ha consisted of more or less pure oak forest (Falinski, 1986, p. 88).

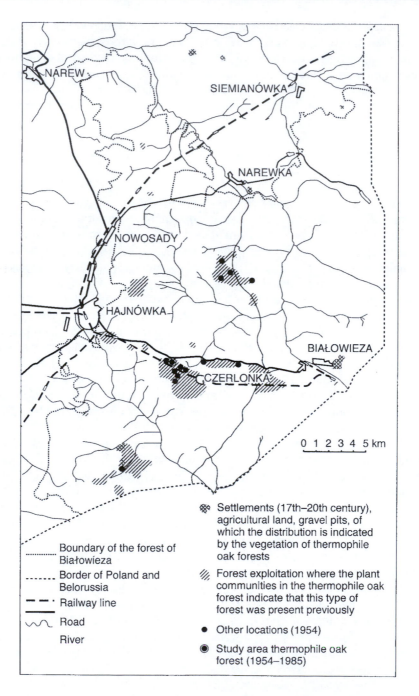

Fig. 5.46. Locations where the thermophile oak forest has recently gradually disappeared from the western (Polish) part of the forest of Białowieza as a result of cultivation and forestry management (from Falinski, 1986, p. 202).

Questions about the origin of this type of forest have given rise to a great deal of discussion. According to one theory, it is a relic from the Boreal and Atlantic periods (9000–5000 BP), and according to another theory, these forests were the result of human intervention. Matuszkiewicz (1977) claims that the results of the pollen survey carried out by Dabrowski (1959) in a few peat bogs in the National Park show that there were herbs in the Boreal–Atlantic period (9000–5000 BP) which are also found in the thermophile oak forest. Therefore it was not possible to exclude the possibility that this type of forest is a relic from that period.

However, as the appearance of the thermophile oak forest in and outside the forest of Białowieza is always correlated to wood-pasture, Matuszkiewicz (1977) also considered that it was possible that it had an anthropogenic origin. This type of forest could have resulted from cutting down trees and burning vegetation so that cattle could graze (Derkman and Koop, 1977, p. 12; Matuszkiewicz, 1977). The fact that there is a correlation between the thermophile oak forest in the forest of Białowieza and the grazing of cattle is evident, in so far as the characteristic vegetation of light-demanding herbs completely disappeared from the reserve at Czerlonka and the forest was gradually colonized by shade-tolerant hornbeam, after grazing stopped there in 1960 (Falinski, 1986, p. 201, *et seq.*; Kwiatkowska and Wyszomirski, 1988; 1990; Jakubowska-Gabara, 1996; Kwiatkowska *et al.*, 1997; H. Koop, Wageningen, 1997, personal communication). During the period 1950–1960, grazing was even fairly intensive (a biomass of 665 kg ha^{-1}) (Falinski, 1986, p. 205). The species diversity is now developing towards a lime–hornbeam forest. Therefore, the thermophile oak forest is identified as a replacement community created by humans of the original lime–hornbeam forest (Falinski, 1986, p. 212; Jakubowska-Gabara, 1996).

On the edges of the thermophile oak forests, and more in general, of the lime–hornbeam forest, there is widespread mantle and fringe vegetation (*Prunetalia*). Falinski (1986) noted that it is striking that these were missing or very rare in the thermophile oak forests in the forest of Białowieza (Falinski, 1986, p. 401). These communities are only found in the forest of Białowieza in the vicinity of settlements, where they are now increasing (H. Koop, Wageningen, 1997, personal communication). Elsewhere, they are widespread in the Polish countryside. In the 1960s, these communities appeared fairly suddenly in the forest of Białowieza in places where they had not been seen for a long time. For example, there was an invasion of hawthorn, wild pear and wild apple as well as several species of roses (*Rosa* spp.) on an old railway track (Falinski, 1986, pp. 373, 425). This phenomenon shows that these species can grow in the forest of Białowieza, but are missing for one reason or another. Another reason could have been the lack of light, although this is not probable in the thermophile oak forest, in view of the many species of light-loving plants which are found there. Another possibility could have been the lack of pasture land.

Grazing was prohibited in more and more areas of the forest of Białowieza in the 17th century and later (Hedemann, 1939; Peters, 1992, p. 25). Because of the decline in the area of pasture land, the commoners may have decided to cut down the thorny scrub in the remaining pasture land or to burn it so as to increase the area of grassland. They would have left trees – particularly oaks – standing, because felling these was strictly prohibited under the regulations at the time, the 'ius forestis'. The extremely high density of cattle which was present in the thermophile oak forests in the 1960s also indicates a lack of grassland. As we noted above, this amounted to 665 kg ha^{-1} (= 1.9 cow ha^{-1}) in 1960. In the most important thermophile oak forest reserve at Czerlonka, there was no regeneration of oaks according to Falinski (Falinski, 1986, p. 228). There are oak trees, while there are no indications that young oaks have been planted. If these were never planted, the oaks must have become established spontaneously. This means there was regeneration. Elsewhere in Poland, there is no more regeneration of oak in thermophile forests as well, because grazing has been stopped (Jakubowska-Gabara, 1996). When there was grazing, this regeneration would have taken place in unshaded scrub, as we saw in Chapters 2 and 4.

As described in Chapter 4, thorny scrub was removed in historical times because cattle could not graze where thorns grew (see Hobe, 1805, p. 124). The wood of hawthorn and blackthorn burns very well and, as we noted in Chapter 4, it was therefore traditionally very popular for firewood. As it burns so well, it would have been easy to burn the thorny scrub. The oaks would have survived this sort of fire because of the nature of their bark. As we saw in Chapter 4, cowherds started fires in the forests and woods to increase the area of grassland for their cattle. At the beginning of this century fires were still lit in the forest of Białowieza to create grassland for cattle (Escherich, 1927, p. 111). If there was any regeneration of oak in thorny scrub in the thermophile oak forests, and the thorny scrub was burnt to increase the area of grassland, the regeneration of oak in particular, and of trees in general, will not have taken place since then because of the fact that the seedlings were no longer protected against the cattle, and were destroyed by being eaten and trampled, or by fire. Young trees which had already grown out of reach of cattle, i.e. where they could no longer be eaten, will have grown into trees and formed the crown layer of the thermophile oak forest. This theory explains the presence of oaks and of species of plants which require light, as well as the lack of regeneration of oak in this type of forest.

Because of the clear correlation between grazing and thermophile oak forests, it is easy to assume that the presence of these forests goes back to the establishment of settlements in the forest, as Matuszkiewicz (1977) supposed. The oldest settlements date from the 17th century. They are located in places where there were thermophile oak forests until recently (Falinski, 1896, pp. 372–373). Figure 5.46 shows the location of this type of forest and the settlements.

5.13.5 Is the thermophile oak forest or the lime–hornbeam forest most similar to the original vegetation?

As noted above, the origin of the thermophile oak forests has been a matter of some discussion. If it is a relic of the prehistoric vegetation, as Matuszkiewicz

(1977) supposes, thermophile oak forests or analagous vegetation would also have directly preceded human settlements in the forest. There are indications that this type of vegetation was present at the time.

For example, in the 14th century, the Polish hunter-king Jagello built a hunting lodge, or possibly even a small hunting castle, in the centre of the area, in Białowieza Stara (Falinski, 1986, p. 26). If the area was one of the last virgin forests, as is generally assumed, the central part will certainly have been a closed forest at the time of King Jagello. You might wonder what the king was thinking of when he built a lodge in the middle of the forest in the 14th century. Did he hack his way through the forest to create a large open space in the middle and build the hunting lodge there so that he could hunt aurochs and bison? In the Middle Ages, these were the most popular game for the royal hunt (Szafer, 1968; Schama, 1995, p. 52). In the dark virgin forest round about, there will have been hardly any wild animals, such as aurochs and bison, because the closed forest is not the biotope of these species. The fact that there are few or no bison living in the National Park nowadays where the archetype of the original vegetation is believed to have developed, is highly significant in this sense (see Krasinski, 1978; Falinski, 1986, p. 165; Krasinska and Krasinski, 1998). Why would a hunter king go to all sorts of lengths to go miles into the dark virgin forest if there was nothing to hunt there? One logical explanation for the fact that he did so is that there was not a closed and inaccessible dark virgin forest at that time at all, but vegetation comparable to wood-pasture, i.e. a mosaic of grasslands and groves which was reasonably accessible. The king could have penetrated miles into this without too much difficulty, to enjoy the pleasures of hunting the large numbers of wild animals that could have been expected in this landscape, including aurochs and European bison, far from the inhabited world and in complete peace and quiet.

A map of the forest of Białowieza published in 1830 (Eichwald, 1830) shows the contours of this type of landscape of groves and grasslands (see Fig. 5.47). The shape of the groves must have been the result of the concentrically expanding thorny scrub of blackthorn where there was regeneration of trees, as described in Chapters 2 and 4. This sort of original vegetation explains why grazing of cattle and hay-making could have been widespread activities in this area in the 16th century, as Eichwald (1830, p. 249) noted in historic sources. The park-like landscape then developed and was maintained for thousands of years as a result of the grazing of wild herbivores such as aurochs, European wild horse and wild boar, as well as the effect on the vegetation caused by European bison, red deer, roe deer and elk.

In that case, the thermophile oak forests are (were) indeed relics from the Boreal–Atlantic period (9000–5000 BP), as Matuszkiewicz (1977) suggested. The fact that grasslands are important for bison, as well as for specialized herbivores such as aurochs, is clear from the fact that the European bison graze widely in pastures specially made for them in the forest which are regularly mown (Krasinski, 1978). In this case, the cattle with which the thermophile oak forests are associated do not mean that the thermophile oak forest consists

Fig. 5.47. Part of a map of the forest of Białowieza, published by Eichwald (1830). The map shows that the forest was composed of concave-shaped groves with open areas between them, which were almost certainly grassland (photograph, K. Peters).

of vegetation with an anthropogenic origin, but that the cattle are a modern analogy of the wild cattle, the aurochs, and that the related vegetation is a modern analogy of the original vegetation.

There are further indications of the presence of thermophile oak forests before the arrival of humans. For example, in the 17th century, the tsar built a hunting lodge where the present village of Białowieza was established in 1696 in the middle of the wilderness (see Fig. 5.37). Until recently there was a thermophile oak forest there. Some of the centuries-old oak trees still remain. They have a short trunk and a large broad crown, showing that they grew in an open landscape. If this vegetation dated from before the time that humans kept cattle, the vegetation must have been maintained by the grazing of the large herbivores at the time, as noted above. This would also have resulted in the regeneration of trees, in particular of oak. If this is the case, it not only explains the presence of certain light-demanding species of plants in the thermophile oak forest. It also explains, as the pollen diagrams and historic sources show, that oak and hazel survived without interruption for thousands of years next to lime, hornbeam and elm, not only up to the time that humans settled in the forest in the 16th and 17th centuries, but also afterwards when cattle grazed there, because the regeneration of trees takes place where cattle graze as we saw in Chapters 2 and 4. In that case, the cattle would have been a modern analogy of the aurochs. Evidence for this is that since cattle have stopped grazing and the wild species of cattle, aurochs and European wild horse have become extinct, the thin forests have changed into closed forests and there is no longer any regeneration of oak. All the colonists had to do in the case that there were thermophile oak forests or analogous vegetation after they settled, was to make use of the natural grassland and groves that were there; the grasslands for cattle and the groves (with oaks) for the pannage of pigs and the collection of firewood and timber. The aurochs and European wild horse were replaced by the cattle which took over the role of their wild ancestors with their grazing.

As noted above, the process of regeneration of trees in general, and oak in particular, would have continued in thorny scrub. If there was no measure to thin out the oaks in the thorny scrub, the oaks could also have grown up together in groups as well as far apart. In that case they experience more competition among themselves for the light and the trunks will be free of branches for a longer length than when they grow in more open conditions. This can explain why nowadays there are oaks with a broad crown, as well as oaks with relatively tall branchless trunks, in the forest of Białowieza. This can also explain why in the past, the trunks of the oaks that were shipped from the Baltic harbour towns to Western Europe (Buis, 1985, pp. 502, 505, 509–514) were free of branches over a longer length than the oaks in the west of Europe (Simpson, 1998). The reason that the young oaks were not thinned out could be that there was sufficient wood for fuel and therefore the underwood did not have to be protected against the shade of the trees and no coppice-with-standards was developed.

When the aurochs disappeared completely from Poland in the 16th century, the other wild herbivores were virtually eradicated in the 18th and 19th centuries, as historic sources show, and the grazing of cattle was prohibited in increasingly large areas, the grasslands in the forest could have been colonized by hornbeam, small-leaved lime and other shade-tolerant species, as in the National Park and the thermophile oak forest. In this way, the open park-like wood-pasture gradually and increasingly changed into a closed forest. Because these shade-tolerant species grew relatively closely together, they developed the long, straight, branchless trunks which are now so characteristic of the trees which are several hundred years old, for example, in the National Park. In this way, an increasingly large area of the thermophile oak forest or an analogous type of vegetation could have changed into a closed lime–hornbeam forest in the past few centuries. In that case, the thermophile oak forest is not the community which replaced the original lime–hornbeam forest as a result of human intervention, but the lime–hornbeam forest is the community which replaced the original thermophile oak forest or analogous vegetation as a result of human intervention. The human intervention responsible for this was the termination of grazing of large herbivores such as (aurochs) cattle and (tarpan) horse.

The above theory explains more than the current theory, which states that the original vegetation was a closed forest, with, as an exponent, the closed lime–hornbeam forest. It also gives a more logical explanation of the establishment of settlements in the middle of this area. For example, in 1696, the settlement of Białowieza developed in the middle of what was a closed forest, according to the current theory, but a park-like landscape, according to the above theory. Until recently, there was a thermophile oak forest in the immediate vicinity of the village. If the thermophile oak forest or analogous vegetation was the original vegetation, there was already pasture land available for the cattle, and it was not necessary to first cut down part of the forest. The same applies for the other settlements which were established over the centuries in what is now the forest of Białowieza.

5.14 Other Forest Reserves in Europe

In other forest reserves in Western and Central Europe it has also been found that there is little or no regeneration of oak as in the above cases. For example, this applies to the pedunculate oak in the Lanzhot forest reserve in the Czech Republic (36.6 ha) on the River Morava (March). On the other hand, small-leaved lime and broad-leaved lime, ash and field maple do regenerate well there in the gaps in the canopy. Like pedunculate oak, wild pear does not regenerate there either (Prusa, 1982; 1985, pp. 50, 51, 70, 73; Den Ouden, 1992, pp. 47–58, 98–99; Bönecke, 1993). Data from other river forest reserves along the March, and also along the Rhine and Elbe, show that there is no regeneration of pedunculate oak either, or of species such as wild apple or wild pear, though there is regeneration of field maple, English elm (*Ulmus procera*), small-leaved lime and hornbeam (see Dister, 1980, pp. 65, 66; 1985; Dister and Drescher, 1987; Dornbusch, 1988). As in many other reserves, this failure is attributed to (excessively) high numbers of game. Fencing areas of forest off against game is not considered a solution because, on the basis of experience, it is expected that ash and field maple will replace the young oak trees within the fenced-off areas (see Prusa, 1982; 1985, p. 73; Den Ouden, 1992, p. 66; Bönecke, 1993).

As we noted in Chapter 4, pedunculate and sessile oak no longer regenerate in wood-pastures in England when grazing stops. This was observed, *inter alia*, in the Mens, Epping Forest, Staverton Park and Birkhamsted Common (Tittensor, 1978, pp. 351–352, 556, 372; Rackham, 1980, pp. 175, 202, 294–296, 326, 327, 356). According to Rackham, oak always regenerated well in Staverton Park when there was grazing, but this stopped completely when grazing stopped.[16] In Epping Forest, the pastures were taken over by beech and

---

[16] In this respect, Rackham commented on Staverton Park: 'The decline of grazing was followed by a great increase of holly and birch, relatively sentitive trees. But oak, which had regenerated moderately freely under grazing, ceased to do so after grazing ended' (Rackham, 1980, p. 294). In addition, he wrote: 'Oak regenerates embarrasingly well in the less-wooded parts of most wood-pastures' (Rackham, 1980, p. 295). With regard to the absence of regeneration of oak, he also noted: 'Since about 1850 something has happened to prevent this turnover [regeneration], but without affecting the ability of oak to reproduce outside woodland. By the 1910 this was a widely-recognized problem, the subject of a classic study by A.S. Watt [1919]' (Rackham, 1980, p. 296). Grazing came to an end at about this time in virtually all the 'commons', because grazing rights were withdrawn (see Rackham, 1980, p. 6).

hornbeam. The beech trees grew out above the oaks, which then died off. Apart from oak, wild apple also declined there because of the increased shade of beech and hornbeam. With regard to hazel, Rackham also found that regeneration has come to an end, and sees this as a parallel to the lack of regeneration of oak.[17]

The Draved Skov in the south of Denmark is a special reserve, because it has been used in palynology as a reference forest for the prehistoric situation (see Iversen, 1958; Andersen, 1970). It is a former wood-pasture (Iversen, 1958; Aaby, 1983). Based on pollen diagrams going back to the Atlanticum and in which the percentages of pollen of grasses and light-demanding herbs are low, Iversen (1958) concluded that the untouched vegetation was a closed and shady forest. Wood-pasture would have had little or no influence on the forest because the percentage of pollen from grasses and herbs was so low. Due to the similarity between the composition of species in the pollen diagrams and the current forest, Iversen (1958) believed this forest would very soon be practically identical to the original primeval forest if forestry was stopped. He also believed there was a far-reaching degree of similarity in the fact that the oak would survive beside the lime and the beech there.

In the use of Draved Skov as reference forest, the pollen diagrams of this forest have been interpreted based on the current vegetation. Based on this, correction factors have been calculated, as explained in Chapter 3, for the percentages of pollen of the various species in the pollen diagrams. The percentages determined in the sediments are adjusted using these correction factors, so that the pollen diagrams correspond more to the present vegetation in the forest. The premises here are that the original vegetation was a closed forest, that the various species that are found in the pollen analyses could continue to exist beside one another in a closed forest and that after the wood-pasture ended, the forest returned to its original state.

The following data, unless stated otherwise, come from Aaby (1983). The written history of the forest goes back to AD 1173. In the 16th century, pigs were driven there in good mast years. That means that the oak occurred there, if not with a prominent place, as evidenced by the relatively high percentage of oak in the pollen spectrum, in addition to the lime, beech, birch and hazel (see Iversen, 1964; 1969). Afterwards, the number of pigs and amount of mast would have greatly decreased. After 1600, a considerable number of cattle and horses grazed there. In 1785, grazing ended and more rational forestry techniques were introduced. This is in accordance with what happened in other parts of Europe (see Chapter 4). After this, inventories were made at various times. These show that the last cohort of oaks dates from about the time that grazing ended. For example, according to an inventory there were about 150 old oaks

---

[17] With regard to the absence of regeneration of hazel, Rackham noted: 'Bearing in mind the parallel case of oak, I hesitate to counsel alarm without having studied the problem in depth, but appearances suggest that the reproduction in hazel is even more problematical than in other heavy-seeded trees' (Rackham, 1980, p. 210).

present in 1952. However, the most striking thing is that, analogous to the other forest reserves discussed before, there are no smaller diameter classes of the oak, i.e. young trees present (see Fig. 5.48). There are such trees of the shade-tolerant beech. Hence, here again there is an image of an ageing, dying out population of oaks. The premise at the basis of the calculation of the correction factors, namely that the oak will continue to be part of the closed forest, is therefore incorrect. The joint presence of beech, oak, lime and hazel is the result of the history of the forest reserve. This illustrates that certainly in the case of such long-living plants as trees, history matters as regards the succession of the vegetation. The data of the current situation are therefore not sufficient for developing models for the succession, because events that happened in the past continue to have an effect in the current situation (Van Hulst, 1979a,b; 1980). In this case, this is the presence of oak in a closed forest with shade-tolerant species like beech and lime.

5.15 Spontaneous Succession in Forest Reserves in the East of the United States

In the east of the United States, there are forest reserves and so-called old growth forests, which are assumed to have had little influence from felling or grazing by livestock since the arrival of the colonists. In many cases, a number

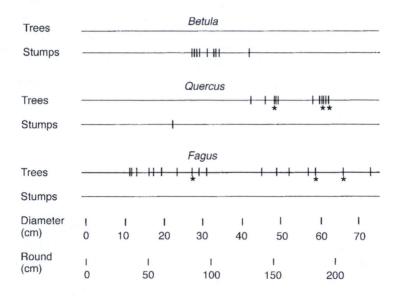

Fig. 5.48. The distribution of diameters of trees and stumps at 1.3 m and 0.2 m high, respectively, in Draved Skov. The trees whose age was determined using tree rings are marked with an asterisk.

of trees in the canopy level date from the period before colonization by Europeans, the so-called pre-settlement period. In this section, I will discuss the succession of a number of these forest reserves, whereby I follow the premise that the possible patterns in this succession can be derived from a certain species (McCook, 1994). As such, I have selected the white oak (*Q. alba*) because I consider this species of oak as a representative of the light-demanding species of oak in the east of the United States. Other members include scarlet oak (*Q. coccinea*) and black oak (*Q. velutina*). However, the white oak is the most common oak and occurs in the east of the United States in every type of deciduous forest and on every type of soil (Barnes, 1991; Abrams, 1996). These include areas where, as a result of the amount of precipitation and moisture levels in the soil, shade-tolerant species occur beside the white oak, such as American beech (*Fagus grandifolia*), sugar maple (*Acer saccharum*), silver maple (*A. saccharinum*), red maple (*A. rubrum*), American elm (*Ulmus americana*), basswood (*Tilia americana*), hemlock (*Tsuga canadensis*), white ash (*Fraxinus americana*) (see Cottam, 1949; Curtis, 1970; Ehrenfeld, 1980; Nowacki and Abrams, 1992).

In accordance with the theories on succession and climax vegetation, these forests would develop in spontaneous succession to an analogy of the vegetation originally present. That means that species, which are known from pollen analyses to have survived for millennia after the last Ice Age in the east of the United States without human intervention, survive in those forests. This regards both light-demanding species of oak (*Quercus*) and the shade-tolerant genera *Acer*, *Fagus*, *Fraxinus*, *Ulmus* and *Tsuga*.[18] As noted in Chapter 3, the oak is particularly well represented in the pollen diagrams. It should be stated here that no difference is made here between the more or less light-demanding species of oak and that the east of the United States contains 30 species of oak.

The term 'old growth' regularly appears in publications on the forest reserves. In America there is a tacit implication that old growth forests are primary, native, natural and virgin. Usually, old growth is seen as a late stage in the development of the succession, but it is also common to apply the term to managed forests (Peterken, 1996, p. 17). As an example, I will now discuss the spontaneous developments in Goll Woods and Sears-Carmean Woods in the north-west of Ohio; in the Hutchington Memorial Forest, also known as Mettler's Wood, in New Jersey; in Sophia's Wood in Pennsylvania; and in Duke Forest in North Carolina.

Goll Woods and Sears-Carmean Woods are two old growth forests in the north-west of Ohio. The data come from Cho and Boerner (1991). Goll Woods is 146 ha in size and Sears-Carmean Woods 71 ha. They are situated in an area that was characterized by oak forests and oak savannas at the time of the arrival of the first colonists. Surveyors have found a high percentage of oaks in the area. A number of large oaks still occur in these forests. However, maple and beech have the upper hand. The population structure shows that no smaller diameter classes of oak are present (see Fig. 5.49). It is the image of an ageing, dying out

---

[18] See Davis (1967; 1984), Wright (1971), Watts (1979), Delcourt and Delcourt (1987; 1991, pp. 90–91), Webb (1988), Roberts (1989, pp. 72–74).

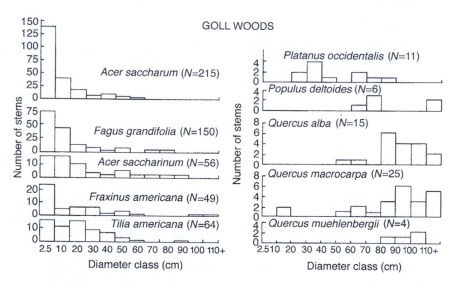

Fig. 5.49. The distribution in 10 cm diameter classes of the various tree species over the thickness classes at chest height in Goll Woods in Ohio. The two bur oak (*Quercus macrocarpa*) trees in the 10–20 cm diameter class were both in a large hole in the canopy (redrawn after Cho and Boerner, 1991, p. 12).

population. In contrast, the population structure of shade-tolerant species such as sugar maple, American beech, basswood, white ash and silver maple show the image of a regenerating population. The image is analogous to that of Dalby Söderskog in the south of Sweden (see Fig. 5.32). Regeneration took place in gaps in the canopy. However, growing seedlings of beech and sugar maple had a much larger share in gaps in the canopy. Most gaps were filled by sugar maple. According to Cho and Boerner (1991), both species will continue to increase and in the current disruptive regime, they will in time replace the oak.

The second example is the Hutchington Memorial Forest, also known as Mettler's Wood, in New Jersey, where the canopy consisted of 80% white oak. The forest is 26 ha in size. The data presented here come from Monk (1961a,b), Sulser (1971) and Davison (1981). The forest is seen as 'the nearest approach to climax' in the area. The oldest oaks were 200–250 years old (Monk, 1961b). In addition to white oak, there were black oak, northern red oak (*Q. rubra*) and red hickory (*Carya ovalis*). Further species are white ash, red maple, black gum (*Nyssia sylvatica*), American elm, pin oak (*Quercus palustris*), swamp oak (*Quercus bicolor*) and the exotic Norway maple. There was advance regeneration of all species. Seedlings of white oak and black oak were common in the forest. However, they usually died within a few years. Therefore saplings were rare. There was no white oak at all <25.5 cm at chest height. This is not true for shade-tolerant species such as red maple, Norway maple and sugar maple. Their seedlings were common and, in the case of red maple and Norway maple, so were saplings. There were few sugar maple saplings, but there were still more than white oak saplings. Sugar maple colonized the forest from a nearby river plain. Storms in 1950 and 1956 blew over 300 large trees. In the resulting gaps in the canopy, oak or hickory would eventually grow only here and there. Dogwood (*Cornus florida*) also regenerated in gaps in the canopy. Outside the gaps made by the wind, shooting sprouts was the most important form of regeneration of the red maple. The beeches in the forests usually formed clones. Hence, the white oak and the black oak did not penetrate the higher thickness classes. The reproduction of oak did not keep up with its disappearance. The expectation was, therefore, that under the current circumstances, the white oak and the black oak would in time disappear and that the shade-tolerant species would replace them.

A third example is Sophia's Wood in Pennsylvania. It is currently a 40 ha old forest that was considered untouched in 1786. Unless stated otherwise, the data on this forest are from Abrams and Downs (1990). In 1923, the forest contained 300-year-old oak. At that time, the canopy consisted of 50% white oak and 40% northern red oak, scarlet oak and black oak. The scarlet oak was in poor condition. The species was dying out. Between 1930 and 1940, selective felling would have taken place that would have initiated the increase in shade-tolerant species. There were still two oaks of <300 years old. Of all the trees that formed the canopy, 90% was >120 years old. The canopy consisted of many American beech, tulip poplar (also called yellow poplar) (*Liriodendron tulipifera*), red maple, sugar maple with very few oaks left. American beech, tulip poplar and red maple dominated. A density (no. per hectare ± SE) of seedlings (plants <10 cm diameter at breast height and lower than 1.37 m) and saplings (trees <1.37 m tall) of 23,400 (±300) and 50 (±2) were found for red maple; for beech the figures were 1100 (± 15) and 100 (± 5) respectively; for sugar maple 5100 (± 61) and 100 (± 3) respectively; for white oak 3000 (± 82) and 0 (± 0) respectively; for black oak 100 (± 5) and 0 (± 0) respectively; and for scarlet oak 0 and 0 respectively. That meant that the seedlings did not succeed in growing. Other species did succeed, as is evidenced by the fact that there were many young beeches and tulip poplars in the canopy storey. This was in contrast with the pre-settlement period, when 40% of the witness trees (see Chapter 4) were white oak, and beech and maple were only a small minority. The conclusion was that the oak is being replaced in these forests by the shade-tolerant species beech, red maple, sugar maple and the light-demanding tulip poplar. This last species is a formidable opponent for the oak, because it grows quickly, can become up to 60 m high and can become as old as an oak (>500 years) (Peterken, 1966, p. 149).

Duke Forest is located in North Carolina. The data come from Christensen (1977). It was considered a relatively undisturbed forest in which only dead oaks and larger oaks of 120 years of age would have been removed. Most of the oaks were 120 years old and were white oaks. During a period of 23 years (1951–1952 to 1974), the density of oak decreased by 40–50%; the white oak showed the greatest decrease, namely from 512 to 300. This is an average annual decrease of 2.3%. The majority of deaths of white oak were in the small year classes. They disappeared in a period of 10 years. The distribution of white oak in diameter classes changed during the period between the two samples from an inverse J curve into a bell curve/normal curve distribution, the curve that is characteristic of an ageing, dying out population. One species, the post oak (*Q. stellata*) had no descendants after 5 years. This is compared with an enormous increase of red maple of 230%, while the species was rare in the canopy. The red maple increased in almost every size class.

Similar-sounding developments were determined in old growth forests elsewhere, including Ohio (Whitney and Somerlot, 1985), New Jersey (Botkin, 1990), Pennsylvania (Lutz, 1930; Morey, 1936; Nowacki and Abrams, 1992; 1994; Abrams and Ruffner, 1995), Maryland (McCarthy and Bailey, 1996), Indiana (Parker *et al.*, 1985), Virginia (Ware, 1970), Massachusetts (Ogden III, 1961; Lorimer, 1984), New York (Whittaker and Woodwell, 1969; Lorimer, 1984) and Tennessee (Barden, 1981; McGee, 1984). Abrams (1992) concluded that a scarcity of young oak trees is typical of many forests in eastern North America, and this has been used as evidence for predicting oak replacement by shade-tolerant species. All in all, there is a large amount of literature that indicates that the oak is being suppressed, even if the ground is regularly covered with thousands of seedlings.[19]

According to the current theory, regeneration would have to take place in the gaps in the canopy but this appears not to happen.[20] In those conditions, the oak has no chance. Nor does heavy windthrow as a result of hurricanes, whereby trees were blown to the ground over hundreds of hectares, offer any solace for light-demanding species like the oak to successfully regenerate. In that case it is mainly young trees from the advance regeneration of shade-tolerant species that were present under the canopy. They were 'released' and grew up. The species that can suppress the oak, dominate the vegetation (Spurr, 1956; Hibbs, 1983; Peterson and Picket, 1995). Large-scale disruptions will therefore not assist the oak if the storey of shade-tolerant species remains intact. Such disruptions will actually speed up the replacement of oak (Nowacki and Abrams, 1992). This is confirmed by empirical experiments in forestry, where clear felling does not lead to successful regeneration of the oak, even if the oak is well represented in the advance regeneration, i.e. up to thousands of seedlings per hectare. Without further human assistance in the form of combating shade-tolerant species, the herb storey, or the fast growing light-demanding tulip poplar, upcoming young light-demanding oaks like the white oak, black oak and chestnut oak will disappear in a few decades.[21] Only in conditions unique for the oak forests in the east of the United States, such as dry, poor sites where

---

[19] See further Korstian (1927), Adams and Anderson (1980), Barden (1981), Host *et al.* (1987), Barnes (1991), Nowacki and Abrams (1992), Crow *et al.* (1994), Abrams (1996), Dodge (1997).

[20] See also Ehrenfeld (1980), Barden (1981), McGee (1984), Ward and Parker (1989), Clinton *et al.* (1993), Lorimer *et al.* (1994).

[21] See Gammon *et al.* (1960), Loftis (1983), Beck and Hooper (1986), Host *et al.* (1987), Lorimer *et al.* (1994), Norland and Hix (1996), Arthur *et al.* (1998), Barnes and Van Lear (1998), Cook *et al.* (1998), Brose (1999), Brose *et al.* (1999).

there are no species that can suppress the oak, does the oak seem to be able to perpetuate its dominance in the canopy (McCune and Cottam, 1985; Ross *et al.*, 1986; Orwig and Abrams, 1994). This phenomenon is advanced as an argument that the occurrence of oak will be limited to extreme edaphic or climatic conditions (Abrams, 1992; Nowacki and Abrams, 1992; Hodges and Gardiner, 1993), but that contradicts the data from the pollen analyses.

5.16 Conclusions and Synthesis

This chapter examines the spontaneous succession in forest reserves in the lowlands of Central and Western Europe and in the east of the United States. It looked in particular at whether pedunculate and sessile oak, and hazel can survive in these areas. The most important findings and comments on these findings are presented below.

> • There are no new young generations of oak trees in any of the forest reserves in the lowlands of Central and Western Europe that were examined.

Oak seedlings are regularly found in all the forest reserves, sometimes even in large numbers. They all disappear within a few years. As a result, the old oak trees which are still present are no longer replaced, and as the old oaks die, there is a gradual decline in the number of oak trees in the reserves. Everything indicates that oak seedlings tolerate little shade and therefore die. The seedlings of other species of trees, such as small-leaved lime, hornbeam, wych elm, smooth-leaved elm, sycamore, ash and beech appear to tolerate shade better. They are virtually always present under the canopy. When a gap appears in the canopy so that the light improves, they seem to profit immediately and quickly grow taller.

> • The rate at which oaks are disappearing is not determined by the age of the oaks, but by the time which other species need to grow up above the oaks. This usually happens in a period of less than one century.

In contrast with oaks, species such as beech, small-leaved lime, hornbeam, sycamore, ash, smooth-leaved elm and wych elm do regenerate successfully in the closed forest in gaps in the canopy. They grow up into the crown layer, and then grow past mature oaks, which die as a result. The result is a progressive

replacement of oaks by these other species. This happens in all the forest
reserves throughout the lowlands of Western and Central Europe with every
type of soil. Therefore it is a general phenomenon.

> • The history of the forest reserves and an examination of the annual
>   rings of trees shows that the old oaks all date from the time when
>   there was grazing in the forests concerned.

There is a clear relation between the end of grazing by typical grass-eaters such
as cattle, horse and sheep, and the stagnation of the regeneration of oak.

> • In the presence of herbivores such as red deer, roe deer, elk and
>   European bison, which are not specialized grazers, there is no
>   regeneration of oak.

Where there are high densities of wild animals which are not specialized grazers, the regeneration of all the species of trees in the forest reserves is greatly
restricted, as shown by fencing off areas of forest and the developments in the
forest of Białowieza in the period 1880–1915. When there is a reduction in the
density of wild large herbivores, trees such as small-leaved lime, hornbeam,
beech and wych elm regenerate successfully, but oak does not. Therefore, a high
density of wild herbivores is not the explanation for the failure of oak to regenerate. Moreover, the New Forest has shown that oak can regenerate when there
are high densities of wild animals, though not in a closed forest (see Chapter 4).

> • The lack of regeneration of oak in the forest reserves must be
>   attributed to the fact that the forest reserves all developed to become
>   closed forests.

Pedunculate and sessile oak do not survive in closed forests. These species cannot regenerate in gaps in the canopy in accordance with Watt's gap phase
model (1947) or Leibundgut's cyclical model (1959; 1978). If these species do
succeed in becoming established in a gap in the canopy, they still lose out in the
competition from shade-tolerant beech, hornbeam, elm and lime, etc. These
species even grow under the oaks. Nor do hazel and wild fruit trees survive in
closed forests. A similar explanation seems likely for the absence of regeneration
of these species in the shade-intolerant closed forests.

> • If the vegetation in prehistoric times consisted of closed forests with different species of trees, it follows from the above that these cannot have included oak trees.

Pollen diagrams from Central and Western Europe show that oak trees have been present in Central and Western Europe without interruption since the Boreal period (9000 BP). If there was already a closed forest in Central and Western Europe consisting of birch, Scots pine and hazel shrubs in the Boreal period, when the oaks arrived after the end of the Ice Age, as the current theory suggests, it is not clear how oak could have become established there in the first instance. Hazel would also have disappeared from the forest after a short time, although the pollen diagrams do not show this to be the case. In contrast, compared with species which tolerate shade better such as elm, lime, beech and hornbeam, hazel and oak were generally represented by the highest percentages of pollen in the pollen diagrams throughout prehistoric times.

> • The developments in the National Park of Białowieza provide a good insight into the relationship between grazers and the regeneration of trees and shrubs, because the original fauna of large herbivores was present until the 16th century and was then replaced by grass-eating domesticated cattle.

According to pollen analyses, oak, lime and hornbeam have been present together in this area, without interruption, since the Boreal–Atlantic period. The regeneration of oak took place until the 19th century, i.e. during the period that the specialized grass-eaters of the original fauna of large herbivores, aurochs and European wild horse were present, and even after, when they had disappeared and were replaced by grazing domesticated cattle. The most recent developments in the vegetation of the National Park of Białowieza show that the grazing by herbivores, such as cattle in particular, forms a determining factor for the regeneration of oak. When the grazing came to an end, there was no more regeneration of oak, not even when the density of the other herbivores was subsequently greatly reduced. Small-leaved lime and hornbeam did profit from these developments; species which can regenerate in a closed forest if the seedlings in the forest are not all eaten and trampled by large ungulates. It is fairly certain that the reason for the lack of regeneration, e.g. of small-leaved lime in the forest of Białowieza in the period of high densities of wild herbivores, was that the seedlings were eaten by these animals, particularly as large herbivores have a great preference for lime leaves.

> • The hypothesis that the thermophile oak forests in the forest of
>   Białowieza developed as a result of human intervention in the
>   lime–hornbeam forest, after settlements had been established, should
>   be reversed; in fact the settlements were established in places where
>   the thermophile oak forest already existed.

Colonists who settled in the area in the 16th and 17th centuries were able to
graze their cattle and make hay in this sparse forest. These forests must therefore have been naturally present in the form of a park-like landscape as a result
of grazing by aurochs, European bison, European wild horse and red deer which
lived in the area at the time. Like the regeneration in the New Forest, the trees
must have regenerated in thorny scrub in the presence of these wild herbivorous fauna. This explains why oak, lime, hornbeam and other species were present from the Boreal–Atlantic period without interruption, as shown by the
pollen diagrams and historic sources of this area. All the colonists had to do was
to use the wood-pastures and groves that were naturally present. The aurochs,
European bison, European wild horse, red deer and other species of wild animals were driven out and replaced by cattle.

According to this hypothesis, the whole of the present forest of Białowieza
must originally have consisted of thermophile oak forest and other types of forest which owe their existence to grazing by large herbivores. An exception will
be the areas that were impassable for the animals, as a result of high ground
water levels, or weak ground. The fact that there was regeneration of all the
species of trees, including oak and hazel, in the area at the time that it was
grazed by cattle and wild herbivores means that the regeneration of trees took
place in thorny scrub, as in the New Forest (see Chapter 4), and that groves
developed from these. Because of the absence of measures such as thinning out
young oak trees in the thorny scrub, a measure which was introduced, as we
saw in Chapter 4, to produce mast oaks or create coppices, many young oaks
could have grown relatively close together in the thorny scrub so that they had
fairly straight trunks. This explains why there are both so many oak trees with
broad crowns and oak trees with fairly straight, branchless trunks in the forest.

When the grazing of cattle came to an end, the wood-pasture turned into a
closed forest with all the characteristics of the lime–hornbeam forest; a change
which has recently taken place in the last thermophile oak forests in the forest
of Białowieza. Because these shade-tolerant species of trees grew in large concentrations relatively close together, they developed the tall, straight, branchless trunks which are now so characteristic of the trees which are several
hundreds of years old, such as those in the National Park.

To summarize, the pollen analyses and the spontaneous developments in the
forest reserves show that closed forests such as those in the forest reserves cannot
be seen as modern analogies of the prehistoric vegetation. The results of the examination of spontaneous developments in the forest reserves show that there is no

regeneration of pedunculate and sessile oak anywhere. Oak disappears from a closed forest in just one or a few generations as a result of the absence of the regeneration of oak and the death of old oaks which become overgrown by other species. The same applies to hazel. The spontaneous developments in forest reserves which are seen as natural processes did not in any of the cases examined lead to a situation which corresponded to the palynological facts, which provide the only indication of the composition of the original natural vegetation. The developments in the forest reserves show that oak cannot successfully become established even in the largest gaps in the canopy. On the other hand, the composition of species in the reserves wholly corresponds to the picture outlined in Chapters 2 and 4, of large numbers of young oaks which germinate and grow in wood-pastures on the fringes of thorny scrub. The occurrence of thorny bushes, hazel and wild fruit trees in the forest reserves show that this process actually took place there. As there is no more grazing in the former wood-pastures, the oaks will eventually disappear. Table 5.11 summarizes the conclusions with regard to spontaneous succession in the forest reserves.

Table 5.11. Summarizing conclusions on the regeneration or failure to regenerate of the most important species of trees in former wood-pasture and former coppices, where grazing by herbivores such as cattle and horses, came to an end, and a non-intervention style of management was introduced.

	Oak	Beech	Lime	Hornbeam	Elm
La Tillaie	−/−	+	n.a.	n.a.	n.a.
Le Gros-Fouteau	−/−	+	n.a.	n.a.	n.a.
Neuenburger Urwald	−/−	+	n.a.	+	n.a.
Hasbrucher Urwald	−/−	+	n.a.	+	n.a.
Sababurg	−/−	+	n.a.	n.a.	n.a.
Rohrberg	−/−	+	n.a.	n.a.	n.a.
Priorteich	−/−	+	n.a.	+	n.a.
Kottenforst	−/−	n.a.	+	+	n.a.
Chorbusch	−/−	n.a.	+	+	n.a.
Geldenberg	−/−	+	n.a.	n.a.	n.a.
Rehsol	−/−	+	n.a.	n.a.	n.a.
Johannser Kogel	−/−	+	n.a.	+	n.a.
Krakovo	−/?	n.a.	n.a.	+	n.a.
Unterhölzer	−/0	?	n.a.	n.a.	n.a.
Dalby	−/−	+	n.a.	n.a.	+
Vardsätra	−/−	n.a.	n.a.	n.a.	+
Białowieza N.P.	−/−	n.a.	+	+	+
Białowieza thermophile oak forest	−/0	n.a.	n.a.	+	n.a.

−/−, no regeneration and a decline in the number of old trees; −/0, no regeneration, but no decline in the number of old trees; +, regeneration and an increase in the number of trees; ?, no information or insufficient information available; n.a., the species concerned is not found in the reserve, or no attention has been devoted to it in the study.

- The above shows that Draved Skov has incorrectly been used as a reference for the untouched vegetation originally present in the lowlands of Central and Western Europe.

Spontaneous developments in the various forest reservations covered, including Draved Skov itself, show that the presence of oak beside shade-tolerant species such as lime, beech, hornbeam and elm cannot be interpreted as evidence of the presence of these species in each other's company in a prehistoric, untouched closed forest. The correction factors used to correct percentages of pollen from the various types of trees from pollen samples are corrected based on the composition of species in this forest, from the assumption that this forest with oak and shade-tolerant species side by side is a modern analogy of the prehistoric, untouched vegetation. This is incorrect and therefore the calculated correction factors are also incorrect.

- In spontaneous development in forests in the east of the United States, no successful regeneration of oak takes place in forests where light-demanding oaks – represented by white oak – are part of the canopy. They are replaced by shade-tolerant species like American beech, red maple, silver maple, sugar maple, American elm, white ash, hemlock and the light-demanding, fast growing tulip poplar, which can become as old as an oak.

In the east of the United States, in forests where oaks are part of or dominate the canopy, either in small or in large gaps in the canopy, or in the case of the blow-down over large areas by heavy storms, there is no successful regeneration of oak. Shade-tolerant species and the light-demanding tulip poplar progressively replace them. These data show that it is very improbable, if not impossible, that the prehistoric vegetation, as shown in the species composition of the pollen samples, was a closed forest.

Establishment of Trees and Shrubs in Relation to Light and Grazing

6

6.1 Introduction

The difference in the light required by different species of trees is considered by many authors[1] as the most important factor for regeneration, the competition between trees and succession in forests. On this basis, species are categorized into pioneering species, which require light, and climax species, which tolerate shade. On the basis of models of forest succession, this classification is considered adequate (Botkin, 1993, p. 69). Factors concerning their location, such as soil humidity, temperature and type of soil, are admittedly important for regeneration (Dengler, 1990, p. 24 *et seq.*; Madsen, 1995), but are seen as being subordinate to the amount of light available (Burschel and Schmaltz, 1965a; Röhrig, 1967; Madsen, 1994; 1995; Ziegenhagen and Kausch, 1995).

The question is whether, and to what extent, light is a determining factor for the establishment of pedunculate oak (*Quercus robur*), sessile oak (*Q. petraea*) and hazel (*Corylus avellana*). Both species of oak and hazel are characterized as requiring light (Ellenberg, 1986, p. 82; Dengler, 1992, p. 179). In a number of forests the number of pedunculate and sessile oak seedlings which grow is certainly sufficient to replace the parent plants, despite predation and other factors

[1] Vanselow (1949, p. 103), Burschel and Huss (1964), Huss and Stephani (1978), Bormann and Likens (1979, p. 106), Sumer and Röhrig (1980), Whitmore (1982; 1989), Swaine and Whitmore (1988), Dengler (1990, pp. 273–298), Mayer (1992, pp. 317–320), Madsen (1995), Peters (1995), Zoller and Haas (1995), Kuiters *et al.* (1997).

which prevent the germination of acorns.[2] Light does not play a role in the germination as such (Sanderson, 1958, p. 38; Andersson, 1991). Experience of forestry has shown that pedunculate and sessile oak seedlings need a lot of light relatively quickly if they are to grow.[3] As we saw in Chapter 4, young oaks have to be provided with full daylight after 5–10 years by successively felling the standing parent trees, if this 'natural' regeneration is to be successful.[4] For beech, the duration of 'natural' regeneration extends to a period of 30 years. Obviously, beech tolerates a great deal more shade than oak.[5] As the hypothesis states, and spontaneous developments in forest reserves confirm, pedunculate and sessile oak and hazel regenerate little or not at all in closed forests in gaps in the canopy, while beech (*Fagus sylvatica*), hornbeam (*Carpinus betulus*), small-leaved lime (*Tilia cordata*), broad-leaved lime (*T. platyphyllos*), field maple (*Acer campestre*), Norway maple (*A. platanoides*) and various species of elm (*Ulmus* sp.), do regenerate in these conditions.

In this chapter, I will examine to what extent light required, or the tolerance to shade of seedlings of pedunculate and sessile oak, broad-leaved and small-leaved lime, beech, hornbeam and hazel, are determining factors for the establishment of these species in closed forests. I will look at the individual species and discover what a reduction in the intensity of light means in terms of the competition of the species as they grow into trees. I will study a number of experiments in which the seedlings of pedunculate and sessile oak, beech, broad-leaved and small-leaved lime and hazel are exposed to different amounts of daylight.

In the experiments described, the reduction of daylight was usually achieved artificially, for example, with nets or blinds. The reduction in the amount of daylight is measured and given as a percentage of full daylight in the test design. This method is seen as the most practical method for experiments of a relatively short duration, which must be reproduced (Dohrenbusch, 1987). In order to compare the relative amounts of daylight, the light intensities must be determined at the same time in the open field (100% daylight) and in shady places (Anderson, 1964). This was done in these experiments. The use of this method does mean that the percentages of daylight in tests in different places in Europe cannot simply be compared. Equivalent percentages of daylight in

---

[2] See Watt (1919), Vanselow (1949, p. 222), Doing-Kraft and Westhoff (1958), Sanderson (1958, p. 38), Krahl-Urban (1959, p. 233), Ovington and McRae (1960), Mellanby (1968), Shaw (1968a,b; 1974), Rackham (1975, pp. 108–109; 1980, pp. 296–297), Mayer and Tichy (1979), Newbold and Goldsmith (1981, pp. 9–21), Tendron (1983, p. 58), Lemée (1987), Wolf (1982; 1988), Dengler (1990, pp. 58, 63, 273–276), Anderson (1991), Gurnell (1993).
[3] See Bühler (1922, pp. 312, 331), Vanselow (1926, pp. 63, 86–87; 1949, p. 131), Tangermann (1932), Reed (1954, pp. 84–87), Turbang (1954), Klepac (1981), Hausrath (1982, pp. 75–76), Tendron (1983, p. 58 e.v.), Leibundgut (1984b, p. 86), Raus (1986), Dengler (1990, p. 63), Mayer (1992, p. 345).
[4] Bühler (1922, pp. 312, 331), Vanselow (1949, p. 131), Reed (1954, pp. 84–85), Turbang (1954), Tendron (1983, p. 59), Mayer (1992, p. 345).
[5] See Vanselow (1926, p. 215; 1949, pp. 103, 133, 203), Reed (1954, pp. 86–87), Leibundgut (1984b, p. 86), Dengler (1990, pp. 275, 278), Mayer (1992, p. 341).

different places in Europe do not actually mean that there is the same amount of light. Amounts of light can only be compared in different places in Europe on the basis of absolute amounts measured (Anderson, 1964). Further information about the design of the tests in different experiments is given in Appendix 12.

As most experiments in this field were for a duration of only one or a few years, I will also look in this chapter at the empirical information obtained from forestry practice. In addition, I will see whether the information about seedlings provides any confirmation of the alternative hypothesis, i.e. that the original vegetation in the lowlands of Western and Central Europe was a vegetation which consisted of alternating grasslands, scrub and groves in which large herbivores played a determining role in the succession. I will see whether this sort of vegetation could also have developed with the wild herbivorous fauna after the end of the Ice Age, when trees migrated from their refuges in the south and south-east of Europe to the north and north-west of Europe, as a result of climate changes. In this chapter I will also explore whether the data that are presented in this chapter about the lowlands of Europe are also relevant to the east of the United States.

6.2 Pedunculate Oak (*Q. robur*) and Sessile Oak (*Q. petraea*)

6.2.1 The reaction of pedunculate and sessile oak seedlings to reduced amounts of daylight

Pedunculate and sessile oak seedlings which were exposed to reduced amounts of daylight in different experiments in the first growing season demonstrated a reduction in the total dry weight (Ovington and McRae, 1960; Jarvis, 1964; Ziegenhagen and Kausch, 1995). The plants invested relatively more dry matter in the above ground parts, the stem and leaves, which are concerned with assimilation. This was at the expense of the root system, which lost weight in relation to the stem. This development manifested itself in an increase in the stem–root ratio (see Tables 6.1, 6.2 and 6.3 and Fig. 6.1). Figure 6.2 shows that with reduced amounts of daylight compared with full daylight, pedunculate oak seedlings invest a great amount of starch in the root, despite a reduction in their growth. Even with the lowest light intensities, this amount was still considerable in comparison with full daylight. Morphologically, the seedlings reacted with an increase of the specific leaf area (SLA) and the leaf area/weight ratio (LAR).[6] With a reduction in the amount of daylight, the average height also increased because of the etiolation of the stem. Table 6.3 shows that this

---

[6] The specific leaf area (SLA) is the area of the leaf per dry weight of leaf, or cm^2 leaf per gram of leaf (cm^2 g^{-1}). The leaf area ratio (LAR) is the relation between the area of the leaf and the dry weight of the whole plant, cm^2 per gram of plant (cm^2 g^{-1}).

Table 6.1. Effect of a reduced amount of daylight on sessile oak seedlings in the first growing season (weights are dry weights) (from Ovington and McRae, 1960, p. 552).

	Percentage of daylight			Significance of difference between treatments
	85	45	15	
Weight of remnant of acorn (g)	0.73	0.7	0.69	0
Weight of leaves (g)	0.61	0.57	0.51	**
Weight of stem and branches (g)	0.57	1.16	1.06	0
Weight of roots (g)	2.26	1.29	1.1	***
Weight of seedling (g)	4.27	3.72	3.36	***
Ratio of stem–root	0.52	2.12	1.43	—
Length of stem (cm)	12.4	17.5	17.6	***
Height of seedling (cm)	11.6	16.6	17.1	***
Number of leaves	6.8	7.5	6.8	**
Surface area of leaves (cm²)	97.0	120.0	125.0	***

— not statistically evaluated; 0, not significant; * Significant $P < 0.05$; ** significant $P < 0.01$; *** significant $P < 0.001$. The authors gave the ratio of root–stem; to be consistent with the other data, it was converted to a stem–root ratio.

Table 6.2. Growth of sessile oak seedlings in the first growing season with reduced amounts of daylight. Weights are dry weights (from Jarvis, 1964, p. 549).

	Percentage of daylight				Significance of difference between light intensity ($P < 0.05$)
	100	56	34	20	
Weight of cotyledons when acorn is planted (C1) (mg)	836	836	836	836	—
Weight of cotyledons when harvested (C2) (mg)	110	95	104	95	0
Weight of root (R2) (mg)	1351	1197	977	755	*
Weight of stem (S2) (mg)	359	412	348	418	0
Weight of leaf (L2) (mg)	449	427	426	423	—
Surface area of leaf (cm²)	60.1	69.9	80.4	91.2	—
R2 + S2 + L2 = W2 (mg); total weight of plant	2159	2036	1751	1596	*
S2/R2; stem–root ratio	0.26	0.34	0.36	0.56	*
Height (cm)	11.0	12.8	13.3	15.6	*

Significance in accordance with the key of Table 6.1.

Table 6.3. Effects of the reduction in the amount of daylight on the growth of pedunculate oak seedlings in the first growing season (data are averages ± standard deviation) (from Ziegenhagen and Kausch, 1995, p. 101).

Percentage of daylight	Length of stem (cm)	Number of leaves	Total surface area of leaves (cm^2)	Total dry weight of stem (g)	Total dry weight of leaves (g)
100	13.8 ± 0.73	7.5 ± 0.40	82 ± 7.3	1.45 ± 0.17	0.63 ± 0.06
50	15.3 ± 0.83	7.0 ± 0.45	112 ± 10.7	1.34 ± 0.10	0.61 ± 0.04
25	18.8 ± 0.98	6.3 ± 0.30	122 ± 8.7	1.24 ± 0.10	0.53 ± 0.03
10	15.8 ± 0.95	5.6 ± 0.45	109 ± 8.3	0.78 ± 0.06	0.37 ± 0.03

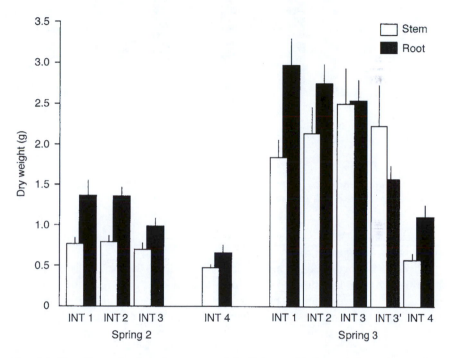

Fig. 6.1. The effect of reduced amounts of daylight on the dry weight (with SD) of the root and the stem of 1- and 2-year-old pedunculate oak seedlings. The light intensities are 100% (INT 1), 50% (INT 2), 25% (INT 3) and 10% (INT 4). The intensity INT 3' is also 25% daylight, but concerns plants which grew with INT 4 (10% daylight) for the first growing season, and INT 3' (25%) in the second growing season (redrawn from Ziegenhagen and Kausch, 1994, p. 103).

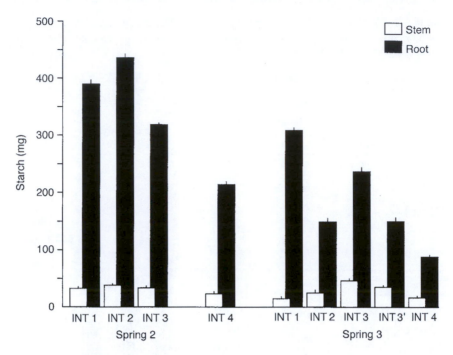

Fig. 6.2. The effect of decreasing amounts of daylight on the distribution of starch between the root and the stem of 1- and 2-year-old pedunculate oak seedlings. For the light intensities, see Fig. 6.1 (redrawn from Ziegenhagen and Kausch, 1994, p. 103).

etiolation did not continue with extremely low amounts of daylight, but that there was a decrease in the average height.

In general, pedunculate and sessile oak seedlings developed, in the case of a reduction in the amount of daylight, in a way that is general for seedlings exposed to reduced amounts of light, namely by investing more in the assimilating parts above ground, at the expense of the part of the plant under the ground, the root system (see Brouwer, 1962a, b; Brouwer and Kuiper, 1972, pp. 130–131, 207–211; Bannister, 1976, pp. 81–82; Packham and Harding, 1982, p. 42; Van Hees, 1997). This pattern continued in pedunculate and sessile oak in the second growing season, though there was even poorer growth when there was a reduction in the amount of daylight. This is shown in Tables 6.4, 6.5 and 6.6, as well as Fig. 6.1. In the second growing season, Ziegenhagen and Kausch (1995) found that, as Table 6.6 indicates, the relative growth rate (RGR) of the stem of pedunculate oak and the net assimilation rate (NAR)[7] were higher with a reduced amount of daylight than with full

[7] The relative growth rate (RGR) is the increase of weight of dry material per gram of dry weight of the plant per day (RGR = LAR x NAR); meaning mg plant per gram plant per day (mg g^{-1} day^{-1}). The net assimilation rate (NAR) is the mg plant per cm^2 leaf per day (mg cm^{-2} day^{-1}).

Table 6.4. Effect of reduced amounts of daylight on sessile oak seedlings after two growing seasons. Weights are dry weights (from Ovington and McRae, 1960, p. 552).

	Percentage of daylight			Significance of the difference between treatments
	85	45	15	
Weight of leaves (g)	0.75	0.79	0.67	*
Weight of stems and branches (g)	0.88	1.35	1.27	***
Weight of roots (g)	3.62	2.80	1.89	***
Weight of seedlings (g)	5.25	4.94	3.83	—
Stem–root ratio	0.45	0.76	1.03	***
Length of stem (cm)	23.8	29.5	30.3	***
Height of seedling (cm)	17.2	26.0	27.8	***
Number of leaves	13.5	9.8	8.5	***
Surface area of leaves (cm^2)	167.0	187.0	169.3	0
Number of branches	2.4	1.6	1.4	***
Average length of branches (cm)	4.6	5.3	5.7	—

See Table 6.1 for significance levels.

Table 6.5. The effects of a reduction in the amount of daylight on the growth of pedunculate oak seedlings after two growing seasons (data are averages ± standard deviation). INT 3′ are seedlings which grew with 10% daylight (INT 4) in the first growing season, and 25% daylight (INT 3) in the second growing season (from Ziegenhagen and Kausch, 1994, p. 101).

Percentage of daylight	Length of stem (cm)	Number of leaves	Total surface area of leaves (cm^2)	Total dry weight of stem (g)	Total dry weight of leaves (g)
100	46.8 ± 3.0	51.5 ± 5.1	583 ± 57.8	9.74 ± 0.7	4.15 ± 0.46
50	53.4 ± 3.1	50.2 ± 6.7	830 ± 69.0	15.06 ± 1.4	4.40 ± 0.48
25	74.3 ± 4.3	48.0 ± 9.4	899 ± 136.4	16.72 ± 1.9	3.75 ± 0.61
10	40.0 ± 3.3	13.8 ± 1.9	306 ± 44.9	2.91 ± 0.4	1.08 ± 0.16
INT 3′	60.6 ± 4.5	29.3 ± 1.7	737 ± 66.9	9.00 ± 1.0	3.19 ± 0.39

Table 6.6. Effects of differences in amounts of light (100–10%) and changes in this on the relative growth rate (RGR), leaf area ratio (LAR) and net assimilation rate (NAR) for pedunculate oak after two growing seasons (data are averages with ± standard deviation). INT 3′ are seedlings which grew with 10% daylight in the first growing season (INT 4), and 25% daylight (INT 3) in the second growing season (from Ziegenhagen and Kausch, 1994, p. 101).

Percentage of daylight	RGR	LAR (dm^2 g^{-1})	NAR (g m^{-2})
100	0.84 ± 0.02	0.61 ± 0.05	149.2 ± 15.2
50	0.91 ± 0.01	0.59 ± 0.06	174.0 ± 22.1
25	0.91 ± 0.01	0.59 ± 0.10	209.5 ± 42.9
10	0.68 ± 0.05	1.19 ± 0.17	79.7 ± 12.5
INT 3′	0.90 ± 0.01	0.88 ± 0.14	130.1

Table 6.7. The effects of decreasing amounts of daylight on the growth and survival of sessile oak seedlings which were exposed to this for eight growing seasons from their second year (from Shaw, 1974, p. 170).

% daylight	1965	1966	1967	1968	1969	1970	1971	1972	Max. height in 1972 (cm)	Av. weight in 1969 (g) of surviving plants
85%										
Height (cm)	6.28	10.75	14.37	16.69	20.39	23.79	32.95	35.35	150.0	7.04
Survival (%)	100	94	92	82	74	72	72	72		
43%										
Height (cm)	7.09	10.15	12.33	13.90	15.71	17.12	20.44	20.25	32.0	1.28
Survival (%)	100	98	96	86	68	56	42	42		
31%										
Height (cm)	6.79	9.63	11.50	13.55	18.09	21.37	22.46	25.95	41.0	1.02
Survival (%)	100	92	84	76	32	24	18	16		
19%										
Height (cm)	7.67	10.30	12.28	12.75	15.95	16.38	18.19	19.39	31.0	1.32
Survival (%)	100	86	86	78	52	48	32	32		
15%										
Height (cm)	6.91	8.83	11.12	12.20	14.20	15.40	16.46	17.71	22.5	0.97
Survival (%)	100	100	86	78	48	40	18	18		

daylight. Young pedunculate oaks produced the largest quantity of starch in the tap root in full daylight, as shown in Fig. 6.2. With light intensities lower than full daylight, pedunculate and sessile oak therefore still grew relatively well in the second growing season. However, in a longer series of observations, Shaw (1974) observed that after a few years there were great differences in the development of the sessile oak seedlings (see Table 6.7). The seedlings exposed to very little light not only grew much less, but were also prone to a much higher mortality rate. The greatest differences became apparent only after a few years, when the greatest distinction was visible even with a relatively slight reduction in the amount of daylight, i.e. a reduction in the light intensity from 85 to 43%. Oosterbaan and Van Hees (1989) also found a higher mortality rate with less daylight in sessile oak. After a period of four growing seasons, the mortality in their study with crown cover of 90–100% (undisturbed canopy), 70% (slightly thinned out canopy), and 40–50% (greatly thinned out canopy) amounted to, respectively, 85%, 58% and 47% (Oosterbaan and Van Hees, 1989, p. 12).

One explanation for the improvement in growth in the first years of life, when there is a reduction in the amount of daylight, could be that the seedlings have reserve nutrients in the cotyledons in the acorn. Jarvis (1964) found that in the first year of life, the cotyledons reduced by almost 90%, irrespective of the light intensity (see Table 6.2). Brookes *et al.* (1980) found that for pedunculate and sessile oak seedlings growing in full daylight, the weight of the cotyledons decreased as the growth of the root and stem of the seedling increased. The total weight of the seedlings (stem, root and cotyledons together) remained constant.

This situation continued until the second growth shoot was completed, i.e. when the second crown of leaves unfolded. Figures 6.3 and 6.4 show this development. The second crown of leaves has fully unfolded in growth stage 6.

The endogenous rhythm of pedunculate and sessile oak seedlings produced two growth shoots in the first year of life, each with one crown of leaves; i.e. a total of two crowns of leaves. (Jones, 1959; Hoffmann, 1967; Shaw, 1974; Brookes *et al.*, 1980; Harmer, 1990; Alaoui-Sossé *et al.*, 1994). Brookes *et al.* found that at this stage there was no net increase in the dry weight of the seedling (Figs 6.3 and 6.4). This shows that the formation of the first and second shoot in the first growing season, with the first and second crown of leaves respectively, takes place to an important extent with the help of the reserves in the cotyledons. For the second growth shoot, the seedlings use sucrose from the leaves and starch from the stem of the first growth shoot, as well as reserves from the cotyledons (Alaoui-Sossé *et al.*, 1994). It is only after

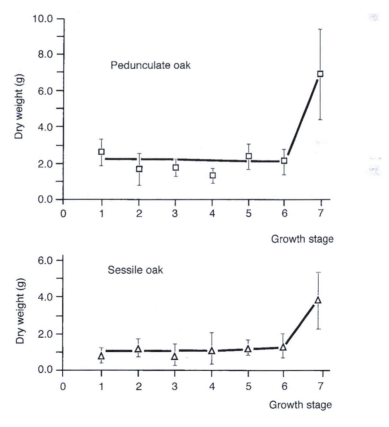

Fig. 6.3. The total dry weight of pedunculate oak □ and sessile oak seedlings △ at different stages of growth. The average values and standard deviation are shown. The first crown of leaves unfolded in stage 4, the second in stage 6 (redrawn from Brookes *et al.*, 1980, p. 169).

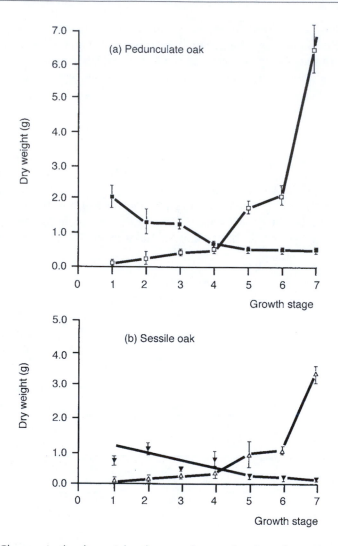

Fig. 6.4. Changes in the dry weight of root and stem of pedunculate oak and sessile oak, during growth in the first growing season. The average values and standard deviation are shown. Pedunculate oak: ■ dry weight of cotyledons; □ dry weight of root and stem. Sessile oak: ▼ dry weight of cotyledons; △ dry weight of root and stem of pedunculate oak (redrawn from Brookes *et al.*, 1980, p. 170).

the crown of leaves of this second growth shoot has unfolded that the pedunculate and sessile oak seedlings in the experiment of Brookes *et al.* grew significantly, as shown in Figs 6.3 and 6.4. The reduction in the weight of the cotyledons shows that this growth was achieved wholly by means of assimilation by the parts of the seedling that are above ground, and by absorbing nutrients through the roots (Brookes *et al.*, 1980; Alaoui-Sossé *et al.*, 1994; see Fig.

6.4). After the completion of the second crown of leaves, i.e. after the first growing season, the cotyledons obviously do not play a significant role any longer if the plants grow in full daylight (Brookes *et al.*, 1980). However, the effect of the reserves in the cotyledons on growth does extend into the second growing season because the first growth shoot has formed with the help of the reserves, and to some extent, the second growth shoot (Alaoui-Sossé *et al.*, 1994). This was also found for other species of oaks (see Triplathi and Khan, 1990). The cotyledons seem to be important in the first year if the growth conditions for the seedlings are unfavourable, such as is the case if there is grazing and when competion of grasses and herbs plays a role. However, even then the importance is limited (Frost and Rydin, 1997b).

From the beginning, pedunculate and sessile oak reveal a growth of the root system unparalleled by any other species of tree in Europe (Jones, 1959; Krahl-Urban, 1959, pp. 39–40; Shaw, 1974; Newbold and Goldsmith, 1981, p. 28; Van Hees, 1997). More than 90% of the total weight of the root is invested in a deep-rooting tap root (Shaw, 1974). The stem–root ratios vary from 0.5 to 0.1 (Jarvis, 1964; Shaw, 1974; Newbold and Goldsmith, 1981, p. 28). The stem–root ratio can be 3–6 times as large in oaks as in other species of trees (Jones, 1959). With the exception of Jarvis's experiment (1964), these levels were not found in the experiments described here. With full daylight and 56% daylight, Jarvis found a stem–root ratio of 0.25 for a short period. The development of the root system also indicates that in the first growing season the reserves in the cotyledons provide a significant proportion of the nutrients. The large amount of starch which the seedling stores in the root in relation to the stem, as indicated in Fig. 6.2, cannot be acquired by assimilation, because the storage takes place while the assimilation system is being developed. The roots are actually a sort of 'sink' during this period (Alaoui-Sossé *et al.*, 1994). In 3-year-old pedunculate oak where the cotyledons were exhausted, and which were exposed to a greatly reduced amount of daylight by Hoffmann, there was, in particular, a significant reduction in the growth of the roots compared with the specimens growing in full daylight (see Fig. 6.5). The growth of the root of the shaded plants was not only greatly reduced overall, but was also restricted to a few periods of the year; on the other hand, in young pedunculate oak growing in full daylight, the growth of the root system continued throughout the year (Hoffmann, 1967; Brookes *et al.*, 1980; Harmer, 1990; Alaoui-Sossé *et al.*, 1994). It is known that if the light conditions do not improve, oak seedlings will still vegetate for some time and eventually disappear after 4–5 years (Seeger, 1930; 1938; Lemée, 1987).

The fact that reallocation of nutrients from the cotyledons is mainly responsible for the growth of the seedlings in the first year of life, and assimilation is mainly responsible for growth in the second and subsequent years, is also clear from transplanted seedlings which grew with 10% daylight in their first year and were given 25% daylight in their second year. They achieved higher levels of the dry weight and absolute quantities of starch in the tap root than the seedlings which were exposed to 10% daylight in the first and second years of

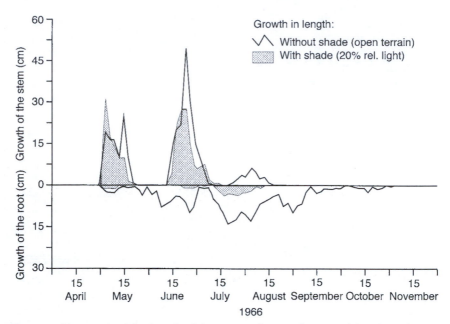

Fig. 6.5. The growth of the length of the root and stem of 3-year-old pedunculate oaks (*Quercus robur*) in full daylight and with relative amounts of daylight from 20% (redrawn from Hoffmann, 1967, p. 746).

life (see Table 6.5 and Figs 6.1 and 6.2). The fact that light does play a role in the growth, even in the first growing season, is clear from the fact that the transplanted seedlings do not reach the levels of those which grew with 25% daylight during the first and second growing seasons. Sonessen (1994) also showed that assimilation, and therefore the amount of light, plays a role in the first growing season in pedunculate oak seedlings which grew in full daylight. Plants from which the cotyledons had been removed after the stem was full-grown, i.e. after the first crown of leaves had unfolded, did not reveal any difference in the dry weight at the end of the second growing season, compared with plants which still had the cotyledons, irrespective of whether they grew in poor or rich soil (Sonessen, 1994). This is shown in Fig. 6.6. Therefore, in principle, the seedlings can continue to develop independently of the cotyledons after the completion of the first crown of leaves in the first growing season.

On the basis of his data (Table 6.7), Shaw (1974) concluded that to achieve reasonable growth oak needs more than 50% daylight. This sort of percentage of daylight is found in thin oak forests or in gaps in the canopy of oak forests with an area of 0.01 ha, and in beech forests with an area of 0.2 ha (Shaw, 1974; Von Lüpke, 1987). Therefore, pedunculate and sessile oak seedlings can grow in large gaps in the canopy in the first years. As we saw in Chapter 5, developments in forest reserves show that seedlings do grow into young trees in the gaps in the canopy, but that in these conditions, there is no development into mature trees which form part of the canopy.

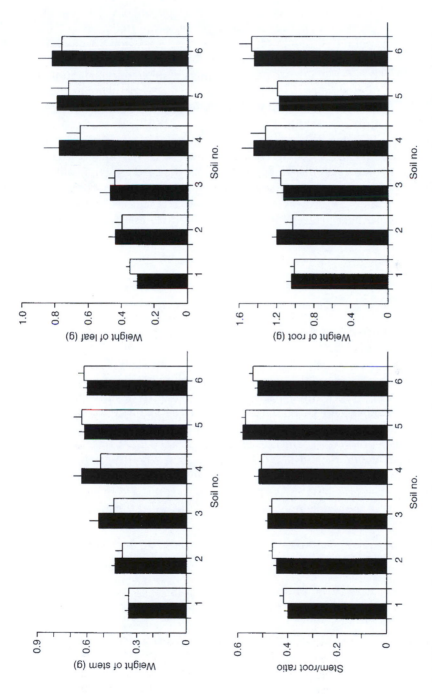

Fig. 6.6. The average values with the standard deviation of pedunculate oak seedlings which have grown with and without cotyledons in full daylight on six different types of soil during two growing seasons. The black bars are plants which have grown with cotyledons, the white bars are those of plants in which the cotyledons were removed in the first growing season after the first crown of leaves was completed (redrawn from Sonessen, 1994, Figures f, g, h and i, p. 67).

To summarize, it may be concluded from the information presented above, that young sessile and pedunculate oaks are relatively tolerant of shade for a few years because of the reserves in the cotyledons (see also Van Hees, 1997). This explains the low compensation levels of respectively 5.9% and 8% daylight which Jarvis (1964) found in 1-year-old pedunculate oak seedlings in two experiments. As the further growth of the seedlings with a reduced amount of daylight shows, this tolerance to shade is not representative of their physiological capacity. Other experiments, in which oak seedlings grew with reduced amounts of daylight, confirmed this (see Cieslar, 1909; Fabricius, 1929; Hoffmann, 1967). This was also shown by the empirical data obtained from forestry practice. In forests, seedlings of these two species of oak survive from 1 to at most 3–4 years under the canopy.[8] Therefore, the good growth results of pedunculate oak seedlings in the first and second years of life cannot simply be extrapolated to later ages. This is also clear from the difficulties encountered when attempts are made to regenerate oak 'naturally' in forests by means of the shelterwood system, the selection system and the group selection system.[9]

6.2.2 The establishment of both species of oak in relation to the alternative hypothesis

As we saw in Chapter 4, pedunculate and sessile oak are widespread in wood-pasture (also see Rackham, 1980, p. 293). Some authors[10] attribute this to human activity, such as the protection provided for oaks when they produced the mast, and the related fact that young oaks were planted. However, Chapters 2 and 4 described in detail how oaks particularly become established in grazed areas in grassland, brushwood and fringes of thorny scrub without any such measures.[11] In the commons in the Chilterns, Watt (1934b) found that oaks and other species of trees became established in scrub consisting of juniper, blackthorn and hawthorn, and developed into oak groves. Both blackthorn and one-styled hawthorn are species which require light and which only become established in open grassland.[12] As Chapter 4 shows, thorns were widespread in wood-pasture in the Middle Ages. Oaks became established when the grazing

[8] See Forbes (1902), Watt (1919), Seeger (1930; 1939), Tansley (1953, p. 295), Doing-Kraft and Westhoff (1958), Rackham (1980, p. 297), Newbold and Goldsmith (1981, p. 36), Wolf (1982; 1988), Lemée (1987), Dengler (1990, p. 63; 1992, p. 170).
[9] Bühler (1922, pp. 310, 331, 433, 566), Vanselow (1926, pp. 63, 87–88; 1949, pp. 117, 128), Boden (1931), Seeger (1938), Krahl-Urban (1959, p. 146), Hausrath (1982, p. 76).
[10] Doing-Kraft and Westhoff (1958), Krahl-Urban (1959, pp. 21, 56, 140), Schubart (1966, pp. 14, 51, 56–58, 68–70, 74), Rackham (1980, p. 300), Koop (1981, p. 92), Mantel (1990, p. 444), Harris and Harris (1991, p. 36), Jahn (1991, p. 395).
[11] See Forbes (1902), Adamson (1921; 1932), Watt (1924; 1925; 1934a,b), Nietsch (1939, p. 117), Fenton (1948), Tansley (1953, pp. 130–133, 489), Peterken and Tubbs (1965), Rackham (1980, pp. 289, 293, 294–297, 300), Tubbs (1988, pp. 153–157), Pott and Hüppe (1991, pp. 23–27).
[12] Puster (1924), Watt (1924; 1934b), Ekstam and Sjörgen (1973), Rackham (1980, pp. 352, 355), Smith (1980, p. 352), Ellenberg (1986, p. 95), Tubbs (1988, p. 157; 1987).

was such that thorny scrub developed, and spread when the thorn bushes spread. The rate at which oak groves spread is the same as the rate at which the front of the thorny scrub advances into the grassland (Watt, 1934b). Therefore, the presence of oak in wood-pasture is related to the grazing of large herbivores.

In this respect, it is striking that, according to certain publications (including Watt, 1919; Peterken and Tubbs, 1965; Rackham, 1980, p. 293; Tubbs, 1988, pp. 153–157), oak is extremely predominant in wood-pasture. For example, Watt (1925) noticed that young oaks particularly grew in grassland bordering on a beech forest. The jay (*Garrulus glandarius*) buries acorns, which then germinate to produce seedlings (Mellanby, 1968; Bossema, 1979; Kollmann and Schill, 1996). The question is whether the jay plays a role in the predominance of oak in wood-pasture.

From mid-September to mid-November, jays collect acorns of pedunculate and sessile oaks (see Fig. 6.7). They carry these in their gullet, throat and beak. They bury them at some distance from the tree where they collect the acorns. This distance varies from a few score metres to several kilometres (Schuster, 1950; Chettleburgh, 1952; Bossema, 1979, pp. 32–33). Jays do not distinguish between pedunculate and sessile oak. They select healthy, ripe acorns which

Fig. 6.7. Jay collecting acorns. An acorn, which has been swallowed, is visible at the back of the throat. The jay takes one to six acorns at a time, holding one, the largest or longest in its beak; the other acorns are carried in the gullet or throat (photograph, J. Korenromp).

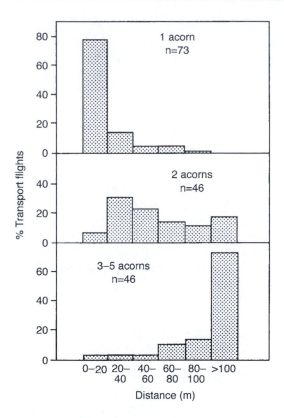

Fig. 6.8. The percentage distribution of distances over which jays move a particular number of acorns (redrawn from Bossema, 1979, p. 31).

have not been affected by parasites and are not too small (Bossema, 1979, pp. 23–28). They take one to six acorns on every flight, carrying one of these – the largest or longest – in their beak, and the others in their throat and gullet (Schuster, 1950; Chettleburgh, 1952; Bossema, 1968; 1979, pp. 20–26). The more acorns a jay carries at a time, the larger the distance it travels to bury them (see Fig. 6.8). Usually, several birds collect acorns from a particular place at the same time. Chettleburgh (1952) calculated that in mid-October, at the height of the collection, 35–40 jays removed and buried 63,000 acorns in a period of 10 days. Schuster (1950) came to the conclusion that in 4 weeks, approximately 65 birds dispersed at least 500,000 acorns throughout the area.

When they bury the acorns, jays have a clear preference for open terrain, such as large open spaces in forests and grasslands and fields (Kollmann and Schill, 1996). In this open terrain, they have a preference for a transitional area of short to long grass or brushwood (the periphery of fields), the outer edge of hedges and the fringes of thorny scrub and mantle vegetation of forests, as shown in Figs 4.2, 4.3, 4.5 and 6.9 (Chettleburgh, 1952; Bossema, 1979, pp. 35, 45–47, 51, 54, 69; Vullmer and Hanstein, 1995; Rousset and Lepard,

Fig. 6.9. On the left, the mantle and fringe vegetation of a grove is visible in the Borkener Paradise, Germany. The jay plants the acorns at the very edge, on the transition to grassland, the fringe (see also Fig. 4.5). On the right of the photograph, there is an oak growing in a hawthorn bush. Here the jay has planted an acorn at the foot of the hawthorn bush. At the back, on the left of the photograph, there is an oak growing in the middle of a clump of hawthorn (photograph, F.W.M. Vera).

1999). These are the structures which develop in grasslands when there is grazing. Research by Kollmann and Schill (1996) also showed the preference for open terrain, in particular the mineral soil therein, but not for the horizontal and vertical structures. Chettleburgh (1952) saw jays hiding acorns in a common with thorny shrubs, such as hawthorn. He observed a jay flying down into a hawthorn bush and burying an acorn at the foot of the bush. This observation explains the phenomenon of oaks which seem to grow entwined with hawthorn in wood-pasture (personal observation and see Fig. 6.9). In addition, jays like to bury acorns in places where the soil is loose and they can easily push the acorns into the ground (Bossema, 1979, pp. 44, 49).

When a jay has several acorns it buries them all separately. The distance between two acorns varies from 0.2 to 15 m, but is generally between 0.5 and 1 metre (Bossema, 1979, p. 32). The jay pushes the acorn into the ground. If it will not go in far enough, the jay hammers it further into the ground. Then it covers over the hole with a few sideward movements of its beak, and camouflages it with leaves, little lumps of earth and stones (Chettleburgh, 1952; Bossema, 1979, pp. 32, 36–37, 78). The jays easily find the acorns they have buried. The vertical structures for which jays appear to have a clear preference when they bury the acorns seem to serve as a beacon (Bossema, 1979, pp. 57–58). The jays then dig up the acorns, peel them *in situ* and eat the contents or fly off with the unpeeled acorn in their beak to eat it elsewhere. On the other hand, a jay can never find an acorn buried by another jay in its absence (Bossema, 1979, pp. 57–58, 62–65). Burying the acorns seems to greatly reduce their predation by mice and squirrels.[13] According to Bossema, mice find acorns hidden by jays only in exceptional cases (Bossema, 1979, pp. 78–82). This is shown, for example, by the large number of pedunculate and sessile oak seedlings which grow in open grasslands, abandoned fields, roadside verges, the periphery of hedges and fringes of scrub and forest mantle at a great distance from the old fertile oaks (see Figs 4.2, 4.5 and 6.9 and Forbes, 1902; Watt, 1919; Mellanby, 1968; Bossema, 1979, pp. 55, 70).[14] Mellanby (1968) found 2000 pedunculate oak seedlings per hectare in an area where trees had been felled, up to at least 150 m from the nearest fruit-bearing pedunculate oak, and almost 5000 per hectare at a distance of 200 m.

Jays eat the acorns they hide throughout the year. However, in the period from April to August, they do so far less than in the rest of the year (Bossema, 1979, p. 16). It is during this period that the seedlings appear. The stem appears in May and the seedling is complete in June, i.e. the first crown of leaves has unfolded (Bossema, 1979, pp. 67, 70). As indicated above, the seedlings can develop independently of the cotyledons from that time. This is very important for the seedling, as in June, the jays start to look for seedlings which have grown from the acorns they buried, together with their young. When a parent bird finds a seedling, it takes hold of the stem with its beak and lifts the plant. This raises the acorn above the ground, or the soil that is brought up shows where it is hidden under the ground. In that case, the jay will dig it up. The acorn is removed from the seedling and peeled and the parent bird then gives it to its young (Bossema, 1968; 1979, pp. 17, 68). The young birds imitate their parents by yanking up all sorts of plants at random. In contrast with their parents, they do not (yet) make a distinction between different species (Bossema, 1979, p. 77). The adult jays select seedlings on the basis of size and the colour of the stem. It has to be green because this characterizes a seedling which is still attached to an acorn. Older seedlings without an acorn lose their green colour as their stem becomes woody; these are ignored (Bossema, 1979, pp. 71–72, 76). Young green seedlings which have grown from acorns and have not been

---

[13] Watt (1919), Shaw (1968b), Bossema (1979, pp. 80–90), Jensen (1985), Nilsson (1985), Lemée (1987), Kollmann and Schill (1996).
[14] Watt (1919) wrote:

> among the low growing shrubs such as Hawthorn (*Crataegus monogyna*), Blackthorn (*Prunus spinosa*), and Brambles (*Rubus spp.*), among the stems and fronds of bracken (*Pteridium aquilinum*) or among bilberry (*Vaccinium myrtillus*) and heather (*Calluna vulgaris*) there we find numerous seedlings coming up. Again among grass such as meadow grass, acorns germinate in great numbers – a fact noted by Forbes and also by Warming (Watt, 1919, p. 186).

Forbes (1902), cited by Watt (1919), wrote: 'A grassy surface, however, seems the natural seed-bed of oak, for very successful examples may often be seen on rough pasture adjoining woods, which, for some reason or other, has been allowed to lie waste, or is only lightly stocked with cattle during the summer' (Forbes, 1902, p. 256).

hidden by the jay concerned are *also* ignored. Bossema (1979) found this to be the case in birds which he kept in captivity for his research. They left 'strange' green seedlings alone. It was only when the researchers had shown the jays that these were attached to acorns that they started to inspect 'strange' seedlings for acorns themselves after a while, by lifting them up (Bossema, 1979, p. 77).

In many of the thriving oak seedlings, Bossema (1979) found the scar on the stem where the jay had picked it up with its beak (Bossema, 1979, pp. 68–69). It was also determined in other research (Vullmer and Hannstein, 1995; Rousset and Lepart, 1999) that the cotyledons had been removed from young oaks. The development of the seedlings revealed that they were not hampered in their growth by the removal of the cotyledons (Bossema, 1979, pp. 70, 76, 98). As we noted earlier, Sonessen (1994) also found this to be the case when he removed the cotyledons of seedlings growing in full daylight, after the first crown of leaves had been completed (see Fig. 6.6). The results of research by Andersson and Frost (1996) confirmed this once again. The conditions of Sonessen's experiment were met in the oaks planted by the jays. The jays inspect the seedlings in May and June when the first crown of leaves unfolds, and they hide the acorns in light, open places. Seedlings can be hampered in their growth, or even die when they are uprooted. This could happen if they were lifted up a few times in succession. This does not happen, because the jays inspect only the seedlings which develop from the acorns they planted themselves. Therefore each seedling is only lifted up once. The chance that the young oak will be uprooted is small, because seedlings which grow in full daylight form an extremely extensive root system with a long tap root immediately after germination.[15] This root system ensures that the seedling is securely anchored and not easily uprooted by being inspected by a jay. Therefore, the disadvantage of the inspection is offset by the advantage of growing in extremely light conditions. It is only extremely young seedlings which have therefore grown late in the season that are occasionally pulled out of the ground with their roots by a jay. The fact that Bossema (1979) found young oaks in open terrain, but not in the forest, led him to the conclusion that the chances of survival of young oaks is high in places where jays like to hide the acorns (Bossema, 1979, p. 98). This also explains the successful regeneration of oaks in wood-pasture, in the fringes of unshaded scrub, in grasslands and the mantle of forests, while there is no regeneration of oaks in the adjacent forests at the same time, as several researchers have found.[16] As the seedlings of pedunculate and sessile oak are found mainly in open terrain such as

[15] Jones (1959), Krahl-Urban (1959, p. 94), Ovington and McRae (1960), Jarvis (1964), Hoffmann (1967), Newbold and Goldsmith (1981, p. 28), Harmer (1990), Ziegenhagen and Kausch (1995).
[16] Fenton (1948), Peterken and Tubbs (1965), Rackham (1980, pp. 295–296). Two remarks by Rackham (1980) illustrate this. The first is: 'Oak regenerates embarrassingly well in the less-wooded parts of most wood-pastures'; the second: 'there are many places where oak regeneration begins only a few feet outside a wood boundary' (Rackham, 1980, pp. 295, 296). See also Chapter 4.

abandoned fields, grazed meadows and meadows which are no longer grazed, on verges, abandoned railway tracks and in heathland, the oak is attributed pioneering characteristics (see Tansley, 1953, pp. 293–296; Shaw, 1974; Rackham, 1975, pp. 110; 1980, p. 291; Koop, 1981, p. 46; Newbold and Goldsmith, 1981, p. 36). However, the capacity to become established in this sort of open terrain should be seen not so much as a characteristic of the oak as a characteristic of the jay.

Bossema (1979) correctly characterized the relationship between the oak and the jay as symbiotic, as both species profit from the relationship. The oak provides the jay with a significant proportion of its diet, while the oak profits from the jay because the latter 'plants' the acorns in places where the seedlings and young trees can thrive. Furthermore, the jay selects healthy and relatively large acorns which produce strong seedlings (see Jarvis, 1963). In combination with the dispersal by the jays, this increases the chances of successful establishment. In addition, burying the acorns over a widespread area restricts the chances of predation (Bossema, 1979, pp. 94–111).

Several authors (Watt, 1919; Tansley, 1953, p. 293; Ashby, 1959; Tanton, 1959; Harmer, 1990) consider the predation of acorns as the most important factor in the failure of oaks to regenerate. The relationship between the jay and both species of oak shows that the predation of seeds is not in itself a negative phenomenon. For the oak, it is even an important condition for becoming successfully established (Mellanby, 1968), possibly more important than a good mast. The success of regeneration in forest depends to a great extent on full masts, because acorns lying spread out on the forest floor are easy prey for predators. In open landscapes, this direct relationship is not so simple. For example, in a particular open site, Mellanby (1968) found 50 times as many oak seedlings when the production of acorns was 10 times less, because the conditions for jays to hide acorns in the ground had improved as a result of the ground being dug up. The presence of jays and good conditions for burying acorns, such as those arising from cultivation, therefore appear to have a greater determining effect on the successful growth of seedlings than the actual production of acorns. The digging up of the soil can be seen as being similar to the rooting of (wild) boar (*Sus scrofa*), which means that in natural conditions the rooting of wild boar, for example, in grasslands, has a facilitating effect for jays (see Treiber, 1997). In wood-pasture with livestock, domestic pigs sent out to pannage will have facilitated the burying of acorns by jays, and therefore the regeneration of oak. In addition, the grasslands in which the (wild) boar root about develop as a result of the large grazing herbivores as well as the fringes of fields and thorny scrub to which jays are strongly attracted to bury acorns.

In addition to jays, mice also play a role in the distribution of acorns, albeit over distances of less than tens of metres (Jenssen and Nielsen, 1986; Rousset and Lepart, 1999). A limiting factor for this activity by mice is that short graze, which occurs through grazing, is not a mouse biotope and the animals do not therefore move around there (Jenssen and Nielsen, 1986; Vullmer and

Hanstein, 1995; Kollmann and Schill, 1996). As a result of jays and mice bury-ing acorns in open grassland, young oaks shoot in these open types of terrain (Kollmann and Schill, 1996; Frost and Rydin, 1997a; Rousset and Lepart, 1999). Whether these young oaks survive in grazed grassland depends to a great extent on whether they are protected by a thorny bush or other species not eaten by grazers, as a consequence of the acorn being deposited there (Rousset and Lepart, 1999).

Young oaks which grow in the fringes of hawthorn and blackthorn scrub in grazed areas are at risk of being nibbled by large herbivores. However, both species of oak have a great capacity for survival and recovery which they prob-ably owe to the tap root, from which reserve nutrients can be mobilized (Shaw, 1974). Following an examination of the annual rings, badly eaten oaks, less than 0.5 m tall, and which were in full daylight, proved to be 25 years old. A few years with a reduction of grazing are enough for these trees to grow up out of reach of the browsing animals within 2–3 years (Shaw, 1974). Mellanby (1968) found that pedunculate oak seedlings growing in grasslands survived being mown down once a year for 4 successive years. Mowing twice or more times per year killed the seedlings in the space of 2 years. In forestry, the capacity for oak to recover is used by cutting the young planted oaks just above the first dormant bud at the foot after a few years. The following spring, this produces a shoot which often grows to a height of 1 m after only 1 year. Nine-year-old oaks (it is not specified which species) which were cut back 4 years after being planted, produced shoots in the following growing season with an average height of 70 cm and a maximum height of 1.3 m (Rümelin, 1926). As a result of this sort of measure, young oaks can grow above the grass which can blot out the light, or above the browse line of deer or livestock in a much shorter time (Rümelin, 1926; Seeger, 1938; Frank, 1939). Oaks which grow in full daylight are therefore well protected against browsing (Mellanby, 1968; Shaw, 1974; Rackham, 1980, p. 297). The net result is that oaks regenerate well in grazed, semi-open landscapes. The predation of acorns by jays and grazing by large herbivores are no obstacle to the establishment of oak. On the contrary, they both actually contribute to the oak's successful establishment.

Jays do not bury acorns in the deciduous forest unless there is an extremely large open space (Bossema, 1979, p. 98). When seedlings grow in gaps in the canopy in forest reserves, there is always an old fruit-bearing oak in the imme-diate vicinity (see also Chapter 5). As jays prefer places with a relatively large amount of light, they sometimes hide acorns in gaps in the canopy of forests of Scots pine which had already changed into open forests as a result of storms. This is clear from the fact that oak seedlings grow here at large distances from the old oak trees (Gross, 1933; Kuper, 1994, p. 64; Van Hees, 1997).

6.3 Beech (*F. sylvatica*)

6.3.1 The response of beech seedlings to a reduction in the amount of daylight

The average dry weight in growing beech seedlings decreased with a decline in light intensity in their first year of life. At the same time, there was a relative increase in the assimilating part of the plants above the ground. This was expressed in an increase in the specific leaf area (SLA) and the leaf area–weight ratio (LAR). As for the pedunculate and sessile oak, this development was at the expense of the roots. Tables 6.8 and 6.9 and Figs 6.10, 6.11 and 6.12 show this. Therefore, in principle, beech responds in the same way to a reduction in the amount of daylight as the two species of oak and other seedlings (see also Van Hees, 1997). A gradual difference between beech, on the one hand, and

Table 6.8. The stem–root ratio of beech with a reduction in the amount of daylight in two places with different fertility of the soil; Gahrenberg is poor in nutrients and Dransfeld is rich in nutrients (from Burschel and Schmaltz, 1965, p. 204). The original table shows the root/stem ratio. For the sake of consistency in the presentation of data in this chapter, these values have been converted into the stem/root ratio. The conclusions are based on the original values and significant difference at $P < 0.05$ and $P < 0.01$. These significant differences are also shown in this table and are indicated with *. The plants were planted as 1-year-old seedlings and harvested after two growing seasons, i.e. when they were 3 years old.

Percentage of daylight	Plants from the nursery		Plants growing wild	
	Gahrenberg	Dransfeld	Gahrenberg	Dransfeld
100	1.05	1.10	0.96	1.06
77	0.93	1.23	0.97	1.05
28	1.45	1.35	1.22	1.11
12	1.39	1.43	1.25	1.08
1	1.39	1.61	1.52	1.27
Significant difference at:				
$P < 0.05$	0.18*	0.12*	0.14*	—
$P < 0.01$	0.26*	0.16*	0.20*	—

Table 6.9. Leaf area–biomass ratio (LAR; $cm^2\ g^{-1}$) in beech with a reduction in the amount of daylight in two places with different fertility of the soil; Gahrenberg is poor in nutrients and Dransfeld is rich in nutrients (from Burschel and Schmaltz, 1965. p. 204). The plants were planted as 1-year-old seedlings and harvested after two growing seasons, i.e. when they were 3 years old.

Percentage of daylight	Gahrenberg		Dransfeld	
	Plants from nursery	Plants growing wild	Plants from nursery	Plants growing wild
100	38.9	30.5	34.2	34.7
77	48.8	37.2	34.0	41.5
28	54.6	49.0	54.5	48.8
12	80.0	60.6	73.0	64.9
1	156.0	200.0	119.0	129.8

pedunculate and sessile oak on the other hand, is that when there is a reduction in light intensity, the reduction in the growth of the root system is less marked in beech (also see Neemann and Stickan, 1992). Moreover, the height of the stem stays virtually the same (also see Brown, 1953, p. 4). Therefore, Figs 6.10, 6.11 and 6.12 also clearly show that the plants grow better, and therefore tolerate more shade, in richer soil with certain reductions in the

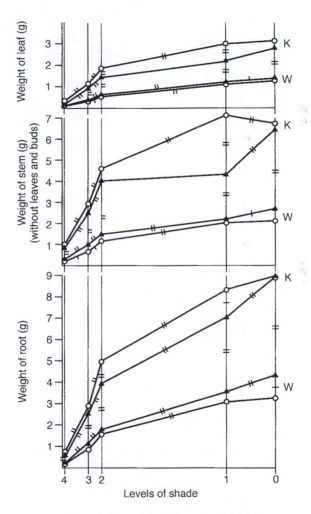

Fig. 6.10. Development of the different parts of beech seedlings under the effect of reduced amounts of daylight. The light levels are: 0 = 100%; 1 = 77%; 2 = 28%; 3 = 12%; and 4 = 1% daylight. ▲, Gahrenberg; ○, Dransfeld; K, Cultivated plants; W, Wild plants, ⫻, the difference is significant; ⫻, the difference is highly significant. All values are averages per plant. The significance of the differences are shown as: significant: $P < 0.05$; highly significant: $P < 0.01$ (redrawn from Burschel and Schmaltz, 1965, pp. 201, 204).

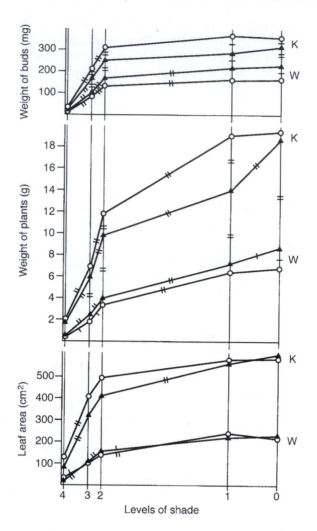

Fig. 6.11. Development of the different parts of beech seedlings under the effect of reduced amounts of daylight. See Fig. 6.10 for explanation (redrawn from Burschel and Schmaltz, 1965, pp. 201, 204).

amount of daylight. This was also found to apply for other species of trees (Grubbs *et al.*, 1996; H. Koop, Wageningen, 1997, personal communication). This distinction disappears as the light intensities are reduced.

A reduced amount of daylight also affects the survival of seedlings; Suner and Röhrig (1980) found a positive correlation between the number of surviving seedlings and the amount of daylight (see Table 6.10). Oosterbaan and Van Hees (1989) also found this sort of positive correlation. Under a canopy that was, respectively, undisturbed (90–100% crown cover), slightly thinned out (70% crown cover), or greatly thinned out (40–50% crown cover), the

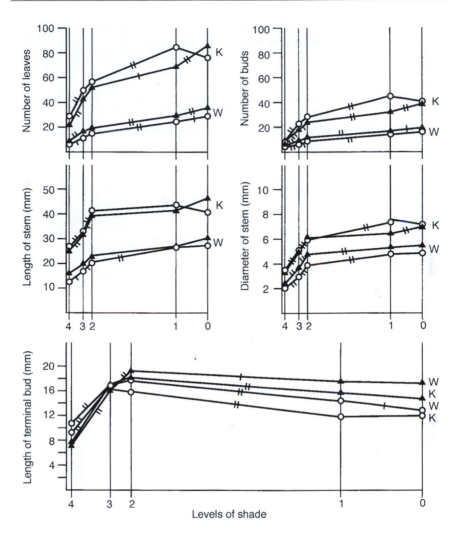

Fig. 6.12. Development of the different parts of beech seedlings under the effect of reduced amounts of daylight. See Fig. 6.10 for explanation (redrawn after Burschel and Schmaltz, 1965, p. 207).

mortality rate of beech seedlings after four growing seasons was, respectively, 87%, 74% and 47% (Oosterbaan and Van Hees, 1989, p. 10).

In the third growing season of beech seedlings, Suner and Röhrig (1980) observed the phenomenon that there was a marked reduction of growth of the height, while on the other hand, the growth in the length in all the shoots together was greatest. This phenomenon is caused by a marked branching development which gives the seedlings a broad shape. This type of growth is characteristic of beech seedlings growing in shade and semi-shade (see e.g. Fricke, 1982, pp. 89–90, 102). This develops as the result of the formation of

Table 6.10. Mortality of beech seedlings with reduced amounts of daylight under 50%. The number of stems m^{-2} is shown (redrawn from Suner and Röhrig, 1980, p. 146).

Percentage of daylight	Number of plants				Percentage of mortality from early summer 1977 to autumn 1979
	Spring 1977	Autumn 1977	Autumn 1978	Autumn 1979	
49.0	304	206	212	120	60
31.1	364	348	296	0	19
30.4	376	348	276	268	29
24.6	392	328	268	224	43
21.8	280	232	164	140	50
18.3	258	—	178	160	38
16.9	348	304	244	204	41
15.5	366	316	236	108	70
12.1	388	312	148	140	64
11.5	541	445	292	211	61
8.0	416	352	244	192	54
7.6	357	286	190	131	63
6.5	472	404	192	84	82
5.5	313	250	105	60	81
4.6	322	240	59	21	93
Correlation coefficient r	−0.292 −	−0.61 +	−0.73 ++	−0.11 −	0.66 +

Significance: −, none; +, $P < 0.05$; ++, $P < 0.01$.

lateral, leafy, so-called exploitation shoots; a morphological adaptation to create a surface which catches as much light as possible (Dupré *et al.*, 1986; Neemann and Stickan, 1992).

6.3.2 A comparison of beech and sessile oak seedlings

In a beech forest, Von Lüpke (1982; 1987) compared the growth of 2-year-old beech seedlings and sessile oak seedlings from a nursery (i.e. grown in full daylight), seedlings which had grown up together from the third year of life in an area where all the trees had been felled (100% daylight), and seedlings in a large gap in the canopy (45.5% daylight) and under a canopy that had been thinned out for the first time to promote 'natural' regeneration (11.3% daylight) (Von Lüpke, 1987). Half of the seedlings grew without any ground cover for the first 4 years, and with ground cover for the next 4 years; the other half grew with

Table 6.11. The development of the dry weight of the stem (without leaves) (g) of sessile oak and beech from a nursery (where they grow in full daylight), which were transplanted at 2 years and exposed to different amounts of daylight from their third year: 100% in the open field, 45.5% in large gaps and 11.3% under the canopy. Results followed by the same letter do not differ significantly (from Von Lüpke, 1982, pp. 59, 66).

Time	Open field		Large gap		Canopy	
	With veg.	Without veg.	With veg.	Without veg.	With veg.	Without veg.
Sessile oak						
Initial value 1977	± 2	± 2	± 2	± 2	± 2	± 2
Final value 1980	60 a	98 b	45 a	93 b	4 c	4 c
Beech						
Initial value 1977	± 3	± 3	± 3	± 3	± 3	± 3
Final value 1980	16 a	29 b	10 ac	36 b	6 ac	8 ac

ground cover for 8 years. The test sites were fenced off against wild animals (Von Lüpke, 1982, pp. 15–18; 1987).

The data in Table 6.11 show that in both sessile oak and beech, the weight of the stem declined with reduced amounts of daylight. There was little difference in the growth of these species whether they grew in an open space or in a large gap. A comparison of the two species revealed large differences in growth in the open field and the large gap. The sessile oak was ahead of the beech for the first 4 years of growth.

Both species responded to the presence of ground cover with a reduction in growth. For sessile oak, the dry weight of the stem without leaves in the sectors with ground cover in the open field was 61% of the level of those in sectors with no ground cover. In the large gap, it was 48% of that of the stems in the sector without ground cover. Under the canopy, there was no difference between the two types of treatment (Von Lüpke, 1982, p. 59). This was almost certainly caused by the shade of the canopy, so that there was only very slight ground cover. This was the same for beech. The dry weight of the stem without leaves in the sectors with ground cover in the open field was 55%, and in the large gap, 28% of the weight of the seedlings which grew with no ground cover. As for the oak, there was no difference between the two types of treatment for the beech under the canopy (Von Lüpke, 1982, p. 66). The young trees of both species therefore responded strongly to the presence of ground cover. At a certain point, the average weight of the stem of sessile oak fell below that of beech when there was a reduction in the amount of daylight, for both treatments. In this respect, it is striking that with the lowest light intensity, there was a reduction to 4% of the weight achieved by sessile oak plants in full daylight in the most favourable conditions (i.e. without ground cover), while for beech, there was a reduction to 28% of the weight achieved by young plants in full daylight. Therefore the reduction for sessile oak was seven times higher than for beech. For the average height of the stem, the levels for the sessile oak also fell below those of beech at a certain point, when there was a reduction in light intensity (see Fig. 6.13). In

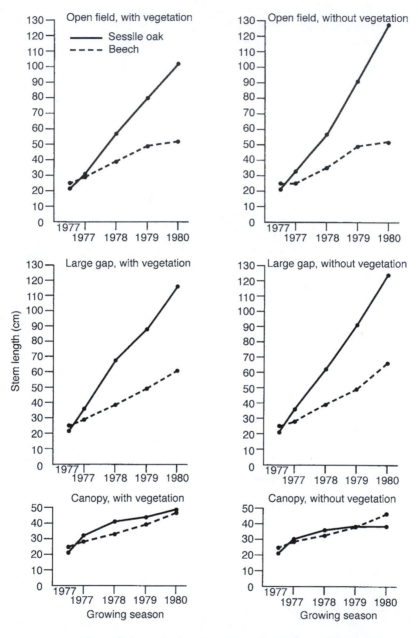

Fig. 6.13. The progress of the length of the stem (cm) of transplanted, 2-year-old sessile oak and beech seedlings from a nursery (where they grow in full daylight), which grew with different amounts of daylight from the third year, i.e. 100% (open field), 45.5% (large gap) and 11.3% (under the canopy). For the level of significance, see Table 6.12, which shows the values on which these graphs are based (from Von Lüpke, 1982, pp. 57, 65).

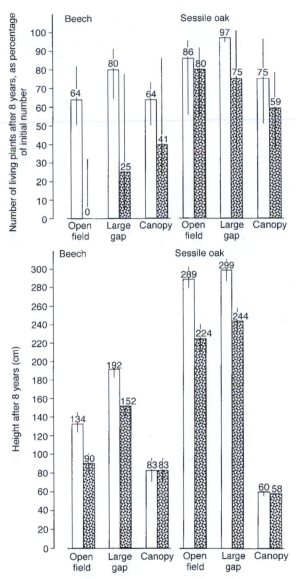

Fig. 6.14. The number of surviving sessile oak and beech seedlings (top) and the average height (bottom) of seedlings which grew for 2 years in a nursery (where they grew in full daylight) and were exposed to different amounts of daylight for eight growing seasons from the third year (redrawn from Von Lüpke, 1987, p. 21). Open bars = fields kept free of ground cover from the beginning of the experiment; shaded bars = fields with ground cover from the beginning of the experiment. The relative amounts of daylight were 100% (open field), 45.5% (large gap) and 11.3% (under the canopy). The average height is calculated on the basis of the surviving plants. The values above the columns are the values on the *y* axis. They are shown as averages with a reliability interval of 95% (*P* = 0.05). The values are averages of different sectors. As all the beech trees in the open field with vegetation died in several sectors, the 0 value was very widespread.

Table 6.12. The development of the length of the stem (cm) of 2-year-old sessile oak and beech from a nursery, which were exposed in the third year to different amounts of daylight: 100% in the open field, 45.5% in large gaps and 11.3% under the canopy. Results followed by the same letter do not differ significantly (see Von Lüpke, 1982, pp. 59, 66).

Time	Open field		Large gap		Canopy	
	With veg.	Without veg.	With veg.	Without veg.	With veg.	Without veg.
Sessile oak						
Initial value 1977	± 21	± 21	± 21	± 21	± 21	± 21
Final value 1980	102 a	128 b	116 a	124 b	49 c	42 c
Beech						
Initial value 1977	± 25	± 25	± 25	± 25	± 25	± 25
Final value 1980	52 a	52 b	61 b	66 b	47 c	47 c

view of the development of the levels, beech therefore grows better than sessile oak with very low levels of daylight, and sessile oak responds much more strongly to a reduction in the amount of daylight. Sessile oak still grows well when the reduction in the amount of daylight is not too great. The average length of the stem of sessile oak was no different after four growing seasons in the open space, and the large gap, in situations without ground cover, but was different in situations with ground cover (see Table 6.12). For beech, the stems were tallest after four growing seasons in the large gap. However, they were much less tall than those of sessile oak in the large gap, as Fig. 6.13 clearly shows. In contrast with the average dry weight, there was no difference in the average length of beech between the specimens growing with and without ground cover.

After eight growing seasons, the differences in the average height increased, as shown in Fig. 6.14. For sessile oak, the height under the canopy is still only 20–24% of that in the large gap, and therefore clearly less than that of beech under the canopy. In the open field, oak was much taller than beech, as it was in the large gap, although the differences were less there. Beech reached its maximum height in the large gap in the sector without ground cover but was still considerably shorter than sessile oak. For both species, the effect of ground cover on growth is still clearly visible both in the open field and in the large gap.

The effect of the ground cover could also be seen in the mortality, particularly of beech. In the open field, the mortality of seedlings in the sector with ground cover was extremely high; almost 100% (see Fig. 6.14). According to Von Lüpke (1987), the most important cause was the nibbling of common vole (*Microtus agrestis*). For this species, rough ground cover is an excellent biotope. Sessile oak suffers much less from being nibbled by voles, although the oaks were not spared this entirely.

Beech also suffered a great deal of damage in the open field as a result of late frosts (Von Lüpke, 1987). Beech is much more sensitive to this than pedunculate or sessile oak (Watt, 1923; Newbold and Goldsmith, 1981, pp. 34–45;

Dengler, 1990, pp. 59–60; Mayer, 1992, p. 94). For sessile oak, the difference in mortality between the sectors with and without ground cover was only significant in the large gap (Von Lüpke, 1987). In this respect, it was striking that the mortality of sessile oak with the lowest amount of daylight, i.e. the conditions under the canopy, did not differ very clearly from that with the two other light regimes. Very probably the reason for this is that the seedlings had grown in full daylight in the nursery for the first 3 years of their life, and were therefore able to develop strongly. This is confirmed by the observations of Oosterbaan and Van Hees (1989, pp. 10, 12), who found that the mortality of sessile oak and beech seedlings which had been exposed to the same reduced amounts of daylight from the moment of germination was virtually no different. The young oaks had a reserve of nutrients in the acorn and were therefore able to achieve relatively strong growth with low amounts of daylight.

In terms of the development of the stem, the ground cover therefore has a clear effect on the growth of both beech and sessile oak. As we saw before, a reduction in the amount of light in the first instance results in a greater length of the stem as a result of etiolation. This means that the poorer growth of the plant is not directly reflected in a reduction of its length. The fact that there was a reduction in the length of the stem after eight growing seasons under the canopy can therefore be seen as a reflection of a greatly reduced condition of the plant as a whole.

This experiment shows that in the open field and in large gaps in the canopy, sessile oak can, in principle, remain ahead of beech with regard to the growth in length, while the plant as a whole also grows well. A very long-term practical experiment in forestry, in which the growth of beech and oaks of the same age was followed for a period of more than a century, provides an insight into the way in which oak and beech continue to develop in relation to each other.

In 1891, Schwappach started an experiment in the Westerwald at Wiesbaden in Germany, where there were stands of oak and beech of the same age, 68 years old, 52 years old and 40 years old (Wiedemann, 1931). He did not indicate which type of oak was concerned. In these stands, Schwappach selected so-called 'future' oaks. These were identified by means of coloured rings on the trunk. Some of these oaks were completely freed from beech. This meant that any beech trees which could have been a threat in any way for the oaks concerned, were felled. In the rest of the stand only poor specimens of beech were kept to provide shade on the forest floor so that no grasses and herbs could grow. Directly under the oaks themselves, Schwappach spared the poorly growing beech trees to provide shade for the oak trunks and prevent the oak trunk from forming epicormic shoots. Another section of 'future' oaks were freed to a lesser extent. The large beech trees were spared, as well as poorly growing beech trees under the oaks, to provide shade for the oak trunks (Schwappach, 1916; Wiedemann, 1931; Bonnemann, 1956a). Oaks which had not been selected were left to their fate, i.e. no measures were taken to protect them from the beech trees. The measures which were taken to free the oaks were repeated

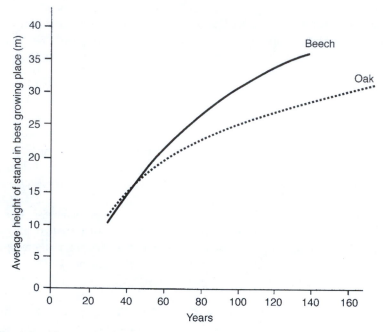

Fig. 6.15. The average height for different ages of oak and beech in the best growing areas (redrawn from Mayer, 1992, p. 200).

every time recordings were made, or otherwise every 6 years (Schwappach, 1916; Bonnemann, 1956).

Developments showed that oaks were ahead of beech in height, or grew at the same rate for the first 50 years. After 80–100 years (1930 records), the beech trees had grown taller than the oaks in all the sectors. The largest beech trees were 2–4 m taller than the oaks (the sources do not say anything about the heights). The oaks which had been strongly protected from beech were less tall and had a thicker trunk than those which were less protected against the beech. The latter had narrower crowns which were tightly surrounded by the crowns of adjacent beech trees (Wiedemann, 1931). At the age of 100–130 years (1952 records), the beech trees were up to 5 m taller than the oaks which had been protected. The oaks which had been left to their fate had virtually all disappeared (Bonnemann, 1956).

The fact that beech trees grow past oaks is generally known in forestry practice. This happens both with mature oaks and with young growing oak trees.[17] This is illustrated in Fig. 6.15. It means that in the case of the 'natural' regeneration of oaks in forests where there are also beech trees, after thinning out the canopy (shelterwood system) or the creation of gaps in the canopy

[17] See *inter alia* Schwappach (1916), Vanselow (1926, pp. 63, 67), Boden (1931), Wiedemann (1931), Seeger (1938), Hesmer (1958, p. 57), Krahl-Urban (1959, pp. 146, 165, 188, 212), Dengler (1990, p. 224).

(group selection cutting), if no measures are taken to protect the young oaks, the regeneration of oaks leads to the regeneration of beech because the young beech trees win out in the competition against the young oaks.[18] Pedunculate and sessile oak seedlings can grow in the presence of young beech trees only if the latter are constantly trimmed.[19] At a later age, the beech trees also have to be kept in check because they will otherwise grow taller than the oaks and replace them.

6.3.3 The establishment of beech in relation to the alternative hypothesis

As explained in the preceding paragraphs, in principle, beech grows well in full daylight, unless the seedlings are nibbled by mice or there are late frosts. Therefore, in principle, beech can regenerate in a grazed, more open, park-like landscape such as that postulated by the alternative hypothesis as being the natural vegetation. One difference between beech and oak is that under these light conditions, oak develops a much stronger root system. In relation to beech, this could mean a competitive advantage for oak. This also applies with regard to the consequences of the browsing of large herbivores in such a landscape. Because of the extensive root system, oak can recover better from this browsing, because more reserve nutrients can be reallocated from the root to the stem. In addition, the cotyledons of both species of oak are under the ground, while they are above the ground on the stem in beech seedlings (see Fig. 6.16). Pigeons, small birds and rodents eat the cotyledons of beech before they have developed into leaves, because of the carbohydrates and fats which they contain (see Jensen, 1985). As they graze, the large herbivores also bite off pieces of young beech trees standing on the fringes of thorny scrub, and therefore also eat the cotyledons (Watt, 1923; 1925). This is impossible with oak. Therefore in park-like landscapes grazed by large herbivores, beech is at a disadvantage compared with oak. This advantage is further increased because there are also all sorts of herbs and long grasses in the fringes of the mantle vegetation, where the young trees can grow, protected from being eaten by large herbivores.[20] These are a suitable biotope for mice and voles. Beech is much more sensitive to the nibbling of rodents than oak, and the mortality of beech is higher (Watt, 1923; Burschel *et al.*, 1964, p. 39; Von Lüpke, 1987). Another factor which has an adverse effect on beech seedlings is the sensitivity to late frosts. This occurs more often in such open situations with no cover, than in a forest (see Von Lüpke, 1987). Apart from this, an important – if not the most important – factor which is

[18] See *inter alia* Schwappach (1916), Bühler (1922, pp. 310, 433), Vanselow (1926, pp. 63, 67, 86–87, 92–93, 110–111), Endres (1929b), Wiedemann (1931), Seeger (1938), Frank (1939), Dannecker (1955), Bonnemann (1956a), Hesmer (1958, p. 57), Krahl-Urban (1959, pp. 146, 165, 188, 212, 219).
[19] Bühler (1922, p. 433), Vanselow (1926, pp. 87–88), Meyer (1931, pp. 356–357), Krahl-Urban (1959, pp. 146, 165, 188, 212, 219).
[20] See *inter alia* Tüxen (1952), Dierschke (1974), Westhoff and den Held (1975, pp. 115–125), Rodwell (1991, pp. 33–351), Oberdorfer (1992a, pp. 81–106).

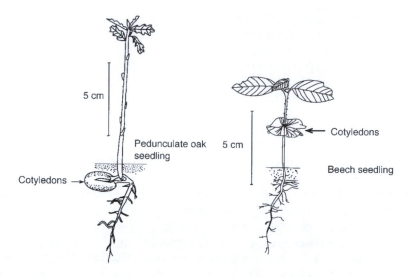

Fig. 6.16. (Pedunculate) oak and beech seedlings and the place where the cotyledons are on the seedling, i.e. under the ground for the oak, and above the ground for the beech (redrawn from Packham and Harding, 1982, pp. 91, 119).

responsible for oaks being much more predominant in park-like landscapes is the jay.

There is no equivalent species for spreading the nuts of beech trees (Ashby, 1959; Nilsson, 1985). It is only in exceptional cases that jays take beech nuts, for example when there are few or no acorns (Bossema, 1979, p. 29; Nilsson, 1985). Rodents do move beech nuts and hide them, but they do not move them large distances (Watt, 1925; Jenssen, 1985). Usually they eat the nuts after a while. In contrast with the acorns hidden by the jays, beech nuts hidden by rodents therefore germinate much less often. In so far as they do germinate, being moved by mice appears to be much less important for the dispersal of the species. Rodents rarely carry the nuts further than 10 m (Jenssen, 1985). Therefore, in general, the dispersal of beech is limited to places where the nuts fall from the crown straight on to the ground. There may be some sidewards displacement by the wind, but in general, this is limited to about 10 m (Watt, 1923; 1924; 1925; Brown, 1953, pp. 36–37; Newbold and Goldsmith, 1981, p. 13).

It is only the nuts of free-standing beech trees that may land at a few score metres from the parent tree as a result of strong winds (Watt, 1925). Probably these factors, and the presence of the jay, are responsible for the fact that relatively few beech seedlings are found in grasslands and on the fringes of thorny scrub, though all the more oaks are found there, even in the immediate vicinity of a beech forest (Watt, 1925).[21] In certain situations, beech can successfully

[21] Watt (1925) said about this: 'Nothing is more striking than the local frequency of young oaks colonizing the grassland adjoining the beech associes and the comparative infrequency of the beech' (Watt, 1925, p. 30).

regenerate in grazed grasslands outside thorny scrub. This is clear from the phenomenon of so-called solitary 'Weidbuchen' (pasture-beeches) (Eickstedt, 1959; Schwabe and Kratochwil, 1986; 1987). The regeneration of beech in grazed grasslands took place with a grazing density of up to 1 LU (livestock unit) per hectare (= 350 kg ha^{-1}) (Schwabe and Kratochwil, 1986). Possibly, short grass contributes to this success, because this means that it is not a good habitat for rodents. These solitary beech trees usually develop from several trunks. The trunks are the result of young trees being chewed or broken off by the horns of cows, and then regrowing (see Fig. 6.17). A young beech which grows in open grassland is initially nibbled by browsing cattle. When it is turned into a bush in this way, it is known as a 'Kuhbusch' (cow-bush) (Schwabe and Kratochwil, 1986). The beech can remain at this stage of growth for a long time as a result of the browsing cattle. Schwabe and Kratochwil (1986) found one bush 1.4 m tall and with a diameter of 1 m, which proved to be 220–230 years old. At a certain point, one or more shoots can grow out of reach of the cattle in the centre of the 'Kuhbusch', and develop into a tree (Schwabe and Kratochwil, 1986). If several shoots grow at the same time, this can result in a 'Weidbuche' (pasture-beech) with several trunks. A number of different trunks from one or more specimens can also grow together into a beech with a single trunk (Schwabe and Kratochwil, 1987, pp. 36–39).

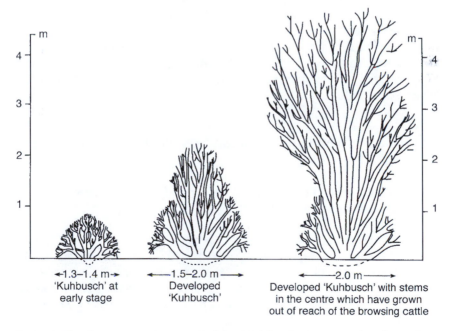

Fig. 6.17. The three successive stages of the 'Weidbuchen' (pasture-beeches). The rounded shape with stems growing out is characteristic of the last stage (redrawn from Schwabe and Kratochwil, 1986, p. 29).

The proportion of beech trees in wood-pasture that still exists today, and information about the history of wood-pasture, shows that beech regenerates in park-like landscapes with thorny scrub.[22] At first, beech only becomes established on the fringe of scrubland, because the light intensity in the centre of unshaded scrub is less than 1% daylight (Dierschke, 1974), where beech cannot grow. Therefore it does not become established there (Salisbury, 1918; Watt, 1924; Tüxen, 1952; Tubbs, 1988, p. 154). If the scrub becomes thinner as a result of the shade of trees growing on the fringe of the scrub, such as oak and ash, beech seedlings can become established under the canopy of these trees.[23] The shade provided in these situations, which allows the growth of beech, prevents the profuse growth of plants and grasses, so that the young trees are not much bothered by rodents. The only question is whether they can survive with the densities of large herbivores which can live in wood-pasture. The developments in the Ornamental Woods in the New Forest, described in Chapter 4, indicate that in situations where trees regenerate outside the forest in unshaded scrub, there is no regeneration in the grazed forest at the same time. Experiments with fencing show that the browsing and trampling of large herbivores are partly responsible for this. Seedlings have a chance of growing successfully only if they are protected, for example, by holly (Morgan, 1991). In that case, the small amount of daylight which penetrates through the evergreen holly will still prevent the growth of the seedlings. For example, beech regenerates between the branches of holly if the holly grows in a gap in the canopy of a forest dominated by oak. This was observed in wood-pasture where grazing had come to an end (Rackham, 1980, pp. 175, 294).

To summarize, the above shows that beech cannot be considered as a token sort of forest. This species has characteristics which allow it to become established and regenerate in park-like landscapes very easily, as the alternative hypothesis postulates.

6.4 Broad-leaved Lime (*Tilia platyphyllos*) and Small-leaved Lime (*T. cordata*)

6.4.1 The response of broad-leaved and small-leaved lime seedlings to reduced amounts of daylight

Paice (1974) subjected small-leaved lime seedlings to four different light levels in their first year. The response of the seedlings in terms of dry weight and leaf area are shown in Tables 6.13 and 6.14. The seedlings responded in the same way as other species described up to now, i.e. with a reduction in the dry weight, and initially an increase in the leaf area with a reduction in the amount of light.

[22] See Watt (1923; 1924), Eickstedt (1959), Rackham (1980, pp. 359, 323–325), Harding and Rose (1986, pp. 20–21), Koop (1989, pp. 106–107), Pott and Hüppe (1991, p. 292).
[23] Watt (1924; 1934b), Puster (1924), Ekstam and Sjörgen (1973) , Rackham (1980, pp. 342–325), Ellenberg (1986, p. 95), Tubbs (1988, p. 157; 1987).

Table 6.13. The increase in the total dry weight (g) of small-leaved lime seedlings with reduced amounts of daylight. The intensities are: low = 3.1; medium = 11.3; high = 48.2, and very high = 82.6 W m^2 (from Paice, 1974, p. 188).

Age (weeks)	Amount of light			
	Low	Medium	High	Very high
2	0.006 ± 0.001	0.011 ± 0.002	0.028 ± 0.003	0.025 ± 0.004
4	0.008 ± 0.001	0.033 ± 0.005	0.111 ± 0.015	0.119 ± 0.007
6	0.013 ± 0.001	0.083 ± 0.013	0.281 ± 0.039	0.255 ± 0.057
8	0.018 ± 0.003	0.119 ± 0.022	0.589 ± 0.080	0.728 ± 0.123
10	0.020 ± 0.003	0.210 ± 0.030	0.873 ± 0.104	0.826 ± 0.107

Table 6.14. The development of the leaf area (cm) of small-leaved lime seedlings with reduced amounts of daylight (as in Table 6.13) (from Paice, 1974, p. 188).

Age (weeks)	Amount of light			
	Low	Medium	High	Very high
2	2.07 ± 0.27	3.83 ± 0.83	8.03 ± 0.98	7.61 ± 1.07
4	3.52 ± 0.62	12.29 ± 2.19	23.54 ± 3.69	23.94 ± 1.09
6	4.95 ± 0.67	28.46 ± 3.88	50.79 ± 7.24	43.06 ± 7.98
8	5.14 ± 0.78	21.61 ± 3.48	76.72 ± 10.32	64.88 ± 8.77
10	5.71 ± 0.89	29.84 ± 4.19	103.07 ± 10.74	70.82 ± 8.00

Figure 6.18a shows that the leaf area ratio (LAR) increased with decreasing light intensity.

In contrast with the other species described up to now, the net assimilation rate (NAR) of the small-leaved lime increased as the light intensity decreased (see Fig. 6.18b). After 8 weeks, there was therefore no significant difference left in the relative growth rate (RGR) between the different levels of light intensity, although this difference was initially very large (see Fig. 6.18c). This shows that small-leaved lime seedlings can adapt to relatively small amounts of light, and can therefore continue to grow with very low light intensities. Thus Röhrig (1967) found that broad-leaved and small-leaved lime seedlings, which had grown with considerably reduced amounts of daylight, did not differ much in terms of performance, except for those plants which had grown with extremely low amounts of daylight of 1% (see Tables 6.15, 6.16 and 6.17). It has also been found that the relative growth rate (RGR) of small-leaved lime under a canopy did not differ at 6% and 2.4% daylight (Pigott, 1991). Röhrig's experiment showed that small-leaved lime seedlings did not even grow so much worse with 1% daylight than those which were subjected to higher levels of daylight (except in the case of full daylight). Röhrig found the same to apply for the separate parts, such as leaves, stems and roots. Lyr *et al.* (1965) also found that with greatly reduced amounts of daylight, small-leaved lime seedlings still grew well (see Fig. 6.19). This finding, as well as other observations in the field (Paice, 1974, p. 259; Pigott, 1975; 1991), shows that small-leaved lime can continue

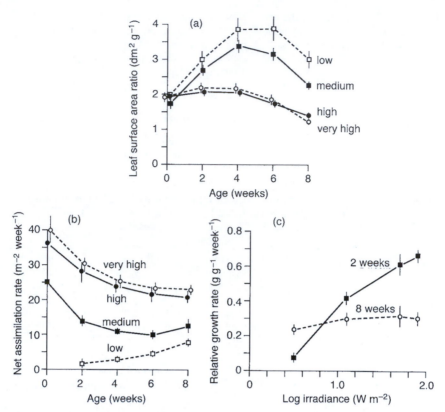

Fig. 6.18. The leaf area ratio (LAR) (a), the net assimilation rate (NAR) (b) and relative growth rate (RGR) (c) of small-leaved lime seedlings which were collected in the wild in Shrawley Wood, England, and subsequently grew in the laboratory with four different light intensities for a period of 8 weeks. The intensities are shown in (a) and (b): 3.1 □, 11.3 ■, 48.2 ● and 82.6 ○ W m^{-2} PAR (photosynthetic active radiation). The relative growth rate (RGR) is shown in (c) against these light intensities. The line with symbols ■ shows the differences in the second week of the experiment, and the line with symbols ○, the differences after 8 weeks. The averages with the standard deviations are shown (redrawn from Paice, 1974, pp. 190, 191, 192).

to grow slowly but steadily in deep shade for many years (certainly up to 5 years).

6.4.2 A comparison of broad-leaved and small-leaved lime seedlings and pedunculate and sessile oak seedlings

Röhrig (1967) grew broad-leaved and small-leaved lime and pedunculate and sessile oak under the same light conditions. Compared with pedunculate and

Table 6.15. The average length of the stem (cm) and dry weight of 2-year-old broad-leaved and small-leaved lime with different reduced amounts of daylight (from Röhrig, 1967, p. 234).

Relative amount of daylight (%)	Length of stem (cm)	Dry weight			
		Leaves	Stem	Root	Total
Small-leaved lime					
100	14.96	0.58	0.52	1.60	2.70
78	17.20	0.59	0.73	1.24	2.56
24	10.18	0.26	0.34	0.50	1.10
8	12.99	0.33	0.40	0.56	1.29
1	7.44	0.10	0.12	0.14	0.36
Broad-leaved lime					
100	17.24	1.02	1.00	1.51	3.53
78	15.39	0.63	0.69	1.13	2.45
24	13.16	0.39	0.50	0.81	1.70
8	13.67	0.39	0.52	0.75	1.66
1	8.03	0.14	0.13	0.18	0.45

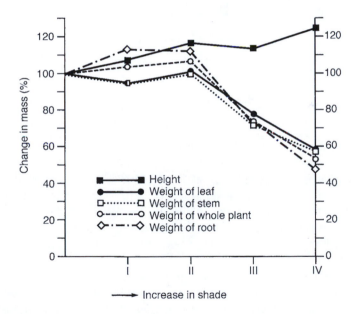

Fig. 6.19. The relative change in weight of the individual parts of small-leaved lime with a reduction in the amount of daylight. The values are expressed as a percentage of the values achieved in full daylight. Three-year-old plants were used for the experiment. The relative amounts of daylight were: I, 68%; II, 35%; III, 12%; IV, 1%. It was only for levels III and IV that the average values for the root, the leaf and the total weight of the plant differed significantly from the values achieved in full daylight. For the stem, this only applied for the value with IV (redrawn from Lyr *et al.*, 1965, p. 322).

Table 6.16. The difference in total dry weight of 2-year-old broad-leaved and small-leaved lime grown with a reduction in the amount of daylight expressed as a percentage of the average value reached with full daylight (redrawn from Röhrig, 1967, p. 234).

	Percentage of daylight			
	78	24	8	1
Small-leaved lime 100	95−	41+	48+	13++
Broad-leaved lime 100	69−	48+	47+	13+++

The differences were evaluated with a variance analysis. The significance of the differences is as follows: − not significant; + significant $P < 0.05$; ++ highly significant $P < 0.01$; +++ most significant $P < 0.001$.

Table 6.17. Differences in the dry weight of the leaves (L.), the stem (St.) and the root (Ro.) in 2-year-old small-leaved lime and broad-leaved lime, with a reduction in the amount of daylight expressed as a percentage of the average value reached with full daylight (from Röhrig, 1967, p. 234).

	Percentage of daylight								
	24			8			1		
	L.	St.	Ro.	L.	St.	Ro.	L.	St.	Ro.
Small-leaved lime 100	45+	41+	31++	57−	49−35++		17++	15++	9++
Broad-leaved lime 100	38+	50+	54−	38+	52+	50+	13++	13++	12++

For the significance of the differences, see Table 6.16.

sessile oak, the broad-leaved and small-leaved lime did not, in principle, respond differently to a reduction in the amount of daylight. This is shown in Fig. 6.20. Compared with the performance of plants growing in full daylight, the two species of lime did reveal a smaller reduction in growth than pedunculate and sessile oak. Therefore both species of lime thrived better in shade than both species of oak.

One aspect which can play a role in the development of small-leaved lime seedlings in the competition with the two species of oak, is the fact that small-leaved lime grows in width when there is shade. In this respect, small-leaved lime is similar to beech. One of the differences with beech is that the young small-leaved lime seedling also grows upwards with small amounts of daylight, from 10 to 50%. The seedling develops a central stem which grows upwards (see Fig. 6.21). With amounts of light below 10% daylight, the central stem grows horizontally (Pigott, 1988). With more than 50% daylight, as in open areas, the seedling forms a few side shoots just above the ground, which eventually become thicker than the central stem, so that the plant develops a broom-like shape. Small-leaved lime seedlings which are planted in an open terrain

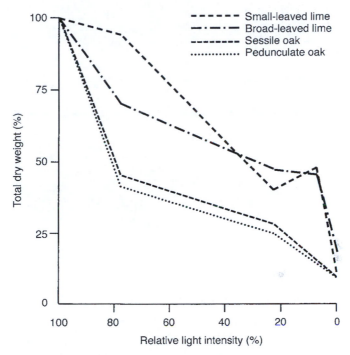

Fig. 6.20. The effect of different relative amounts of daylight on small-leaved and broad-leaved lime seedlings and sessile and pedunculate oak seedlings (redrawn from Mayer, 1992, p. 218, based on Röhrig, 1967).

where trees have been felled, retain this shape until they have formed a closed plant layer (Pockberger, 1963; 1967, p. 40; Hesmer, 1966; Hesmer and Günther, 1966; Belostokov, 1980; Pigott, 1988). This cover usually forms in the fifth to the seventh year, then a central stem grows up from the plant (Hesmer and Günther, 1966). Thus a small-leaved lime seedling which is exposed to a great deal or very little light when it is very young, does not grow vertically, but grows horizontally.

In forestry, the great tolerance of lime seedlings to shade is well known from empirical data, as is the rapid growth in height when the light conditions are favourable for the growth of a central stem. In a forest there can be large numbers of small-leaved lime seedlings under a closed canopy. This is partly the result of the fact that the small-leaved lime forms fruit every year, and the seedlings which develop from these can grow with relatively low amounts of daylight. When there are oaks and small-leaved lime in a forest, the small-leaved lime will predominate, unless measures are taken against it. The central stem grows taller much more quickly for the first 20–50 years than that of an oak or beech (Erteld, 1963; Böckmann, 1990, pp. 61–63, 105–109; Pigott, 1991). For example, lime seedlings, after the canopy of the forest is thinned out or gaps are made in it that announce the 'natural' regeneration of shelterwood or

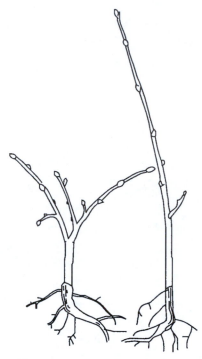

Fig. 6.21. Small-leaved lime seedlings of Polish origin at the end of the second year of life, after growing respectively in full daylight (left) and with 50% daylight (right) (redrawn from Pigott, 1988, p. 30).

selection cutting, will quickly grow high as a result of the increased amount of daylight.[24] In doing so, they compete with oak seedlings, which disappear.[25] There are reports from Russia of oak forests eventually changing into lime forests as a result of this (Morosow, 1927, p. 143). Thus both species of lime can easily be regenerated in forests by making gaps in the canopy (see Keiper, 1916; Pockberger, 1963; Rühl, 1968; Freiherr von Berlepsch, 1979; Hocker, 1979). The rapid growth, which is the response to the increase in the amount of light in a gap in the canopy, also applies to species described as pioneers, such as birch (*Betula* sp.) and rowan (*Sorbus aucuparia*) (Pockberger, 1963; 1967, p. 69). Because of the seedlings' high tolerance to shade on the one hand, and the rapid growth when there is more light, on the other hand, broad-leaved lime and small-leaved lime are described in forestry as semi-shade species.[26]

[24] Mantyk (1957), Hesmer (1958, pp. 47, 285–286; 1966), Krahl-Urban (1959, p. 156), Pockberger (1963), Rühl (1968), Davies and Pigott (1982), Koss (1982), Namvar and Spethmann (1986), Pigott (1991).
[25] Keiper (1916), Reed (1954, pp. 116–117), Mantyk (1957), Erteld (1963), Pockberger (1963), Rühl (1968), Freiherr von Berlepesch (1979), Davies and Pigott (1982), Böckmann (1990, pp. 60–62), Pigott (1991).
[26] Keiper (1916), Mantyk (1957), Pockberger (1963; 1967, pp. 40, 69), Namvar and Spethmann (1986), Dengler (1990, pp. 184–186), Mayer (1992, p. 108).

The fact that lime rapidly grows taller and can win out in the competition against oak is also revealed by the practice of planting limes under oaks. In this practice, the limes provide shade for the trunks of the oaks to prevent dormant buds on these trunks from growing, and to ensure that the trunk grows as straight as possible. Beech is also used in this way. As a result, the branchless trunks of oaks provide valuable wood for veneer.[27] The small-leaved lime tree grows very rapidly under the oaks for the first few decades, and will grow past the oak if no measures are taken to oppose it (Erteld, 1963; Hesmer, 1966; Hesmer and Günther, 1966; Böckmann, 1990, pp. 100, 132, 133). Even pedunculate oak with a head start of 30 years can be caught up by small-leaved lime (Fricke *et al.*, 1980). To prevent this, limes are cut back several times at intervals of a few years. If this is not done, the oaks disappear as a result of the competition (Hesmer and Günther, 1966).

To summarize, it can be concluded that the available data clearly show that both species of lime regenerate well in closed forests and therefore survive well because the seedlings tolerate a great deal of shade. The regeneration takes place in gaps in the canopy. At an early stage, the limes become significant competitors for the seedlings of both species of oak, because lime seedlings, in contrast with oak seedlings, will continue to grow for a long time when there is little light, and grow taller more quickly than oak when there is more light.

6.4.3 The establishment of both species of lime in relation to the alternative hypothesis

Both species of lime also grow in mantle and fringe vegetation in wood-pasture[28] (see Fig. 3.7). In Anglo-Saxon sources, lime is even described as a hedgerow tree (Rackham, 1980, p. 248). In England, lime is rare nowadays in wood-pasture (Rackham, 1980, p. 242; Harding and Rose, 1986, p. 23). On the other hand, in Slovensky Kras in Slovakia, lime (broad-leaved lime) is widespread in grazed, park-like landscapes (Jakucs and Jurko, 1967; see Fig. 3.7). In comparison with oak and beech, the two species of lime are relatively rare throughout Central and Western Europe. The reason for this is probably that in previous centuries, measures were taken against limes because the wood was considered useless and the trees took up the place of trees which provided better wood.[29] This means that in many places there are hardly any lime trees left to serve as a source of seed.

As small-leaved lime only grows taller when there is 50–10% daylight (I do not have any data for broad-leaved lime), this species will not grow tall in full

[27] Mantyk (1957), Hesmer and Günther (1966), Rühl (1968), Fricke (1982, pp. 2–4), Koss (1982), Jahn (1987).
[28] Trepp (1947, pp. 106, 109), Jakucs and Jurko (1967), Pockberger (1963; 1967, p. 40), Rühl (1968), Rackham (1980, pp. 242–250), Smith (1980, p. 353), Koss (1982), Pigott (1991), Oberdorfer (1992a, p. 148), personal observation.
[29] See Keiper (1916), Pockberger (1967, pp. 30, 39), Streitz (1967, p. 63), Rühl (1968), Müsall (1969, p. 97), Pigott (1981; 1991), Koss (1982).

daylight on the fringes of scrub. It is only when there is some shade from the foliage of shrubs or other species of trees growing in the scrub, that the lime will grow taller. A lime seedling may wait for shade while staying low to the ground, and then grow taller. As the amount of light increases it can then grow taller with characteristic speed. As the seeds of both species of lime have wings, and can therefore be dispersed by the wind, they can land in either grassland or scrubland. Small-leaved lime also germinates in grassland (Pigott, 1991). In view of its morphological reaction to full daylight, the seedling will not grow upwards there unless thorny scrub becomes established.

Livestock and wild animals like to eat both species of lime.[30] The trees are not necessarily harmed by this, because both species have an enormous capacity for regeneration (Pigott, 1991). Although limes are known to be resistant to mice and voles, the bank vole (*Clethrionomys glareolus*) nibbles lime seedlings and can badly damage them. Nevertheless, the plants continue to grow after being nibbled several times (Pigott, 1985). Limes are very sensitive to being frayed by roe deer and red deer, which can kill the young trees (Hesmer, 1960). They are exposed to this in forests with a closed canopy because the protective thorny scrub is missing there. In scrub, they are spared this. Another advantage of both species of lime in locations outside the forest is that they are not very sensitive to late frosts (Hesmer, 1960; Dengler, 1990, p. 184; Mayer, 1992, p. 109).

6.5 Hornbeam (*Carpinus betulus*)

6.5.1 The response of hornbeam seedlings to a reduction in the amount of daylight

I do not have any data about the growth of hornbeam seedlings with different percentages of daylight. However, forestry experience shows that seedlings tolerate shade to the extent that they can grow under a closed canopy for some years (Eichhorn, 1927; Schrötter, 1964; Bezacinsky, 1971). Hornbeam seedlings respond to gaps in the canopy like small-leaved lime seedlings, by growing taller very quickly (Eichhorn 1927; Bezacinsky, 1971; Reif *et al.*, 1998). In morphological terms, the hornbeam seedlings respond to reduced amounts of daylight in a similar way to beech and small-leaved lime, with horizontal growth. Up to the 15th year, they form branching so-called sylleptic shoots in the leaf axilla of the annual shoots. This means that the seedling develops a broad, relatively dense crown, low down to the ground (Hesmer, 1958, p. 260). This can mean that it is very competitive for oak seedlings. An even more threatening aspect for oak is that the seedlings can thrive in the shade for a long time, and then quickly grow taller when there is an increase in the amount of light. Therefore young oaks have virtually no chance (Eichhorn,

---

[30] Keiper (1916), Edlin (1964), Koss (1980), Pigott (1988), Dengler (1990, p. 184), Mayer (1992, p. 109).

1927; Bezacinsky, 1971). As a result, measures for the 'natural' regeneration of oaks in oak forests which also contain hornbeam, such as thinning out the canopy and creating gaps in the canopy, result in the large-scale growth of young hornbeams which replace the young oaks.[31] This large-scale growth is further stimulated because hornbeam bears fruit every year, and the seeds disperse far and wide because they are winged. Hornbeam seedlings survive shade better than those of oak (Le Duc and Havill, 1998; Reif *et al.*, 1998). The tolerance of the seedlings to shade means that they can wait for better light conditions, for longer than the oak seedlings. Thus, when these conditions arrive, there are already a lot of young hornbeams ready to grow. In addition, these young hornbeams grow taller much more quickly than oak seedlings and also more quickly than beech seedlings during the first decades of their life (Eichhorn, 1927; Hesmer, 1958, pp. 260, 269; Bezacinsky, 1971; Von Lüpke and Hauskeller-Bullerjahn, 1999). Figure 6.22 illustrates this.

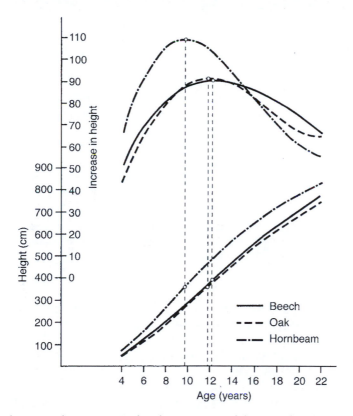

Fig. 6.22. The increase in height per year and the growth in height of beech, oak and hornbeam up to the age of 22 years, in mixed oak, beech and hornbeam forest (redrawn from Bezacinsky, 1971, p. 11).

[31] See Eichhorn (1927), Hesmer (1958, pp. 260–262), Krahl-Urban (1959, p. 167), Bezacinsky (1971), Röhle (1984), Pigott (1991).

The failure to take measures against hornbeam with the 'natural' regeneration of oak in French, German and Slovakian forests have resulted in a 'natural' regeneration of hornbeam (Hesmer, 1958, p. 261; Bezacinsky, 1971; Pigott, 1991). Cutting back the young hornbeam only exacerbates the situation for oaks because of the enormous growth of shoots on the stool of the hornbeam (Bezacinsky, 1971). When this happens, the hornbeam can only be kept down with very vigorous measures such as digging or ploughing up the young trees. In this case regeneration of oak is only possible if young oaks are planted (Eichhorn, 1927; Hesmer, 1958, p. 268).

Hornbeam is very sensitive to being nibbled by deer and field vole (*Microtus agrestis*). They can be nibbled down to the ground (Hesmer, 1960; Abs *et al.*, 1999). The fact that hornbeam regenerates well in forests despite this is partly because it produces so much fruit (Hesmer, 1960; Fricke, 1982, pp. 106 *et seq.*, 184–185; Reif *et al.*, 1998). In view of the above, hornbeam must be considered as a species which regenerates well in closed forests.

6.5.2 The establishment of hornbeam in relation to the alternative hypothesis

As regards the alternative hypothesis, hornbeam also regenerates well in scrub consisting of hawthorn and blackthorn in grazed areas.[32] They can even be very numerous in wood-pasture (Rackham, 1980, pp. 221, 223, 225). Hornbeam probably owes this capacity to a combination of characteristics which are very similar to those of the two species of lime, namely a great tolerance to shade when it is a seedling, so that it can grow in scrubland, and the fact that it rapidly grows taller, once the plant has reached a height above the scrub. The seeds have wings so that they can be dispersed by the wind. Other characteristics which allow it to regenerate in unshaded scrub are its great capacity for regeneration after being eaten by livestock (Rubner, 1960, p. 63; Bezacinsky, 1971) and its resistance to late frosts (Rubner, 1960, p. 13).

6.6 Ash (*Fraxinus excelsior*), Field Maple (*Acer campestre*), Sycamore (*A. pseudoplatanus*) and Elm (*Ulmus* sp.)

6.6.1 The response of seedlings of ash, field maple, sycamore and elm to reduced amounts of daylight

Other species of trees also have the characteristics of both species of lime and of hornbeam, namely, a high tolerance to shade in the seedling and rapid growth in height, when the seedling is given more light. These species include

---

[32] See Salisbury (1918), Watt (1925; 1934b), Adamson (1932), Jakucs and Jurko (1967), Burrichter *et al.* (1980), Smith (1980, p. 353), Pott and Hüppe (1991, p. 292), Oberdorfer (1992, p. 92).

sycamore, field maple, wych elm (*Ulmus glabra*), European white elm (*U. laevis*), common elm (*U. minor/carpinifolia*) and ash (*F. excelsior*).[33] Therefore, like broad-leaved and small-leaved lime, they are identified in forestry as semi-shade species (Dengler 1992, pp. 178–186). The great tolerance to shade of wych elm seedlings is clear from the fact that they survive surrounded by stinging nettles (see Pigott, 1975; Derkman and Koop, 1977, pp. 23–33 and Chapter 5). Ash seedlings can survive up to 15 years or longer under the shade of a canopy, as long as they are not overgrown by plants, such as perennial dog's mercury (*Mercurialis perennis*). If a gap appears in the canopy above the ash seedlings, they respond by rapidly growing taller (Wardle, 1959; Gardner, 1975; Von Lüpke, 1989; Tapper, 1992; 1993). The fact that the seedlings of these species can survive under a forest canopy for years, gives them a head start on both species of oak, which therefore do not have a chance from the outset.

6.6.2 Ash, field maple, sycamore and elm in relation to the alternative hypothesis

All these species are sometimes very common in grazed, park-like landscapes and scrub consisting of thorny shrubs, next to, for example, pedunculate oak, sessile oak and hazel.[34] As in the case of the two species of lime and of horn-beam, the seeds are dispersed over large distances by the wind because they have wings. All these species regenerate well when they are eaten. Fluttering elm and smooth-leaved elm (*U. minor/carpinifolia*) also develop spring from the roots (Boeijink *et al.*, 1992). Clones of the common elm have been known to advance at a rate of 0.65 m year^{-1} with a radius of 210 m (Rackham, 1980, p. 269).

6.7 Hazel (*Corylus avellana*)

6.7.1 The response of hazel seedlings to reduced amounts of daylight

Hazel seedlings respond in a similar way to reduced amounts of daylight to the seedlings of other species of plants, including the trees described up to now. They invest relatively more in the parts above the ground engaged in assimilation, at the expense of the roots. Tables 6.18 and 6.19 show this. Seedlings which grew in full daylight in their first year, and were exposed to reduced amounts of

---

[33] Watt (1925), Wardle (1959), Röhrig (1967), Gardner (1975), Thill (1975), Namvar and Spethmann (1985), Spethmann and Namvar (1985), Von Lüpke (1989), Dengler (1992, pp. 179, 181, 183), Mayer (1992, pp. 100, 103), Tapper (1992; 1993).

[34] See Salisbury (1918), Watt (1924; 1925; 1934a, b), Tüxen (1952), Jakucs and Jurko (1967), Pott and Hüppe (1991, pp. 289, 293, 298), Rodwell (1991, pp. 334–345), Oberdorfer (1992a, pp. 82–83), personal observation. Watt (1924) remarked: 'For the most part the protection of scrub is a necessity to the initial establishment of ash and oak, and in consequence their rate of advance is limited by the rate of invasion of the former' (Watt, 1924, p. 161).

Table 6.18. The differences in the growth of hazel seedlings in their first year of life, with a reduction in the amount of daylight (average values ± standard deviation). No. 2 replicates the test site (from Sanderson, 1958, table 59).

Relative amount of daylight (%)	Height (cm)	Dry weight of) total plant (g)	Surface area of leaf (cm^2)	LAR (cm^2 g^{-1})	Stem–root ratio
100	1. 8.4 ± 2.1	0.79 ± 0.07	40 ± 8	50.6	1.2 ± 0.4
	2. 7.5 ± 1.0	0.61 ± 0.06	36 ± 10	59.0	1.8 ± 0.4
50	1. 13.4 ± 2.8	0.78 ± 0.34	96 ± 45	123.0	2.7 ± 1.1
	2. 13.2 ± 5.6	0.50 ± 0.04	62 ± 21	124.0	2.3 ± 0.1

Table 6.19. The average growth of hazel seedlings in their second year of life (average values ± standard deviation). The seedlings grew up in full daylight in the first year and were exposed to a reduction in the amount of daylight in the second year (from Sanderson, 1958, table 60).

Relative amount of daylight (%)	Height (cm)	Dry weight (g)	Stem–root ratio
100	10.1 ± 3.9	0.54 ± 0.34	1.5 ± 0.3
15–20	10.3 ± 2.6	0.21 ± 0.05	3.3 ± 1.4
10	11.9 ± 2.0	0.13 ± 0.06	

daylight in the second year, did respond with a decrease in the total dry weight, but not with a reduction in height, as indicated in Table 6.19. The stems of seedlings in deep shade (15–20% and 10% daylight) grew so poorly that it was doubted that they would survive a second year with the same light intensity (Sanderson, 1958, p. 98).

For seedlings which grew in full daylight in the first growing season, and were then exposed to reduced amounts of daylight for the next two growing seasons, Sanderson found a reduction of the dry weight (see Table 6.20). With a light intensity of 10% daylight, the seedlings reached 20–30% of the weight of plants which had grown in full daylight. According to Sanderson, the plants growing with the lowest levels of light were not healthy (Sanderson, 1958, pp. 98–99).

It is striking that the dry weight of hazel seedlings was no different in the first growing season when the amount of daylight was halved (see Table 6.18). As Sanderson (1958) mentioned, the reason for this could be that the plants were not exposed to the same reduction in the amount of daylight all year round. For a part of the year, they were exposed to more than twice as much light as the rest of the year. The removal of daylight was achieved by a hedge of beech trees, which allowed much more daylight through before the leaves came out, while the hazel already had leaves (Sanderson, 1958, pp. 96, 98–99). The reserve nutrients in the hazel nut may also have contributed to the relatively strong growth in the first growing season. As Table 6.21 shows, Sanderson (1958) found that the underground cotyledons were exhausted by the end of the first growing season. According to the data in this table, a reduction in the amount of daylight does not affect this (Sanderson, 1958, p. 89). Table 6.22

Table 6.20. The average growth of hazel seedlings which grew in full daylight in the first year and were then exposed for 2 years to a reduction in the amount of daylight to 10% (average values ± standard deviation). No. 2 is a replication (from Sanderson, 1958, table 61).

Relative amount of daylight (%)		Height (cm)	Increase in height in 2nd year (cm)	Dry weight of total plant (g)	Stem–root ratio
100		22.0 ± 1.7	6.5 ± 2.6	2.99 ± 3.4	1.00 ± 0.22
25	1.	24.0 ± 2.2	8.7 ± 3.2	2.93 ± 7.0	0.86 ± 0.22
	2.	17.6 ± 3.7	5.9 ± 2.3	1.13 ± 1.1	0.86 ± 0.65
10	1.	20.0 ± 4.5	6.0 ± 2.9	0.88 ± 0.36	1.60 ± 0.10
	2.	17.1 ± 2.6	4.1 ± 1.9	0.62 ± 0.36	1.30 ± 0.46

Table 6.21. The decline in the weight of the cotyledons of hazel seedlings with 100% and 25% daylight. No. 2 is a replication (from Sanderson, 1958, table 69).

Time of sample		Number of seedlings with cotyledons from an initial number of five		Average dry weight of cotyledons (g)	
		100% daylight	25% daylight	100% daylight	25% daylight
29 June	1.	5	5	0.0506 ± 0.017	0.0461 ± 0.015
	2.	5	5	0.0645 ± 0.026	0.0764 ± 0.036
2 August	1.	1	5	0.0475 —	0.0473 ± 0.020
	2.	2	2	0.0309 ± 0.005	0.0363 ± 0.005
1 September	1.	0	0	0	0
	2.	0	0	0	0

Table 6.22. The percentage survival of artificially damaged hazel seedlings (from Sanderson, 1958, table 48).

Damage caused	Stage at which damage was caused	Number of plants treated	Survival (%)
Cotyledons removed	Rad. 1 cm; no plumula	5	0
Cotyledons removed	Rad. 5 cm; plumula erect	5	20
Cotyledons removed	Later roots	5	20
Radicula removed	Rad. 0.5 cm	25	88
Radicula removed	Rad. 3–5 cm	25	92
Plumula removed	Plumula erect	20	100
Tip of plumula removed	Plumula erect	20	100

shows that hazel seedlings survive the removal of the cotyledons if this happens after the stem has appeared above the ground and has fully developed, i.e. when the first crown of leaves is completed. If it happens earlier, the seedling dies (Sanderson, 1958, pp. 67, 70). These observations show that after the first growing season, the seedling depends on assimilation for further growth. It is only then that the effects of a reduction in the amount of daylight become clearly visible in the development of the plant (see Tables 6.19 and 6.20).

In its first year, the hazel develops a tap root (Sanderson, 1958, pp. 101–103). The extent to which this happens appears to depend not on the type of soil, but on the amount of light (Sanderson, 1958, pp. 101–104). As Table 6.20 illustrates, the stem–root ratio increases as the amount of daylight

Table 6.23. Average stem–root ratio of hazel seedlings in the second growing season on different types of soil (average values ± standard deviation) (from Sanderson, 1958, table 64).

Time of sample	Humus soil	Limestone soil	Sandstone soil
April	0.51 ± 0.1	0.43 ± 0.1	0.50 ± 0.27
May	1.56 ± 0.4	2.00 ± 0.7	2.30 ± 0.3
June	2.90 ± 0.9	3.30 ± 1.8	2.00 ± 0.7
August	3.00 ± 1.2	2.60 ± 1.1	1.90 ± 0.9
September	—	1.90 ± 1.09	2.49 ± 0.5

decreases. In full daylight, the stem–root ratio dropped to very low values after the first growing season, showing that the hazel invested a great deal of energy in the root system. The increase was greater in the course of the second growing season, because of the development of leaves and the plants growing thicker. Table 6.23 shows this development. This table also shows that the ratio again increased in the second growing season, reflecting the growth of the stem.

I do not know of any experiments in which hazel grew side-by-side with species of trees such as pedunculate and sessile oak or beech, with decreasing intensities of daylight, and consequently it is not possible to make a direct comparison with these species of trees. There are no data available either on the effect of the long-term exposure of hazel to reduced amounts of daylight. There are only incidental observations and empirical data from forestry practice. For example, in forests, Sanderson (1958) found only 1-year-old seedlings (Sanderson, 1958, pp. 77–78). This shows that for 1 year, hazel can survive on the reserve nutrients in the hazel nut, but will not regenerate successfully in closed forest or hazel scrub. Observations elsewhere also show that hazel does not achieve generative regeneration in closed forests or hazel scrub (see Dister, 1980, p. 71). With regard to hazel, Rackham (1980) noted that it does not regenerate nowadays, and sees this as a parallel to the oaks' failure to regenerate. In view of the centuries of exploitation of hazel as coppiced wood from coppices-with-standards, hazel does survive with some shade for a long time (Rackham, 1980, pp. 205–206; Dengler, 1990, p. 262; Savill, 1991, p. 31; Boeijink *et al.*, 1992; Peterken, 1992). For this reason, hazel is characterized as being tolerant of shade (see Weeda *et al.*, 1985, p. 100; Savill, 1991, p. 31; Boeijink *et al.*, 1992, p. 69). However, in coppices-with-standards, the tolerance to shade concerns only vegetative regeneration, i.e. the new growth from the stool. This is vegetative and not generative reproduction. Furthermore, the canopy of the coppice-with-standards should not be too thick or the hazel will disappear (Evans, 1992). The canopy of coppices-with-standards is thin, so thin that thorny bushes such as hawthorn and blackthorn can also grow there (Schubart, 1966, pp. 106, 172). The number of standards should not be too great, as the coppice and hazel will otherwise disappear.[35] The crown cover in

[35] See Hart (1966, p. 168), Rackham (1980, pp. 145, 207–208), Hausrath (1982, p. 31), Evans (1992), Fuller (1992), Peterken (1992).

the coppice-with-standards should not be more than 20–30% (Cotta, 1865, pp. 136–140; Warren and Thomas, 1992). As we saw in Chapter 5, hazel eventually disappears when there is a closed canopy (Nietsch, 1939, p. 28; Schubart, 1966, p. 168; Malmer *et al.*, 1978; Rackham, 1980, p. 203; Hytteborn, 1986). In this respect, the honey fungus (*Armillaria* sp.) and other related species also play a role. They affect the root system of shaded hazel trees, which die as a result (Rackham, 1980, p. 208).

As regards the survival of hazel in a closed forest, another important factor is that it is a shrub which is lower than the trees. We have already noted that there are indications that hazel cannot become established in a closed forest. Even if hazel could become established in a forest in gaps in the canopy, it is always at a disadvantage compared with trees, because these will always eventually overgrow the hazel. Moreover, as undergrowth in forests, hazel does not flower, or hardly flowers at all (Bertsch, 1929; Borse, 1939; Sanderson, 1958, pp. 76, 128, 153, 160, 253; Rackham, 1990, p. 203). It does not even flower under a canopy of sessile oak and hairy birch with 30% daylight (Sanderson, 1958, p. 128). Therefore it does not form any fruit, and generative reproduction becomes impossible.

6.7.2 Hazel in relation to the alternative hypothesis

Sanderson (1958) found that the presence of hazel seedlings was limited to unshaded grasslands and the mantle and fringe vegetation in the transitional area from grassland to woodland. Figure 6.23 shows one of his recordings, which demonstrates this. Young hazel grew there, together with thorny bushes such as hawthorn and blackthorn (Sanderson, 1958, pp. 72–87). Hazel was found little or not at all in closed forest, but it was widespread in grazed, park-like landscapes. It grew as a solitary bush and as part of the scrub and mantle vegetation, but also as virtually closed scrub (see Figs 3.6 and 3.7).[36] As regards grazing, hazel is very resistant to the browsing of livestock when it is young. It responds with strong new growth from the roots (Sanderson, 1958, pp. 82, 88, 120–122, 127). As noted above, hazel invests a great deal of dry material in the root system, which could explain its capacity for regeneration. Hazel not only recovers from browsing livestock, but responds to this with vegetative regeneration through underground runners. In this way, one seedling can expand into scrub of several square metres (Sanderson, 1958, pp. 120, 123). When it is older, hazel can also cope with heavy grazing, both as a solitary shrub and as part of scrub including hawthorn or blackthorn (Bär, 1914, pp. 341–343; Sanderson, 1958, pp. 119, 122, 127; Jahn, 1991, p. 445).

---

[36] Salisbury (1918), Adamson (1921; 1932), Watt (1924; 1925; 1934a, b), Borse (1939), Tüxen (1952), Tansley (1953, pp. 259–260, 473–475), Sanderson (1958, pp. 119, 123), Müller (1962), Dierschke (1974), Westhoff and Den Held (1974, pp. 239–241), Smith (1980, p. 474), Harding and Rose (1986, p. 25), Pott and Hüppe (1991, p. 292), Rodwell (1991, pp. 333–351), Oberdorfer (1992a, pp. 81–106).

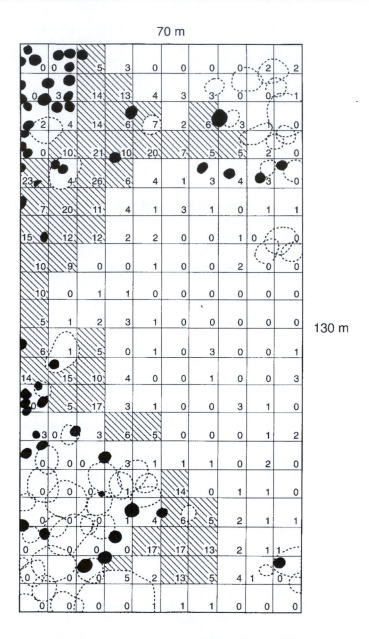

Fig. 6.23. Hazel seedlings in grassland in Lathkill Dale, England. The crowns of hazel bushes which produced nuts are shown in black. The crowns of trees are indicated with dotted lines. The numbers at the bottom right of the squares show the number of hazel seedlings in every square (7 m × 7 m). The squares with diagonal lines contain five seedlings or more. As this figure shows, these are in the immediate vicinity of groves and in mantle and fringe vegetation (from Sanderson, 1958, Figure 21).

As regards the dispersal of nuts, these are eaten in large numbers in the forest. This predation is reduced when the nuts are buried, and when they are taken outside the forest into grassland (Sanderson, 1958, pp. 52–54). Jays bury hazel nuts, but prefer acorns, if they are available (Bossema, 1979, pp. 20, 22, 29). Rodents also hide hazel nuts, although they do not seem to play an important role in grazed grasslands, because the short grass is not a suitable biotope for voles and mice. Most hazel nuts are hidden by nuthatches (*Sitta europaea*), which push and hammer the nut into the ground, and then cover up the spot, as jays do for acorns (Sanderson, 1958, p. 48; personal observation). It is not clear whether they have a preference for the edges of fields and other transitional areas in open vegetation, such as grassland. On the one hand, this is not likely as the nuthatch is strongly dependent on tall trees, while on the other hand, tall trees are always in the immediate vicinity of the mantle and fringe vegetation in grazed, park-like landscapes, where the hazel bushes grow. The mantle and fringe vegetation borders on the grasslands where the hazel nuts can be hidden (see Fig. 3.6). In view of the occurrence of hazel bushes in grasslands and on the fringes of scrub, it is therefore quite possible that the nuthatch plays a role there.

6.8 Wild Fruits, Wild Apple (*Malus sylvestris*), Wild Pear (*Pyrus pyraster*) Wild Cherry (*Prunus avium*) and members of the rowan family (*Sorbus* sp.)

6.8.1 The establishment of wild fruit trees

In addition to hazel and the two species of oak, wild apple (*M. sylvestris*), wild pear (*P. pyraster*), wild cherry (*P. avium*), and rowan (*S. aucuparia*), are also identified as species which require light.[37] The wild service tree (*S. torminalis*), whitebeam (*S. aria*) and true service tree (*S. domestica*) are characterized as species which require light, or which tolerate semi-shade, and are therefore also included among the species of trees which require more light (Namvar and Spethmann, 1985b; Mayer, 1992, pp. 119–120). They are all found in coppices-with-standards and wood-pasture, vegetation with plenty of light.[38] They disappear from these when they develop into closed forests (Dister, 1980, p. 67; Rackham, 1980, pp. 356–357; Buttenschøn and Buttenschøn, 1985; Namvar

[37] See Bindseil (1958), Beck (1977; 1981), Gottwald (1985), Kausch von Schmeling (1985), Namvar and Spethmann (1985; 1986), Pryor (1985; 1988, p. 6), Heymann and Dautzenberg (1988), Diez (1989), Schalk (1990), Savill (1991, p. 55), Mayer (1992, pp. 110, 120–121), Kleinschmit (1998), Wagner (1998; 1999).
[38] Endres (1888, p. 82), Bühler (1922, p. 65), Vanselow (1926, pp. 22, 290), Grossmann (1927, p. 23), Meyer (1931, p. 293), Bock (1933), Peterken and Tubbs (1965), Hart (1966, p. 276), Schubart (1966, pp. 16, 89–93, 106, 138–142), Flörcke (1967, pp. 67, 99), Streitz (1967, pp. 37–38, 53), Beck (1977; 1981), Dister (1980, p. 67), Rackham (1980, pp. 355–356), Jahn and Raben (1982), Kausch von Schmeling (1985), Namvar and Spethmann (1985), Tubbs (1988, pp. 139, 141), Dornbusch (1988).

and Spethmann, 1985; Dornbusch, 1988). In closed forests, they are found virtually only on the fringes.[39] A number of these species are in danger of extinction. For example, the true service tree is threatened with extinction in parts of Germany (Dagenbach, 1981). In the east of England, there were only five wild pear trees left in the 1980s (Rackham, 1980, pp. 356, 358).

There are no quantitative data available about the ecology of the seedlings of wild apple, wild pear, wild cherry, rowan, wild service tree, whitebeam and true service tree, or about their regeneration in general. The reason for this is undoubtedly that they are of no interest to the timber industry. Wild cherry, which produces valuable wood for the furniture industry, is an exception to this.[40] Forestry practice has shown that wild cherry seedlings easily germinate under a canopy (Pryor, 1988, p. 20; Spiecker and Spiecker, 1988; Diez, 1989). Thus, as for the two species of oak, germination does not require any light. However, the seedlings die off if they are not exposed to a lot of light very quickly (Schalk, 1990). Experience has shown that for healthy growth, they must be in full daylight for between 3 and 5 years. The last standards must be removed by this time, creating an open space with only young wild cherry trees (Diez, 1989). If wild cherry is to be regenerated with group selection cutting, relatively large gaps must be made in the forest canopy.

The 'danger' for wild cherry, as for the two species of oak, is above all, that seedlings of species which tolerate more shade, such as beech, hornbeam and sycamore, may already have been present under the canopy for a number of years. Because it requires a great deal of light, wild cherry is not able to regenerate in advance in this way (Diez, 1989). This means that species which tolerate shade have a head start on wild cherry when the light conditions improve because of a gap in the canopy. Therefore wild cherry seedlings must have a head start of 3–5 years' growth on such shade-tolerant species, or they will disappear as a result of the competition (see Beck, 1977; Jahn and Raben, 1982; Diez, 1989; Schalk, 1990). Because of the growth rate, which is not paralleled by any other species of tree, wild cherry can stay well ahead of other shade-tolerant species in very light conditions, if it has such a head start.[41] The wild cherry's high requirements as regards light, and its rapid growth rate, means that it is sometimes identified as a pioneer (Binseil, 1958; Pryor, 1988, p. 6; Diez, 1989).

If the empirical data about wild cherry available from forestry sources are representative of other light-tolerant species and indicative of the species which require light or semi-light, the seedlings of these species have light requirements very similar to those of the two species of oak. This certainly applies to rowan, which, like wild cherry, is identified as a pioneer (Namvar and Spethmann, 1985). With their characteristics as a pioneer, rowan and wild cherry would be

[39] Lindquist (1935, pp. 121, 131), Bindseil (1958), Rackham (1980, pp. 351, 355–356), Namvar and Spethmann (1985), Diez (1989), Röös (1990, p. 148).
[40] Beck (1981), Gottwald (1985), Kausch (1985), Pryor (1985; 1988, p. 5), Otto (1987), Diez (1989), Noffke (1989), Röös (1990, pp. 7, 134), Schalk (1990), Savill (1991, p. 85).
[41] Bindseil (1958), Beck (1977), Kausch (1985), Pryor (1988, p. 6), Spiecker and Spiecker (1988), Schalk (1990), Savill (1991, p. 83).

able to survive in small numbers in closed forests in gaps in the canopy (Jahn and Raben, 1982; Koop, 1982). However, wild cherry particularly suffers there from being frayed by roebucks. Wild cherry and rowan, as well as wild apple, wild pear, wild service tree, whitebeam, and true service tree, are all affected by being nibbled by voles, mice and other wild animals.[42] Wild apple and wild pear do have thorns at the end of the twigs (Buttenschøn and Buttenschøn, 1985; Namvar and Spethmann, 1986; Heymann and Dautzenberg, 1988; Boeijink *et al.*, 1992, pp. 119, 152), but they do not afford sufficient protection from being eaten by wild animals, because the trees are nowhere near as thorny as blackthorn and hawthorn bushes.

6.8.2 Wild fruit and members of the rowan family in relation to the alternative hypothesis

Wild cherry, wild apple, wild pear, rowan, wild service tree, whitebeam and true service tree are more common in grazed, park-like landscapes than in closed forests. They grow there in unshaded, thorny scrub.[43] Birds particularly contribute to their establishment. Hawthorn and blackthorn bushes serve as a place for the birds to rest or roost. The seeds and fruits of the wild fruit trees are particularly popular with songbirds (Passeriformes) (Namvar and Spethmann, 1985; Snow and Snow, 1988). They land in the scrub in the birds' droppings, or in the grass around the shrubs. Therefore the shrubs serve as a concentration point for the wild fruit trees. The fruits of wild cherry, wild apple, wild pear, rowan, wild service tree, whitebeam and true service tree are eaten by rodents, domestic cows, wild boar, deer, badgers (*Meles meles*) and foxes (*Vulpes vulpes*), as well as by birds (Müller-Scheier, 1977; Buttenschøn and Buttenschøn, 1978; 1985; Sorensen, 1981; Herrera, 1984; Kollmann, 1992). On average, 91 seeds and 15 seed pods of wild apple were found per kg of cattle faeces. The seedlings grew in groups in the faeces. In comparison with ungrazed areas, it seems that wild apple trees benefit greatly for their regeneration from the presence of large herbivores, such as cattle. On the other hand, there are no wild apple seedlings in areas grazed by sheep (Buttenschøn and Buttenschøn, 1985).

Once they have germinated and are growing, wild fruit trees and members of the rowan family are resistant to the browsing of large herbivores. They all form shoots from a stool, have new growth from roots, or display both characteristics.[44] Only wild apple does not seem to have these qualities. The

---

[42] Beck (1977), Dister (1980, p. 66; 1985), Kausch von Schmeling (1985), Namvar and Spethmann (1985b; 1986b), Spiecker and Spiecker (1988), Diez (1989).

[43] Salisbury (1918), Adamson (1921; 1932), Tansley (1922), Watt (1925; 1934a, b), Tüxen (1952), Jakucs (1961), Jakucs and Jurko (1967), Schubart (1966, pp. 138–141), Dierschke (1974), Buttenschøn and Buttenschøn (1978; 1985), Rackham (1980, pp. 353, 355–357), Smith (1980, pp. 310, 353), Namvar and Spethmann (1986), Röös (1990, p. 148), Pott and Hüppe (1991, pp. 299, 297), Rodwell (1991, pp. 316–321), Oberdorfer (1992, pp. 98, 148), personal observation).

[44] Rackham (1980, pp. 349, 350), Dagenbach (1981), Namvar and Spethmann (1985), Pryor (1988, p. 6), Diez (1989), Schalk (1990), Mayer (1992, p. 120).

stool of the wild cherry tree produces so many shoots, and so much spring grows from the roots, that in Germany this species is sometimes known as the forest weed ('forstliches Unkraut') (Diez, 1989, p. 781). The new growth from the roots of the wild cherry tree results in the development of groups of trees (clones). Wild pear also has the property of forming clones (Rackham, 1980, pp. 349, 356).

Ancient written sources show that all these species survive in grazed landscapes. These date back to the early Middle Ages. As we saw in Chapter 4, they were referred to, together with oaks, as fruitful trees, and were protected because of the fruit, the mast, which was used for pannage for pigs in the autumn. They could be removed only with the express consent of the lord or, in the case of free commons, with that of the commoners' meeting.[45]

To summarize, it may be concluded that wild fruit trees and members of the rowan family (*Sorbus* sp.) flourish as well in open, grazed, park-like landscapes as they struggle to survive in closed forests.

6.9 The Establishment of Blackthorn, Hawthorn, Juniper and other Thorny Shrubs in Relation to Grazing and Diversity

6.9.1 Germination conditions

Little or nothing is known in a quantitative sense about the conditions in which thorny species, such as hawthorn and blackthorn, juniper (*Juniperus communis*), roses (*Rosa* sp.), brambles (*Rubus* sp.) and gorse (*Ulex europaeus*), become established in grasslands. The available information is mainly of a qualitative nature. They are all known to require light (Ellenberg, 1986, pp. 94–95; Coops, 1988; Grubb *et al.*, 1996). They owe this qualification to the fact that they become established in open vegetation, such as grasses and coppices,[46] and disappear when the crowns of trees form a closed canopy.[47] In order to germinate in grassland, blackthorn and juniper certainly require bare, mineral soil. This is found in grassland because of earthworms and moles, which dig up the earth and form molehills. In addition, it is brought about by the hooves of cattle, sheep and horses, scraping the soil, and the rooting of (wild) pigs (Ahlén, 1975: Hard, 1976, p. 99; Miles and Kinnaird, 1979b; Smith, 1980, p. 328).

Songbirds particularly contribute to the establishment of these species in grassland, because they eat the fruits and then excrete the seeds undigested

---

[45] Endres (1888, p. 82), Bühler (1922, p. 65), Vanselow (1926, pp. 22, 190), Grossmann (1927, p. 23), Meyer (1931, p. 293), Schubart (1966, pp. 48–49, 89–93, 138–142, 192), Streitz (1967, pp. 37–38), Rackham (1980, pp. 355–357).
[46] See Gradmann (1901), Bernçtsky (1905), Watt (1924; 1934b), Grossmann (1927, pp. 113–114), Tansley (1953, pp. 130, 474–476), Schubart (1966, p. 106), Knapp (1971, pp. 16–17, 248–250, 288, 427–432), Ahlén (1975), Rackham (1980, pp. 353, 355), Ellenberg (1986, pp. 43, 60–61), Dengler (1990, p. 262), Peterken (1992).
[47] See Puster (1924), Watt (1924; 1934b), Ekstan and Sjörgen (1973), Rackham (1980, p. 352), Ellenberg (1986, p. 95), Tubbs (1988, p. 157); Coops (1988).

(Snow and Snow, 1988, pp. 96–99). There is a period between the time when the fruits are eaten and the seeds are excreted, when the birds can cover smaller or larger distances. As many species of fruit-eating songbirds, such as thrushes (Turdidae) live in a park-like landscape (Smith, 1980, pp. 346–347, 369–390; Tubbs, 1988, pp. 159–162; Cramp, 1992; Hondong *et al.*, 1993, pp. 146–147; Schepers, 1993), this also involves flying over grasslands, and consequently the seeds can land there. Mammals also eat the fruits and excrete the seeds undigested (Herrera, 1984; Buttenschøn and Buttenschøn, 1985). Apart from the thorny species, other species of shrubs with fleshy fruits are also dispersed by birds, such as guelder rose (*Viburnus opulus*), elder (*Sambucus nigra*) and dogwood (*Cornus sanguinea*) (Adamson, 1921; Tansley, 1922; Snow and Snow, 1988, pp. 65–68, 70–71, 75–76; Kollmann, 1992). Because birds sometimes roost in thorny shrubs, the seeds of these species of shrubs can be dispersed in their droppings on the fringes or in the neighbourhood of this scrub.

Grassland is maintained by specialist grass eaters, such as cattle, horses and sheep (Ahlén, 1975). Although thorny shrubs are protected by their thorns against being eaten, this does not mean that they are wholly spared. Blackthorn bushes and juniper are browsed by cows and sheep (Buttenschøn and Buttenschøn, 1978; Bokdam, 1987). The annual shoots of blackthorn are not spiny (Heukels and Van der Meijden, 1983, p. 199), and are therefore eaten. It takes up to 3 weeks for the thorns of hawthorn seedlings to harden. Up to that time they are vulnerable to grazing livestock (Rackham, 1980, p. 355), and can disappear from the grassland as the result of grazing. For example, it has been observed that rose seedlings (*Rosa* sp.) disappear where there is heavy grazing by sheep (Buttenschøn and Buttenschøn, 1985). Nevertheless, thorny shrubs successfully become established in grazed grasslands and spread there, as we saw in Chapters 2 and 4. In parts of Europe, thorny shrubs are seen as weeds (Grossmann, 1927, pp. 113–114; Ellenberg, 1986, p. 95; 1988, pp. 20–21). They develop in grasslands which are undergrazed, or with all-year-round grazing when the animals are left to graze outdoors all summer and winter.[48] All-year-round grazing was the original form of grazing by livestock in the commons and 'forestes'. As we saw in Chapter 4, the commoners could graze as many animals on the common or in the 'forestis' as they were able to keep throughout the winter with fodder they had collected there themselves. They could not bring any livestock or fodder in, or take them out, so that the numbers were regulated on the basis of the ecological capacity of the common or 'forestis'.

Thorny scrub can advance rapidly when grazing is reduced for a number of growing seasons, or if it suddenly stops altogether.[49] Blackthorn is the only

[48] Hobe (1805, p. 124), Bernçtsky (1905), Sloet (1913, p. 360), Grossmann (1927, pp. 113–114), Hilf (1938, p. 10), Meyer (1941, pp. 105, 106), Tansley (1953, pp. 130, 475–476), Knapp (1971, pp. 16–17, 248–250, 288, 427–432), Rackham (1980, p. 353), Ellenberg (1986, pp. 43–44, 60–61). In German this is called 'Abweiden', or 'Hut- oder Triftweide' and in English 'rough grazing'.

[49] Tansley (1922), Sheail (1980, pp. 2–3), Smith (1980, p. 328), Buttenschøn and Buttenschøn (1985), Weeda *et al.* (1987, p. 104), Schreiber (1993).

species which advances in a clonal fashion with runners (Watt, 1934b; Rackham, 1980, p. 349; Weeda *et al.*, 1987, p. 104; Tubbs, 1988, pp. 153–157). For the successful establishment of juniper, a reduction in grazing, or an interruption even seems to be a prerequisite.[50] Thus young juniper grows rapidly after grazing has been interrupted (Buttenschøn and Buttenschøn, 1985). The most spectacular examples of advancing thorny scrub occurred in England in areas where the population of rabbits collapsed as a result of myxomatosis (Sheail, 1980, pp. 2–3; Smith, 1980, p. 328).

In grasslands where there has been no grazing for a long time, shrubs and trees often do not become established for decades.[51] Seedlings do not become established in ungrazed grasslands because the seeds remain lying on the litter, dry out and therefore do not germinate. Even if they do germinate they cannot become (sufficiently) rooted in the mineral soil because of the thick layer of litter so that the seedlings dry out. If they do succeed, the seedlings die from lack of light because of the shade of the tall grasses and plants, or they die because they are nibbled by rodents, particularly the bank vole (*M. agrestis*), as the layer of litter is an ideal biotope for rodents.[52] The development of scrub and groves is therefore not so much the result of a gradual process of the successive establishment of species of trees and shrubs in grasslands where grazing has ended, it is much more the result of their initial establishment, after grazing has ended.[53] Even the clonal expansion of blackthorn in grassland comes to an end with the prolific growth of grasses and plants. The young stems which grow from the rootstocks degenerate in the tall vegetation and disappear (Hard, 1972; 1975; 1976, p. 1986; Buttenschøn and Buttenschøn, 1985; Schreiber, 1993).[54] The thick layer of litter which forms as a result of the lack of grazing in the grassland disappears after a few years when grazing, e.g. by cows, restarts. The litter disappears because it is trampled, as well as by grazing (Van Wieren, 1988, pp. 27–31). As a result, there is once again an environment in which shrubs and trees can germinate. In addition, compared with the situation where there is no grazing, the density of rodents is considerably lower in grazed grassland (Tubbs and Tubbs, 1985; Hill, 1985, cited by Putman, 1986, pp. 166–168; Putman *et al.*, 1989).

---

[50] Vedel (1961), Fitter and Jennings (1975), Miles and Kinnaird (1979b), Ward (1981), Rosén (1982, pp. 85–86), Rodwell (1991, p. 321), Hillegers (1994).

[51] Persson (1974, p. 29), Hard (1972; 1975; 1976, pp. 184–185), Smith (1980, pp. 302–303, 313), Schreiber (1993). German foresters have a saying: 'Wo Gras wächst, wächst kein Holz' (Hard, 1976, p. 185).

[52] Fabricius (1897), Watt (1943a), Tansley (1953, pp. 140–141), Krahl-Urban (1959, p. 130), Burschel and Schmaltz (1965b), Pigott (1975), Miles and Kinnaird (1979a, b), Huss and Stephani (1978), Mayer and Reimoser (1978), Smith (1980, pp. 307, 329), Koop (1981, p. 30), Von Lüpke (1982), Röhl (1984), Dengler (1990, pp. 58–59, 120–121, 163), Hill *et al.* (1992).

[53] Eglar (1954), Morris (1967, cited by Smith, 1980, p. 294), Hard (1975; 1976, pp. 184–185). Persson (1974, p. 29), Buttenschøn and Buttenschøn (1985); Schreiber (1993).

[54] Because the expansion of the blackthorn after grazing stops is shown in the average number of metres per year over the whole period of observation, the process of fast initial expansion and subsequent dying out is obscured (see Hard, 1975; 1976, pp. 185–187). That is why literature creates the idea that the blackthorn is continuously expanding in grazed grasslands, whereas this does not have to be the case.

The above shows that an interruption in grazing leads to the establishment and advance of scrub and groves into grassland. This interruption occurs in two ways when there is grazing all year round. First of all, grazing is interrupted locally because the animals do not cover the whole area they graze in winter during the growing season of plants. Compared with winter, there is less grazing in summer. Secondly, there are large-scale interruptions when the numbers of animals are decimated, for example, as the result of a lack of food or the spread of a disease.

6.10 The Establishment of Trees and Shrubs at the End of the Last Ice Age

6.10.1 Grazing and the establishment of trees and shrubs in prehistoric times

If the alternative hypothesis about the establishment of trees and shrubs in grazed, park-like landscapes is correct, this should also be able to explain the establishment of trees and shrubs in prehistoric times. As we saw in Chapter 3, palynological studies have shown that hazel, oak, lime and elm reached their furthest points in the north and north-west of Europe, between 10,000 BP and 6000 BP. Beech and hornbeam only came much later, respectively in about 4000 BP and 2000 BP (Huntley and Birks, 1983, pp. 143, 155–159, 175–177, 192–193, 205–208, 216–217, 361–363, 399–410, 421–436; Delcourt and Delcourt, 1987; Birks, 1989; Bennett *et al.*, 1991). Large herbivorous ungulates such as aurochs, European bison, red deer, roe deer (*Capreolus capreolus*), elk (*Alces alces*), tarpan (*Equus przewalski gmelini*), and the omnivorous wild boar, had been present in this region long before, since the Allerød period (13,000–10,800 BP).[55] At that time 'modern' temperatures already prevailed in north-west and Central Europe (see Fig. 6.24). Depending on the place, the average July temperature was 15°–18°C.[56] In the Allerød period, genera of plants which are characteristic of temperate climate conditions appeared, such as *Typha, Filipendula, Urticularia, Plantago* and *Rumex*.[57] In this period (13,000–10,800 BP), there were few or no trees yet. The vegetation still consisted mainly of grasses or herbs. The question is how trees became established under these climate conditions in the presence of large herbivores. In Chapter 5, I suggested that domestic livestock in the 'forestes' could be considered as a modern analogy of the original fauna. In that case, the trees could have become established

[55] Degerbøl and Iversen (1945), Degerbøl (1964), Degerbøl and Fredskild (1970), Grigson (1978), Söffner (1982), Birks (1986), Aaris-Sørensen *et al.* (1990), Cordy (1991), Current (1991), Housley (1991), Street (1991), Stuart (1991).
[56] Coope (1977; 1994), Van Geel *et al.* (1980/81), Amman *et al.* (1984), Lemdahl (1985), Atkinson *et al.* (1987), Berglund *et al.* (1987), Dansgaard *et al.* (1989), Ponell and Coope (1990), Kolstrup (1991).
[57] Van der Hammen (1949), Iversen (1954), Polak (1959), Klostrup (1980), Pennington (1970), Berglund *et al.* (1984), Van Geel *et al.* (1989), Bohncke *et al.* (1987).

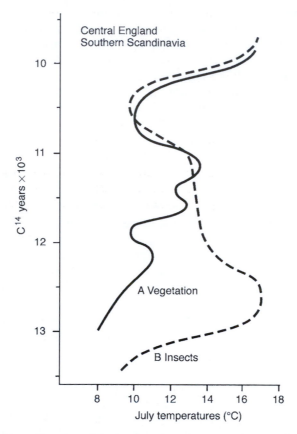

Fig. 6.24. Reconstruction of the increase in temperature in central England and southern Scandinavia, after the end of the last Ice Age, 14,000 BP (from Berglund *et al.*, 1984, p. 27). The unbroken line shows the reconstruction which has been accepted for a long time. It is based on the appearance of the first pollen from trees known to require warmth. The broken line is of a more recent date, and is based on the discovery of insects, mainly beetles (Coleoptera). The difference between the two lines is believed to be due to the different responses of trees and beetles respectively to changes in the climate. Insects react more rapidly because they are much more mobile. Therefore they are considered to be more reliable indicators of changes in temperature resulting from changes in the climate (Coope, 1994). The reconstruction on the basis of insects corresponds to the findings based on trapped gas bubbles in the icecap of Greenland (see Dansgaard *et al.*, 1989).

and regenerated in the same way that the process of regeneration took place in wood-pasture, i.e. in thorny scrub. What are the arguments for this?

On the basis of their food preferences, wild ruminants can be divided into three categories: 'concentrate feeders' (which eat buds, leaves and twigs of shrubs and trees); 'intermediate feeders' (species which alternately eat grass and herbs, on the one hand, and bark, leaves and twigs of trees and shrubs, on the other hand), and 'grass/roughage eaters' (specialized grass eaters) (Van de Veen, 1979, pp. 131–134; Van de Veen and Van Wieren, 1980; Van Soest, 1982, p. 338; Hofmann, 1985; 1989; Van Wieren, 1996). The 'concentrate feeders' include elk and roe deer, the 'intermediate feeders' include European bison and red deer and the 'grass/roughage eaters' include the aurochs. The tarpan, the wild version of the domestic horse, was a non-ruminant, though it was a specialized grass eater (Van de Veen and Van Wieren, 1980; Van Soest, 1982, p. 338). The omnivorous wild boar also includes a considerable amount of grass in its diet (Briedermann, 1990, pp. 175–176; Groot-Bruinderink *et al.*, 1994).

Figure 6.25 shows the wild animals and livestock, divided on the basis of their food preferences. The similarity between livestock and the wild herbivores is that domestic cattle, domestic horses and domestic pigs are domesticated versions of, respectively, the aurochs, tarpan and wild boar. They do not differ in terms of the food they prefer. Goats and sheep are not indigenous in the lowlands of Central and north-west Europe. However, the diet of the goat is similar to the food preferences of the European bison, red deer, roe deer and elk, while the diet of sheep falls under that of aurochs and tarpan. As regards food strategies, goat and sheep do not differ from the wild fauna. Nevertheless, different species of herbivores may have different effects on the vegetation with the same feeding strategies, when there is a difference of emphasis. For example, cattle and sheep both eat grass, but sheep are more selective than cows. Sheep prefer the young, soft parts of grasses and plants, and crop the plants much lower down to the ground, compared with cattle. Sheep can remove the growing ends and runners of grasses which are above the ground, and even pull plants out of the ground. As a result, the grass cover is reduced and bare patches develop (Klapp, 1971, p. 433). On the other hand, extremely large numbers of seedlings of trees and shrubs grow in grassland grazed by sheep. Yet they often do not become successfully permanently established, because sheep browse and strip woody plants, including blackthorn, more than cattle (Marks, 1942; Buttenschøn and Buttenschøn, 1978; 1985; Coops, 1988; Mitchell and Kirby, 1990; Van Wieren, 1996). The sheep's narrower head seems to make stripping and cutting young trees and shrubs easier, because it can poke it through the branches more easily. For this reason, many more trees and shrubs grow in grassland grazed by cattle than in grassland grazed by sheep (Buttenschøn and Buttenschøn, 1978).

As discussed earlier, this explains why Watt (1924) found that the thorny scrub advanced and was accompanied by an increase in the regeneration of trees when areas formerly grazed by sheep were subsequently grazed by cattle; a phenomenon which he interpreted as a phase in the process being a result of the end of grazing (see Chapter 2). It also explains why restrictions were imposed on sheep, though not on cattle, in the course of the 17th century. It was particularly the increase in the number of sheep at that time as a result of the emergence of a trading economy and an increasing demand for wool from the textile industry (Mantel 1990, p. 439; Bieleman, 1992, pp. 80, 84), that

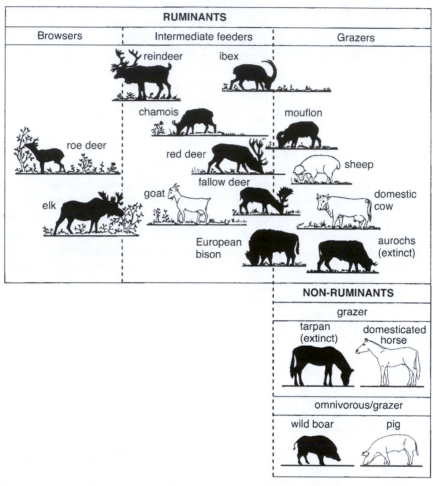

Fig. 6.25. The different species of large herbivores indigenous in Europe, as well as the omnivorous wild boar, classified according to their feeding strategy. The domesticated species are shown in white. The indigenous species of the lowlands of Central and Western Europe include the aurochs, tarpan, bison, red deer, elk, roe deer and wild boar (redrawn from Hofmann, 1973; 1976; 1985; Van de Veen and Van Wieren, 1980).

must have had a marked effect on the vegetation. The exorbitantly high densities of sheep which eventually resulted were described in Chapter 4.

Therefore, in terms of food preferences, the wild fauna did not differ essentially from the livestock which were introduced by humans in Central and Western Europe only 8000 years after the arrival of wild herbivores. The livestock and the related browsing of species of plants were therefore not a new phenomenon for the vegetation of Central and north-west Europe. All-year-round grazing by livestock is comparable to the grazing of wild herbivores.

After all, wild fauna must survive all year round with the food provided in the environment, and the reserves of fat which they can build up from these for the seasons when no more food is produced. In the Middle Ages, this also applied to livestock, which in principle grazed outside (all year round). It was possible to give them additional fodder, but only with fodder collected in the common or the 'forestis'. Both the numbers of wild herbivores and the numbers of livestock were regulated by the food supply. For livestock, this applied because the local communities were self-sufficient as regards the food supply for livestock, as we saw in Chapter 4. For the vegetation, this means that the establishment of trees and shrubs which takes place in the presence of livestock can be seen as a modern analogy of the process which took place in the past in the presence of wild ungulates. The fact that hazel and oak do not regenerate in closed forests without large herbivores, though they do regenerate in park-like landscapes with grasslands grazed by large herbivores, means that the establishment of trees and shrubs must have taken place in a similar way, as otherwise it would not be possible to explain the continuous presence and high frequencies of oak and hazel pollen in the pollen diagrams of Central and Western Europe over a period of more than 9000 years. The shape of the growth of about ten sub-fossilized oaks by the Waal near Nijmegen, of which one specimen is dated 8420 BP (Boreal period), also indicates that some trees grew in open conditions such as those which are found in a park-like landscape. A thick oak, approximately 12 m tall and with a diameter of 1.5 m, had the first thick branch at a height of approximately 2.5 m which means that this tree grew in an open field.

The presence of large predators, such as wolf (*C. lupus*), lynx (*L. lynx*) and brown bear (*Ursus arctos*), must have resulted in densities of wild ungulates that were approximately the same as those of livestock that did not receive supplementary feeding in combination with the wild animals that lived in the wood-pastures in the Middle Ages and shortly afterwards. For otherwise the permanent presence of oak and hazel in the vegetation from before the arrival of livestock cannot be explained. Whatever influence the large predators had, the densities that are required for the regeneration of oak and hazel must have been the result. Fossils will not be able to give a definite answer on the densities of large herbivores. First of all, the chance of fossilization depends on special conditions and is therefore very small (Davis, 1987, pp. 22–46). Secondly, the chance of finding fossilized bones is small. The greatest chance is in human settlements, because there was a concentration of bones as waste there. In that case, these bones are the remains of a human selection from the animals in the surrounding area. Even based on bones found in a settlement, it is difficult, if not impossible, to determine densities. For example, Davis (1987, p. 27) mentions the case of Fort Lignier between Carlisle and Pittsburg in the United States. From 3 September 1758 until the spring of 1766 (2364 days), 4000 soldiers made their camp in the fort. The daily meat ration per soldier was 1 lb (453 g). The bones found near the fort and analysed by Guilday (1970, cited by Davis, 1987) suggested a total meat consumption of no more than 4000 lb

(1800 kg) for the entire period. That was sufficient to feed all the men for 1 day or two men for the entire period that the army was present in the fort.

After the end of the last Ice Age, juniper may have prepared the way for trees to become established in the presence of large herbivores. All the pollen diagrams of Central and Western Europe show that after 13,000 BP, there was a marked increase in the frequency of the pollen of this shrub, prior to the increase of birch and pine pollen (see Huntley and Birks, 1983, pp. 128–131, 243–248, 311–314). Birch and pine, as well as all other species of trees and hazel, regenerate in juniper scrub.[58] Songbirds eat the juniper seeds and the ripe berries are available for (migrating) birds throughout the winter (Snow and Snow, 1988, pp. 96–99; Garcia *et al.*, 1999). After the end of the Ice Age, migrating birds, which breed in the steppes and tundra and spend the winter in regions further south, could have been responsible for the northern advance of the juniper. The species which are most likely to have done this are those which breed in the steppes and tundra today, such as the shore lark (*Eremophila alpestris*), the Lapland bunting (*Calcarius lapponicus*), and the snow bunting (*Plectrophenax nivalis*) (Cramp, 1988, pp. 210–225; Jonnson, 1993, pp. 358, 365, 538). Judging from the rate at which juniper colonized Central and north-west Europe after the Ice Age, in comparison with birch and Scots pine, with their winged seeds which are dispersed by the wind, the dispersal of juniper berries by birds seems to be the only explanation.

Hazel also advanced relatively rapidly (Erdtmann, 1931; Huntley and Birks, 1983, pp. 167–184). In view of the weight of hazel nuts, this is difficult to explain without allowing for the possibility of their dispersal by birds. Thus, hazel nuts are preferred by the nutcracker (*Nuccifraga caryocatactus*), a species which currently has a boreal habitat (Swanberg, 1951; Turcek, 1966; Turcek and Kelso, 1968; Löhrl, 1970). The nutcracker transports seeds over distances from 10 to 15 km (Swanberg, 1951; Sutter and Amann, 1953; Reimers, 1958, cited by Turcek and Kelso, 1968; Müller-Schneider, 1977). A similarly rapid colonization applied to the oak, of which the acorns were transported by birds (especially jays). Species such as elm, lime and ash, of which the seeds are dispersed by the wind, migrated more slowly than oak.

As remarked above, grazing creates the short grass and plant vegetation required for juniper to become established, while a relaxation in the grazing gives the seedlings an opportunity to grow out of the reach of browsing animals.[59] This means that fluctuations in the numbers of wild herbivores contribute to the establishment of this species. Considering the enormous mortality

---

[58] Watt (1934a), Konigsson (1968, pp. 20, 154), Smith (1980, p. 253), Rodenberg (1988, pp. 82–83), Buttenschøn and Buttenschøn (1985), Rùsen (1988, p. 96), Rousset and Lepart (1999).
[59] Vedel (1961), Fitter and Jennings (1975), Miles and Kinnaird (1979a, b), Ward (1981), Rùsen (1982, pp. 85–86), Buttenschøn and Buttenschøn (1985), Rodwell (1991, p. 321), Hillegers (1994).

which can occur among large wild herbivores as a result of disease or sudden shortages in the food supply, potentially up to more than 50% of the total population (Clutton-Brock *et al.*, 1991; Prins and Van der Deugd, 1993; Young, 1994), it is almost certain that such fluctuations took place naturally and resulted in the advance of thorny species. In Africa, the establishment of thorny acacia is related to the mortality rate of large herbivores (Prins and Van der Deugd, 1993). A sudden decline in the number of herbivores could also have contributed to the establishment of birch and Scots pine. The grazing and trampling of these large herbivores create the conditions for the germination and establishment of these two species, in the form of mineral soil and short grass. A sudden decline in the number of herbivores means that a number of young trees have several years to grow out of reach of browsing animals. Juniper becomes established and advances as a potentially protective scrub under similar conditions, as we saw earlier in this chapter. Although birch and Scots pine seedlings are adversely affected by grazing and being trampled, they cannot survive without these large ungulates, because of the combination of factors required for the establishment of both species.[60] The solution to this dilemma is a temporary reduction in the number of large herbivores, for example, as a result of a sudden high mortality rate because of lack of food or disease. This gives juniper the chance to become established, and in addition, birch and Scots pine seedlings are able to grow out of reach of browsing herbivores.[61] An additional factor is that large ungulates do not like to eat birch and Scots pine. This means that in the case of grazing, they grow relatively well unprotected (Olberg, 1957; Van Wieren, 1996; Jorritsma *et al.*, 1999). In all probability, fire does not play a role in the regeneration in the sense that it is necessary for regeneration (Olberg, 1957).

It is very difficult, or impossible, to establish when blackthorn and hawthorn arrived in Central and north-west Europe after the Ice Age, because of the unlikelihood of finding pollen of these plants, which are pollinated by insects. As the berries of both species are eaten by birds and large ungulates and then excreted (Herrera, 1984), they may have advanced very rapidly. They could have arrived in Central and north-west Europe with the ungulates in the Allerød period (13,000 BP) after the end of the Ice Age. From the point of view of the climate, these species could have become established at that time (see Fig. 6.24). Birds could also have been involved in the dispersal of seeds, as well as ungulates. The way in which blackthorn has become established in the steppe–forest zone in the Ukraine and southern Russia shows how blackthorn could have advanced rapidly in the open former steppe tundra after the Ice Age. In the Ukraine and southern Russia, the shrub became established as a vanguard in the steppes, far in advance of the trees (Mayer-Wegelin, 1943, p. 14;

---

[60] See Børset (1976), Buttenschøn and Buttenschøn (1985), Ellenberg (1986, pp. 366–367), Rodenberg (1988, pp. 82–83), Rùsen (1988, p. 96), Van Wieren (1988, pp. 35–360, Atkinson (1992).
[61] Chard (1953), Kinnaird (1974), Miles and Kinnaird (1979a, b), Newbold and Goldsmith (1981, p. 96), Ellenberg (1986, pp. 366–367).

Fig. 6.26. Park-like landscape with trees growing from thorny scrub, showing what prehistoric vegetation may have looked like. The photograph was taken in the Borkener Paradise, Germany (photograph, F.W.M. Vera).

Fig. 6.27. An overview of how the landscape in Central and Western Europe would have looked under the influence of grazing and browsing by the wild fauna, including specialized grazers such as aurochs and tarpan. The photograph was taken in Slovenski Kras, Slovakia. Mountains, like those in the background, could have been covered in closed forest, if they were less accessible for the aurochs and the tarpan (photograph, F.W.M. Vera).

Fig. 6.28. Trees die of old age, and sometimes as a result of persistent drought in the centre of a grove, as shown here in Denny Wood in the New Forest, England. In this way, grassland develops in the central part of the grove. Several grasslands which have developed in groves as a result of regeneration may join together so that the area is considerably increased. In this way, grasslands in a grove turn into larger scale grasslands (H. Koop, Wageningen, 1997, personal communication) (photograph, H. Koop).

Fig. 6.29. Thorny bushes establish themselves in the open grassland that arises as a result of grazing in the grove and subsequently trees shoot up in these bushes. These in turn form a grove that in time changes to grassland again.

Walter, 1974, pp. 131, 149, 152–153). Trees then became established in this blackthorn scrub (Mayer-Wegelin, 1943, p. 14; Leimbach, 1948; Walter, 1974, pp. 131, 149, 152–153). This is possible because of the moisture conditions resulting from the localized deposits of snow in an environment which is otherwise hostile to trees, so that trees can become established (Leimbach, 1948). In the case that blackthorn and hawthorn arrived later, only juniper will initially have served to protect trees and shrubs. Subsequently, blackthorn may have become established of its own accord, or in the juniper scrub, and have replaced the juniper, as Watt (1934b) observed in the Chilterns. Later, groves must have developed from the scrub as a result of the effect of grazing by the wild herbivorous fauna. As a result of the presence of the large ungulates and the shade there will be no more regeneration of trees (see Fig. 4.9). The result is a mosaic of grasslands, scrub and groves (see Figs 5.47, 6.26 and 6.27). After a while, the centre of the grove degenerated as the trees grew old, because regeneration in the groves was obstructed by shade and wild ungulates (see Chapter 4, and Figs 4.10, 4.11 and 6.28). This resulted in grasslands, where groves then developed once again because thorny scrub was the first to grow in the grassland (see Figs 6.28 and 6.29). Chapter 3 shows how this sort of landscape could accord with the pollen diagrams.

There is a relatively large body of information available from research about the effects of cattle, horses, red deer, roe deer and elk on vegetation. This does not apply to European bison. The European bison in the forest of Białowieza are fed liberally with extra hay in winter to prevent them from stripping the bark off the trees and therefore damaging the forest (Borowski and Kossak, 1972; Krasinski, 1978; Pucek, 1984; Falinski, 1986, p. 165). There is clear evidence that without this additional feeding, there is extensive stripping of trees. For example, six adult bulls which did not go to the feeding place from December 1969 to March 1970, the period when extra food is provided, stripped almost 3000 trees, mainly ash, in an area of 12 ha (Borowski and Kossak, 1972).

As regards the European bison's biotope, it is striking that they avoid virtually the whole of the National Park (Falinski, 1986, p. 31; Jedrzejewska *et al.*, 1994; Krasinka and Krasinski, 1998). According to the prevailing theory, this contains the most natural vegetation, and therefore also the natural biotope of European bison. This is clearly not the case. Other data (Heptner *et al.*, 1966, pp. 491–493, 505) show that European bison prefer a park-like landscape, and that the presence of European bison in the forest of Białowieza should be seen as a throwback, on the basis of which it cannot be concluded that a closed forest is the biotope of this species. The fact that European bison make wide use of the feeding pastures created for them in the forest, which are regularly mown for them (Krasinski, 1978), shows that grass is also an important part of their diet, and that grasslands must therefore have formed part of their natural biotope. In addition to grass, a considerable proportion of the European bison's diet consists of the bark of trees, particularly oak. It also strips the bark of hornbeam, ash, elm, wild apple, wild pear and hazel, and eats the fruit of wild apple and wild pear (Heptner *et al.*, 1966, pp. 507–508; Borowski and Kossak, 1972; Van Wieren, 1985; Falinski, 1986, p. 167; Wallis de Vries, 1998).

For the aurochs, a specialized grass eater, it is even clearer that the closed forest cannot have been its natural biotope. However, the aurochs is described as a forest dweller, in the sense that the disappearance of the last primeval forests in Europe in the 15th century is given as the reason for the extinction of this species (Green, 1989; Vereshchagin and Baryshnikov, 1989, p. 510). Aurochs were found up to the Middle Ages in the region where the 'forest' of Białowieza is believed to have been one of the last primeval forests. In view of the aurochs' diet, it cannot have been a closed forest; it must have contained grassland.

The grazing of grasses by specialized grass eaters, such as cattle, initiates the regrowth of grass. Cattle, as well as (wild) horses, can live on a diet of grass with a high fibre content (Janis, 1975; Sinclair, 1979a; Van Soest, 1982; Wallis de Vries, 1994, pp. 125–138; Van Wieren, 1996, pp. 82–101). By cropping the grass, they stimulate new growth of young grass, with a higher content of soluble carbohydrates and proteins (Drent and Prins, 1987), which is easily digested by species of herbivores which do not digest the fibre in old grass so easily, such as European bison and red deer (Van de Veen and Van Wieren, 1980; Vera, 1986; 1988; 1989; Gordon, 1988; Wallis de Vries, 1994, pp. 125–138). As cattle and horses are in a sense responsible for the regrowth of grass, the wild cattle and wild horses will, in the natural situation, have provided the European bison and red deer with the grass they can digest. Therefore wild cattle and wild horses acted as facilitators for red deer and European bison (Gordon, 1988; Groot Bruinderink *et al.*, 1999).

Grasslands which are now maintained for European bison by mowing them, must have been kept in a condition that was suitable for European bison, by aurochs and tarpan in the natural situation. After the introduction of livestock, domestic cattle and domestic horses must have taken over this role. However, the introduction of livestock also created competition for aurochs and tarpan, which eventually disappeared (Szafer, 1968; Van de Veen and Van Wieren, 1980; Vera, 1986). As domestic cattle and horses have now also disappeared from the forest of Białowieza, mowers have to take over the role of the aurochs, cattle and (wild) horses. Mowing is necessary because the European bison do not graze in the grasslands in the forest which are not mown (Krasinski, 1978).

On the basis of these considerations, it can be said that a park-like landscape with grasslands as a natural situation would explain the presence of aurochs and European bison, as well as other wild herbivores, such as red deer and tarpan, throughout the Holocene period up to the early Middle Ages, much better than a closed forest. When European bison strip trees in a park-like landscape, this could have accelerated the degeneration of groves into grassland. The stripping can result in the death of a tree. As trees die, the canopy becomes thinner, so that more light penetrates to the forest floor, which stimulates the growth of grass and plants. In this way, European bison can have speeded up the transformation of groves into grassland. Grazing by aurochs and tarpan would then have contributed to the creation of grassland in the same way that grasslands developed in groves in the degenerating ornamental woods in the New Forest because of the grazing of cattle and horses. European bison and red deer, which also eat a great deal of grass[62] could also have contributed to this, although grazing by cattle and horses is necessary, as we saw above, for ensuring that the grass is suitable fodder for red deer and bison, which are less able to digest coarse fibre.

6.10.2 Continuity in diversity

All-year-round grazing is the natural form of grazing as practised by large herbivores. As we noted above, with this form of grazing, not all the parts of plants that are above the ground are removed in some parts of the habitat during the growing season, because the animals do not graze there, or graze there only very little. It is only in winter, when the growth of plants stagnates, that these parts are consumed. With all-year-round grazing, there are species such as field scabious (*Knautia arvensis*), greater yellow rattle (*Rhinanthus serotinus*), creeping bell flower (*Campanula rapunculoides*), meadow clary (*Salvia pratensis*), wild parsnip (*Pastinaca sativa*), hedge bedstraw (*Galium mollugo*) and false oat grass (*Arrhenatherum elatius*) (Smith, 1980, p. 301; Helmer *et al.*, 1995). These species are familiar from hay fields in farmland. A hay field is an artificially created analogy for part of the vegetation which develops with the natural grazing of wild herbivores. It is the part where the herbivores graze little or not at all in summer, but where they do graze in winter. In areas which are grazed all year round, there are also birds which are characteristic of agricultural hayfields, such as the corncrake (*Crex crex*) (de Hullu, 1995).

Park-like landscapes which have developed and are maintained with all-year-round grazing, contain all the species of grasses and plants which are now found only in various types of agricultural grasslands (see, *inter alia*, Wollinger and Plank, 1981; Hillegers, 1986; Ellenberg, 1986, pp. 615–690, 714–776). Because of the variety of types of vegetation, such as grasslands and groves, as well as the transitional systems of the mantle and fringe vegetation, these park-like landscapes have an extremely large diversity of species of plants and animals.[63] In addition to all the species of trees and shrubs found in the mantle and

---

[62] Heptner *et al.* (1966, pp. 504–505), Borowski and Kossak (1972; 1975), Van de Veen (1979, pp. 212–213, 218), Falinski (1986, pp. 163–170), Hofmann (1986), Cornelissen and Vulink (1996a, b), Van Wieren and Wallis de Vries (1998), Groot Bruinderink *et al.* (1999).

[63] See *inter alia* Salisbury (1918), Adamson (1921; 1932), Tansley (1922), Watt (1924; 1925; 1934a, b), Müller (1952), Tüxen (1952), Sjörgen (1973; 1988), Dierschke (1974), Rose (1974), Rose and James (1974) cited by Tubbs (1988, pp. 148–149), Smith (1980, pp. 380, 318–319, 349, 353), Rosén (1988), Tubbs (1988, pp. 25–26), Pott and Hüppe (1991, pp. 289–299), Rodwell (1991, pp. 319–321, 334–361), Kollmann (1992), Oberdorfer (1992a, pp. 87–105; 1992b, p. 148), Anonymous (1993), Hondong *et al.* (1993), Pietzarka and Roloff (1993).

fringe vegetation, it is also extremely diverse in terms of other species of plants (Smith, 1980, pp. 326, 369; Hondong *et al.*, 1993, pp. 126–140; Pietzarka and Roloff, 1993, and see Appendix 8). Historical sources as well as the current situation show that wild fruit is widespread in wood-pastures (see Chapter 4). In various Central and Western European countries, wild apple (*M. sylvestris*) and wild pear (*P. pyraster*) are among the endangered species threatened with extinction (Rackham, 1980, p. 356; Kleinschmit, 1998; Wagner, 1998; 1999).

Apart from the diversity of species of plants, grazed, park-like landscapes are also characterized by a large diversity of species of invertebrates, including insects (Darlington, 1974; Morris, 1974; Tubbs, 1988, pp. 25–26, 157–159; Hondong *et al.*, 1993, pp. 148–164; Alexander, 1998). More than 50% of all the species of insects found in the whole of Great Britain live in the New Forest alone (20,000 hectares) (Tubbs, 1988, pp. 25–26). The New Forest and Windsor Forest are the richest areas in England. The cause of this is the presence of the very old trees in particular (Alexander, 1998). Of all the European species of butterflies, 80% live in a habitat combining grasslands, scrub and groves with mantle vegetation (Bink, 1992, pp. 88, 142, 168–457). The oak has a special place as a host for insects. There is no other species of tree in Europe associated with so many species of insects (Darlington, 1974; Morris, 1974). As we noted above, the oak plays a prominent role in wood-pasture. Furthermore, there is an enormous variety of species of birds in grazed, park-like landscapes (Smith, 1980, pp. 346–347, 369, 388–390; Tubbs, 1988, pp. 159–162; Cramp, 1988; 1992; Hondong *et al.*, 1993, pp. 146–147; Schepers, 1993). These include the nightingale (*Luscinia megarhynchos*), whitethroat (*Sylvia communis*), lesser whitethroat (*Sylvia cuorruca*), garden warbler (*Sylvia borin*), red-backed shrike (*Lanius collurio*), song thrush (*Turdus philomelos*), all the species of woodpeckers, and many birds of prey, including the common buzzard (*Buteo buteo*), goshawk (*Accipiter gentilis*), hobby (*Falco subbuteo*) and imperial eagle (*Aquila heliaca*). Many species of birds, particularly songbirds, are dependent on the combination of grassland, scrub and groves. Grazed, park-like landscapes are even the last places in Europe where the imperial eagle breeds (Cramp, 1980, pp. 226–227; Voous, 1986, pp. 93–95). In their turn, the birds contribute to the diversity in grazed landscapes, as noted above. Apart from the jay, whose role in the establishment of oak was discussed in detail earlier in this chapter, songbirds are particularly important for the establishment of species of plants with fleshy fruits, such as wild fruit trees, hawthorn and blackthorn.

When grazing comes to an end, the diversity of species of plants and animals is in time reduced. As we saw in Chapter 4, the grasslands become overgrown with shade-tolerant trees, and the grasses, plants, shrubs and trees which require light, including both species of oak and hazel, finally disappear. In the first instance, oak and hazel also establish themselves in abandoned grassland. In fact, this is a result of the presence of the grassland on the location, since once a forest has established itself, they cannot survive (Kollmann and Schill, 1996; Bodziarczyk *et al.*, 1999; see Chapters 2 and 4). In grasslands which are not immediately colonized by trees and shrubs, there is also a great reduction in the diversity of grasses and plants (Bakker, 1987; 1989, p. 215; Van Wieren, 1991). When grazing stops, the grasslands eventually disappear, along with the mantle and fringe vegetation, and in their wake, all the species of plants and animals, which are dependent on the transition between, and the combination of grassland and groves. Therefore grazing contributes to diversity to a very great extent.

Because of this enormous diversity in a single interrelated system, wood-pastures are therefore sometimes seen as relics of some of the plant communities which were originally much more widespread. The epiphytic lichen flora in the New Forest in England forms one example of this. These flora are the richest of the whole of the lowlands of Western Europe (Rose, 1974; 1992; Tubbs, 1986, p. 149). There are 278 species, of which two have never been found anywhere else in Great Britain and a total of 312 species of epiphytic bryophytes and lichens, meaning that the New Forest houses the richest number of species of epiphytic flora in Europe (Rose, 1992). The groves richest in these flora contain $130-178$ species km^{-2}. By way of comparison, there is no forest area known in the lowlands of the continental part of Western Europe which contains more than 150 species (Rose and James, 1974, cited by Tubbs, 1986, pp. 148–149). The blackthorn scrub also contains flora of characteristic beard moss (*Usnea* sp.). The large majority of lichens require light, and are found mainly on the fringes of groves (Rose, 1974; 1992). The two species of oak are particularly rich in epiphytes, and up to 150 species of epiphytes can be found on them (Rose, 1974). Because these flora contain species which do not spread easily, and which are found, in so far as it is possible to check this, only in places where there has been a continuous cover of trees, Rose and James (1974) believe that their presence goes back to the Atlantic primeval forest (Rose and James, 1974, cited by Tubbs, 1986, p. 148). The character of the groves in the New Forest as a throwback to these forests could be because there has always been forest there, though not elsewhere (Tubbs, 1985, p. 150). However, in view of the importance of the two species of oak for these flora, and the fact that these epiphytes are mainly found on the periphery of forests, it could also be maintained that this character indicates the historical continuity of a park-like landscape, and the factors responsible for this, i.e., the large herbivores. Rose (1974) came to this conclusion on the basis of the requirements of many of these species, with regard to the habitat, including the fact that they require light. In fact, this indicates that there cannot have been a uniform closed forest there in prehistoric times. Grasslands would have had to be part of the natural vegetation at the time, possibly kept in an open state by the wild herbivores which were there (Rose, 1974).

With regard to the alternative hypothesis, it should be noted, in relation to the above, that the condition of the continuity of a cover of trees in combination with very light conditions is met by the presence of many fringes of forests. Their continued existence is ensured in park-like landscapes because trees are constantly growing on the fringes of the groves. Therefore the regeneration of trees takes place next to the older trees which are already present, and where the lichens grow. The colonization of a new periphery has to take place only over a very small distance. The fact that the colonization of plantations took place in the vicinity of the grazed groves with a rich flora of lichens in the New Forest, shows that this process is taking place.

What was said above about the epiphytic flora also applies for species of plants which serve to indicate the presence of primeval forests. Many of these species from old forests easily colonize new forests from the edges of the old forest (Peterken and Game, 1984). In grazed landscapes, this group of plants is therefore able to accompany the advancing front of the clonally spreading blackthorn where the groves develop. Palynological research has shown that in the course of history, these species of plants colonize open terrain where trees have become established (Day, 1993; Willis; 1993).

6.11   A Park Landscape in the East of the United States?

In previous chapters, data were presented that led to questioning the theory that the original vegetation in Europe and the east of North America was a closed forest. Just as is the case in Europe for the light-demanding pedunculate and sessile oak, in the east of the United States light-demanding species of oak, such as the white oak (*Quercus alba*), the black oak (*Q. velutina*) and the scarlet oak (*Q. coccinea*), were replaced in closed forests by shade-tolerant species. An exception is the light-demanding, fast growing tulip poplar (*Liziodendron tulip-ifera*) that can grow very old. This tree survives well in closed forests, but this is also at the cost of the oaks (Loftis, 1983; Brose, 1999a, b; Brose *et al.*, 1999). The lowlands of Europe have no equivalent of this species of tree. As in Europe, this replacement leads to a situation that contradicts the palynological data. These show that the oak was always part of the originally present, untouched vegetation. So that cannot have been a closed forest. In addition, there are historical descriptions of park-like landscapes, greatly dominated by oaks, in the east of the United States. From those landscapes date the oaks, so-called open grown oaks, that are now in forests surrounded by younger forest-grown trees (Marks, 1942). Park-like landscapes and open grown oaks were and are still found in Europe. Further, there is the fact that grazing by cattle did not obstruct regeneration either in Europe or in the east of the United States. Finally, it was pointed out in this chapter how the need for light, or the tolerance for shade of species of trees and bushes in the lowlands of Central and Western Europe offer a link to the alternative hypothesis, namely that the original vegetation consisted of vegetation of grasslands, scrub and groves, resulting in a park-like landscape. Based on all this data, the question arises: can this alternative hypothesis for the original vegetation in Europe shed some light on the situation in the east of the United States?

In the east of the United States, species occur in the edges of the forests as seedling and young trees that do not have an opportunity to establish themselves in the forest proper (Ranney and Bruner, 1981). These species include the white oak and the black oak. Forest edges even favour the white oak, which can be very numerous there (Ranney and Bruner, 1981; Whitney and Runkle, 1981). For example, in old growth forests there were saplings of white oak in the edges, while at the same time they were missing from the interior of the forest (Whitney and Runkle, 1981). In addition, a large number of the species from the forest do occur in the edges, but the opposite is not true. Some species, including the hawthorn, even appear only in the edges (Ranny and Bruner, 1981). The result is that the composition of species in the edges clearly differs from the interior of the forest (Matlack, 1994). It is striking that the composition of species of an open terrain that is becoming overgrown by forest, is very similar to that of the forest edge. For establishment in open terrain, it is not of decisive importance whether a species is light demanding or shade-tolerant (Myster, 1993). In other words, shade-tolerant species of tree that can establish themselves in a closed forest, can also do so in open terrain, while the opposite is not true for light-demanding species.

From the forest in the direction of the grazed grassland, a transitional zone appears that is characterized by scrub with species of bush like hazel (*Corylus* sp.) and dogwood (*Cornus* sp.) and young trees, including light-demanding species such as white oak, black oak and northern red oak (*Q. rubra*) (Marks, 1942; Whitney and Runkle, 1981). Oaks (including the black oak and the white oak) also occur on open terrain, such as fallow fields and grasslands, including lawns and grasslands grazed by cattle (Scot, 1915; Darley-Hill and Johnson, 1981; Harrison and Werner, 1982; Jokella and Sawtelle, 1985, cited in Crow *et al.*, 1994; Crow *et al.*, 1994), in a number of cases even profusely (Jokella and Sawtelle, 1985; Crow *et al.*, 1994). This is due to the blue jay (*Cyanocitta cristata*) (Darley-Hill and Johnson, 1981; Harrison and Werner, 1982; Johnson and Atkinson, 1985). This jay 'plants' acorns at sites structurally suitable for the colonization of oak, namely in short grassland, edges of forests and in disturbed soil (Fowels, 1965, cited by Darley-Hill and Johnson, 1981; Johnson and Webb, 1989). According to Darley-Hill and Johnson (1981), the distribution of nuts from the forest to edges and open disturbed sites increases the chance of survival of light-demanding species of tree. A blue jay moves up to several thousands of acorns as far as 4 km from the tree where it collected the acorns (Darley-Hill and Johnson, 1981). Mice also move acorns, but only over a distance of a few tens of metres (Johnson and Webb, 1989). The blue jay follows the line-shaped elements in the landscape when it distributes the acorns. The bird itself is a typical so-called edge sort, i.e. prefers to be in areas with transitions from a close vegetation like a forest to open terrain like grassland or fields. In addition, this jay also broods in prairies with few trees (Johnson and Webb, 1989). The blue jay also moves and plants beech nuts (Johnson and Atkinson, 1985). This is a difference with the jay in Europe.

Protection of unpalatable bushes in grassland grazed by cattle leads to survival and growing into a tree for light-demanding species that hardly regenerate in the forest, if at all, like the oak, prairie crab apple (*Malus ieonsis*) and the exotic European apple (*Malus* sp.) (Scot, 1915; Marks, 1942; McCarthy, 1994; Stover and Marks, 1998). For example, there are more than 300 species of hawthorn alone, of which most are thorny (see Britton and Brown, 1947), compared with two in Europe. Furthermore, there are the species which propagate by root stock: smilax and gooseberry (*Ribes* sp.) (see Britton and Brown, 1947). In the east of North America there are, furthermore, a large number of thorny species of bush or other species that animals do not eat. In grazed grasslands these bushes shoot up, such as roses, hawthorns, junipers, gooseberry (*Ribes*), prickly ash (*Zanthoxylum americanum*) and the clonally expanding species like sumach (*Rhus* spp.) (Scot, 1915; Bromley, 1935; Marks, 1942; Den Uyl, 1945; 1962; McAndrews, 1965; Parker *et al.*, 1985; Stover and Marks, 1998). In thorny bushes and sumach scrub, young trees grow, including oak (Scot, 1915; Marks, 1942; Wistendahl, 1975; Werner and Harbeck, 1982). There is even one species of hawthorn called Oakers thorn. In this light, it is striking that in areas in Ohio where there was heavy grazing until 1940, the white oak was present in all year classes (Whitney and Somerlot, 1985). None of the trees and bushes can regenerate in a forest if it is grazed by cattle or other specialized grass-eating ungulates, because no seedlings grow (Den Uyl, 1945; 1962; Steinbrenner, 1951; Ward and Parker, 1989). An image arises of stag-headed trees (Whitney and Somerlot, 1985). That means that thanks to grazing a forest develops into grassland, in which in turn trees can regenerate. If grazing ends in a grazed forest, then under the canopy, seedlings of the shade-tolerant species grow up at once (Marks, 1942). For example, Den Uyl (1962) determined that after grazing ended in a forest with a canopy dominated by oaks, 225,000 to 2,000,000 sugar maple seedlings shot up per hectare. These will then suppress the oaks.

To summarize, it appears that in the east of the United States all components of a park landscape analogous to the wood-pastures in Europe (Barnes, 1991) are present. These are:

- Considerable percentages of pollen in pollen diagrams of oak beside those of shade-tolerant species of tree that currently suppress oaks.
- Historical descriptions of park-like landscape, in which the oak dominated.
- Oaks not regenerating in closed forests, while they do in grazed grassland and in scrub and in mantle and fringe vegetation.
- A species of jay that plants acorns at a distance at sites that are structurally suitable for establishment of oak.
- An indigenous large ungulate that is a specialized grazer analogous to cattle (see Fig. 6.30).

What is missing, is the entire image as it is still present in Europe. Given the similarities in the jigsaw puzzle pieces that make up the image, I nevertheless consider it extremely probable that in the east of the United States, in places where the light-demanding white oak occurred in historic times, there was a landscape analogous to the park landscape in Europe. The analogy of the specialized grazer from Europe, cattle, was the bison in the east of the United States (see Reynolds *et al.*, 1982; Truett, 1996). It should be noted here that this hypothesis holds for areas where the oak occurs, i.e. up

GRAZERS

1 Reedbuck 75 kg
2 Oryx 200 kg
3 Wildebeest 250 kg
4 African buffalo 750 kg
5 Aurochs/cattle 600–700 kg
6 Plains bison 1000 kg

1	10	100	1000 kg

Fig. 6.30. The American bison is a grazer in terms of food selection and feeding strategy, making it ecologically analogous to European cattle (redrawn from Hofmann, 1973; 1976; 1985; Van de Veen and Van Wieren, 1980).

to a height of approximately 1000 m (Barnes, 1991). Furthermore, the areas should be accessible and passable for the specialized grazer, the bison, i.e. flat areas, slightly rolling hill landscape and valleys in mountain areas.

Due to the fact that the east of the United States has a more continental climate than Europe (Barnes, 1991), fire as a result of lightning could have played a role. There are differences of opinion as to whether fire is selective or not. It is important to note that the assumption that fire works selectively comes from the search for a solution for oaks being suppressed by shade-tolerant species.[64] I recall the statement by Abrams and Seischab (1997, p. 374): 'What disturbance factor other than fire could historically have prevented these species from replacing oak?'. My answer to this is: grazing.

In the distribution area of the white oak, a number of light-demanding pines also occur, such as white pine (*Pinus strobus*), ponderosa pine (*P. ponderosa*) and loblolly pine (*P. taeda*). These species often no longer regenerate either (Morey, 1936; Ware, 1970). A hypothesis of grazing for the regeneration of oaks does not contradict the continued existence of these species. On the contrary, it is known that in grazed grasslands, white pine shoot up (Bromley, 1935; Marks, 1942). According to Bromley (1935), there is no regeneration of white pine, unless there is grazing. The specialized grazer, cattle, avoid pines (Bromley, 1935). Grazing even changed mixed forests into pine forests (Whitney and Davis, 1986). Grazing reduces the chance of fire, because the amount of flammable material decreases. As stated earlier in Chapter 2, this increases the establishment of woody species. Nevertheless, grazing does not exclude fire. Like in Europe, fluctuations will have occurred in the numbers of large herbivores in North America, as result of disease or lack of food. In the case of a sudden reduction, areas will have remained ungrazed. Plant material will have remained there, in which fire could occur. Fire does not work selectively among young trees.[65] Several generations of young trees may have been destroyed by fire. That

---

[64] See Smith (1962), Raup (1964), Whittaker and Woodwell (1969), Botkin (1970), Jokela and Sawtelle (1985), Abrams (1992), Abrams (1996), Abrams *et al.* (1995), Abrams and Seischab (1997), Dodge (1997), Ross *et al.* (1986), Brose *et al.* (1999a,b).
[65] Korstian (1927), Whitney and Davis (1987), Johnson (1992), Huddle and Pallardy (1996), Arthur *et al.* (1998), Barnes and Van Lear (1998), Brose *et al.* (1999a,b), Chambers *et al.* (1999).

could explain why stands of the white pine and ponderosa pine have an irregular regeneration (Hough and Forbes, 1943; Covington and Moore, 1994). Fire would in that case occur irregularly and have eliminated a number of generations of young trees.

6.12 Conclusions and Synthesis

This chapter looked at the establishment of trees and shrubs in relation to the important abiotic factor, light, and the biotic factor, grazing. In view of the hypothesis of this study, we looked particularly at pedunculate and sessile oak and hazel. Many of the findings shown below are direct results of experimental research and will not be further elucidated in this section.

- In comparison with the seedlings of beech, hornbeam, broad-leaved and small-leaved lime, and other so-called shade or semi-shade species, pedunculate and sessile oak seedlings require a relatively large amount of daylight to grow.

- Because of the reserve nutrients in the acorn, pedunculate and sessile oak tolerate greatly reduced amounts of daylight relatively well in the first 2 years of life.

When the reserves in the cotyledons have been used up, the seedlings clearly tolerate shade less well. The fact that oak invests in the root system relatively strongly compared with other species of trees, also plays a role. If the light conditions do not improve, they continue to vegetate for some time and disappear within 3–4 years.

- Like oak, hazel seedlings have the property that for the first year, they feed on the reserve nutrients in the cotyledons, and invest relatively strongly in the root system while they grow.

As for oak, these properties can be seen as an indication that hazel also tolerates shade badly after a few years, and eventually disappears from closed forests, as observations in the field have shown.

- Hazel cannot regenerate in closed forests.

Even in relatively open forests, the shrub does not flower, and therefore does not develop pollen and seed, so that it cannot regenerate generatively, and therefore cannot form a permanent shrub layer in the forest. This clearly deviates from the theory that in prehistoric vegetation, hazel formed a shrub layer in the closed forest, because the pollen from this species is well represented in the pollen samples.

> • With low levels of light, beech has a competitive advantage over oak.

Like oak, beech grows best in full daylight. With reduced levels of light, the growth of beech is also reduced, like that of oak, but it is still better than that of oak. In this way, beech can successfully compete against oak at the stage when they are young plants.

> • Broad-leaved lime and small-leaved lime are the most pronounced shade-tolerant species of all the trees.

The seedlings of both species of lime continue to grow steadily with very low light intensities. The seedling of the small-leaved lime has been shown to adapt physiologically to low light intensities. In the small-leaved lime, a certain reduction of daylight appears to be necessary for the seedling to grow taller. In full daylight, the small-leaved lime does not grow taller. Therefore small-leaved lime seedlings require some shade to grow into a tree.

> • Under the closed canopy of a forest, the seedlings of beech, hornbeam and broad-leaved lime and small-leaved lime survive longer than those of pedunculate and sessile oak. This means that there are already a large number of seedlings of these species waiting for light conditions to improve.

Beech, lime and hornbeam grow in such a way that they are able to catch more of the reduced amount of daylight when they are in the shade. Because they also produce fruit nearly every year, they have a twofold advantage over oak with regard to regeneration.

> • The seedlings and young trees of hornbeam and lime, and other so-called shade and semi-shade species, respond to improved light conditions by growing taller more quickly than oak seedlings and young oak trees.

When gaps form in the canopy because trees die, hornbeam, lime, field maple, sycamore, Norway maple, ash and the various species of elm have an extra advantage over oak. This means that when gaps appear in the canopy in a closed forest, the fast-growing seedlings of these species close the gap in a very short space of time. Even if young lime trees are several decades younger than oaks which are already established, they can grow taller than the oaks in a few more decades, and successfully compete against them.

> - If oak and beech seedlings are growing side by side in gaps in the canopy of a closed forest, the oak initially has a head start on the beech. However, this head start is lost within a century, and the oak loses the competition and disappears.

Initially, oak grows better than beech in gaps in the canopy, and for a long time also remains taller. However, in less than a century, oaks are surpassed by beech trees of the same age, and then the oaks die because of the shade cast by the beech. Both experiments and spontaneous developments in forest reserves show that beech always surpasses and replaces oak. This applies both for beech of the same age, and for beech which grows later under the canopy of the oaks. An exception is soil where the beech grows poorly and therefore does not become high enough from below the oak to pass it and suppress it.

> - Even in ungrazed open terrain, where the light conditions for oak are most favourable, compared with the other species of trees described, oak cannot survive because of the properties of these other species.

Both experiments and the spontaneous developments in forest reserves show that species such as beech, lime, hornbeam etc., can become established under an oak canopy, and eventually successfully compete against the oak. Therefore, for oak to survive, certain factors are needed to compensate for the disadvantageous competitive position of oak in relation to other species.

> - Oak successfully regenerates and survives in grazed, park-like landscapes and is even dominant in this environment.

Clearly, the competitive disadvantage of oak in relation to other species of trees has been removed in these situations. Grazing produces an open terrain which is a good environment for oak to grow in. It also provides an environment for thorny scrub to become established, where oak seedlings can grow, protected from browsing animals. However, where there is grazing, this thorny scrub also forms a good environment for the establishment and growth of other species which compete with oak. These species also all regenerate in grazed, park-like landscapes.

However, in relation to these competitive species, oak has the advantage, as regards its establishment, that it has a unique dispersal mechanism in the form of the jay. Jays plant large numbers of acorns on the fringes of thorny scrub, the best situation for oak to grow. As oak is less sensitive to the nibbling of rodents and late frosts than its competitors, it has a competitive advantage over these species in these fringes. In addition, on the fringes of thorny scrub, oak has the advantage that it grows taller more quickly than its competitors in the first years, in full daylight. Because of the jay, oak also has the advantage in terms of numbers because the competing species do not have a comparable facilitator. Their dispersal takes place randomly by the wind, over very short distances, compared with oak.

When oaks have grown into groves from the thorny scrub, and the thorny scrub has disappeared under the shade of the canopy, they continue to retain their advantage in terms of numbers, because of the presence of large herbivores. This is because seedlings of other competitive species of trees growing under the oaks disappear when they are eaten and trampled by large herbivores. Therefore in groves, oaks do not always lose out in the competition and disappear. The oak survives there with shade-tolerant species.

> * Hazel regenerates well and survives in grazed, park-like landscapes. In comparison with oak, the regeneration is largely restricted to the fringes of the mantle vegetation of the groves and the grassland in the immediate vicinity.

The analogy between oak and hazel also seems to apply as regards their establishment. Like oak, hazel seems to have its own facilitator in Central and Western Europe in the form of the nuthatch. Given the ecology of the nuthatch, this also seems to explain why hazel becomes established mainly on the fringes, and much less at great distances in the open field.

> * In contrast with the closed forest, where herbivores do not play a significant role, all species of trees and hazel shown to be present after the last Ice Age by palynological studies, can regenerate and survive in grazed, park-like landscapes, where large herbivores play a steering role in the succession.

The diversity of park-like landscapes, with large herbivorous mammals, can be traced back to the end of the Ice Age, as regards the establishment of species which can be shown to have been present by palynological studies. The advantage of trees and shrubs that were spread by birds, is also shown by the much more rapid colonization of these species, from their refuges in the Ice Age, towards the north, compared with species of which the seeds are transported by the wind. The dispersal and establishment of species, both as regards their colonization after the Ice Age, and as regards regeneration, not only depend on facilitating factors such as birds and grazing by large herbivores, but are also strongly influenced by fluctuations in the numbers of herbivores as a result of high mortality rates caused by disease or lack of food.

To summarize, it may be said that the data collected in this chapter on the ecology of different species of trees and shrubs, and the findings of the previous chapters, lead to the conclusion that the original vegetation must have been a park-like landscape. In this park-like landscape, the succession of species of trees was governed by large herbivorous mammals and species of birds, such as the jay, which acted as facilitators for certain species of trees. It was a landscape characterized by a large diversity of species of plants and animals. Some of these species have survived since prehistoric times in wood-pasture, even after the disappearance of the original large herbivores, because livestock, and the way in which it was kept, was a modern analogy of the original, mainly grass-eating herbivorous species, namely aurochs and tarpan living together with the remaining species of wild types of ungulates mentioned. This analogy certainly applied in Central and Western Europe up to the end of the 18th century, even after aurochs and tarpan had disappeared as a result of the competition with their domesticated descendants in particular. In the course of the centuries, farming has developed in a way that is less and less like the natural processes. As a result, the diversity of species which were naturally present in a single interrelated system, together with large herbivorous mammals eventually became fragmented and divided over different types of agricultural areas.

- Based on the conditions under which light-demanding oaks established themselves and grew in the east of the United States, the hypothesis that the original vegetation was a park-like landscape in which large herbivores had a leading role is also applicable as an alternative for the hypothesis that the original vegetation was a closed forest.

Given the large number of similarities in the way light-demanding oak establish themselves and grow in Europe and in the east of the United States, it is extremely probable that in the east of the United States in the natural conditions, there was also regeneration of trees in grazed grasslands in analogy to the European park-like landscapes.

- Fire did not play a decisive role in the regeneration and permanent presence of oak.

Due to a more continental climate in the east of the United States, fire probably occurred more often in natural conditions than in the lowlands of Central and Western Europe. Given the fact that fire is barely selective, it was not the determining factor for the continued existence of oaks in the presence of shade-tolerant species of tree and the light-demanding tulip poplar.

Final Synthesis and Conclusions   7

7.1 The Null Hypothesis and the Alternative Hypothesis

In this concluding chapter, I summarize in stages the conclusions and synthe-
ses formulated in the preceding chapters in relation to the central hypothesis
formulated in the introduction, and the null hypothesis and alternative hypothesis formulated alongside this. This chapter concludes with the synthesis of the
findings of the various chapters and draws the final conclusions on that basis.
I will start with Europe. At the end I will present the conclusions for the east of
the United States, which are partly based on the conclusions for Europe.

In the organization of this study of the literature, I looked first of all at the
current theories on succession in the vegetation of the lowlands of Central and
Western Europe. I included the vegetation in the east of the United States here,
because this is considered to be an analogy of the vegetation in Europe. At the
same time, I examined whether large herbivores were assigned a role in these
theories. Subsequently, the research results from the palynology of prehistoric
vegetations and the influence of humans on these was related to this. The next
building block was to investigate how humans influenced the wilderness in
Europe since the early Middle Ages by the use made of it. This concerns, in particular, the exploitation of wood and the grazing of livestock. The first synthesis
produced by these building blocks was compared with the spontaneous development of the vegetation in 'modern' forest reserves without any human intervention, such as the exploitation of wood and the grazing of livestock. This was
examined particularly in relation to the possible influence of large herbivores
on the developments of the vegetation, especially the influence of specialized
grass eaters. This evaluation produced additional building blocks for the synthesis. In the final step of the study, the whole range of findings was examined
from the perspective of the autecology of the most important species of plants,
the relationships between these and the influence, in particular, of grazing by
specialized grass eaters. At one stage in the study, the findings from autecology
and the relationship with herbivores were compared for consistency with the
facts that are known about the establishment of species of trees and the development of the vegetation in prehistoric times.

The *null hypothesis* in this study is that pedunculate and sessile oak and hazel survive in a closed forest and regenerate in gaps in the canopy in accordance with Watt's gap phase model (1947) and Leibundgut's cyclical model (1959; 1978). Large herbivores present in the natural state are dependent on the developments of the vegetation. According to this hypothesis, they do not have an influence on the course of the succession and regeneration of forests.

The *alternative hypothesis* is that the natural vegetation consists of a mosaic of large and small grasslands, scrub, solitary trees and trees growing in groups (groves), in which the indigenous fauna of large herbivores is essential for the regeneration of species of trees and shrubs which are characteristic in Europe. According to this hypothesis, wood-pasture should be seen as the closest modern analogy of this landscape.

7.2 The Findings

The study of existing theories produced the following findings:

- The hypothesis that grazing leads to a retrogressive succession from forest to grassland must be rejected.
- The assumption that wild herbivores do not have a decisive influence on the succession in the original vegetation is based on observations of succession *without* large wild herbivores. As there were large herbivores present in the original system, this assumption is based on a circular argument.
- The assumption that wild herbivores do not have a decisive influence on the succession in the original vegetation leads to the view that where there are wild animals, they can only occur in low densities, because they do not have an influence on the vegetation. This assumption is, in itself, another circular argument.
- The assumption that virgin mountain forests can be seen as an analogy of the original vegetation of the lowlands of Central and Western Europe is incorrect, because these mountain forests are at altitudes where the characteristic species of oaks found in the lowlands cannot grow and there are no large herbivores (aurochs and tarpan) (which might be) relevant for the vegetation, which were originally present in the lowlands.

The research into the theories in palynology and the results of pollen analyses produced the following findings:

- Palynology extrapolated the prevailing theories and concepts back to prehistoric times, so that data from prehistoric times were interpreted as a confirmation of the prevailing theories. In this way, palynology added a new circular argument to the circular argument mentioned above.
- The decline in the relative frequency of tree pollen in the pollen diagrams is explained in palynology as being the result of the introduction of farming (the 'Landnam theory'). By clearing open spaces in the forest and then graz-

ing livestock, the forest was believed to have become more and more open. For palynologists, this demonstrated that grazing livestock leads to a retrogressive succession from forest to grassland and heathland, once again confirming the prevailing theory.

- The high relative percentages of hazel pollen in sediments in Central and Western Europe are consistently left out of the total sum of tree pollen or added to the percentage of the trees by palynologists. This strongly contributes to the view that the original vegetation was a closed forest. When the hazel pollen is presented together with the pollen grains of grasses and herbs as representatives of open terrain in a single sum, this leads to quite a different picture. After including hazel pollen in this way in the total sum, the pollen spectrum can be clearly explained by a grazed park-like landscape. In a certain sense, the pollen grains from the oak should be considered a separate category. On the one hand, it is a species whose permanent presence indicates open terrain, because oaks only grow successfully into trees in open conditions. On the other hand, as a blooming tree it can be part of groves, although individual trees are also possible.
- Theories on forest as the original vegetation have been extrapolated to the past based on pollen analyses, after which the subsequent image of the past served as proof of the validity of these theories.
- The results of research into modern pollen sedimentation, in landscapes varying from entirely covered with forest to very open, show that the percentage of non-arboreal pollen (NAP) in pollen spectra cannot simply be taken as a linear measure for the actual openness. If they are interpreted using current criteria, very open landscapes appear in these spectra as closed forest.
- In terms of the species diversity and the relative representation of tree pollen, modern pollen spectra of park-like landscapes grazed by large herbivores reveal great similarities with the pollen spectra of prehistoric times, which are interpreted as indicating a closed forest.

The study of written sources on the use of the wilderness by humans since the early Middle Ages, and the changes which have resulted from this produced the following findings:

- The meaning of the terms 'silva', 'forest', 'forêt', 'Forst', 'Wald', 'woud', 'bos', 'Busch', 'wood' as derived from the context in which these terms were used in the Middle Ages, is that they did *not* have the meaning of a closed forest at that time. All these terms related to the uncultivated wilderness which consisted, according to the whole range of uses, of a mosaic of grasslands, scrub, solitary trees and groves.
- The texts of the regulations on the grazing of livestock dating from the Middle Ages and afterwards show that the regeneration of trees took place *with* grazing. The grazing of livestock was not regulated for the benefit of generative regeneration of trees *in* the forest but only for the benefit of the vegetative regeneration of scrub. The livestock became a threat to the young trees only

when humans exposed the seedlings of trees growing in scrub to large browsing herbivores when the thorny scrub was cut down to collect firewood.

- In the 18th century, a change in the demand for wood led to the development
  of the tree forest where the aim was the regeneration of trees in the forest
  itself. Large areas of closed tree forest did not actually develop until after
  1700. Grazing livestock in this tree forest was harmful because the livestock
  destroyed the seedlings of the trees in the forest, and therefore made the generative regeneration of trees impossible in the forest. As the tree forest is seen
  in the prevailing theory as a reference to the natural vegetation, the grazing
  of livestock is considered harmful for the survival of the original forest.
- The arguments derived by plant geographers and ecologists, such as Moss,
  Tansley and Watt, and the palynologist Iversen, from forestry and historical
  texts, showing that the grazing of livestock was traditionally seen as a threat
  to the original forest, are incorrect.
- Neither the type of forest that developed in the 18th century, nor the way in
  which this forest was regenerated, can be seen as analogies of the vegetation
  and regeneration of trees present at the time when the first regulations on
  grazing livestock were issued in the Middle Ages.
- The historical data do not show that forest disappears as a result of the grazing of livestock. They do show that the grazing of livestock leads to the development of groves. Therefore these data do not provide any arguments for the
  theory of retrogressive succession, nor can they be used as arguments for the
  theory that the original vegetation was a closed forest.

The following findings were produced with regard to the spontaneous development of the vegetation in forest reserves:

- There are no new young generations of oak trees in any of the forest reserves
  that were studied. This means that the oak is becoming extinct in forest
  reserves. Old oaks die because they become overgrown by competing species.
  The rate at which oaks disappear is not determined by the age of the oaks,
  but by the time other species need to grow taller than the oaks. This usually
  happens in a period of less than a century.
- The old oaks all date from the time that the forest reserves concerned were
  grazed. The absence of the regeneration of oak in the forest reserves must be
  attributed to the fact that after grazing came to an end, they all developed
  into closed forest. A significant reduction in the density of herbivores such as
  red deer, roe deer, elk and European bison, i.e. non-specialized grass-eaters,
  does not lead to the regeneration of oak, but does result in the regeneration
  of competing species.
- The developments in the National Park of Białowieza provide a good insight
  into the relationship between grass-eating herbivores and the regeneration
  of trees and shrubs because the original fauna of large grass-eating herbivores, aurochs and tarpans, were present there until the 16th century and
  were then replaced by grass-eating domestic livestock.
- The hypothesis that the thermophile oak forests in the forest of Białowieza

developed as a result of human intervention in the lime–hornbeam forest, after settlements were established, should be reversed; in fact, the settlements were established in places where the thermophile oak forests had become established naturally. The lime–hornbeam forests developed from the thermophile oak forest after grazing by specialized grass-eaters came to an end.

- Spontaneous developments in the forest reserves and pollen analyses show that closed forest, as it has developed in forest reserves, cannot be viewed as a modern analogy of the prehistoric vegetation.
- The composition of species in former wood-pastures that are now forest reserves and where the light-demanding oak and hazel occur beside shade-tolerant species such as lime, elm, beech and hornbeam, is incorrectly seen as a modern analogy of the untouched prehistoric vegetation based on pollen diagrams. The combination of light-demanding and shade-tolerant species in the current vegetation is a result of the history of the forests. The light-demanding species date from the time of the use as coppice-with-standards and/or wood-pasture; the shade-tolerant species from that period and the period after this use ended. The shade-tolerant species mainly appeared then.
- Correction factors used to modify the representation of the various species of tree in pollen samples are based on the presence of tree species in forest reserves that are considered to be a modern analogy of the original vegetation. These correction factors are based on an incorrect premise and lead to circular reasoning.
- The results of the study of the spontaneous developments in forest reserves show that there is no regeneration of pedunculate and sessile oak anywhere. Oak disappears from a closed forest in one or just a few generations as a result of the lack of regeneration and the death of old oaks which become overgrown by other species of trees. The same applies to hazel.
- Spontaneous developments in the forest reserves which are seen as natural processes do not, in any of the cases studied, lead to a situation which corresponds with the palynological facts, which provide the only indication of the composition of the original natural vegetation.

The study of the autecology of the most important species of trees and hazel, and an evaluation of its consistency with the known facts about the establishment of species of trees and the development of the vegetation in prehistoric times produced the following findings:

- Pedunculate and sessile oak seedlings require relatively high levels of daylight to grow in comparison with the seedlings of beech, hornbeam, broadleaved and small-leaved lime and other so-called shade or semi-shade species.
- Because of the reserve nutrients in the acorn, pedunculate and sessile oak tolerate greatly reduced amounts of daylight relatively well for the first 2 years of their life.
- In common with oak, hazel seedlings have the property that they survive for

the first year on the reserve nutrients in the cotyledons and invest relatively strongly in their root system during growth.

- Under the closed canopy of a forest the seedlings of beech, hornbeam and broad-leaved and small-leaved lime survive longer than pedunculate and sessile oak seedlings. This means that there are a large number of seedlings of the former species waiting for the light conditions to improve. The seedlings and young trees of hornbeam and lime and other so-called shade and semi-shade species respond by growing taller more quickly than oak seedlings when the light conditions improve.

- When oak and beech seedlings grow together in the gaps in the canopy of a closed forest, the oak initially has a head start over the beech. However, this head start is caught up within a century and the oak disappears as a result of the competition. Even in ungrazed open terrain where the light conditions are the most favourable for oak in comparison with the other species described, oak will not be able to survive because of the capacity of the other species to grow under the oaks and then replace them.

- Oak appears to regenerate successfully and survive in grazed park-like landscapes and is even dominant in this landscape. Hazel also regenerates well and survives in grazed park-like landscapes. In contrast with oak, the regeneration of this species is largely restricted to the fringes of the mantle vegetation of groves, and the grassland in the immediate vicinity.

- All the species of trees and hazel which were shown to be present without interruption after the end of the last Ice Age by palynological studies, regenerate and survive in grazed park-like landscapes where large herbivores have a determining effect on the succession; this contrasts with the situation in closed forests where herbivores do not play a significant role, and where there is *no* regeneration of the two species of oak and hazel.

Research into the situation in the east of the United States has led to the following conclusions. These conclusions are based in part on the above conclusions about the lowlands of Central and Western Europe. A number of these conclusions are practically identical. The reason for this is that the same starting points and premises were used for both these areas when defining the theories on the original vegetation. For the sake of clarity, identical conclusions are repeated here.

- Forests in the east of the United States that develop spontaneously cannot be considered an analogy of the original vegetation in the lowlands of Central and Western Europe.

- The assumption that wild herbivores in the original forest vegetation did not influence the succession is based on research into spontaneous succession of vegetation on old fields and abandoned grassland, without involving the original specialized grazer, the bison, in this research.

- The assumption that grazing by cattle introduced by colonists led to a retrogressive succession of the original forest to open park landscape because cattle prevent the regeneration of trees, is in general incorrect.

- Analogous to the situation in Europe, regeneration of trees occurs in open grassland when there is grazing by cattle. Seedlings grow to trees protected by thorny bushes. It is very probable that this process also took place in the case of grazing by bison, since both species can be considered to be ecologically analogous.
- Based on palynological research, the theories about forest as original vegetation are extrapolated to the past, after which the subsequent image of the past is introduced as evidence for the validity of these theories.
- The results of research into modern pollen sedimentation in landscapes varying from entirely covered with forest to very open show that the percentage of NAP in pollen spectra cannot simply be seen as a linear measure for the actual openness.
- Colonists from Western and Central Europe took their mental models with them to the east of North America. Terms like 'forest', 'forêt', 'Forst', 'Wald', 'bosch' and 'woud' in historical texts must therefore be interpreted in the light of their European meaning at that time. That means that they did not mean forest in the modern sense of the word.
- European colonists were very familiar with park-like landscapes through their European background. Historical descriptions that place such landscapes in the east of the United States should therefore be considered reliable.
- In the 18th and 19th centuries, the images from forestry that appeared in Europe on the effect of cattle on the regeneration of trees will also have been transmitted to the United States.
- After the bison or the cattle introduced by the Europeans had disappeared, closed forest developed on a large scale without these specialized grazers.
- Analogous to Europe, there is no regeneration of light-demanding oaks in the closed forest via the gap phase model nor due to large-scale windthrow. Here, light-demanding oaks are suppressed by shade-tolerant species.
- Based on the conditions in which oaks regenerate in the east of the United States, namely in grazed grassland, in scrub in grazed grassland, in mantle and fringe vegetation of a forest bordering on grazed grassland and in the edges of forests, it is likely that there were park-like landscapes in which the process of regeneration of trees took place analogous to the European park-like landscapes. The establishment of oak in these vegetations can be explained here by the fact that blue jay buried acorns.
- Old oaks in the east of the United States date from the time when a specialized grazer, the bison, was present. A relationship between grazing by bison and the presence of oaks is most probable, analogous to the relationship between grazing by cattle and the presence of oaks in Europe.
- Due to a more continental climate, fire may have occurred more often in the east of the United States than in Europe. The role ascribed to fire in relationship to the regeneration of the light-demanding oak in the presence of shade-tolerant species of tree is mainly a result of not considering grazing as an explanatory model.
- The fact that fire mainly affects young trees and is not very selective in terms

of species, means that in the presence of shade-tolerant species fire is not advantageous for the light-demanding oak. However, grazing, open grassland in combination with groves and the presence of the blue jay are advantageous for the light-demanding oak. This forms a simple explanation for the permanent presence of oaks beside these species without fire during the Holocene.

- The occurrence of fire is related to the lack of specialized grazers, since the lack of grazing leads to the accumulation of flammable material. The occurrence of fire in historical times may therefore also have been the result of a specialized grazer like the bison being wiped out locally.
- Fire will be able to destroy generations of young trees. Naturally open landscape could have become more open in this way.
- The burning of so-called underwood by native Americans involved the burning of bushes and scrub in grassland and the mantle and fringe vegetation of groves bordering open grassland. This opened up the landscape and increased the accessibility and views of wild animals. It would also have prevented or slowed down the regeneration of trees. The burning of the native Americans therefore would gradually have opened up a landscape further than was already naturally opened through grazing and fire. Cutting firewood outside the villages would have contributed to openness in their vicinity too.

7.3 Final Conclusion

The synthesis of the findings leads to the conclusion that the original vegetation in the lowlands of Europe is a park-like landscape where the succession of species of trees is determined by large herbivorous mammals and birds such as the jay, which act as facilitators for certain species of trees. As a result of the grazing of specialized grass-eaters, such as wild cattle and wild horses, grasslands develop where thorny shrubs become established, eventually evolving into thorny scrub. The rooting of wild boar also contributes to the establishment of the thorny shrubs. Subsequently, seedlings of all the species of trees and all the other species of shrubs also grow on the fringes of this thorny scrub. These are then protected against the browsing of the large herbivores by the thorny scrub. Birds, such as the jay and the nuthatch, play an important role in the establishment of oak and hazel. They respectively collect acorns and hazel nuts and bury them on the fringes of the scrub and in the grassland. Eventually in time only those trees will survive that are protected by thorny or spiky bushes or plants that are otherwise not attractive (i.e. inedible) for the large ungulates. The crowns of the trees which grow out of the thorny scrub may join together, resulting in a grove. The grove advances into the grassland as the thorny scrub advances. Blackthorn does this concentrically by means of underground suckers. If the protective bush or plant does not expand clonally, like the hawthorn, then a single or a few trees can grow in this way (see Fig. 4.3). In the grove, there is no further regeneration of trees because of the shade of the canopy and because of the browsing and trampling by the large herbivores. As no new generations of trees grow, the forest (grove) eventually degenerates into grassland. This process of natural degeneration may be accelerated by processes such as the stripping of the trees by large herbivores such as European bison, and by 'catastrophes', such as droughts and storms, which result in the closed forest turning into large areas of open terrain (see Figs 4.11 and 6.28). In this way, increasingly large areas of grassland develop *in* the grove. Eventually the grove changes to open grassland (see Fig. 6.29). The process of the establishment of thorny scrub and trees then starts all over again in these open grasslands.

Thus, at a certain point in an area there is grassland first, followed by thorny scrub or other unattractive (i.e. inedible) species of plant, then forest (grove) and finally back to grassland. The system described is a non-linear system (C. Geerling, Driebergen, 1998, personal communication). Grazing is dominant in the system. It consists of three modules. Each module is in itself the result of an irreversible development brought about in the system by grazing. The first module is the grassland, where as a result of grazing by specialized grazers, bushes shoot up in which trees can grow protected from being eaten (see Figs 3.6, 4.2 and 4.5). The grazers cannot stop this formation of forest. On the contrary, they facilitate it by offering bushes and trees places to establish themselves, including the oak with the jay as vector. The second module is the formed grove, in which the bushes, in which the trees grew, disappear as a result of the shade of the canopy. Due to the presence of the large ungulates, there is no more regeneration of trees and bushes. A grove arises that only has a canopy storey (see Fig. 4.9). The composition of the canopy does not change, because shade-tolerant species do not get a chance to establish themselves below the canopy and penetrate it, because of the large ungulates. Light-demanding trees in the canopy, such as oak, therefore cannot be replaced by shade-tolerant species. In the groves, the trees in the centre are the oldest, because they established themselves first in the scrub that was present at the time, which subsequently expanded concentrically. The trees become younger towards the edges of the grove, up to the mantle and fringe vegetation that is still in full daylight where the youngest generations are present. Regeneration only takes place there. The composition of species in the grove is therefore determined in the edges of the forest, in the mantle and fringe vegetation outside the forest (grove) (see Whitney and Runkle, 1981) where all species of tree grow successfully in full daylight.

The third module is where the canopy in the centre of the grove becomes more and more open, due to trees decaying through age, possibly in combination with storms, drought and fungal damage, without being replaced by shade-tolerant trees. As a result of the increased openness, more light reaches the ground, so that grasses and herbs can establish themselves. The grasses and herbs in turn attract the specialized grazers among the large ungulates. Due to the lack of protective bushes, there is no successful establishment of young trees as a result of this grazing (see Figs 4.9, 4.10, 4.11 and 6.28). In this way, the grove changes to open grassland over time. Eventually, the surface area of the grassland becomes so large that light-demanding thorny bushes establish themselves there again and young trees can grow in among them (Fig. 6.29). This closes the cycle. At a certain point in time, all the stages of this cycle of succession are present in one place in a large area. Therefore all the biotopes are always present, though not always in the same place. I term this theory about the natural vegetation in Central and Western Europe and the processes responsible for maintaining it: *the theory of the cyclical turnover of vegetations.*

This non-linear succession interrelates two well-known phenomena related to grazing in a single system. The first phenomenon is that scrub shoots up and advances in grassland grazed by specialized grazers like cattle. Subsequently, trees appear here that form a forest. The prevailing theory says that the forest returns as a result of reduced grazing. The second phenomenon is that there is no regeneration of trees in the forest if there is grazing, because no seedlings grow. The prevailing theory says that the forest then degrades to grassland, in other words becomes grassland via retrogressive succession.

The landscape that developed in this way in the wilderness was characterized by a large diversity of species of plants and animals. Some of these species have survived since prehistoric times, even after the disappearance of the original large herbivores in wood-pasture, because livestock and the way in which it was kept was comparable to the original wild animals, mainly grass-eating herbivorous species, like the aurochs and the tarpan. This analogy certainly applied in Central and Western Europe up to the 18th century, when the aurochs and tarpan disappeared as a result of competition with livestock. However, over the centuries, farming practice has developed in a way that is increasingly different from the natural processes. The species diversity which was naturally found in a single interrelated system with large plant-eating mammals therefore gradually became fragmented and distributed in all sorts of different types of agricultural land.

In view of the above, I conclude that for the situation in the lowlands of Central and Western Europe, the null hypothesis must be rejected in favour of the alternative hypothesis, that the natural vegetation is a mosaic of grasslands, scrub, trees and groves in which large plant-eating mammals play an essential role in the process of the regeneration of trees and have a determining effect on the succession of species of trees. In addition, there was closed forest in certain places, namely those inaccessible to the large ungulates. There are few to no data available on the processes in park-like situations in the east of the United States. Nevertheless, there are various pieces of jigsaw puzzle that correspond to pieces from the image in Europe, which together form an image of a park-like landscape with large ungulates as keystone species in the subsistence of the landscape. Based on these similarities, I believe that in the east of the United States, in places where light-demanding oak traditionally occur and that were accessible for specialized grazers like the bison, the original vegetation was a mosaic of grasslands, scrub, forests and groves. Compared with the lowlands of Central and Western Europe, the landscape could have been more open due to fire.

7.4 Epilogue

The rejection of the null hypothesis in favour of the alternative hypothesis has consequences for the conservation of nature, because the frame of reference that is used for this is based on the null hypothesis. As the alternative hypothesis indicates, species diversity in Europe is *not* a result of the introduction of farming. The species diversity is the result of natural processes which were responsible for a large diversity of biotopes and landscapes. Therefore, farming has not led to the creation of new biotopes, such as grasslands and scrub; nor has it led to the creation, for example, of mosaic landscapes, as nature conservationists tend to assume.

As discussed in previous chapters, cultivating nature, for example through agriculture and forestry, means changing nature. Agriculture has existed for 10,000 years; forestry for only 200 years. Compared with nature, agriculture and forestry are recent phenomena. Agriculture and forestry mean producing food and other products or wood using a limited number of plant and animal species. This is accompanied by deliberate changing of the composition of species of the natural system. In this way, humans have even taken some plant and animal species from their natural environment. They have been *domesticated*, from the Latin *domesticus*, meaning belonging to the house (*domus*). This happened to only a few of the indigenous species of mammal in the lowlands of Central and Western Europe, such as wild cattle (aurochs), wild pig (wild boar) and wild horse (tarpan). They have now become domesticated cattle, pigs and horses. In addition, humans have brought a number of non-indigenous species to Europe during the last 10,000 years, such as sheep and goats, derived from the wild sheep (probably *Ovis orientalis* and *Ovis vignei*) and the Bezoar goat (*Capra aegagrus*) from the Middle East, where agriculture, which spread across Europe between 10,000 and 5000 BP, was invented. Annual grasses that form our crops, the grains, also come from there (see Davis, 1987, pp. 126–168).

Agriculture is selection. All over the world, during the last 10,000–12,000 years, about 40 species of the estimated total of approximately 50,000 naturally occurring species of birds and mammals have been domesticated: that is 0.08%! Of the more than 50,000 edible wild plant species in the world, a few hundred have been selected as food plants. Only 15 species of crop (0.03%) provide 90% of the global energy intake. Two-thirds of the energy intake comes from three species of crop (0.006%), namely rice, maize and wheat. For the last 10,000 years, these domesticated species of animals and plants have been favoured over wild species that were *not* selected. This is also true for certain species of tree in forestry, such as the beech. During the past thousands of years, an enormous area has been cleared using the plough, the axe and fire for the small number of selected species, at the cost of the space for the wild forms of the selected, domesticated species, as well as of the species that were *not* selected and *not* domesticated.

The consequence of domestication is that two forms appear of one species: the domesticated form and the wild form. Both need the same living area; both are susceptible to the same diseases. The difference is that the wild form can usually resist disease better than the domesticated form. The result is that everywhere in the world the wild form was and is considered a competitor and potential danger for the domesticated form. In Europe, for example, the wild form of the domesticated pig, the wild boar, is seen as a potential danger for domesticated pigs, because the wild boar can be a carrier of the virus that causes swine fever. As a result of competition for food and space, the wild forms of the domesticated species have either become extinct or brought close to extinction. In Europe, the aurochs became extinct due to suppression by livestock and by hunting. In 1627, the last specimen, a cow, died in Poland (Szafer, 1968). The last tarpan was captured around 1860 in the Ukraine and died in 1887 in Moscow Zoo (Pruski, 1963; Volf, 1979; Diamond, 1989; Veresshchagin and Baryshnikov, 1989). Elsewhere in the world, wild forms of domesticated species have become extinct as well, such as the wild dromedary (*Camelus ferus*) and the wild donkey (*Equus asinus*), while other species are under serious threat, such as the banteng (*Bos javanicus*) in South-east Asia, the wild form of Bali cattle, and the wild camel (*Camelus bactrianus*). Of these, approximately 800 still live in the Gobi desert, on both sides of the China–Mongolia border.

Because the biotope of the approximately 40 species of mammal and bird selected for agriculture can never cover the entire range of the other 49,960 *non*-domesticated species, cultivation has been at the cost of biodiversity. A number of *non*-domesticated species of plant initially survived in the wilderness put to use for livestock pastures and timber, the 'forestis', 'Wald', 'wold' or 'weald'. As discussed in Chapters 4 and 6, livestock in Europe functioned as modern analogies of their wild ancestors until after the Middle Ages and thereby kept situations in place for a long time that were analogous to the original situation. The originally present flora and fauna could continue to exist there, with the exception of a number of large ungulates that were seen as competitors of the livestock and large predators like the wolf, which were seen as a threat to the livestock, sheep in particular. There are still areas used for agriculture that are very rich in species of wild flora and fauna. In addition to the wood-pastures, which have already been discussed in detail, this is also true for other areas where integrated trees and grassland appear, such as the dehesas in Spain and Portugal, for example (see Bangs, 1985; Joffre *et al.*, 1988; Lieckfeld, 1991, pp. 16–36; Beaufoy *et al.*, 1995; Rackham, 1998). However, as was shown in history, humans have always followed the road of specialization and are still doing so. In the Middle Ages, the mantle and fringe vegetation of groves in the wood-pasture were fixed in place and turned into pieces of coppice with a regulated felling cycle. In the 18th and 19th centuries, for the sake of wood production, the wood-pasture was split into pastures to feed the livestock and forests in the modern sense of the word with only trees for wood production. A few species of wild ungulates were tolerated, but in unnaturally low densities, because they were not allowed to damage the timber production and the regeneration of trees *in* the forest, as it has been applied in the lowlands of Central and Western Europe since the 18th and 19th centuries. Due to this, the 'forestes' became closed forests (see Fig. 4.19), in which there has been selection of a species and growth patterns (tall and straight) of trees for the sake of timber production and partly for this reason more and more exotic species have appeared. The mantle and fringe vegetations disappeared. Thorny hedges, which initially served as a separator between pastures, could continue to function as analogies of thorny scrub for species of bird and plant. However, in the past century the hedges have been replaced by barbed wire.

There are still species of wild flora and fauna that survive in land cultivated for agriculture, because the conditions there are similar to those where they naturally occur. The number of species of wild flora and fauna within the context of agriculture is, however, decreasing further and further. The cause is that the pressure of selection is also present in the biotope of the selected domesticated species, such as the pastures for domestic cows, where fertilization, dividing and sowing have focused on a few productive species of grass, such as perennial ryegrass (*Lolium perenne*). Between 1932 and 1984, the surface area of species-rich rough grassland decreased on livestock farms in England and Wales from 7.2 million ha to 0.6 million ha. That is a reduction of more than 90%. Decreases of a similar or even greater order are known of in countries in the European Union (Wolkinger and Plank, 1981; Baldock, 1990; Anonymous, 1996; Bignal and McCracken, 1996). To summarize briefly: agriculture concentrates on a very small number of species, at the cost of the large majority (Scherf, 1995). Agriculture and forestry create a limited monster from the total of biodiversity.

Selection has continued within the species selected for agriculture. For example, several species of cattle that produce more beef and milk replace species bred less or not at all for these characteristics. This threatens more than half of all European livestock species with extinction (Scherf, 1995). Within the highly productive species, there is further selection on certain characteristics, using artificial insemination and embryo transplantation. This makes an ever smaller number of animals the genetic basis for all livestock. For example, in the Netherlands in 1996, 80–90% of the dairy cattle were descendants of 60–80 bulls. Each bull had an average of between 45,000 and 67,500 female descendants (Anonymous, 1996). That figure is an average and large deviations are possible. For example, one bull, Sunny Boy, has approximately 590,000 female descendants. There is an equally extreme narrowing of the genetic basis of plants. For example, since 1900 75% of the genetic variety in crops has disappeared and the number of varieties of rice in India has decreased from 30,000 to 30–50. The result is an enormous loss of genetic material (Loftas, 1995).

Therefore, agriculture strives for uniformity at the cost of biodiversity. An ever smaller area remains for species of wild flora and fauna that are not used or no longer used. Agriculture competes with diversity, with nature, for space. The pressure of selection by agriculture on nature is present all over the world and leads to simplification and depletion of natural ecosystems everywhere.

The image of nature has changed simultaneously with the change in nature in the lowlands of Central and Western Europe. The reason is that there is no more untouched nature that can serve as a baseline. In Europe, people have become separated from their calibration point of nature. Species of wild flora and fauna remain only within the context of cultivated land and there are theories on where they come from. Due to the lack of a baseline and an incorrect theory, cultivated landscapes are seen as the highest ideal for nature conservation. The wild flora and fauna there are the only remaining link with nature. Based on the incorrect theory, the untouched nature is considered to be a relatively monotonous closed forest, poor in species, where large ungulates have no influence on succession. In the name of nature conservation, in many forests in the lowlands of Central and Western Europe a number of very competitive shade-tolerant species of tree, such as the beech, have taken the upper hand, at the cost of the diversity that was present there in the natural conditions. This means that species from open terrain and the mantle and fringe vegetations and the oak, the wild fruit and the hazel can no longer occur in a natural way. At the same time, agriculture is considered to be the only option to retain the desired biodiversity, because according to this theory only humans can provide openness in nature by intervening through agriculture. History teaches us that agriculture changes continually and therefore so does the context for the species of wild flora and fauna. It is a very dynamic function and cannot be otherwise in an ever-changing and developing society. The first precondition for agriculture is to produce food in a way that is safe and acceptable to society and it manipulates nature with this objective. The objective in forestry is to produce timber. Within these restrictions, there can be space for the continued existence of wild flora and fauna. The space available should be used to the utmost, if only due to the enormous areas involved in agriculture and forestry. Because agriculture and forestry are based on selection from nature, it is impossible for the original nature, from which agriculture and forestry arose, to continue to exist within these restrictions. It can only continue to exist beside agriculture and forestry. Therefore, the presence of wild flora and fauna within the context of agriculture and forestry is not enough. This is why more space must be made for this nature, beside agriculture and forestry.

European nature conservation focuses mainly on wild flora and fauna within the context of agriculture and forestry, because people believe that humans have enriched nature, i.e. from the monotonous forest, poor in species: an incorrect theory. If the problem is that human intervention in nature continues to decrease the number of species of wild flora and fauna, then the theories, premises, calibration points, images of nature and strategies used will have to be reconsidered. Only then will nature conservation be able to realize its objective: to strive to allow nature and its variety of species of wild flora and fauna to continue to exist in the long term (Baerselman and Vera, 1995). In view of the aims of nature conservationists to retain the natural diversity, it will, therefore, be necessary to retain the natural processes, such as grazing and browsing by indigenous large herbivores living in the wild, or to redevelop them.[1] This means that the interdependence and the interactions between large herbivores and the vegetation will have to be restored (see Fig. 7.1). Cattle, horses, bison, red deer, elk, roe deer and wild boar will have to be able to operate as wild animals once again (De Bruin *et al.*, 1987; Vera 1988; Nilsson, 1992; Baerselman and Vera, 1995). Without these ungulates the survival of the natural diversity is impossible in the long term. This means that cattle and horses will have to be rehabilitated as species living in the wild. As the wild horse, the tarpan, and the wild cattle, the aurochs, are extinct, their role will have to be taken over by their domesticated descendants. It is particularly the breeds that have not been bred for particular production purposes, the so-called primitive breeds that are most suitable for this. Because they have not been bred much and have also often been kept in semi-wild conditions, it may be assumed that of all the breeds, they have retained the most genes of their extinct wild ancestors.

Keeping these breeds in the wild also means that the highest common denominator of the gene pool of the aurochs and tarpan that is stored in these breeds, is secured for future generations of people. This is done in a way in which the aurochs and tarpan have always survived as a gene pool, until they were

Fig. 7.1. Wild-living cattle (Heck cattle) and wild-living horses (konik horses) in the Oostvaardersplassen, The Netherlands. This nature reserve of 6000 ha is one of the areas in which it is being tested whether cattle and horses living in the wild together with other ungulates living in the wild, such as roe deer and red deer, facilitate the establishment of various species of wild flora and fauna, including thorny bushes and trees. In 1999 there were about 500 Heck cattle, 450 konik horses and 400 red deer in the reserve (photograph F.W.M. Vera).

[1] See De Bruin *et al.* (1987), Vera (1988; 1998), Baerselman and Vera (1995), Helmer *et al.* (1995), Vulink and Van Eerden (1998), Groot Bruinderink *et al.* (1999), Krüger (1999). This is currently taking place, for example, in the Oostvaardersplassen, a nature area of 6000 ha in the province of Zuid Flevoland in the Netherlands and in a large number of nature areas along the rivers Rhine, Maas, Waal and Ijssel.

domesticated by humans about 10,000 years ago. Restoring the natural processes therefore also serves an agricultural purpose. The importance of cattle and horses in such natural areas for the redevelopment of natural processes is that they are key species, together with oak and hazel. Examples are known from around the world that prove that so-called primitive races of cattle and horses have excellent success returning to the wild.[2] However, for the east of the United States it means that the bison should take the place of cattle in places where the subsistence of flora and fauna characteristic for this location has top priority. Experiments with the re-introduction of horses and cattle as animals living in the wild in Europe and bison in the east of the United States could test the alternative hypothesis.

For the policy on agriculture, as well as the policy on nature conservation, this means that the highest priority must be to allow the wilderness to develop once again. In view of the processes required for this, there must be large uninterrupted natural areas which should not be used in the way that led to the disappearance of this wilderness, i.e. farming and forestry. This does not mean that people cannot make use of these areas. In contrast, apart from serving as a gene pool, these areas can be used to experience nature as a source of inspiration for technology to recreate it as a living and working environment. By introducing such natural areas to maintain the biodiversity, in the form of both domesticated and undomesticated species of plants and animals, an important step is taken towards sustainable development. Allowing the wilderness to develop once again is very important for cultural conservation, as well as for safeguarding biodiversity. After all, the wilderness shows us the framework within which our cultural landscape developed. *It is only by knowing the wilderness that we can understand our cultural landscape.*

---

[2] Hall and Moore (1987), Davidson (1989), Hall (1989), Clutton-Brock *et al.* (1991), Guintard and Tardy (1995), Anonymous (1995).

Appendix 1

Structuring the Use of the Wilderness

The Creation of Communities of Users

The kings who used the term 'forestis' descended from the Franks. Like other
'Germanic' people, such as the Angles, the Saxons, the Friesians, the Alemans
and the Lombards, they settled in Western and Central Europe at the time of
the great migration (AD 350–450) (see Van Es, 1994a, p. 81; 1994b,
pp. 82–85). They adopted an ancient system of law passed down by word of
mouth, which meant that anything outside the farm and the fields could be
used by any member of the local community to meet their needs (Meyer,
1931, pp. 300, 392; Mantel, 1990, p. 151).[1] This included firewood and timber for building, honey, and fodder for the livestock. In the course of the 13th
and 14th centuries, local communities on the European mainland started to
reserve the traditional common land strictly for use by their own community
(Grossmann, 1927, pp. 23–24; Buis, 1985, pp. 32, 120, 205; Mantel, 1990,
p. 151). The use of this land was restricted to the so-called commoners, i.e. to
people who had a right to the use of a particular piece of wilderness traditionally used by the local community, on the basis of hereditary customary law
(Streitz, 1967, p. 36; Buis, 1985, p. 102). The customary law was passed on
by word of mouth every year at the so-called commoners' meeting
(Grossmann, 1927, p. 20; Buis, 1985, pp. 102–103, 218).

---

[1] A record dating from 1279 expressed it as follows: 'so wann sie die welde offent, so sin sie
inen allen offen' (Bodeman 1819, quoted by Endres, 1888, p.7).

The criterion for the use of the common land was what was necessary to meet a household's needs, so-called 'own' needs ('eigenen notdurft', 'zur notturfft', 'des Hauses Notdurft').[2] The concept of need was inextricably linked to the way in which the medieval economy was organized. It was mainly based on local self-sufficiency (autarchy). Therefore a local community had to have access to natural resources to provide wood, and a place where they could graze livestock so that they could meet their own needs for meat, milk, skins, manure for the fields where they cultivated cereals (Buis, 1993, pp. 31–32). In this respect, there were neighbouring communities which also had to meet their needs from the wilderness (Endres, 1888, p. 12). The individual local communities therefore eventually made boundaries in the wilderness to indicate which area was considered as being for their exclusive use. The common was such a community. In Western and Central Europe such communities of commoners developed for the use of the so-called common land, for communal use: these were variously known as the 'marke', 'gemeynte', 'meente', 'buur', 'buurschap', 'Gemeinde' or 'Allmend'.[3] The original meaning of the word 'marca' is mark, sign, border or an area surrounded by a border (Buis, 1985, p. 32). The oldest reference to 'marca' in the Netherlands is in a document dating from 792–793 (Buis, 1985, p. 35).

Probably the communities arose as a result of the increasing pressure on the use of the wilderness by a growing population (Buis, 1985, pp. 30, 177, 180, 205; Slicher van Bath, 1987, p. 176). In the west of Germany, the population seems to have doubled between 900 and 1100, and even quadrupled by 1200 (Mantel, 1990, p. 58). In Europe as a whole, the number of people increased markedly after 1150 up to 1300 (Slicher van Bath, 1987). The increase in the population was accompanied by a decline in the area of wilderness available for grazing and wood because of the cultivation and the creation of fields, while at the same time the pressure on the remaining wilderness increased (Grossmann, 1927, p. 23; Buis, 1985, pp. 30, 177, 180, 205; Slicher van Bath, 1987, p. 176; Mantel, 1990, p. 61).[4]

The written records of customary law are seen as the start of the organization of commons (Buis, 1985, p. 23). Writing down the customary law resulted in so-called commoners' shares, also known as 'Slagen', 'scharen', 'loten' or 'laden' (Buis, 1985, p. 261). Usually this entailed a vote at the com-

---

[2] In a record from the common of Wellingen, near the Moselle, dating from 1582, need was defined for the legal norm as 'Reich und Arm, nach Notdurft ihrer Haushaltung in s. Peterswald Brennholz zu hauen, als Haseln, Hainbuchen und Windfälle, und weiter nichts' (Endres, 1988, p. 36). See *inter alia*: Endres (1888, pp. 7, 8, 36), Grossmann (1927, p. 61), Hilf (1938, pp. 125, 160), Kaspers (1957, pp. 124, 149, 185, 205, 214), Rubner (1960, p. 52), Hesmer and Schroeder (1963, pp. 104, 145), Buis (1985, p. 150), Tack *et. al.* (1993, pp. 32, 175).
[3] Hilf (1938, p. 125), Hesmer and Schroeder (1963, pp. 103–105), Schubart (1966, pp. 9–11), Streitz (1967, p. 36), Buis (1985, pp. 31, 42, 121), Mantel (1990, p. 151).
[4] Grossmann (1927, p. 17), Meyer (1931, p. 298), Rodenwaldt (1951), Kaspers (1957, p. 182), Hesmer (1958, pp. 123, 412–415), Hesmer and Schroeder (1963, pp. 105–106), Reed (1954, p. 29), Schubart (1966, pp. 11, 43), Buis (1985, pp. 28, 125 e.v.), Mantel (1990, p. 153).

moners' meeting (Buis, 1985, pp. 230, 261). In addition to rights, the duties of the individual commoners were also laid down, and regulations were imposed on the use of the common (Hesmer and Schroeder, 1963, p. 273; Buis, 1985, p. 23). These regulations were drawn up, depending on the extent to which the lord or landowner could insist on the rights entailed in the concept of the 'forestis' *vis-à-vis* the local communities which had traditionally used the wilderness. They were by no means always able to do so.[5] On the European mainland, this resulted in three sorts of communities.

The free common

There was the so-called free common (Buis, 1985, p. 205). In this case, the lord or landowner could not successfully impose his rights. The commoners themselves decided on the use of the wilderness, and therefore actually exercised the rights which should have gone to the lord according to the concept of the 'forestis'. The land was therefore, in fact, the common property of the commoners (Buis, 1985, p. 165). Jurisdiction on the use of the common land was presided over by a judge of the common who was elected from the commoners, or whose office could eventually become hereditary (Hesmer, 1958, p. 414; Hesmer and Schroeder, 1963, p. 104; Buis, 1985, pp. 105, 221, 250). The regulations for the common use were decided upon at the meeting of commoners, the 'holtsprake' or 'holthinge'. They were then laid down in 'wilkeuren' and drawn up in books on the common (Buis, 1985, pp. 31, 245).[6]

The unfree common

Apart from the free common, there was also the unfree common. Even in the 15th century, the commoners' shares in the common could be more or less

[5] Even in the early Middle Ages, the Merovingian kings tried to control the uncultivated land of local communities which they had declared to be their forestis (forestis nostrum) (Grossmann, 1927, p. 117). Apart from the king, people and authorities which had been loaned a forestis or parts of a forestis by the king, the lords or landowners, also tried to gain control over the use of their land. These people were lords if they had been granted authority by the king, such as dukes, counts and prince-bishops. If they had no authority, they were landowners. They included the nobility and abbeys. A lord could also be a landowner if he owned private land, i.e. did not have authority in his capacity as a lord (Buis, 1985, p. 165). In view of the importance of the land for the local communities, they did not suffer these attempts gladly. There are many descriptions in the literature reflecting the tough struggle between lords and landowners trying to establish their authority in the organizations of the commons on the one hand, and the local communities opposing them, on the other hand (see Meyer, 1931, p. 298; Kaspers, 1957, p. 182; Hesmer, 1958, p. 412; Schubart, 1966, p. 11; Buis, 1985; pp. 125–131; Mantel, 1990, p. 156).

[6] Free commons were found throughout Europe wherever an organization of commoners developed (Grossmann, 1927, pp. 17–19; Rodenwaldt, 1951; Hesmer, 1958, pp. 121–123, 412–415; Hesmer and Schroeder, 1963, p. 126; Streitz, 1967, pp. 35–36; Buis, 1985, pp. 165, 206, 233, 409).

freely traded so that they sometimes fell into the hands of lords and landowners, who then gained an influence in the organizations of commons (Grossmann, 1927, p. 17; Hesmer and Schroeder, 1963, pp. 106, 108; Schubart, 1966, p. 29; Buis, 1985, p. 73). A lord or landowner could acquire the majority of shares, and with the help of the votes attached to these, gain a dominant influence in the annual meeting of commoners when the regulations on the use of the common were decided upon (Buis, 1985, p. 232). The common land then actually passed into the possession of the lord or landowner. The commoners made use of the land and had rights (Grossmann, 1927, pp. 17–18; Hesmer, 1958, pp. 414–415; Hesmer and Schroeder, 1963, pp. 104–106; Buis, 1985, pp. 266–273; 1993, pp. 60–62). The lord could also gain an influence in the organization of the common if a representative acquired the chairmanship of the meeting of commoners, which might be an inherited position (Hesmer, 1958, p. 414; Hesmer and Schroeder, 1963, pp. 104–106; Buis, 1985, pp. 71, 225, 250, 266; 1993, pp. 43–44, 53, 60–62). In this case, the regulations on the use of the common determined by the lord and the commoners together were known as 'wysdommen' ('Weistümer' in the German literature) (Streitz, 1967; p. 36; Buis, 1985, p. 31; Mantel, 1990, p. 156).[7]

The 'gemeynten'

If the lord was able to exercise the rights on the use of uncultivated land, included in the concept of the 'forestis', the local communities inherited the right of use. The lord converted the use traditionally made of the wilderness by the local inhabitants into the right of use of his 'forestis', in return for payment of money or in kind, such as some of the pigs which went out to pannage there.[8] Drawing up the regulations on the use of the common took place in the 'geding', the 'vorsthinc', 'holzgedinghe' or 'waltgedinghe', under the chairmanship of the 'magister forestariorum', who was later known as the 'waldgraaf' (Kaspers, 1957, pp. 40, 46, 70, 95, 137). The legal norm was that the local inhabitants should always be able to meet their own needs with regard to firewood, timber, grazing and pannage for pigs (Kaspers, 1957, p. 185).[9] In

---

[7] There were unfree commons or commons with a lord throughout Western and Central Europe, i.e. in south-west, west and north-west Germany, Lower Austria, Steiermark, Kärnten, Salzburg, Tirol, and a large part of Switzerland (Bühler, 1922, p. 65; Grossmann, 1927, p. 19; Hesmer, 1958, p. 158; Hesmer and Schroeder, 1963, p. 108; Schubart, 1966, p. 10; Buis, 1985, p. 164, 205, 233–234).

[8] See Endres (1888, pp. 2–3, 79, 172), Grossmann (1927, pp. 17, 19), Endres (1929), Rodenwaldt (1951), Reed (1954, pp. 32–33), Kaspers (1957, pp. 126, 149, 184, 193, 232), Hesmer (1958, pp. 158–159, 410–411, 414–415), Hesmer and Schroeder (1963, p. 106), Schubart (1966, pp. 41, 69), Buis (1985, pp. 29, 31, 102, 165–167; 1993, pp, 39–40), Mantel (1990, pp. 100–102, 154–156, 164–165), Iben (1993).

[9] A record dating from 1574 reads: 'der da bawen wolt, der soll die herrn bittn um das holz, so sollen die herrn ihme daz nit versagen' (Endres, 1888, p. 41).

this way, the local communities inherited the right to the use of a 'forestis'. The difference between 'gemeynten' and commons was that in the 'gemeynten', the lord laid down the regulations (Buis, 1985, pp. 31, 165), while in the commons they were determined jointly, following mutual consultation, with a vote linked to the shares in the common (Buis, 1985, p.102). The regulations laid down by the lord were drawn up in ordinances, bills, 'Wald' or 'Forstordnungen' (Buis, 1985, p. 165).[10]

The manor

As on the European mainland, the Anglo-Saxon kings loaned land to the nobles (estates) in return for fealty and services (Aston, 1958; Page, 1972, pp. 45, 53–54). In Anglo-Saxon documents, these nobles were referred to as 'earls', 'gesithes', 'thanes' or 'lords' (Aston, 1958; Page, 1972, p. 53). They owned a manor, the administrative unit into which Medieval England was divided (Hart, 1966, p. 8; Page, 1972, pp. 88–90; Rackham, 1980, p. 112; Cantor, 1982a,b). The administrative structure and settlements (villae or villages) were also loaned or granted by the king to the lords who were then confronted with the traditional right of the common use of the 'waste' by the local inhabitants or 'commoners'. This right had to be respected.

Therefore the manor comprised land which was traditionally used by the inhabitants of local settlements jointly. They inherited the right to the common use of the wilderness of the 'manor' to meet their own needs as regards firewood, timber, wood for building and grazing, pigs, cows, sheep and goats (rights of estovers, common and pannage) (Hart, 1966, p. 8; Page, 1972, pp. 88–90; Cantor, 1982a,b). Therefore a manor consisted of cultivated land, the fields and hay fields, and the uncultivated wilderness which was described as the 'waste', 'wealde', 'wold' or 'headh' (Krause, 1892; Rübel, 1914; Borck, 1954; Trier, 1963, p. 45).

The use of the wilderness by the commoners was regulated in the so-called 'Common Law' or 'Manorial Law'. This was exercised by local officers who fell under a 'manor-court' (Rackham, 1980, pp. 112, 174–175; Putman, 1986, p. 17). The regulations on the use, drawn up by the lord and the commoners were laid down in 'costumals' (see Peckham, 1925). After 1066, when William the Conqueror conquered England from Normandy, the Anglo-Saxon structure of use changed.

---

[10] This form of community of users was found particularly in the southern Netherlands and France and also in the west of Germany. In the Netherlands they were known as 'gemeynten' and in Germany as 'Gemeinden' (Hesmer, 1958, p. 123; Buis, 1985, pp. 31, 165; Mantel, 1990, p. 163).

The Forest Law

After the Norman Conquest, William the Conqueror imposed the 'Forest Law' on the Anglo-Saxon system of use (Hart, 1966, p. 7; Darby, 1976, p. 55; Cantor, 1982a; Tubbs, 1988, p. 67). The act served to protect the king's sovereign rights to all wild animals. This particularly concerned deer. An area to which the 'Forest Law' applied was known as the 'Forest'.

William the Conqueror applied the 'Forest Law' not only to land belonging to the Crown, but also to land which already had a clear owner, the lord (see Rackham, 1980, pp. 176–177, 184). If the Forest Law was declared to apply to one or more manors, three parties were involved in the use of a Forest: the Crown, the lord or lords and the commoners. In return for the loss of their land, the lords were granted the right to graze their own livestock freely in the forest (Stamper, 1988). Thus, one form of agricultural use continued to be possible although the law prohibited extending or improving farming (Darby, 1976, p. 55; Putman, 1986, p. 17; Tubbs, 1988, p. 68).

The Forest Law applied to all the land used by deer, i.e. not only the land where the deer found food and rested, but also the land they passed through. Like the concept of the 'forestis', Forest Law applied to all sorts of terrain. In 1598, an English lawyer named Manwood defined a Forest as a 'territory of woody grounds and fruitful pastures, privileged for wild beasts and fowls of forest, chase and warren, to rest and abide there in safe protection of the King, for his delight and pleasure' (Tubbs, 1988, p. 67). Therefore, like the terms 'forestis', 'Forst' and 'forêt', the term 'forest' did not have the modern meaning of a forest. It was a wilderness which included not only trees and shrubs but also grasslands, water, peat bogs, marshes and species of wild animals.

Appendix 2

The Use of the Wilderness

In the Middle Ages, the wilderness was an integral part of the agricultural system. The amount of manure available for fertilizing the fields was determined by the number of animals that were kept. In turn, these numbers were determined by the area of uncultivated land which could be grazed outside the fields. Therefore the relation between the field and the wilderness could not be changed to provide more farmland without paying a price, because this could jeopardize the fertilizing of the fields (Endres, 1888, p. 27; Grossmann, 1927, p. 23; Buis, 1985, p. 603). Thus the size of the herd was determined by the area of uncultivated land where the livestock could find food (Grossmann, 1927, p.23: Buis, 1985, pp. 34, 174). If there were neighbouring settlements, the area was limited because the inhabitants of these settlements in their turn had to graze their cattle in the wilderness. The ratio between the area of field and wilderness needed to keep the system of farming going in terms of food supply also depended on the productive capacity of the soil. In the rather poorly productive periglacial sandy areas in Central and Western Europe (see Fig. A2.1), a significant area of wilderness was needed for grazing.

In the 10th century, the system of eternal rye cultivation was 'discovered' there. This involved cultivating rye, year in, year out (Hesmer, 1958, p. 455; Buis, 1985, p. 601; Ellenberg, 1990, p. 34; Pott, 1992). To fertilize the fields, sods of turf were removed from the wilderness and taken to the deep-litter barns where the livestock were put at night. In the barns these sods were mixed with manure and urine. After a while the barns were emptied and the mixture of manure, urine and sods was put out on the fields (Buis, 1985, p. 601; Slicher van Bath, 1987, pp. 15, 281; Ellenberg, 1990, p. 31). Sometimes sods were also moved directly from heathland to fields (Pott, 1992). The relationship between the area of field in relation to the area where sods were cut and livestock grazed

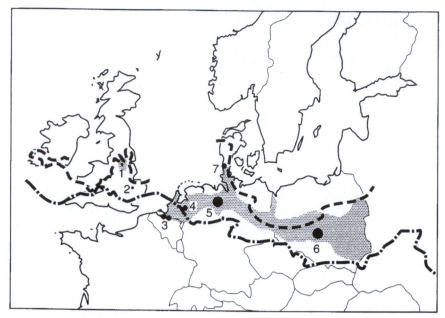

Extent of sand belts

— — Maximum extent of latest (Weichselian) ice sheets

—·— Maximum extent of Pleistocene ice sheets

Scattered dune fields:

· <100 km²

● 100–1000 km²

● 1000–10,000 km²

Fig. A2.1. Spreading of sand dunes and sandy areas from the glacial period in Central and Western Europe. Smaller areas are numbered and the borders are also marked indicating the extent of the ice caps during the last two glacial periods (redrawn from Koster, 1988, p. 70).

on this sandy land was 1–3 to 5–7 (Buis, 1985, p. 602; Slicher van Bath, 1987, p. 17). This meant that no more than 15–25% of the available area was in cultivation. If the wilderness was also used for cutting sods as fuel, cutting heather (to collect humus), and for the collection of sand and loam, the ratio of cultivated land to wilderness was even 1–15 to 1–20 (Buis, 1985, p. 602). The ratios do not include the green land ('prata', 'prés', 'meadows') (Slicher van Bath, 1987, p. 84). On more fertile soil such as loess, a much larger area of land was cultivated in comparison to the wilderness, because the farmers were less dependent on animal manure for produce from the soil (Buis, 1985, p. 174). On loess soil, the ratio between the area of cultivated land and forest/wilderness could be the reverse of that on sandy soil, i.e. 3–6 to 1 (Jansen and Van de Westeringh, 1983, p. 23). The larger area of fields in turn produced a great deal of straw which was mixed with stable manure and used to fertilize the fields.

Appendix 3

The Destruction of the Wilderness

The long list of historical reports given by Hesmer and Schroeder (1963) of the destruction of the commons and in the 'forestes' in the lowlands of Lower Saxony, to the west of the Weser and in the so-called Münster Bocht, illustrates the destruction of the wilderness. I will quote from a number of these in chronological order.

In 1549, a declaration was issued in the Heller mark: 'dair met deselffte so iemmerlich, wie eyn tidtlanck weß her gescheen nicht mochte verhouwen, verdilget vnnd in den grundt entlich verdoruen werdenn'. In 1696 it proved 'daß sich die Heller marckt gantz verwuestet befunden, ob solches aber durch verhawung der interessirten oder vereintreibung des schadtlichen Viehes hergeruhret, darüber würden die interessirte den Besten Bericht abstatten konnen.'

In 1550, there were complaints in the royal commons of the county of Ravensberg that 'trees were being felled destructively'.

In 1580, the 'Heepener Sunder' was 'completely felled'.

In 1595, trees were felled in the Hoppinger mark because: 'Vnde zeigete der subtituirter Holtzrichter dabei ahnn, daß bißannhero Jeder nach gelegennheit seiner Theill oder Wharen zu houwen gewieset vnnd mit der Maellbardenn gewaret, Nhun aber houwe Jederman seines gefallens muthwilliger weise vnngewieset, vnde verdrincken daß Holtz vnnoetiger weise.'

In 1606, there was a report about the Dortmund Forest that: 'Nachdem offenbar am Tage vnd menniglich fur Augen, In was grossen vnwiderbringlichen Abgangh vnd verderb das Dormundische geholtz oder Forst durch das Tagliche niederhawen Auß fuhren vnd tragen des Holtzes ... gerathen vnd kommen ist'.

The report from the prince-bishopric of Münster in 1608 read:

> In diesem Winter ... daß in den Benachpartten Embtern, die besten fruchtbarn Eichen boeme heuffigh niddergehouwen, nach dem Lipström also außerhalb diesem Stifft gefurtt vnd verkaufft wurden ... daß alstan diß Liebe Vatterlandt dadurch inß eußerste verderb gesetzt, vnd die posteri an holtz mangell erleiden vnd sich nit erhalten werden khonnen. Vnd ist auch vor Gott nicht zuuerantwort-ten, daß die fruchtbaren boeme, vmb der lieben mast willen, so Gott der almechtigh bescheret, mit solcher vndanckbarheitt sollen verhouwen vnd verher-get werden.

As early as 1610, it was established in the Wettruper mark, that 'ereignet, daß die Marcke vnnütz verhauen vndt gantz verwüstet worden'.

In 1629, references were made to the Ammeter mark: 'vber das vngebürliche Holttshouwen vnd verderben'.

In 1647, the Hülsberger mark was 'Verwöst- und abhauwungh' and 'verwüst-und schädlich hauwung des gehöltzes'.

In the Amt Ahaus, there were complaints in 1657 about 'verwüst- und schädlich hauwung des gehöltzes' by the royal subjects.

In 1674, the Limburger mark: 'für langen iaren Verhawen gewesen, daß Von Mast Vndt Holtz haw nichts zu bekommen'.

In the 'Staver Wald', a great deal of damage had been done in 1684 'nicht allein daß geholtz seher verhauen sondern mit eintreibung der schaffen vnd hauungh der heide vnd plaggen binnen geholtts' and in 1772, it was apparent that 'daß der Börger und staffer waldt täglich mehr entblößet werden, weil die schaffe undt Viehe den jungen auffschlag abfreßen'.

In 1684, 'das geholtz ... mit vnordentlichen hauen plagen vnd heiden meien verdorben' in the Baumweg.

In 1694, the capital of Cologne, as the 'Holtrichter' of the Oerschen 'Hardt', reported that: 'sich auff etliche stunden erstreckende Ohrer Hardt und Busch, so unlengst mit Trefflichen fruchtbaren Eichen, auch unzehlbahren schlag zum Brandt dienlichen Holtz dergestalt woll besetzt geweßen ... nunmehr durch unmeßig unverandtwordtliche Verhawunge in solchen erbärmlichen abgang gerathen, daß, daß kein eintzig Taugliches Holtz mehr darin zu finden'.

In 1699, there were complaints about the 'Schernen' ('die Malleute', who helped the 'holtrichter', known in Dutch as the 'gezworenen, gecommiteerden, gedeputeerden, vorsters, forestarii, bosbewaarders, or boswachters'; see Buis, 1985, pp. 245 *et seq.*) who 'totally ruined and destroyed these marks'.

In 1705, there were references to an 'unbelievable decline' of the 'Holz' in the county of Oldenburg. This was mainly the result of theft, the absence of supervision, and violations of the provisions by the supervisors themselves, as well of excessive grazing.

In the Wilderloh, it was noted in 1705 that 'a countless selection of the best and most beautiful oak and beech trees were cut down in a few years', and the young seedlings were being destroyed 'by the livestock grazing there'.

In 1744, the Rahder Osterwald 'von Holtz Wachs gantz und gar kahl gehauen'.

In the Dorgerloh, where 146 pigs were still sent out to pannage in 1590, 'gar sehr, wie noch fast täglich geschieht, devastiert' in 1750, so that there was not a single tree left by 1780.

In 1779, it was written about Hombruch that 'Hat verhauenes Eichenholtz und gantz verbissenes Schlagholtz, ist folglich in desolaten Umständen'.

The fate of the 'forest' of the town of Minden is typical of the whole picture outlined by Hesmer and Schroeder (1963, pp. 128–141). In 1460, the 'forest' was described as a beautiful and 'fruitful' 'forest'. However, the foresters appointed by the town of Minden did not look after it. In 1601, the 'forest' was partly destroyed because 'eine Zeit hero vnter wehrender vneinigkeit zur übermaß ... verhauen'. This disagreement was a dispute between the town of Minden and the lord, the prince-bishop of Minden, about the use of the 'forest'. In 1618, they concluded a convention providing, amongst other things,

> Zum Dritten hat man ein Zeittlang hero verspüret, das der Ungemessener freyer gebrauch des Minderwaldts, undt das, wie derselb in guetten Stande zu erhalten, keine bequeme Ordnung gemacht, daruber mit geburenden ernst gehalten, die Sonsten gewöhnliche Holtzungs Gerichte nicht angestellet, die Excedenten fur demselben eingeklaget, weiniger mit geburender straffe belegt, unnd derogleichen Verbleib zu dessen Verödung nicht weinig Ursache gegeben. Weilen nun gleichwoll ... nicht zu verandtwortten stehet, einen solchen großen ortt undt stadlichen Grundt unnd boden also lenger fast Öde unnd wüste liggen zu laßen.

It did not seem to change very much, for in 1650, it was recorded that the 'forest' was destroyed to such an extent that it had to be enclosed for 10 years. However, in 1678, the 'forest' contained only 'very poor wood'. In 1744, the lord told the town of Minden:

Wir haben eine Zeithero mit dem größten Mißvergnügen ver- und wahrnehmen müßen, wasmaßen auf den Uns und unserer Stadt Minden zugehörigen also genandten Minderwald gar wenige acht gegeben, darinnen überall nicht wirtschafftlich verfahren, die Anweiß- Abstamm- und Anpflanzung des Holtzes nicht forstmäßig geschehen, die Deputanten sich ihres habenden Rechts mißgebrauchet, zur aller Jahreszeit Holtz gehauen und nicht angepflantzet noch geheget worden, so daß wir befürchten, es werde diesem sonsten weitlaufftigen und ansehnlichen Geholtze in wenig Jahren des Garaus gemachet werden.

According to a description dating from 1770, 'Der Wald lieget inden letzten Zügen'. In 1782, it was largely destroyed. According to a contemporary source, it was a forest in name only, because

Auf den gantzen großen Reviere befindet zich nichts weiter als nur einige sehr wenige und gantz abgestandene Eichen. In de Gründen siehet man zwar etwas Ellern Unterholz welches aber wegen der darin hergebrachten Hude und Weide ... so ruiniret und von den Viehe abgefressen wird, daß es gar keinem Gebüsche mehr ähnlich, und auch niemahlen fortkommen kann.

Finally, in 1787, the Minder 'forest' was 'fast nichts weiter als eine bloße Heide ... worauf hin und wieder noch etwas Unterholtz und Buschwerk steht, und daher dieses ehedem mit den besten und schönsten Eichen versehene große Revier, in eine wahre, und man mögte fast sagen die Schöpfung entehrende Wüsteney verwandelt' (Hesmer and Schroeder, 1963, pp. 128–141).

The final result of grazing livestock in the 'forests' was described by authors in the 19th century in caustic terms as outright devastation (see Hobe, 1805, p. 124; Hesmer, 1958, pp. 90–98; Mantel, 1990, p. 245). When Von Schwerz saw the Eifel during a 2-year inspection tour through the Rhineland and Westphalia, commissioned by the Prussian government, he wrote in 1816:

Man sollte sehen und weinen! Ein Land wie die Eifel, wo es nicht an Raum fehlt, wo der Boden zum Theil keinen Werth für die übrige Cultur hat, weil es an Dung und Dungmaterial gebricht, da heben die Berge von allen Seiten ihre nackten Schädel, welche kein Gesträuch deckt, und wo kein Vöglien ein Schattenplätzchen ze seinem Neste findet. (Hesmer, 1958, p. 91)

Appendix 4

The 'Hanau-Münzenbergse Forstordnung', the 'Mainzer Forstordnung' and the 'General-Verordnung'

The method of regeneration which developed in the first half of the 18th century was described in some detail in the Hanau-Münzenbergse 'Forstordnung' in 1736 (Bühler, 1922, p. 303; Hausrath, 1982, p. 64; Mantel, 1990, p. 357). This regulation was included verbatim in the 'Mainzer Forstordnung', which was issued in 1744 and related to the Spessart, amongst other places. Vanselow (1926, pp. 222 *et seq.*) reproduced the paragraphs concerned of this regulation verbatim. In my opinion, the technique for regeneration is best described by the texts themselves. Therefore I will cite a number of the paragraphs of the 'Mainzer Forstordnung' from Vanselow (1926, pp. 222–227).

Following the 1744 regulation, the 'General Verordnung' appeared for the Spessart in 1774. I will also quote from this regulation because it clearly shows how the technique developed and what consequences it had, particularly for the grazing of livestock. The two regulations also clearly reveal that there was a certain degree of ambivalence. In fact, keeping oak standards hindered the regeneration of beech, yet they had to be kept to provide the mast for the wild animals and for pigs. Therefore the pole-forest was still part of the coppice with standards. The oak standards were also still identified as fruitful trees while the beech trees were called seed trees. The 'Heeg-Reiser' and 'Heegreiser', referred to respectively in the two regulations, are the standards in the coppice which must be spared so that they can grow into mature trees. The 'Waldrecht' refers to the 'ius forestis' which decreed that oaks, the so-called 'Herrenholz', must be spared. The phrase 'Eichen zu Wald-recht stehen lassen' in paragraphs 10, 12 and 16 of the 'Mainzer Forstordnung' refers to a standard which is spared to provide mast for pigs. They also served as a protective canopy for beech seedlings (Vanselow, 1926, pp. 31, 38). The term 'Berg' in the regulation refers to the coppiced wood to be harvested (Trier, 1952, p. 119). A few paragraphs from the 'Mainzer Forstordnung' follow below.

Paragraph 9

Die gesunde fruchtbare Bäum sollen auf den jungen Schlägen nebst denen gewöhnlichen Heeg-Reiser stehen gelassen werden:

Die gesunde fruchtbare Bäum sollen auf den jungen Schlägen, und darneben auf jeden Morgen die nöthige Heeg-Reiser von Eichen und Buchen, darunter aber sonderlich das Eichen-Holz, so viel zum graden Fortwachs dienlich, stehen bleiben, was aber oben in Wipffeln trucken und dürr, und am Stamm hohl wird, weil es von Jahren zu Jahren abnimmt, und endlich gar niederfallt, mit weggehauen, und was an Handwerks-Holz daran noch tuchtig ausgehauen, und das übrige zu Brennholz geschlagen werden: Wie dann die Forst-Beambte und Forst-Knechte, so solche Heeg-Reiser aushauen und stehen lassen sollen, welche so stark sehn, daß sie von Schnee und Wind nicht unterdruckt werden können; Ebenmäßig sollen auch die junge Schläg wohl in acht genommen damit weder Zaun-Gärten, Lattenstangen, Hopffen- oder Reissstangen daraus gehauen, und die Berg dadurch schändlich verderbet werden.

Paragraph 10

Die Schläg sollen anfänglich nicht zu licht gehauen werden:

Die Schläg sollen anfänglich, damit die Sonn das Erdreich nicht verdruckte und dem jungen Anflug den Nahrungs-Safft entziehe, nicht zu licht gehauen, sondern hin und wieder gesunde Heister und Heeg-Reiser daneben auch alle gute und gesunde Eichen zu Wald-Recht stehen lassen.

Paragraph 11

Wann die erste Ausläuterung vorzunehmen:

Mann alsdann der junge Anwachs in den bereits vorhandenen oder künsstig zumachenden Schlägen eines Knies hoch und drüber erwachsen, und also die Ausdruckung des Erdreichs nicht so mehr zu befürchten ist, so soll alsdann die erste Ausläuterung der stehen gebliebenen haubahren Heister geschehen, und solche ebenfalls nicht hier und dar, sondern dem Schlag nach durchgängig vorgenommen werden.

Paragraph 12

Was hernach ausgeläutert werden kann:

Wann der junge Auffwachs sodann Manns lang erwachsen, gleichwohl aber noch hier und dar zu Wald-Recht etwas stehen gelassen, müssen solche Bäum zu Beförderung des jungen Holzes, wofern es ohne sonderbaren Schaden geschehen kann, was nicht zu werckholß dienlich vollends aus-geläutet, und mit Säuberung des Walds, Auffbindung des Reiß-Holßes, auch sonsten mit der Anführung alles in die Wege gerichtet werden, daß dadurch Kein sonderlicher Schaden geschehe.

Paragraph 13

Wann und wie das junge Holß auszuscheidelen:

Ist es nun soweit damit gekommen, das die Asschneidelung geschehen muß, so sollen Unsere Untertanen, welche sich der Orthen beholßigen oder sonst die Mast und Hute haben, auf Befehl und Ansag Unserer Forst-Bedienten die Aufschneidelung des jungen Holßes nach der Ordnung, wie die Schläge geführet, und sie von Unsern Forst-Bedienten angestellt und angewiesen wer-den, dergestalt verrichten, daß dem jungen Stamm die Aste bis eines Manns hoch genommen, der Stamm selbst aber gänßlich geschonet, und das abge-hauene Reis-Holß zu Säuberung des waldes sogleich auffgebunden werde, weßhalben und daß solches recht geschehe, sollen Unsere Forst-Bediente überall dahen seyn, und wohl zusehen, daß aller Schaden und Mißbrauch in diesem Fall vermieden und abgestellet werde. Die jenige Unterthanen nun, welche, wann sie zu socher Verrichtung begehret, und auffgeforderet, ohne genugsame Ursach zuruck bleiben, sollen von Unsern Forst-Bedienten ohne Nachsehen zur Buß gebracht, und auf Buß-Tag oder Forstgericht dem Befinden nach gestrafft, den andern aber, welche ihre Arbeit wohl verrichtet, das abgehauene und auffgebundene Reiß-Holz ohnentgeldlich geschenckt werden.

Paragraph 14

Was nach der Ausschneidelung im Wald zu thun:

Worauff alsdann ein solcher Wald und Schlag, wann er nicht her-nachmahls von den untüchtigen und unterdruckten Stangen und Krackel-Holß zu säuberen und auszuläutern, zo lang bis er wieder recht haubar worden, in Ruhe gelassen, und nichts außer dem Eichen-Bauholß zur höch-sten Nothdurfft darinnen angewiesen und gefället werden solle.

Paragraph 15

Die verbeißte Schläg und Dornen sollen weggehauen werden:

Desgleichen sollen auch die verbeißte Schläg kahl auf der Erden, und wo Dornen vorhanden, solche in vollem Safte, auf daß sie desto eher vergehen, ausgehauen werden.

Paragraph 16

Von der Heeg und geheegten Orthen:

Damit aber auch die neue Schläge und Gehäu in behöriger Ordnung gehalten, und der junge Ausflug von dem Viehe nicht abgefressen oder verbeißt werde; so wollen Wir zwar geschehen lassen, daß den ersten Sommer über, wann der Schlag in hohen Wäldern, wo das Klaffter-Holß gehauen, im Frühe-Jahr geschehen, solcher annoch mit dem Viehe betrieben werde, weilen dadurch das Erdreich wund getretten, und also des Saamen von denen zu Waldrecht stehen gelassenen Eichen und Buchen desto besser in das Erdreich kommen und wieder auffschlagen kann, jedoch, daß alsdann, wann die Blumen-Huthe, des ersten Sommers, vorbey, und die Maste beginnt reiff zu werden, nicht weniger in denen Waldungen, wo Stamm-Reiß gehauen, gleich von Anfang ein jedweder mit seinem Horn-Viehe (dann das Schweine-Viehe mag wohl nach gefallener Maste ein paar mahl durch getrieben werden), es seye was es wolle heraus, und der Orth solang in Heege und Zuschlag verbleiben, bis das junge Holß dem Viehe wieder aus dem Maul und schier zu Auschneidelung erwachsen, als dann und nicht eher soll solcher Orth von Unsern Forst-Bedienten zur Huthe wieder ohnentgeldlich auffgetan, und jedwederem, so darzu berechtiget, darinnen zu hüthen vergönnet werden.

Paragraph 17

Wann die Unterthanen die Huthe nicht entbehren können, wie daselbst die Wälder zu tractiren:

Da sich aber zutragen sollte, daß an ein- und anderen Orthen die Unterthanen gar nicht, oder doch wenig an ihrer Huth entrathen könnten, so sollen auf den ersterer Fall, wann Unsere Forst-Bedienten denselben mittler weil nichts zur Huthe einzuthun wissen, die Huth-Wälder zwar in Heege nicht gelegt werden, denen Unterthanen aber wird hiemit bey willkührlicher Straff befohlen, alle Jahr ein Stück derselben nach dem andern, welches ihnen Unsere Forstt-Bedienten zeigen sollen, mit jungen Eichen und Buchen, oder nach Gelegenheit Hayn-Buchen, welche ihnen von denen Forst-Bedienten aus Unsern Wäldern, wo es ohne Schaden geschehen kann, gratis hergeben werden sollen, ordentlich nach der Maase, wie hernach folget, zu bepflanßen, und an der ausgehenden Stelle wieder neue zu seßen, auch solche mit Dornen und

Pfählen dergestalt wohl zu bewahren, daß den jungen Stämmen von dem Viehe kein Schaden geschehen kann, wohen dann denen Hirten sonderlich hierdurch eingebunden wird, daß sie das Viehe, soviel immer möglich, von solchen neu bepflanßten Orthen abhalten, oder doch wenigstens allen Schaden mit Sorgfalt zu verhüthen suchen, widrigenfalls aber ohne Nachsehen auf dem Buß-Tage ernstlicher Bestraffung gewärtig zu seyn: Auf den andern Fall aber, soll zwar denen Untethanen, so viel zu ihrer ohnentbehrlichen Huthe vonnöthen, jedesmahl offen gelassen, damit aber auch die Waldungen nicht gantz und gar in Abgang kommen, und dadurch ein schädlicher Holz-Mangel der Posterität zugezogen werde, wollten Wir daß von Zeit zu Zeit ein Stück nach dem andern so viel nehmlich auf einmahl an der Huthe zu entrathen stehet, in scharpsste Heege gelegt, und wann soches, wie oben gemelt, erwachsen, und wieder zur Huthe aufgethan worden, alsdann ein ander Stück, und so ferner bis der gantze Wald wieder arthhafft gemacht, gleichfalls in Zuschlag genommen und geheeget werde.

Paragraph 18

Die Unterthanen sollen die Huthe-Wälder umackern oder hacken, wann sie in Heeg gelegt werden können:

Gleich wie aber die Erfahrung lehret, daß in solchen von langer Zeit her betriebenen Mast- und Huthe-Wäldern der junge Aufschlag sehr schwer und langsam hervor kommt, so sollen Unsere Unterthanen, damit die Orthe nicht allzulang in Zuschlag verbleiben, sondern sobald möglich wieder zur Huthe auffgethan werden können, jedesmahl auf Befehl Unserer Forst-Beambter den neuen in Heeg zu legenden District umackern oder hacken, Unsere Forst-Bediente aber denselben alsdann mit Eicheln und Aeckern im Herbst, und da den Boden zu Auffbringung der Eicheln nicht tüchtig, zu gehöriger Zeit mit Thaunen-Saamen, welchem in Mangel desselben Unsere Forst-Bedienten zu beschreiben haben, ordentlich besäen, der gebühr verpflegen, und in Summa alles das thun, was zu Wiedererzieh- und Arthaffmachung deren Waldungen in allen Stücken nöthig und nußlich seyn möge.

The General-Verordnung of 1744

The 'General-Verordnung' of 1744, i.e. 30 years later, goes even further as regards the measures for regeneration. The decree still referred to fruitful trees forming the canopy ('Heegreiser'), but also to seed trees ('Saamenbäume'), which should be spaced at intervals every 18–20 paces.

The text of the General-Verordnung indicates that the seed trees of the pole-forest should be left to stand with the oaks as far as possible, though when standard oaks made the regeneration impossible because of the shade they cast, the deteriorating oaks should be felled. This still shows the importance attached to oak standards for the mast. It also indicates that the regeneration of beech trees from seed was combined less and less often with keeping oak standards (see Vanselow, 1926, p. 50). The oaks left standing were not meant to obstruct the seed trees in the pole-forest, but, on the other hand, they were still protected because of the mast they produced and were removed only when they really made the regeneration from seed impossible, and were also starting to deteriorate. Thus healthy oaks were simply left standing. When a situation arose in which no oak standards were left because of the age of the trees, the regeneration of the pole forest from seed became the main goal, and consequently other forms of traditional use of the forest by commoners, such as the grazing of horned livestock or the mowing of the undergrowth were completely prohibited. This total prohibition was unique in German history and was undoubtedly inspired by experiences of the previous decades. The decree went one step further than that of 1744. In fact, this was the first step towards separating the wood (the material) from the pasture. The General-Verordnung also extended the cycle from 80 to 140 years. With this sort of cycle, there was a tree forest such as those which exist now.

The above was taken from the following text of the decree:

> hätten dieselben in jedem ausgehauenen Schlage dasjenige haubare Eickenholz, so den in 80 Jahren wieder vorgehenden hieb nicht mehr ohne Beschädigung des hinkünftigen jungen Anflugs oder Stangenholzes auszuhalten, oder wegen seiner Vielheit und allzudicken Stand den Anflug zu verdämpfen scheinet, mit Zuziehung der Jägerei gleich auszusuchen, zu marquiren und die Anzahl der stämme, nebst einem Gutachten, wie solche nach Abzug eigener Nothdurft am besten zu benutzen an Unsere Hofkammer zu weiterer Verfügung einzusenden, jedoch hiebei die Masttragende Bäume, so viel möglich, zu verschonen, und sollten die Buchwaldungen vermischter mit Eichenstämmen bestellt seyn, so wären die Saamenbäume meistens und, so viel thunlich, von solchen Eichstämmen stehen zu belassen, und wenn dieses Eichenholß in einem Schlage allenfalls zu Verdämpfung des hinkünftigen Anflugs zu dicht erstehe, die abständigen Stämme alsdenn hiervon heraus nehmen und noch in Zeiten zu benußen.

> Wäre sogleich nach völlig ausgehauen und gesäuberten Schlägen auf die Wiederaufbringung eines ordentlichen Anflugs ein Hauptaugenmerck zu richten, die genaueste Heege anzulegen, alles Eintreiben des Viehes, Grasen Mähen, und überhaupt was einem Schlage nur schädlich fallen könne, schärfstens zu untersagen. (Vanselow, 1926, p. 226)

Appendix 5

The Development of 'Natural' Regeneration

The 'Dunkelschlag' Method

In 1791, G.L. Hartig gave a systematic description of the regeneration technique from seed in pole-forests described in the Mainzer Forstordnung, in his book, *Holzzucht für Förster* (Bühler, 1922, p. 353; Mantel, 1990, p. 362). In this work, he proposed the so-called 'Dunkelschlag'; a measure which involved felling the inferior wood first in the pole-forest, and sparing the best and strongest trees as seed trees. The trees could still touch each other with the tips of their branches, so that the forest floor remained shaded to prevent drying out and the growth of grasses. Therefore there was some thinning out, although the canopy remained virtually closed (Bühler, 1922, p. 349). Hartig called this method 'Dunklen' or 'Besamungsschlag' (dark- or seed-felling) (Hartig, 1791, cited by Bühler, 1922, pp. 306–307). If there were not yet sufficient seedlings, Hartig said that this part of the forest must be grazed by livestock to trample down the loose soil and prevent undesirable species of plants and trees from growing. Hartig said: 'Der Schlag muß in der dunklen Stellung bleiben, bis er sich größtenteils besamt hat und der Aufschlag $\frac{3}{4}$–$1\frac{1}{2}$ Schuh hochgeworden ist' (the felling must remain in the dark position [in the shade from the crowns of the remaining trees] until it has become seeded for the most part [by the remaining trees] and the seedlings have grown $\frac{3}{4}$–$1\frac{1}{2}$ shoe high (25–40 cm)) (Hartig, 1791, cited by Bühler, 1922, p. 306). Subsequently, more trees were felled in the 'Dunkelschlag'. One seed tree had to be spared every 15–20 paces (10–14 m) to provide seed and to protect the seedlings. New trees then grew in the so-called 'Lichtschlag'. When there were enough seedlings on the entire forest floor and they had reached a height of 2, 3, 4 feet (60, 90, 120 cm), all the trees had to be felled. This was known as the 'Abtriebsschlag' (Bühler, 1922, p. 306). Therefore Hartig divided the harvesting and regeneration of a beech pole-forest into three fellings: the 'Dunkel' or 'Besamungsschlag', followed by the 'Lichtschlag' and concluding the regeneration with the 'Abtriebsschlag' (Mantel, 1990, p. 362). The 'Dunkelschlag' or 'Besamungsschlag' took place when there were sufficient seedlings underneath the canopy of the trees to be harvested, as a result of a good mast year (Bühler, 1922, p. 306; Schubart, 1966, p. 101; Streitz, 1967, p. 73).

However, the 'Dunkelschlag' method did not lead to the regeneration of oaks, which require light (Vanselow, 1926, pp. 63, 87–88; Krahl-Urban, 1959, p. 146). Admittedly, oak seedlings appeared on a vast scale after a good mast, but they were soon overgrown by the equally numerous beech seedlings which flourished better under the virtually closed canopy (Vanselow, 1926, p. 87). At the beginning of the 19th century, the 'Dunkelschlag' method was modified in Germany for oak seedlings (Hausrath, 1982, p. 76). The 'Forstkatechismus', dating from 1806, states that oak should be regenerated like beech. However, the first felling must provide much more light for oak seedlings. Therefore more parent trees have to be felled (Jeitter, 1806, cited by Bühler, 1922, p. 310). In 1819, a technical description was given in Württemberg, stating that for beech, the branches of the trees must still touch each other after the first felling, while oak seed trees must stand 3–4 m apart (Bühler, 1922, p. 331). In 1817, Cotta wrote that oaks must be given more light than beech. Therefore in his view the first felling, i.e. the traditional 'Dunkelschlag', should take place in the first or second winter after the germination of the seeds, and the canopy should be cleared by the second or fourth year so that the oak seedlings would be in full daylight by the second or fourth year (Cotta, 1817, cited by Bühler, 1922, p. 312).

In the period 1816–1820, regulations were drawn up for the natural regeneration of oaks in the Spessart. These stated that when seedlings appeared on a large scale as a result of a good mast, the 'Dunkelschlag', using Hartig's method, must be applied, providing more light than when it is used for the regeneration of beech (Vanselow, 1926, p. 87). However, it became clear that when oak seedlings germinated on a huge scale, and beech trees which produce seed also form part of the crown layer, the young oaks were overgrown and destroyed by the young beech trees. The young beech trees had to be controlled by cutting them back or felling them, if the oak was to survive (Bühler, 1922, p. 433; Vanselow, 1926, pp. 87–88; Meyer, 1931, pp. 356, 357).

This method of regenerating beech and oak continued with a 'trial and error' approach (see Hausrath, 1982, pp. 69–75). The term 'preparation felling' appeared in 1828 in forestry regulations (Vanselow, 1926, p. 97). If there were no seedlings under the canopy (the so-called advanced regeneration), this felling served to provide more light for the parent trees so that they would flower more profusely and produce more fruit. By giving more light to the crown earlier on, it was therefore possible for more seedlings to germinate in the soil. In the mid-19th century, this so-called shelterwood system became common for the regeneration of beech and oak in Germany (Baur, 1842; Hagen, 1866, both cited in Bühler, 1922, pp. 324 and 331, respectively).

This technique of germination was taken over by the Frenchman Lorentz in 1827 and introduced by him to France (Reed, 1954, p. 38; Tendron, 1983, p. 58). By the second half of the 19th century it was applied throughout France (De Gail, 1907, in Woolsey and Greevey, 1920, p. 75). The French do not describe it as shelterwood felling but as progressive felling (coupes progressives) (Woolsey and Greevey, 1920, p. 73). In the second half of the 20th century, this regeneration technique was introduced in Britain (Tsouvalis-Gerber, 1998). This is also apparent from the contents of the article by Forbes (1902), cited by Watt (1919) with regard to the regeneration of oak in grassland (see Chapter 2).

The shelterwood system developed in the way described above and is used in forestry today. In forestry terminology, it is referred to as 'natural regeneration'. I will briefly summarize the technique as it is currently described in forestry handbooks.[1]

Natural Regeneration using the Shelterwood System

The shelterwood method is carried out at the point when the trees are considered ready for felling and a new generation of trees is to be created by means of regeneration. The first measure is to fell some of the trees which form the canopy. This serves to make the canopy thinner, so that the crowns of the remaining trees have more light. When they are thinned out, the cover of the crown is reduced by a maximum of one-third. This felling is known as the preparatory felling. The increased amount of light to which the crowns are exposed means that the trees flower more profusely and consequently produce more seed. These remaining trees will then produce the seed for the next generation. The standing trees form the canopy.

Therefore the canopy produces the seed. When there has been a good mast and the seed has fallen, a second felling is carried out, known as the so-called seedling felling. This serves to provide the young seedlings with sufficient light to develop well. The remaining canopy has to provide a subdued micro-climate and protect the young seedlings from frost and heat. In addition, the shade provided by the canopy prevents the excessive development of grasses and herbs which could otherwise seriously compete with the young seedlings. However, the dilemma is that as the seedlings become older, the shade provided by the canopy also starts to hinder their continued develop-

---

[1] For a detailed description of the shelterwood system I refer to Bühler (1922, p. 323 etc.), Vanselow (1949, p. 103 etc.) and Dengler (1990, p. 273 etc.) and for the oak in particular to Vanselow (1949, p. 82 etc.), Woolsey and Greeley (1920, pp. 73–77 and 472–472), Reed (1954, pp. 84–87), Turbang (1954), Klepac (1981), Tendron (1983, p. 58 etc.) and Krahl-Urban (1959, p. 135 etc.).

ment. For this reason, the canopy is thinned out with fellings at intervals of a few years to ensure that the seedlings always have enough light. Finally, felling the last trees clears the area leaving only a space with young trees. These are then thinned out in the course of the next few years.

The total duration of regeneration, i.e. the period between the preparatory felling and clearing the area, lasts several decades, or as much as 40 years for a shade-tolerant tree, such as beech. On the other hand, for oak, which requires light, the whole process of successive fellings and clearing must take place in less than 10 years. For the regeneration of oak, the first felling takes place after the acorns have fallen, i.e. when the seed has been sown after a good mast. Therefore, there is no preparatory felling which serves to encourage trees to flower more profusely and form more seed. The reason for this is that good masts occur at intervals of many years. Therefore the risk that there will not be a good mast after the preparatory felling is too great, and as a result of the increase in the light penetrating to the forest floor, there could be a luxuriant vegetation of grasses and herbs. If this were to happen, the young oak seedlings which grew under the canopy once there was a mast, would have no chance at all. Therefore it is a matter of waiting for a good mast followed by the seedling felling straightaway. Because of the increased need for light of oak seedlings, the first felling should take place 1 year, or at most 2 years, after the germination of the acorns. Every subsequent felling must take place at intervals of 3 years (Bühler, 1922, p. 218; Tendron, 1983, p. 58). As stated above, the final clearing should take place in less than 10 years. Therefore the canopy is thinned out at a much greater rate for the regeneration of oak than for more shade-tolerant species, such as beech and European silver fir.

At the beginning of the process of regeneration, oak seedlings need a great deal of human intervention to protect them from the competition of grasses, herbs and spontaneously developing seedlings of beech, hornbeam, lime and elm. Like beech, these species tolerate much more shade than oak and the oak seedlings can be destroyed by the competition. Therefore a great deal of human intervention is needed for the 'natural regeneration' of oak.[2]

[2] See Bühler (1922, pp. 218, 295 etc.), Vanselow (1926, p. 27), Tangermann (1932), Hess (1937), Turbang (1954), Hesmer (1958, p. 261), Krahl-Urban (1959, p. 214), Nüßlein (1978), Klepac (1981), Evans (1982), Tendron (1983, pp. 57–63), Raus (1986), Dengler (1990, p. 274).

Appendix 6

The Selection System

In the course of the 18th and 19th centuries, another technique of 'natural regeneration' developed in Europe. This was described above as a so-called 'regulated' or 'orderly' selection system (Mantel, 1990, p. 361). In German, it was referred to as 'Plenterwald', and in French as 'jardinage' (Bühler, 1922, p. 576; Reed, 1954, p. 177; Mantel, 1990, p. 361). For the selection system, one or several (three to five) trees which are ready for felling are removed from the forest, so that seedlings spontaneously develop in the gap in the canopy (Bühler, 1922, p. 570; Leibundgut, 1984a, p. 152; 1984b, pp. 90–94; Dengler, 1990, p. 293 *et seq.*). Thus the wood is removed from the forest in one go, and the conditions allowing for regeneration arise at the same time (Leibundgut, 1984a, p. 152; Dengler, 1990, p. 22; Mantel, 1990, p. 361). The regeneration after the felling is left completely to chance, and the random vagaries of nature (Vanselow, 1957; Dengler, 1990, pp. 22, 259; Mantel, 1990, pp. 295, 361). In the case of the selective system, some care is taken to carry out the felling in an orderly way. When one or several trees have been felled, the forester does not return to that place for a particular period to cut more trees. Therefore there is a fixed cycle for the places where a tree or group of trees has been felled (Cotta, 1865, p. 147). As large trees are constantly felled in different places, followed by regeneration, a forest where the selective system is used consists of a mixture of trees of different ages. In this sort of forest, the different species of different ages are usually in groups (three to five trees) (Dengler, 1990, pp. 293–298; 1992, p. 26).

Special places are chosen for the selective felling, because it has certain advantages, for example, compared with coppicing and clear-cutting. These advantages are that when the wood is harvested, the forest remains standing, so that it affords protection against the wind, soil erosion and avalanches.

Therefore the selection system is often used on mountain slopes. It also protects the young plants, which helps a forest to survive in areas with a harsh climate. In those places, coppicing, coppicing with standards or high forests are not possible because clear-felling is part of the management. This results in the disappearance of the forest climate, and in harsh climates, (micro-)climate conditions occur in the bare, open spaces, such as frost, etc., which prevent regeneration from stools and seed (Cotta, 1865, p. 145).

Historical sources reveal that the selection system developed in the same way as the shelterwood system. It developed from thinning out the forest. For example, in 1764, Zanthier wrote that 'wirtschaftliche Förster' carried out a 'Durchhauung oder Plenterung', consisting of the 'Ausziehung oder Ausläuterung' of dead and dry wood to prevent anything from being lost in the forest (Zanthier, 1764, cited by Bühler, 1922, p. 419). According to Bühler (1922, p. 576), there were many synonyms for the selection system in the German 18th-century literature in the sense of thinning or cutting out, such as: Ausleutern, Ausleuchten, Ausziehen, Ausplündern, Plündern, Plentern, Pläntern, Plendern, Ausplentern, Durchpläntern, Ausbländern, Lüften, Luftmachen, Plätzighauen, Durchhauen, Nachhauen, Femmeln, Fimmeln. The term 'Plenterung' first appeared in 1763. By 1787, it was used throughout Germany (Bühler, 1922, p. 578). At that time, it was not yet an orderly selection system, but consisted merely of thinning out trees in a random fashion. In 1744, one author still opposed this selective system because it did not allow for protection against livestock; a view which was also reflected in the forest decrees of the 15th, 16th and 17th centuries, which opposed the random cutting of wood (scrub and the sprouts from stools) in the 'forestes'. Cotta (1865, p. 147) also remarked that grazing in the forest when a selective system is used is more damaging, because there is young wood everywhere. In his book in which he explained the Dunkelschlag, Hartig (1791) referred to Dunkelschlag itself as 'Hauung, die man Plenterung nennt' (Hartig, 1791, cited by Bühler, 1922, p. 421). It was only in the 18th century that the selection system was used as a targeted method of exploitation next to the Dunkelschlag or shelterwood system.

The selection system is used above all in forests with species which tolerate shade or semi-shade, such as mixed forests of beech, Norway spruce and European silver fir, or forests containing mainly Norway spruce and European silver fir (Vanselow, 1949, pp. 117, 270; Leibundgut, 1984a, p. 154; 1984b, p. 92; Dengler, 1990, p. 294). The European silver fir, which is most tolerant of shade, is most common in forests which are selectively felled (Bühler, 1922, p. 591; Dengler, 1990, p. 295). According to the forestry literature, oak trees cannot be regenerated with the use of this selection system. It does not provide them with sufficient light (Bühler, 1922, p. 566). In the case where oak and beech grow together, selective felling and the shelterwood system ensure that shade-tolerant beech becomes predominant (Boden, 1931; Seeger, 1938).

Many authors[1] have considered that the selection system is a modern version of the felling of odd trees, as described in historical written sources since the Middle Ages. Because of the lack of any form of planned felling, they describe it as 'unregulated' selective felling. The difference between the 'unregulated' selective felling and the 'orderly' selective felling is that in the case of unregulated selective felling, the only reason for removing wood from the forest was to provide firewood from scrub (thorns and hazel) and the resprouting stools, and timber from trees for building, and there was no concern to regenerate the forest as such (Cotta, 1865, p. 146; Buis, 1985, p. 429). In the case of orderly selective felling, this concern does exist.

---

[1] Cotta (1865), Bühler (1922), Vanselow (1925, 1926), Meyer (1931, 1941), Hess (1937), Rodenwaldt (1951), Hesmer (1958), Hesmer and Schroeder (1963), Streitz (1967), Fleder (1976), Buis (1985), Dengler (1990), Mantel (1990).

Appendix 7

The Group Selection System

In the second half of the 19th century, another technique was added to the selection system and shelterwood system by Gayer (1886, 1895), the group selection system. This technique is a sort of combination of the shelterwood system and the selection system (Mayer, 1992, p. 348). For detailed explanation, reference is made to Dengler (1990, pp. 276–285). This is the source of the following data, unless another source is given.

The group selection system means that, as in the shelterwood system, more light is allowed to penetrate the canopy by cutting down some of the trees. However, the trees are felled in an irregular pattern. The forest canopy is therefore opened up here and there, so that there are open gaps with an area of three to five trees, 20–40 m across (a group) or 40–80 m across (a grove) (Dengler, 1990, pp. 276–285; 1992, p. 26). In fact, the group selection system entails the shelterwood system, i.e. opening up the canopy. However, this is not done over a large surface area, as in the case of the shelterwood system, but in different places in the forest, as for the selection system. Subsequently, the different regeneration units are concentrically enlarged, and the following fellings take place in the wake of the expanding concentric circles, which are eventually cleared. Eventually, the growing concentric regeneration units spread towards each other, so that all the old trees are felled in the end and the next generation of trees is established. In French, this method is known as 'jardinage trouée' (Reed, 1954, p. 180).

Like the shelterwood system, this technique is particularly suitable for the regeneration of shade-tolerant species, such as European silver fir, beech and Norway spruce, but not for oak (Seeger, 1938; Frank, 1939; Leibundgut, 1984b, p. 86). As in the case of the selection system and the shelterwood system, the oak loses out as a result of the competition from the beech. This occurred in gaps in the canopy with an area of 100–200 m^2 (Frank, 1939), and even in groves with an area of 1 ha. This became clear in the Spessart, when groves of oaks planted in the 19th century failed to thrive (Vanselow, 1926, p. 111). Young beech seedlings on the perimeter of the groves eventually pushed out all the young oaks in the grove because these young beech trees were not suppressed (Vanselow, 1926, pp. 92–93, 110–111; Endres, 1929a,b, p. 214; Boden, 1931). Thus for the group selection system, as for the shelterwood system, the regeneration of oak requires adaptations of these techniques based on this species' greater requirement of light (Vanselow, 1949, pp. 117, 128). When oaks are regenerated with the group selection system, this only succeeds in high forest consisting purely of oak. As in the shelterwood system, the standards are cleared quickly so that the oak seedlings have enough light to survive and grow (Krahl-Urban, 1959, p. 223; Leibundgut, 1984b, pp. 85–90).

Appendix 8

Species of Trees and Shrubs in Landscapes Grazed by Large Herbivores

Species of trees and shrubs which are found in park-like landscapes grazed by large herbivores and which regenerate in scrub, and in the fringe of the mantle vegetation of groves are listed below. Neither this list, nor the references, are exhaustive.[1]

Acer campestre
Acer pseudoplatanus
Berberis vulgaris
Betula pendula
Betula pubescens
Buxus sempervirens
Carpinus betulus
Clematis vitalba
Cornus mas
Cornus sanguinea
Corylus avellana
Crataegus laevigata
Crataegus monogyna
Euonymus europaeus
Fagus sylvatica
Frangula alnus
Fraxinus excelsior
Hedera helix

Humulus lupulus
Ilex aquifolium
Juniperus communis
Ligustrum vulgare
Lonicera periclymenum
Malus sylvestris
Pinus sylvestris
Populus nigra
Populus tremula
Prunus avium
Prunus padus
Prunus spinosa
Pyrus pyraster
Quercus petraea
Quercus robur
Rhamnus catharticus
Ribes nigrum
Rosa spp.

Rubus spp.
Salix alba
Salix aurita
Salix caprea
Salix cinerea
Sambucus nigra
Sorbus aria
Sorbus aucuparia
Taxus baccata
Tilia cordata
Tilia platyphyllos
Ulmus carpinifolia
Ulmus glabra
Ulmus procera
Viburnum lantana
Viburnum opulus

[1] Data extracted from Salisbury (1918), Adamson (1921; 1932), Tansley (1922), Watt (1924; 1925; 1934a,b), Tüxen (1952), Dierschke (1974), Smith (1980, pp. 380, 318–319, 349, 353), Pott and Hüppe (1991, pp. 289–299), Rodwell (1991, pp. 319–321, 334–361), Kollmann (1992), Oberdorfer (1992, pp. 87–105, 148).

In addition, species of grasses and herbs from grasslands on the one hand, and species of herbs from groves, on the other hand, grow on the fringes of scrub. This means that fringe and mantle vegetation is the most diverse from a botanical point of view, in forests and groves (Smith, 1980, p. 369; Hondong *et al.*, 1993, pp. 126–140; Pietzarka and Roloff, 1993).

Appendix 9

Species of Herbs on the Fringes of Scrub and Groves

Some species of herbs on the fringes of scrub and groves are listed below. The data are taken from Müller (1962), Dierschke (1974) and Smith (1980, p. 326).

Agrimonia eupatoria
Ajuga reptans
Anemone nemorosa
Campanula persicifolia
Campanula trachelium
Centaurea nigra
Centaurea scabiosa
Convallaria majalis
Epipactis helleborine
Gentiana lutea

Geranium sanguineum
Geum urbanum
Inula conyza
Knautia arvensis
Listera ovata
Ophrys insectifera
Orchis militaris
Orchis purpurea
Orchis simia
Origanum vulgare

Phyteuma spicatum
Platanthera bifolia
Polygonatum odoratum
Primula veris
Primula vulgaris
Sonchus arvensis
Stellaria holostea
Viola hirta

Appendix 10
Species Characteristic of the Thermophile Oak Forest

Some species characteristic of the thermophile oak forest in Białowieza are listed below (from: Derkmen and Koop, 1977, p. 12; Falinski, 1986, p. 87).

Angelica sylvestris
Anthyllis vulneraria
Astragalus danicus
Campanula cervicaria
Campanula glomerata
Campanula persicifolia
Carex montana
Carlina aucalis
Centaurea austriaca
Coronilla varia

Digitalis grandiflora
Dracocephalum ruyschiana
Geranium sylvaticum
Gladiolus imbricatus
Lathyrus laevigatus
Lathyrus niger
Lathyrus pratensis
Melittis melissophyllum
Primula veris
Pulmonaria angustifolia

Ranunculus polyanthemos
Succisa pratensis
Trifolium alpestre
Trifolium medium
Trifolium montanum
Trollius europaeus
Vicia sepium
Vicia sylvatica
Vincetoxicum hirundinaria
Viola mirabilis

Appendix 11

Situation and Some Abiotic Features of the Forest Reserves Described in Chapter 5 as They Are Presented in the Publications Mentioned

For a complete description, reference should be made to the relevant publications. Reserves appear in the same order as in Chapter 5. Soil types are quoted literally.

Name of reserve	Location	Soil	pH top layer of soil	Average annual temp. (°C)	Average annual rainfall (mm)	Elevation (m)
La Tillaie Le Gros-Fouteau	France, Forest of Fontainebleau, near Paris	Varies from 'sol brun calcaire' to 'podzol humo ferrugineux' with podsolization, resulting in 'sols lessivés néopodzoliques', 'sols podzoliques' and 'podzols'	3.6–4.8	10.2	697	135–140
Neuenburger Urwald	North-west Germany, south of Willemshafen 53° 24' N lat 7° 58' E long	'Gley Humuseisen podsolen' with humus and iron, 'Gley Braunerden' in sandy soil and 'Gley-pseudogley' in loamy soil	3.4–4.4	8.2	779	7.5–10
Hasbrucher Urwald	North-west Germany between Bremen and Oldenburg	'Gley Braunerde' and 'Gley Pseudogley' in loam or slightly loamy sand	Unknown	8.4	720	25–30
Sababurg (information from Reinhardswald, where it is situated)	Central Germany, near Kassel, 53° 4' N lat and 8° 29' E long	Loam and loamy sand	Unknown	7.8	725–825	350–400
Rohrberg (Spessart)	Central Germany, halfway between Aschaffenburg and Würzberg	'Braunerde' in slightly loamy sand	Slight basic Saturation	7–8	1000	530–540
Priorteich	Central Germany, southern Harz, near Walkenried	Sand with 'Lösslehmauflage'	Unknown	7	850	300

Continued

Name of reserve	Location	Soil	pH top layer of soil	Average annual temp. (°C)	Average annual rainfall (mm)	Elevation (m)
Kottenforst	Western Germany near Bonn	Moderate to strong 'Pseudogley aus Lösslehm'. Mull-humus	3.8–4.2	9.5–10.2	650	170
Chorbusch	Western Germany, Rhine Valley between Cologne and Bonn	'Parabraunerde' in fine to 'schluffig-tönigen Hochflusslehm' with slight pseudogley	3.7–4.3	0.5–0.2	700	40
Geldenberg	West Germany, Reichswald, near Kleve, west of Nijmegen, the Netherlands	'Podsolige Pseudogley, Braunerde in Sandloss'. Rough, humus-like moder to rough humus, very poor in nutrition	2.7	9.5–10.2	780	75
Rehsol	West Germany, Reichswald, near Kleve, west of Nijmegen, the Netherlands	Pseudogley Braunerde with weak to moderately formed 'Pseudovergl.' in Sandloss'. Moder-humus	2.8–3.0	9.5–10.2	780	35
Johannser Kogel	Eastern Austria, Alps near Vienna	Moderately 'frische, feinerdereiche, mittel-bis tiefgründige Barunerden'	Unknown	8–9	700–800	170–380
Krakovo	Slovenia, near the Croatian border	Unknown	Unknown	Unknown	Unknown	150m
Unterhölzer	South-west Germany, Black Forest near Donaueschingen	'Lehmtonige, basenreiche Braunerde'	Rich in bases	6.3	732	670–730

Dalby	Southern Sweden in the province of Skania, near Lund	Heavy clay, rich in calcium, Baltic moraine. Mull-humus	Unknown	Unknown	Unknown	65
Vardsätra	Southern Sweden in the province of Skania, near Lund	Unknown	Unknown	Unknown	Unknown	Unknown
Forest of Białowieza	Eastern Poland on the border of Belarus, south of Białystok 52° 30' N. lat 23° 30'–24° 209 E. long	Loam and loamy sand and coarse-grained sand	Unknown	6.0	641	140–200

Appendix 12

Further Information on the Experiments Quoted on the Growth of Seedlings in Chapter 6

For a complete description of the experiments, reference should be made to the publications mentioned. The representation in the table opposite is in alphabetical order. Soil types are quoted literally.

Authors	Type of tree	Form of the experiment	How shade is produced	Treatments in % daylight	pH soil	Type of soil	If field experiment: inf. on temp., rainfall and elevation
Alaoui-Sossé et al. (1994)	Pedunculate oak	Laboratory experiment, plants grown in greenhouse at 25°C and 16 h photoperiod. Plants in pots	n.a.	n.a.	Unknown	Unknown	n.a.
Burchel and Huss (1964)	Beech	Field experiment, seedlings grown from beech nuts and harvested at the end of the first period of growth	Screens and nylon nets	100, 76, 41, 14, and 12	5.1	'Lössverwitterungsboden' of humus, sand with a lot of loam in the Weser River Valley	Average annual temp. 6.5°C
Burschel and Schmaltz (1965a)	Beech	Field experiment with 1-year-old seedlings, grown in a nursery (n) and in the wild (w). They are harvested after treatment through two growth seasons. The experiments were carried out in the Gahrenberg and Dransfeld forestry areas	Screens	100, 77, 28, 12 and 1	Garenber: 4.2 Dransfeld: 6.6	Garenberg: homogenized soil, poor in nutrition, originally 'Pseudogley-Braunerde' with localized podsolization starting Dransfeld: nitritious homogenized Randzina	Garenberg: ave. ann. temp 6.5°C; ave. ann. rainfall 935 mm; elevation: 340 m Dransveld: ave. ann. temp. 8.5°C; ave. ann. rainfall, 640 mm; elevation: 340 m
Brookes et al. (1980)	Pedunculate and sessile oak	Laboratory experiment. Plants grown in unheated greenhouse in full daylight in pots (diam. 25 cm)	n.a.	100	Unknown	John Innes Seedling compost	n.a.

Continued

Authors	Type of tree	Form of the experiment	How shade is produced	Treatments in % daylight	pH soil	Type of soil	If field experiment: inf. on temp., rainfall and elevation
Hoffmann (1967)	Pedunculate oak	Grown in root case 2 m deep and 1 m square. 2-year-old seedlings from a nursery were used. The root was cut back when they were transplanted	Paper and nets	100 and 20	Dune sand 4.9, loam 8.0 (Hoffmann, 1966)	Mixture of dune sand and loam in a 3:1 proportion (Hoffmann, 1966)	Unknown
Lyr et al. (1964)	Small-leaved lime	Field experiment, seedlings grown in nursery for 3 years were used, the roots were cut back when they were transplanted. Plants grown in trays of 3 × 7 × 1.5 m	Mesh	100, 68, 35, 12 and 1	7.3	2 parts sand, 1 part grass earth	Unknown. Experiment carried out in Eberswalde near Berlin
Jarvis (1964)	Sessile oak	Field experiments in pots with a diam. of 20 cm. Seedlings grown from acorns which were planted and harvested at the end of the first growing season.	Nylon net	100, 56, 20 and 10	4.4	Brown, slightly clayed	Unknown
Von Lüpke (1982; 1987)	Sessile oak and beech	Field experiment in the Göttingerwald. 2-year-old seedlings from a nursery were used. After being exposed to the treatment for 4 years, i.e. when 6 years old, and after 8 years, i.e. 10 years old, the plants were harvested.	Adjacent trees and trees above the seedlings	100, 45 and 11	4	'Sehr schwache psuedovergleyte Braunerde'	Göttingerwald.: Ave. ann. temp. 7.8°C; ave. ann. rainfall: 680 mm; elevation: 280 m

continued

| Oosterbaan and Hees (1989) | Sessile oak and beech | Field experiment on the Veluwe, Central Netherlands, with seedlings which grew spontaneously | Areas of the canopy with 100–200-year-old oaks | 1. Strong lighting 0.4–0.5 cover; 2. Poor lighting, 0.7 cover; 3. untreated; 0.9–1.0 cover | Unknown | Poor loamy holtpodzol soil | Unknown |
| Ovington and McRae (1960) | Sessile oak | Field experiment in cages of 40.5 × 30.4 cm, with 10 acorns planted 1 cm deep. The plants were harvested after the first and second growing season | Nylon net | 96, 83 and 54 | 3.9–4.9 | 10 different types of soil were made. The starting material was mineral soil from an oak forest in the Lake District. The soils were: that soil; only humus from *Picea sitchensis*, *Larix leptosis* or *Quercus petraea* respectively; the humus of each of these species on the mineral soil and each of these types of humus mixed with mineral soil | Unknown |

Continued

Authors	Type of tree	Form of the experiment	How shade is produced	Treatments in % daylight	pH soil	Type of soil	If field experiment: inf. on temp., rainfall and elevation
Paice (1974)	Small-leaved lime	Laboratory experiment. Plants grown in pots with a diam. of 8.7 cm with a photoperiod of 16 hours at 19–21°C and 8 h darkness at 15–16° C	n.a.	Low: 0.18 MJ m^{-2} day^{-1} Medium 0.65 High: 2.8 Very high: 4.8 As an indication for the correspondence with daylight in England: 0.18 MJ m^{-2} day^{-1} = 10% daylight (from Davis and Pigott (1982)	Unknown	John Innes Seedling compost	n.a.
Röhrig (1967)	Pedunculate and sessile oak and small-leaved and broad-leaved lime	Garhrenberg forest area in sections of 2 × 2 m and 80 cm tall. The plants were grown from seed. They were harvested after the second growing season	Screens and nets	100, 78, 24, 8 and 1	4.2	'Schluffiger Lehm'	Ave. ann. temp. 6.5°C; ave. ann. rainfall 935 mm; elevation 340 m

Sanderson (1958)	Hazel	Seedlings grown in pots. Plants cultivated from nuts and harvested after one growing season. 1-year-old plants from a nursery grown on in pots dug into the soil and harvested after 1 and 2 years respectively	Beech hedge (for plants grown from nuts) and a wooden screen	25 and 10	Unknown	Potting compost	Unknown
Schwappach (1916)	Oak and beech	Field experiment in the Westerwald near Göttingen	More or less thinned out forest	Unknown	Unknown	Loam	Ave. ann. temp. 14.1°C; ave. ann. rainfall 870 mm; elevation 380 m
Shaw (1974)	Sessile oak	Field experiment in Coed Cymerau reserve in Wales. 1-year-old plants from nursery were exposed to treatment for 8 years	Unknown	85, 32, 19 and 5	4.2	'Very dark brown loam or moder type'	Coed Cymerau Ave. ann. temp. 10°C; ave. ann. rainfall 1900–2000 mm; elevation 150 m
Suner and Röhrig (1980)	Beech	Field experiment in Grüneplan forestry area in Lower Saxony. Seedlings which germinated after full mast. Seedlings harvested after three growing seasons	Screen of 152-year-old beech trees	49.0–4.6	3.0–3.3	'Podsol Braunerde'	Ave. ann. temp. 8°C; ave. ann. rainfall 800 mm; elevation unknown
Ziegenhagen and Kausch (1995)	Pedunculate oak	Field experiment, acorns planted near Bonn	Nylon nets	100, 50, 25 and 10	7.0	Homogenized	Ave. ann. temp. 9.5–10°C; ave. ann. rainfall 660 mm; (from Wolf, 1988)

References

Aaby, B. (1983) Forest development, soil genesis and human activity illustrated by pollen and hypha analysis of two neighbouring podzols in Draved Forest, Denmark. *Danmarks Geologiske Undersøgelse* II Raekke, Nr. 114 (*Geological Survey of Denmark*, II, Series, No. 114).

Aaris-Sørensen, K. (1980) Depauperation of the mammalian fauna of the Island of Zealand during the Atlantic period. *Videnskabelige Meddelelser fra Dansk Naturhistorisk Forening* 142, 131–138.

Aaris-Sørensen, K., Peterson Strand, K. and Tauber, H. (1990) Danish finds of Mammoth (*Mammuthus primigenius* (Blumenbach)), stratigraphical position, dating and evidence of Late Pleistocene environment. *Danmarks Geologiske Undersøgelse* B 14 (*Geological Survey of Denmark* No. 14), 1–44.

Abe, S., Masaki, T. and Nakashizuka, T. (1995) Factors influencing sapling composition in canopy gaps of a temperate deciduous forest. *Vegetatio* 120, 21–32.

Abrams, M.D. (1992) Fire and the development of oak forests. *BioScience* 42, 346–353.

Abrams, M.D. (1996) Distribution, historical development and ecophysiological attributes of oak species in the eastern United States. *Annales des Sciences Forestières* 53, 487–512.

Abrams, M.D. and Downs, J.A. (1990) Successional replacement of old-growth white oak by mixed mesophytic hardwoods in southwestern Pennsylvania. *Canadian Journal of Forest Research* 20, 1864–1870.

Abrams, M.D. and McCay, M. (1996) Vegetation-site relationship of witness trees (1780–1856) in the presettlement forest of eastern West Virginia. *Canadian Journal of Forest Research* 26, 217–224.

Abrams, M.D. and Nowacki, G.J. (1992) Historical variation in fire, oak recruitment, and post-logging accelerated succession in central Pennsylvania. *Bulletin of the Torrey Botanical Club* 119, 19–28.

Abrams, M.D. and Ruffner, C.M. (1995) Physiographic analysis of witness-tree distribution (1765–1798) and present cover through north central Pennsylvania. *Canadian Journal of Forest Research* 25, 659–668.

Abrams, M.D. and Seischab, F.K. (1997) Does the absence of sediment charcoal provide substantial evidence against the fire and oak hypothesis? *Journal of Ecology* 85, 373–375.

Abrams, M.D., Orwig, D.A. and Demeo, T.E. (1995) Dendrological analysis of successional dynamics for a presettlement-origin white-pine–mixed-oak forest in the southern Appalachians, USA. *Journal of Ecology* 83, 123–133.

von Abs, C., Fischer, J.B. and Falinski, J.B. (1999) Vegetationsökologischer Vergleich von Naturwald und Wirtschaftswald, dargestellt am Beispiel des Tilio-Carpinetum in Waldgebiet von Białowieza/Nordost Polen. *Forstwissenschaftliches Centralblatt* 118, 181–196.

Accetto, M. (1975) Die Natürliche Verjüngung und Entwicklung der Stieleiche und Hainbuche im Urwald-Reservat Krakovo. *Gozdarski Vestnile* 33, 67–85.

Acker-Stratingh, G. (1844) *Over eenige Wilde Dieren, welke vroeger in ons Vaderland geleefd hebben (Eene voorlezing, gehouden in het Genootschap ter Bevordering der Natuurkundige Wetenschappen te Groningen)*, Groningen.

Adams, D.E. and Anderson, R.C. (1980) Species response to a moisture gradient in central Illinois forests. *American Journal of Botany* 67, 381–392.

Adamson, R.S. (1921) The woodlands of Ditcham Park, Hampshire. *Journal of Ecology* 9, 114–219.

Adamson, R.S. (1932) Notes on the natural regeneration of woodland in Essex. *Journal of Ecology* 20, 152–156.

Addison, W. (1981) *Portrait of Epping Forest*. Robert Hale, London.

Ahlén, J. (1975) Winter habitats of moose and deer in relation to land use in Scandinavia. *Viltrevy* 9, 45–192.

Al, E.J. (1995) *Natuur in bossen. Ecosysteemvisie Bos*. Rapport IKC Natuurbeheer, nr. 14, Wageningen.

Alaoui-Sossé, B., Parmentier, C., Dizengremel, P. and Barnola, P. (1994) Rhythmic growth and carbon allocation in *Quercus robur*. 1. Starch and sucrose. *Plant Physiology and Biochemistry* 32, 331–339.

Alexander, K.N.A. (1998) The links between forest history and biodiversity: the invertebrate fauna of ancient pasture-woodlands in Britain and its conservation. In: Kirby, K.J. and Watkins, C. (eds) *The Ecological History of European Forests*. CAB International, Wallingford, UK, pp. 73–80.

van Alsté, H. (1989) *Wisenten in Nederland, feit of fictie? Een onderzoek naar het inheemse karakter van de Wisent (Bison bonasus) in Nederland*. Verslag nr. 1052, Vakgroep Natuurbeheer, Landbouwuniversiteit Wageningen.

Ammann, B., Chaix, L., Eicher, U., Elias, S.A., Gaillard, M.-J., Hofmann, W., Siegenthaler, U., Tobolski, K. and Wilkinson, B. (1984) Flora, fauna and stable isotopes in late-Würm deposits at Lobsigensee (Swiss Plateau). In: Mörner, N.-A. and Karlén, W. (eds) *Climatic Changes on a Yearly to Millennial Basis*. D. Reidel, Dordrecht, pp. 67–73.

van Andel, J. and van den Bergh, J.P. (1987) Disturbance of grasslands. Outline of the theme. In: Andel, J., van Bakker, J.P. and Snaydon, R.W. (eds) *Disturbance in Grasslands. Causes, Effects and Processes*. Dr W. Junk Publishers, Dordrecht, pp. 3–13.

Anderson, J.E. (1991) A conceptual framework for evaluating and quantifying naturalness. *Conservation Biology* 5, 347–352.

Anderson, M.C. (1964) Studies of the woodland light climate. 1. The photographic computation of light conditions. *Journal of Ecology* 52, 27–41.

Andersen, S.T. (1970) The relative pollen productivity and pollen representation of North European trees, and correction factors for tree pollen spectra. Determined by surface pollen analyses from forests. *Danmarks Geologiske Undersøgelse* II. Raekke nr. 96. (*Geological Survey of Denmark* II. Series, No. 96).

Andersen, S.T. (1973) The differential pollen productivity of trees and its significance for the interpretation of a pollen diagram from a forested region. In: Birks, H.J.B. and West, R.G. (eds) *Quaternary Plant Ecology. The 14th Symposium of the British Ecological Society, University of Cambridge, 28–30 March 1972*. Blackwell Scientific Publishers, Oxford, pp. 109–115.

Andersen, S.T. (1976) Local and regional vegetational development in eastern Denmark in the Holocene. *Danmarks Geologiske Undersøgelse (Geological Survey of Denmark)*, 5–27.

Andersen, S.T. (1984) Forests at Løvenholm, Djursland, Denmark, at present and in the past. *Det Kongelike Danske Videnskabernes Selskap Biologiske Skrifter*, 24.

Andersen, S.T. (1989) Natural and cultural landscapes since the Ice Age. Shown by pollen analyses from small hollows in a forested area in Denmark. *Journal of Danish Archaeology* 8, 188–199.

Andersen, S.T. (1990) Changes in agricultural practices in the holocene indicated in a pollen diagram from a small hollow in Denmark. In: Birks, H.H., Birks, H.J.B., Kaland, P.E. and Moe, D. (eds) *The Cultural Landscape. Past, Present and Future*. Cambridge University Press, Cambridge, pp. 395–407.

Andersson, C. (1991) Distribution of seedlings and saplings of *Quercus robur* in a grazed deciduous forest. *Journal of Vegetation Science* 2, 279–282.

Andersson, C. and Frost, I. (1996) Growth of *Quercus robur* seedlings after experimental grazing and cotyledon removal. *Acta Botanica Neerlandica* 45, 85–94.

Andersson, F. (1970) Ecological studies in a Scanian woodland and meadow area, southern Sweden. 1. vegetational and environmental structure. *Opera Botanica. A Societate Botanica. Lundensi Edita* 27, Gleerup, Lund.

Anonymous (1978) Führer durch das Naturschutzgebiet Urwald Sababurg. *Hessisches Forstamt, Reinhardshagen*.

Anonymous in cooperation with Keiper, R. (1985) Paarden van Assateagne Island. Een onuitputtelijke bron van informatie. *Paard en Pony* 14, 16–20.

Anonymous (1988) *Grofwildvisie Veluwe. Ministerie van Landbouw en Visserij*. Ministerie van Landbouw en Visserij, Den Haag.

Anonymous (1993) *A Programme for Nature Conservation on Öland; an Island in the Baltic*. World Wildlife Fund for Nature (WWF), Solna.

Anonymous (1996) *Feiten en Cijfers 1996. Kerngegevens over landbouw, natuurbeheer en visserij*. Ministerie van Landbouw, Natuurbeheer en Visserij, Den Haag.

Anonymous (1999) Natuurontwikkeling langs Limburgse beken. *Natuurhistorisch Maandblad* 88.

Archer, S. (1989) Have southern Texas savannas been converted to woodlands in recent history? *The American Naturalist* 134, 545–561.

Arthur, M.A., Paratley, R.D. and Blankenship, B.A. (1998) Single and repeated fires affect survival and regeneration of woody and herbaceous species in an oak–pine forest. *Journal of the Torrey Botanical Society* 125, 225–236.

Ashby, K.R. (1959) Prevention of regeneration of woodland by field mice (*Apodemus sylavaticus* L.) and voles (*Clethrionomys glareolus* Schreber and *Mocrotus agrestis*). *Quarterly Journal of Forestry* 53, 228–236.

Aston, T.H. (1958) The origins of the manor in England. *Transactions of the Royal Historical Society*, 5th series VIII, 59–83.

Atkinson, M.D. (1992) Biological flora of the British Isles. No. 175. *Betula pendula* Roth (*B. verrucosa* Ehrh.) and *B. pubescens* Ehrh. *Journal of Ecology* 80, 837–870.

Atkinson, T.C., Briffa, K.R. and Coope, G.R. (1987) Seasonal temperatures in Britain during the past 22,000 years, reconstructed using beetle remains. *Nature* 325, 587–592.

Aubréville, A. (1933) La forêt de la Côte d'Ivoire. *Bull. Comm. Afr. Occ. Franc.* 15, 205–261.

Aubréville, A. (1938) La forêt coloniale. Les forêts de l'Afrique occidentale française. *Annales Acaddemis Sciences Coloniales* (Paris) 9.

Auguste, P. and Patou-Mathis, M. (1994) L'aurochs au paléolithique. In: Bailly, L. and de Cohën, A-S. (eds) *Aurochs, le retour. Aurochs, vaches et autres bovins de la préhistoire à nos jours.* Centre Jurassien du Patrimoine, Lons-le-Saunier, pp. 13–26.

Austad, I. (1990) Tree pollarding in western Norway. In: Birks, H.H., Birks, H.J.B., Kaland, P.E. and Moe, D. (eds) *The Cultural Landscape, Past, Present and Future.* Cambridge University Press, Cambridge, pp. 11–29.

Baerselman, F. and Vera, F. (1995) *Nature Development. An Exploratory Study for the Construction of Ecological Networks.* Ministry of Agriculture, Nature Management and Fisheries, The Netherlands, The Hague.

Bakker, J.P. (1987) Diversiteit in de vegetatie door begrazing. In: de Bie, S., Joenje, W. and van Wieren, S.E. (eds) *Begrazing in de natuur.* Pudoc, Wageningen, pp. 150–164.

Bakker, J.P. (1989) Nature management by grazing and cutting. On the ecological significance of grazing and cutting regimes applied to restore former species-rich grassland communities in the Netherlands. Thesis. *Geobotany* 14, Kluwer Academic Publishers, Dordrecht.

Baldock, D. (1990) *Agriculture and Habitat Loss in Europe.* CAP Discussion Paper nr. 3, WWF International.

Baldock, D., Bennett, G. and Clark, J. (1993) *Nature Conservation and New Directions in the Common Agricultural Policy.* Report for the Ministry of Agriculture, Nature Management and Fisheries, The Netherlands. The Netherlands Institute for European Environmental Policy, London.

Bangs, P.R. (1985) Monfragüe: a conservation success in Spain. *Oryx* 19, 140–145.

Bannister, P. (1976) *Introduction to Physiological Plant Ecology.* Blackwell Scientific Publications, Oxford.

Bär, J. (1914) Die Flora des Val Onsernone. *Mitteilungen aus dem botanischen Museum der Universität Zürich* 59, 223–563.

Barber, K.E. (1975) Vegetational history of the New Forest: a preliminary note. *Proceedings Hampshire Field Club and Archaeological Society* 30, 5–8.

Barden, L.S. (1981) Forest development in canopy gaps of a diverse hardwood forest of the southern Appalachian Mountains. *Oikos* 37, 205–209.

van Baren, B. and Hilgen, P. (1984) *Struktuur en dynamiek in La Tillaie, een ongestoord beukenbos in het bosgebied van Fontainebleau.* Doctoraalverslag Vakgroep Natuurbeheer (nr. 702) en Vakgroep Bosteelt (nr. 84–14). Rijksinstituut voor Natuurbeheer, Leersum.

Barker, G. (1985) *Prehistoric Farming in Europe*. Cambridge University Press, Cambridge.

Barnes, B.V. (1991) Deciduous forests of North America. In: Röhrig, E. and Ulrich, B. (eds) *Temperate Deciduous Forest. Ecosystems of the World 7*. Elsevier, Amsterdam, pp. 219–344.

Barnes, T.A. and Van Lear, D.H. (1998) Prescribed fire effects on advanced regeneration in mixed hardwoods stands. *Southern Journal of Applied Forestry* 22, 138–142.

Bartley, D.D., Jones, I.P. and Smith, R.T. (1990) Studies in the flandrian vegetational history of the Craven District of Yorkshire: the Lowlands. *Journal of Ecology* 78, 611–632.

Beaufoy, G., Baldock, D. and Clark, J. (1994) *The Nature of Farming. Traditional Low Intensity Farming and its Importance for Wildlife*. Institute for European Environmental Policy, London.

Beaufoy, G., Baldock, D. and Clark, J. (1995) *The Nature of Farming. Low Intensity Farming Systems in Nine European Countries*. The Institute for European Environmental Policy, London.

Beck, D.E. and Hooper, R.M. (1986) Development of a southern Appalachian hardwood stand after clearcutting. *Southern Journal of Applied Forestry* 10, 168–172.

Beck, O. and Göttsche, D. (1976) Untersuchungen über das Konkurrenzverhalten von Edellaubhölzern in Jungbeständen. *Forstarchiv* 47, 86–91.

Beck, O.A. (1977) Die Vogelkirsche (*Prunus aivum* L.). Ein Beitrag zur Ökologie und wirtschaftlichen Bedeutung. *Forstarchiv* 48, 154–158.

Beck, O.A. (1981) Plädoyer für eine starke waldbauliche Berücksichtigung der Vogelkirsche. *Allgemeine Forstzeitschrift*, 36, 212–213.

Becker, B. (1983) Postglaziale Auwaldentwicklung im mittleren und oberen Maintal anhand dendrochronologischer Untersuchungen subfossiler Baumstammablagerungen. *Geologisches Jahrbuch* A 71, 45–59, Hannover.

Becker, B. and Glaser, R. (1991). Baumringsignaturen und Wetteranomalien (Eichenbestand Guttenberger Forst, Klimastation Wurzberg). *Forstwissenschaftliches Centralblatt* 110, 66–83.

Becker, B. and Kromer, B. (1993) The continental tree-ring record: absolute chronology, ^{14}C calibration and climate change at 11 ka. *Palaeogeography, Palaeoclimatology, Palaeoecology* 103, 67–71.

Becker, B. and Schirmer, W. (1977) Palaeoecological study on the Holocene valley development of the River Main, southern Germany. *Boreas* 6, 303–321.

Becker, B., Kromer, B. and Trimborn, P. (1991) A stable-isotope tree-ring timescale of the Late Glacial/Holocene boundary. *Nature* 353, 647–649.

Begon, M., Harper, J.L. and Townsend, C.R. (1990) *Ecology, Individuals, Populations and Communities*, 2nd edn. Blackwell Scientific Publications, London.

Behre, K.E. (1981) The interpretation of anthropogenic indicators in pollen diagrams. *Pollen et spores* 23, 225–245.

Behre, K.E. (1988) The rôle of man in European vegetation history. In: Huntley, B and Webb, T. III, (eds) *Vegetation History. Section IV: Smaller-scale Studies. Handbook of Vegetation Science*, vol. 7. Kluwer Academic Publishers, Dordrecht, pp. 633–672.

Bell, M. and Walker, M.J.C. (1992) *Late Quaternary Environmental Change. Physical and Human Perspectives*. John Wiley & Sons, New York.

Belostokov, G.P. (1980) Morphogenesis of *Tilia cordata* Mill.; bush-shaped regrowth. *Lesoved* 6, 53–59.

Bennett, K.D. (1983) Devensian Late-Glacial and Flandrian vegetational history at Hockham Mere, Norfolk, England I. Pollen percentages and concentrations. *New Phytologist* 95, 457–487.

Bennett, K.D. (1986) Competive interactions among forest tree populations in Norfolk, England, during the last 10,000 years. *New Phytologist* 103, 603–620.

Bennett, K.D. (1988a) Holocene pollen stratigraphy of central East Anglia, England, and comparison of pollen zones across the British Isles. *New Phytologist* 109, 237–253.

Bennett, K.D. (1988b) Post-glacial vegetation history: ecological considerations. In: Huntley, B. and Webb, T. III (eds) *Vegetation History, Section IV: Smaller-scale Studies. Handbook of Vegetation Science*, vol. 7. Kluwer Academic Publishers, Dordrecht, pp. 699–724.

Bennett, K.D. (1988c) A provisional map of forest types for the British Isles 5,000 years ago. *Journal of Quaternary Science* 4, 141–144.

Bennett, K.D., Tzedakis, P.C. and Willis, K.J. (1991) Quaternary refugia of north European trees. *Journal of Biogeography* 18, 103–115.

Berendse, F. (1990) Organic matter accumulation and nitrogen mineralization during secondary succession in heathland ecosystems. *Journal of Ecology* 78, 413–427.

Berge, K., van den Maddelein, D. and Muys, B. (1993) Recent structural changes in the beech forest reserve of Groenendaal (Belgium). In: Broekmeyer, M.E.A., Vos, W. and Koop, H. (eds) *European Forest Reserves. Proceedings of the European Forest Reserves Workshop, 6–8 May 1992, Wageningen, The Netherlands.* Pudoc Scientific Publishers, Wageningen, pp. 195–198.

Berglund, B.E., (1991) Environment and society in selected areas. Introduction; The Köpinge area; the Bjäresjö area; the Krageholm area; the Romele area. In: Berglund, B.E. (ed.) *The Cultural Landscape during 6,000 Years in Southern Sweden. The Ystad Project.* Ecological Bulletins, Copenhagen 41, pp. 109–112, 167–174, 221–224, 247–249.

Berglund, B.E., Lemdahl, G., Liedberg-Jönsson, B. and Persson, T. (1984) Biotic response to climatic changes during the time span 13,000–10,000 B.P.; a case study from S.W. Sweden. In: Mörner, N.-A. and Karlén, W. (eds) *Climatic Changes on a Yearly to Millennial Basis.* Reidel, Dordrecht, pp. 25–36.

Berglund, B.E. and Persson, T. (1986a) Pollen/vegetation relationships in grazed and mowed plant communities of South Sweden. In: Behre, K.-E. (ed.) *Anthropogenic Indicators in Pollen Diagrams.* A.A. Balkema, Rotterdam, pp. 37–51.

Berglund, B.E., Persson, T., Emanuelsson, U. and Persson, S. (1986b) Pollen/vegetation relationship in grazed and mowed plant communities of South Sweden. In: Behre, K.-E. (ed.) *Anthropogenic Indicators in Pollen Diagrams.* A.A. Balkema, Rotterdam.

Berglund, B.E., Malmer, N. and Persson, T. (1991a) Landscape – ecological aspects of long-term changes in the Ystad area. In: Berglund, B.E. (ed.) *The Cultural Landscape during 6,000 Years in Southern Sweden. The Ystad Project.* Ecological Bulletins, Copenhagen 41, pp. 405–424.

Berglund, B.E., Larsson, L., Leevan, N., Gunilla, E., Olsson, A. and Skansjö, S. (1991b) Ecological and social factors behind the landscape changes. In: Berglund, B.E. (ed.) *The Cultural Landscape during 6,000 Years in Southern Sweden. The Ystad Project.* Ecological Bulletins, Copenhagen 41, pp. 425–435.

Bernadzki, E., Bolibok, L., Brzeziecki, B., Zajaczkowski, J. and Zybura, H. (1998) Compositional dynamics of natural forests in the Białowieza National Park, northeastern Poland. *Journal of Vegetation Science* 9, 229–238.

Bernátsky, J. (1905) Anordnung der Formationen nach ihrer Beeinflussung seitens der menschlichen Kultur und der Weidetiere. *Botanische Jahrbücher für Systematik, Pflanzengeschichte und Pflanzengeographie* 34, 1–8, Leipzig.

Bertsch, K. (1929) Klima, Pflanzendecke und Besiedlung Mitteleuropas in vor- und frühgeschichtlicher Zeit nach dem Ergebnissen der pollenanalystischen Forschung. *Berichte der Römisch-Germanische Kommission des Deutschen Archeologischen Instituts* 18, 1–67.

Bertsch, K. (1932) Die Pflanzenreste der Pfahlbauten von Sipplingen und Langenrain im Bodensee. *Badische Fundberichte* 2, 305–320.

Bertsch, K. (1949) *Geschichte des deutschen Waldes*. Gustav Fischer, Jena.

Best, J.A. (1998) Persistent outcomes of coppice grazing in Rockingham Forest, Northamptonshire, UK. In: Kirby, K.J. and Watkins, C. (eds) *The Ecological History of European Forests*. CAB International, Wallingford, UK, pp. 63–72.

Beutler, A. (1992) Die Großtierfauna Mitteleuropas und ihr Einfluß auf die Landschaft. *Landschafsökologie Weihenstephan*, 6, 25 Jahre Lehrstuhl für Landschafsökologie in Weihenstephan mit Prof. Dr L.C.W. Haber. Festschrift mit Beiträgen ehemaliger und derzeitiger Mitarbeiter (Hrsg. Von F. Duhme, R. Lenz und L. Spandau), pp. 49–69.

Beutler, A. (1996) Die Großtierfauna Europas und ihr Einfluß auf Vegetation und Landschaft. In: Gerken, B. and Meyer, C. (eds) *Wo lebten Pflanzen und Tiere in der Naturlandschaft und der frühen Kulturlandschaft Europas? Natur und Kulturlandschaft. Referate der gleichnamigen Tagung am 22. Und 23. März 1995 in Neuhaus im Solling.* Universität-Gesamthochschule Paderborn, Höxter, pp. 51–106.

Bezacinský, H. (1971) Das Hainbuchenproblem in der Slowakei. *Acta Facultatis Forestalis* 8, 7–36.

Bieleman, J. (1992) *Geschiedenis van de landbouw in Nederland 1500–1950. Verandering en verscheidenheid.* Boom, Meppel/Amsterdam.

Bignal, E.M. and McCracken, D.I. (1992) *Prospects for Nature Conservation in European Pastoral Farming Systems. A Discussion Document.* Joint Nature Conservation Committee, Peterborough.

Bignal, E.M. and McCracken, D.I. (1996) Low-intensity farming systems in the conservation of the countryside. *Journal of Applied Ecology* 33, 413–424.

Bignal, E.M., McCracken, D.I. and Curtis, D.J. (1994) *Nature Conservation and Pastoralism in Europe. Proceedings of the Third European Forum on Nature Conservation and Pastoralism 21–24 July 1992, University of Pau, France.* Joint Nature Conservation Committee, Peterborough.

Bindseil, D. (1958) Die Vogelkirsche als Waldbaum. *Holz-Zentralblatt* 84, 1039–1040.

Bink, F.A. (1992) *Ecologische Atlas van de Dagvlinders van Noordwest-Europa*. Instituut voor Bos- en Natuuronderzoek en Unie van Provinciale Landschappen, Schuyt & Co, Haarlem.

Birks, H.J.B. (1973) Preface. In: Birks, H.J.B. and West, R.G. (eds) *Quaternary Plant Ecology. The 14th symposium of the British Ecological Society, University of Cambridge, 28–30 March 1972.* Blackwell Scientific Publications, Oxford, p. ix.

Birks, H.J.B. (1981) The use of pollen analysis in the reconstruction of past climates: a review. In: Wigley, T.M., Ingram, M.J. and Farmer, G. (eds) *Climate and History*. Cambridge University Press, Cambridge, pp. 111–138.

Birks, H.J.B. (1986a) Numerical zonation, comparison and correlation of Quaternary pollen-stratigraphical data. In: Berglund, B.E. (ed.) *Handbook of Holocene Palaeoecology and Palaeohydrology*. John Wiley & Sons, Chichester, pp. 743–774.

Birks, H.J.B. (1986b) Late-Quaternary biotic changes in terrestrial and lacustrine environments, with particular reference to north-west Europe. In: Berglund, B.E. (ed.) *Handbook of Holocene Palaeoecology and Palaeohydrology*. John Wiley & Sons, Chichester, pp. 3–65.

Birks, H.J.B. (1989) Holocene isochrone maps and patterns of tree-spreading in the British Isles. *Journal of Biogeography* 16, 503–540.

Birks, H.J.B. (1993) Quaternary palaeoecology and vegetation science. Current contributions and possible future developments. *Review of Palaeobotany and Palynology* 79, 153–177.

Björkman, L. (1997) The history of *Fagus* forests in southwestern Sweden during the last 1500 years. *The Holocene* 7, 419–432.

Björkman, L. (1999) The establishment of *Fagus sylvatica* at the stand-scale in southern Sweden. *The Holocene* 9, 237–245.

Björkman, L. and Bradshaw, R. (1996) The immigration of *Fagus sylvatica* L. and *Picea abies* (L.) Karst. into a natural forest stand in southern Sweden during the last 2000 years. *Journal of Biogeography* 23, 235–244.

Björse, G. and Bradshaw, R. (1998) 2000 years of forest dynamics in southern Sweden: suggestions for forest management. *Forest Ecology and Management* 104, 15–26.

Björse, G., Bradshaw, R.H.W. and Michelson, D. (1996) Calibration of regional pollen data to construct maps of former forest types in southern Sweden. *Journal of Paleolimnology* 16, 76–78.

Blink, H. (1929) *Woeste gronden, ontginning en bebossching in Nederland voormaals en thans*. v.h. Moeton & Co., 's-Gravenhage.

Blytt, A. (1876) Essay on the immigration of the Norwegian flora during alternate rainy and dry periods. Cammermeyer, Christiania.

Bobiec, A. (1998) The mosaic diversity of field layer vegetation in the natural and exploited forests of Białowieza. *Plant Ecology* 136, 175–187.

Bock, W. (1932/33) Der Urwald bei Sababurg im Reinhardswald. *Naturschutz, Monatschrift für alle Freunde der Deutschen Heimat* 14, 46–50.

Böckmann, T. (1990) Wachstum und Ertrag der Winterlinde (*Tilia cordata* Mill.) in Niedersachsen und Nordhessen. Dissertation Universität Göttingen, Göttingen.

Boden, I. (1931) Abhandlungen. Die Anzucht und Nachzucht der Eiche im akademischen Lehrrevier Freienwalde. *Jagd und Forstwesen* 63, 185–196.

Bodziarczyk, J., Michalcewicz, J. and Szwagrzyk, J. (1999) Secondary forest succession in abandoned glades of the Pieniny National Park. *Polish Journal of Ecology* 47, 175–189.

Boeijnk, D.E., de Geus, M. and Schalk, B. (1992) *Loofbomen in en buiten het bos, 2*. Dick Coutinho, Muiderberg.

Bogucki, P. (1988) *Forest Farmers and Stockholders. Early Agriculture and its Consequences in North-Central Europe*. Cambridge University Press, Cambridge.

Bogucki, P. and Grygiel, R. (1983) Early farmers of the North European plain. *Scientific American* 248, 104–112.

Bohnke, S., Vandenberghe, J., Coope, G.R. and Reiling, R. (1987) Geomorphology and palaeoecology of the Mark valley (southern Netherlands): palaeoecology, palaeohydrology and climate during the Weichselien Late Glacial. *Boreas* 16, 69–85.

Bokdam, J. (1987) Foerageergedrag van jongvee in het Junner Koeland in relatie tot het voedselaanbod. In: de Bie, S., Joenje, W. and van Wieren, S.E. (eds) *Begrazing in de natuur*. Pudoc, Wageningen, pp. 165–186.

Bönecke, G. (1993) Ein Auen-Urwald in Sudmähren. *Allgemeine Forstzeitschrift* 48, 608–610.

Bonnemann, A. (1956a) Eichen-Buchen Mischbestände. *Allgemeine Forst- und Jagdzeitung* 127, 33–42.

Bonnemann, A. (1956b) Eichen-Buchen Mischbestände. *Allgemeine Forst- und Jagdzeitung* 127, 118–126.

Bonnier, G. and de Layens, G. (1974) *Flore complète portative de la France, de la Suisse et de la Belgique.* Libraire générale de l'enseignement, Paris.

Borck, K.-H. (1954) Zur Bedeutung der Wörter Holz, Wald, Forst und Witu im Althochdeutschen. *Festschrift für Jost Trier*, Meisenheim, pp. 456–476.

Bormann, F.H. and Likens, G.E. (1979a) *Pattern and Process in a Forested Ecosystem.* Springer Verlag, Berlin.

Bormann, F.H. and Likens, G.E. (1979b) Catastrophic disturbance and steady state in Northern hardwood forests. *American Scientific* 67, 660–669.

Borowski, S. and Kossak, S. (1972) The natural food preferences of the European bison in seasons free of snow cover. *Acta Theriologica* 17, 151–169.

Borowski, S. and Kossak, S. (1975) The food habits of deer in the Białowieza Primeval Forest. *Acta Theriologica* 20, 463–506.

Borse, Ch. (1939) Über die Frage der Pollenproduktion, Pollenzerstörung und Pollenverbreitung in ost-preussischen Waldgebieten. *Schriftenreihe d. Physisch-ökonomische Gesellschaft zu Königsberg (Pr.)* 71, 128–144.

Børset, O. (1976) Probleme der Naturverjüngung in den nordischen Wäldern. *Journal Forestier Suisse* 127, 165–181.

Bosinski, G. (1983) Die jägerische Geschichte des Rheinlandes. Einsichten und Lücken. *Jahrbuch des Römisch-Germanischen Zentralmuseums.* Mainz 30, 81–112.

Bossema, J. (1968) Recovery of acorns in the European jay (*Garrulus G. glandarius* L.). *Proceedings Koninklijke Nedederlandse Akademie van Wetenschappen* Serie C, *Biological and Medical Sciences* 71, 10–14.

Bossema, J. (1979) Jays and oaks: an eco-ethological study of a symbiosis. PhD. thesis, Rijksuniversiteit Groningen, Groningen. (Also published in *Behaviour* 70, 1–117.)

Botkin, D.B. (1979) A grandfather clock down the staircase: stability and disturbance in natural ecosystems. In: *Forests: Fresh Perspectives from Ecosystem Analysis. Proceedings of the 40th Annual Biology Colloquium.* Oregon University Press, Oregon, pp. 1–10.

Botkin, D.B. (1990) Oaks in New Jersey: machine-age forests. In: Botkin, D.B. (ed.) *Discordant Harmonies. A New Ecology for the Twenty-first Century.* Oxford University Press, New York, pp. 51–71.

Botkin, D.B. (1993) *Forest Dynamics. An Ecological Model.* Oxford University Press, Oxford.

Botkin, D.B., Janak, J.F. and Wallis, J.R. (1972). Some ecological consequences of a computer model of forest growth. *Journal of Ecology* 60, 849–872.

Bottema, S. (1987) De invloed van de vegetatie op de fauna in Nederland gedurende het Laat Quartair. *Nederlands Bosbouwtijdschrift* 59, 287–294.

Bozilova, E. and Beug, H.-J. (1992) On the Holocene history of vegetation in S.E. Bulgaria (Lake Arkutino, Ropotamo region). *Vegetation History and Palaeobotany* 1, 19–32.

Bradshaw, R.H.W. (1981a) Modern pollen-representation factors for woods in south-east England. *Journal of Ecology* 69, 45–70.

Bradshaw, R.H.W. (1981b) Quantitative reconstruction of local woodland vegetation using pollen analyses from a small basin in Norfolk, England. *Journal of Ecology* 69, 941–955.

Bradshaw, R.H.W. (1988) Spatially-precise studies of forest dynamics. In: Huntley, B. and Webb, T. III (eds) *Vegetation History. Section IV: Smaller-scale Studies. Handbook of Vegetation Science*, vol. 7. Kluwer Academic Publishers, Dordrecht, pp. 725–751.

Bradshaw, R. (1993) Forest response to Holocene climatic change: equilibrium or non-equilibrium. In: Chambers, F.M. (ed.) *Climate Change and Human Impact on the Landscape*. Chapman & Hall, London, pp. 57–65.

Bradshaw, R. and Holmqvist, B.H. (1999) Danish forest development during the last 3000 years from regional pollen data. *Ecography* 22, 53–62.

Bradshaw, R. and Mitchell, F.J.G. (1999) The palaeoecological approach to reconstructing former grazing-vegetation interactions. *Forest Ecology and Management* 120, 3–12.

Bradshaw, R., Gemmel, P. and Björkman, L. (1994) Development of nature-based silvicultural models in southern Sweden: the scientific background. *Forest and Landscape Research* 1, 95–110.

Branch, E.D. (1962) *The Hunting of the Buffalo*. University of Nebraska Press, Lincoln.

Breitenfeld, E. and Mothes, K. (1940) Bestandesgeschichtliche Untersuchungen an mazurischen Wäldern. *Schrift. d. Physisch-ökonomische Gesellschaft zu Königsberg (Pr.)* 71, 240–299.

Briedermann, L. (1990) *Schwarzwild*. 2. bearbeitete Auflage. VEB Deutscher Landwirtschaftverlag, Berlin.

Britton, N.L. and Brown, H.A. (1947) *An illustrated Flora of the Northern United States, Canada and the British Possessions*, vol. II. The New York Botanical Garden.

Brockmann-Jerosch, H. (1936) Futterbäume und Speiselaubbäume. *Berichte der Schweizerische Botanischen Gesellschaft* 46, 594–613.

Broekmeyer, M.E.A. and Vos, W. (1990) Stand van zaken in het Nederlandse bosreservaten-onderzoek. *Nederlands Bosbouw Tijdschrift* 62, 244–347.

Broekmeyer, M.E.A. and Vos, W. (1993) Forest reserves in Europe: a review. In: Broekmeyer, M.E.A., Vos, W. and Koop, H. (eds) *European Forest Reserves. Proceedings of the European Forest Reserves Workshop, 6–8 May 1992, Wageningen, The Netherlands*. Pudoc Scientific Publishers, Wageningen, pp. 9–28.

Broekmeyer, M.E.A., Vos, W. and Koop, H. (eds) (1993) *European Forest Reserves. Proceedings of the European Forest Reserves Workshop, 6–8 May 1992, Wageningen, The Netherlands*. Pudoc Scientific Publishers, Wageningen, pp. 1–2.

Brokaw, N.V.L. (1985) Treefalls, regrowth and community structure in tropical forests. In: Pickett, S.T.A. and White, P.S. (eds) *The Ecology of Natural Disturbance and Patch Dynamics*. Academic Press, Orlando, Florida, pp. 53–69.

Bromley, S.W. (1935) The original forest types of southern New England. *Ecological Monographs* 5, 61–89.

Brookes, P.C., Wigston, D.L. and Bourne, W.F. (1980) The dependence of *Quercus robur* and *Q. petraea* seedlings on cotyledon potassium, magnesium, calcium and phosphorus during the first year of growth. *Forestry* 53, 167–177.

Brose, P., Van Lear, D. and Cooper, R. (1999a) Using shelterwood harvest and prescribed fire to regenerate oak stands on productive upland sites. *Forest Ecology and Management* 113, 125–141.

Brose, P.H., Van Lear, D.H. and Keyser, P.D. (1999b) A shelterwood-burn technique for regenerating productive upland oak sites in the Piedmont region. *Southern Journal of Applied Forestry* 23, 158–163.

Broström, A., Gaillard, M.-J., Ihnse, M. and Odgaard, B. (1998) Pollen-landscape relationships in modern analogues of ancient cultural landscapes in southern Sweden – a first step towards quantification of vegetation openness in the past. *Vegetation History and Archaeobotany* 7, 189–201.

Brouwer, R. (1962a) Distribution of dry matter in the plant. *Netherlands Journal for Agricultural Science* 10 (5 Special Issue), 361–376.

Brouwer, R. (1962b) Nutritive influences on the distribution of dry matter in the plant. *Netherlands Journal for Agricultural Science* 10 (5 Special Issue), 399–408.

Brouwer, R. and Kuiper, P.J.C. (1972) *Leerboek der plantenfysiologie. Deel 3. Oecofysiologische relaties.* Oosthoek, Utrecht.

Brown, J.H. (1991) Methodological advances. New approaches and methods in ecology. In: Real, L.A. and Brown, J.H. (eds) *Foundations in Ecology. Classic Papers with Commentaries.* University of Chicago Press, Chicago, p. 445.

Brown, J.M.B. (1953) Studies on British beechwoods. *Forestry Commission Bulletin* No. 20. Her Majesty's Stationary Office, London.

ten Bruggencate, K. (1990) *Wolters' Woordenboek Engels-Nederlands.* 20ste druk. Wolters-Noordhoff, Groningen.

Bruin, D., de Hamhuis, D., Nieuwenhuijze, L., van Overmars, W., Sijmons, D. and Vera, F. (1987) *Ooievaar. De toekomst van het rivierengebied.* Stichting Gelderse Milieufederatie, Arnhem.

Buckland, P.C. and Edwards, K.J. (1984) The longevity of pastoral episodes of clearance activity in pollen diagrams: the role of post-occupation grazing. *Journal of Biogeography* 11, 243–249.

Bühler, A. (1918) *Der Waldbau nach wissenschaftlicher Forschung und praktischer Erfahrung.* I Band. Eugen Ulmer, Stuttgart.

Bühler, A. (1922) *Der Waldbau nach wissenschaftlicher Forschung und praktischer Erfahrung.* II Band. Eugen Ulmer, Stuttgart.

Buis, J. (1985) *Historia Forestis: Nederlandse bosgeschiedenis.* Deel 1 en 2. H & S Uitgevers, Utrecht.

Buis, J. (1993) *Holland Houtland. Een geschiedenis van het Nederlandse bos.* Prometheus, Amsterdam.

Bunce, R.G.H. (1982) Some effects of man on the structure of Atlantic deciduous forests. In: von Dierschke, H. (ed.) *Struktur und Dynamik von Wäldern. Berichte der Internationale Symposium der Internationalen Verein für Vegetationskunde.* J. Cramer, Valduz, pp. 681–698.

Bunzel-Drüke, M. (1997) Großherbivore und Naturlandschaft. *Schriftenreihe für Landschaftspflege und Naturschutz* 54, 109–128.

Bunzel-Drüke, M., Drüke, J. and Vierhaus, H. (1994) Quaternary Park. Überlegungen zu Wald, Mensch und Megafauna. *Arbeitsgemeinschaft Biologischer Umweltschutz im Kreis Soest e.V.* 17/18, 4–38.

Bunzel-Drüke M., Drüke, J. and Vierhaus, H. (1995) Wald, Mensch und Megafauna – Gedanken zur Holozänen Naturlandschaft in Westfalen. – *LÖBF – Mitteilungen* 4, 43–51.

Burrichter, E. (1977) Vegetationsbereicherung und Vegetationsverarmung unter dem Einfluss des prähistorischen und historischen Menschen. *Natur und Heimat* 37, 46–51.

Burrichter, E., Pott, R., Raus, T. and Wittig, R. (1980) Die Hudelandschaft 'Borkener Paradies' im Emstal bei Meppen. *Abhandlungen aus den Landesmuseum für Naturkunde zu Münster in Westfalen.* Münster 42, Jahrgang 4.

Burrichter, E., Hüppe, J. and Pott, R. (1993) Agrarwirtschaftlich bedingte Vegetationsbereicherung und -verarmung in historischer Sicht. *Phytocoenologia* 23, 427–447.

Burschel, P. (1975) Schalenwildbestände und Leistungsfähigkeit des Waldes als Problem der Forst- und Holzwirtschaft aus der Sicht des Waldbaus. *Allgemeine Forstzeitschrift* 30, 214–221.

Burschel, P. and Huss, J. (1964) Die Reaktion von Buchensämlingen auf Beschattung. *Forstarchiv* 35, 225–223.

Burschel, P. and Schmaltz, J. (1965a) Die Bedeutung des Lichtes für die Entwicklung junger Buchen. *Allgemeine Forst- und Jagdzeitung* 136, 193–210.

Burschel, P. and Schmaltz, J. (1965b) Untersuchungen über die Bedeutung von Unkraut- und Altholzkonkurrenz für junge Buchen. *Forstwissenschaftliches Centralblatt* 84, 230–243.

Burschel, P., Huss, J. and Kalbhenn, R. (1964) *Die natürliche Verjüngung der Buche.* Schriftenreihe der Forstlichen Fakultät der Universität Göttingen und Mitteilungen der Niedersächsischen Forstlichen Versuchsanstalt. Band 34. J.D. Sauerländer's Verlag, Frankfurt am Main.

Bush, M.B. (1993) An 11,400 year paleoecological history of a British chalk grassland. *Journal of Vegetation Science* 4, 47–66.

Bush, M.B. and Flenley, J.R. (1987) The age of the British chalk grassland. *Nature* 329, 434–436.

Buttenschøn, J. and Buttenschøn, R.M. (1978) The effect of browsing by cattle and sheep on trees and bushes. *Natura Jutlandica* 20, 79–94.

Buttenschøn, J. and Buttenschøn, R.M. (1985) Grazing experiments with cattle and sheep on nutrient poor, acidic grassland and heath. IV.: establishment of woody species. *Natura Jutlandica* 21, 47–140.

van Caenegem, R.C. (1967) *De instellingen van de Middeleeuwen. Deel I. De geschiedenis van de westerse staatsinstellingen van de V^{de} tot de XV^{de} eeuw.* Scientia, Gent.

Caesar (51 BC) Book I. *The Conquest of Gaul* (transl. S.A. Handford). Revised with new introduction by J.F. Gardner (1982). Penguin Books, Harmondsworth.

Cajander, A.K. (1909) *Ueber Waldtypen.* J. Simelii Arfvingars Boktryckeriaktiebolag, Helsingfors.

Calcote, R. (1995) Pollen source area and pollen productivity: evidence from forest hollows. *Journal of Ecology* 83, 591–602.

Calcote, R. (1998) Identifying forest stand types using pollen from forest hollows. *The Holocene* 8, 423–432.

Canham, C.D. (1985) Suppression and release during canopy recruitment in *Acer saccharum. Bulletin of the Torrey Botanical Club* 112, 134–145.

Canham, C.D. (1989) Different responses to gaps among shade-tolerant tree species. *Ecology* 70, 548–550.

Canham, C.D. and Marks, P.L. (1985) The response of woody plants to disturbance: patterns of establishment and growth. In: Pickett, S.T.A. and White, P.S. (eds) *The Ecology of Natural Disturbance and Patch Dynamics.* Academic Press, Orlando, Florida, pp. 197–216.

Cantor, L. (1982a) Forests, chases, parks and warrens. In: Cantor, L. (ed.) *The English Medieval Landscape.* Croom Helm, London, pp. 56–85.

Cantor, L. (1982b) Introduction: the English medieval landscape. In: Cantor, L. (ed.) *The English Medieval Landscape.* Croom Helm, London, pp. 17–24.

Casparie, W.A. (1985) The neolithic wooden trackway XXI (Bou) in the raised bog at Nieuw Dordrecht (The Netherlands). BAI, Groningen, *Paleohistoria* 24, 115–164.

Casparie, W.A. and van Zeist, W. (1960) A late glacial lake deposit near Waskemeer (Prov. of Friesland). *Acta Botanica Neerlandica* 9, 191–196.

Caspers, G. (1993) Vegetationgeschichtliche Untersuchungen zur Flussauenentwicklung an der Mittelweser im Spätglazial und Holozän. *Abhandlungen aus dem Westfälischen Museum für Naturkunde* 55, 1.

ten Cate, C.L. (1972) *'Wan god mast gift. ...'. Bilder aus der Geschichte der Schweinezucht im Walde.* Pudoc, Wageningen.

Cermak, (Tschermak) L. (1910) Einiges über den Urwald von waldbaulichen Gesichtspunkten. *Centralblatt für das gesamte Forstwesen* 36, 340–370.

Chaix, L. (1994) L'aurochs d'Evital et les aurochs de Franche-Comté. Dans *Aurochs, le retour. Aurochs, vaches et autres bovins de la préhistoire à nos jours.* Centre Jurassien du Patrimoine, Lons-le-Saunier, pp. 67–75.

Chamber, J.C., Vander Wall, S.B. and Schupp, E.W. (1999) Seed and seedling ecology of Piñon and Juniper species in the Pygmy woodlands of western North America. *The Botanical Review* 65, 1–38.

Chard, J.S.R. (1953) Highland birch. *Scottish Forestry* 7, 125–128.

Chen, S.H. (1988) Neue Untersuchungen über die spät- und postglaziale Vegetationsgeschichte im Gebiet zwischen Harz und Leine (BRD). *Flora* 181, 147–177.

Chettleburgh, M.R. (1952) Observations on the collection and burial of acorns by jays in Hinault Forest. *British Birds* 45, 359–64. (Also further note (1955) *British Birds* 48, 183–184.)

Cho, D.-S. and Boerner, R.E.J. (1991) Canopy disturbance patterns of *Quercus* species in two Ohio old-growth forests. *Vegetation* 93, 9–18.

Christensen, N.L. (1977) Changes in structure, pattern, and diversity associated with climax forest maturation in Piedmont, North Carolina. *American Midland Naturalist* 97, 176–188.

Cieslar, A. (1909) Licht und Schattenholzarten, Lichtgenuss und Bodenfeuchtigkeit. *Zentralblatt für das gesamte Forstwesen* 25, 4–22.

Clark, F.B. (1993) An historical perspective of oak regeneration. In: Loftis, D.L. and McGee, C.E. (eds) *Oak Regeneration: Serious Problems, Practical Recommendations. Symposium Proceedings Knoxville, Tennessee September 8–10, 1992.* Southeastern Forest Experiment Station, Asheville, pp. 3–13.

Clark, J.S. (1997) Facing short-term extrapolation with long-term evidence: Holocene fire in the north-eastern US forests. *Journal of Ecology* 85, 377–380.

Clark, J.S. and Royall, P.D. (1995) Transformation of a northern hardwood forest by aboriginal (Iroquois) fire: charcoal evidence from Crawford Lake, Ontario, Canada. *The Holocene* 5, 1–9.

Clark, J.S., Merkt, J. and Müllers, H. (1989) Post-glacial fire, vegetation, and human history on the northern Alpine forelands, south-western Germany. *Journal of Ecology* 77, 897–925.

Clark, J.S., Royall, P.D. and Chumbley, C. (1996) The role of fire during climate change in an eastern deciduous forest at Devil's Bathtub, New York. *Ecology* 77, 2148–2166.

Clason, A.T. (1967) Animal and man in Holland's past: an investigation of the animal world surrounding man in prehistoric and early historic times in the provinces of North and South Holland. PhD thesis, University of Groningen, J.B. Wolters, Groningen. (Also appeared as: (1967) *Paleohistoria* 13A.)

Clason, A.T. (1977) *Jacht en veeteelt van prehistorie tot Middeleeuwen.* Fibula-van Dishoeck, Haarlem.

Clements, F.E. (1916) *Plant Succession. An Analysis of the Development of Vegetation.* Publication no. 242. Carnegie Institution, Washington, DC.

Clinton, B.D., Boring, L.R. and Swank, W.T. (1993) Canopy gap characteristics and drought influences in oak forests of the Coweeta Basin. *Ecology* 74, 1551–1558.

Clutton-Brock, T.H., Price, O.F., Albon, S.D. and Jewell, P.A. (1991) Persistent instability and population regulation in Soay sheep. *Journal of Animal Ecology* 60, 593–608.

Coard, R. and Chamberlain, A.T. (1999) The nature and timing of faunal change in the British Isles across the Pleistocene/Holocene transition. *The Holocene* 9, 372–376.

Coffin, D.P. and Urban, D.L. (1993) Implications of natural history traits to system level dynamics. Comparisons of a grassland and a forest. *Ecological Modelling* 67, 147–178.

Coles, J.M. and Orme, B.J. (1983) *Homo sapiens* or *Castor fiber? Antiquity* 57, 95–102.

Collins, B.S., Dunne, K.P. and Pickett, S.T.A. (1985) Responses of forest herbs to canopy gaps. In: Pickett, S.T.A. and White, P.S. (eds) *The Ecology of Natural Disturbance and Patch Dynamics.* Academic Press, Orlando, Florida, pp. 217–234.

Connell, J.H. (1978) Diversity in tropical rain forests and coral reefs. *Science* 199, 1302–1310.

Connell, J.H. and Slatyer, R.O. (1977) Mechanisms of succession in natural communities and their role in community stability and organization. *American Naturalist* 111, 1110–1144.

Cook, J.E., Sharik, T.L. and Smith, D.W. (1998) Oak regeneration in the southern Appalachians: potential, problems, and possible solutions. *Southern Journal of Applied Forestry* 22, 11–18.

Coope, G.R. (1977) Fossil coleopteran assemblages as sensitive indicators of climatic changes during the Devensian (last) cold stage. *Philosophical Transactions of the Royal Society of London* Series B 280, 313–340 and discussion.

Coope, G.R. (1994) The response of insect faunas to glacial-interglacial climatic fluctuations. *Philosophical Transactions of the Royal Society of London* Series B 344, 19–26.

Coops, H. (1988) Occurence of blackthorn (*Prunus spinosa* L.) in the area of Mols Bjerge and the effect of cattle- and sheep-grazing on its growth. *Nature Jutlandica* 9, 169–176.

Cordy, J-M. (1991) Palaeoecology of the Late Glacial and early Postglacial of Belgium and neighbouring areas. In: Barton, N., Roberts, A.J. and Roe, D.A. (eds) *The Late Glacial in North-west Europe: Human Adaptation and Environmental Change at the End of the Pleistocene.* CBA Research Report no. 77, Oxford, pp. 40–47.

Cornelissen, P. and Vulink, J.T. (1995) *Begrazing in jonge wetlands.* Flevobericht Nr. 367. Ministerie van Verkeer en Waterstaat. Directoraat-Generaal Rijkswaterstaat. Directie Ijsselmeergebied, Lelystad.

Cornelissen, P. and Vulink, J.T. (1996a) *Edelherten en reeën in de Oostvaardersplassen.* Demografie en terreingebruik. Flevobericht Nr. 397. Ministerie van Verkeer en Waterstaat. Directoraat-Generaal Rijkswaterstaat. Directie Ijsselmeergebied, Lelystad.

Cornelissen, P. and Vulink, J.T. (1996b) *Grote herbivoren in wetlands.* Flevobericht Nr. 399. Ministerie van Verkeer en Waterstaat. Directoraat-Generaal Rijkswaterstaat. Directie Ijsselmeergebied, Lelystad (with English summary).

Cotta, H. (1865) *Anweisung zum Waldbau* (Neunte, neubearbeitete Auflage). Arnoldische Buchhandlung, Leipzig.

Cottam, G. (1949) The phytosociology of an oak woods in southwestern Wisconsin. *Ecology* 30, 271–287.

Covington, W.W. and Moore, M.M. (1994) Southwestern Ponderosa forest structure. Changes since Euro-American settlement. *Journal of Forestry* 92, 39–47.

Cramp, S. (ed.) (1980) *Handbook of the Birds of Europe, the Middle East, and North Africa. The Birds of the Western Palearctic,* Vol. II, *Hawks to Bustards.* Oxford University Press, Oxford.

Cramp, S. (ed.) (1988) *Handbook of the Birds of Europe, the Middle East, and North Africa. The Birds of the Western Palearctic,* Vol. V, *Tyrant Flycatchers to Thrushes.* Oxford University Press, Oxford.

Cramp, S. (ed.) (1992) *Handbook of the Birds of Europe, the Middle East, and North Africa. The Birds of the Western Palearctic,* Vol. VI, *Warblers.* Oxford University Press, Oxford.

Crocker, R.L. and Major, J. (1955) Soil development in relation to vegetation and surface age at Glacier Bay, Alaska. *Journal of Ecology* 43, 427–448.

Cronon, W. (1983) *Changes in the Land. Indians, Colonists, and the Ecology of New England.* Hill and Wang, New York.

Crow, T.R., Johnson, W.C. and Atkinson, C.S. (1994) Fire and recruitment of *Quercus* in a postagricultural field. *American Midland Naturalist* 131, 84–97.

Current, A.P. (1991) A Late Glacial Interstadial mammal fauna from Gough's Cave, Somerset, England. In: Barton, N., Roberts, A.J. and Roe, D.A. (eds) *The Late Glacial in North-west Europe: Human Adaptation and Environmental Change at the End of the Pleistocene.* CBA Research Report no. 77, Oxford, pp. 48–50.

Curtis, D. and Bignal, E. (1990) *The Conservation Rôle of Pastoral Agriculture in Europe.* A discussion Document. Scottish Chough Study Group, Peterborough.

Curtis, D.J., Bignal, E.M. and Curtis, M.A. (eds) (1991) Birds and pastoral agriculture in Europe. *Proceedings of the Second European Forum on Bird and Pastoralism. Port Erin, Isle of Man, 26–30 October 1990.* Scottish Chough Study Group. Joint Nature Conservation Committee, Peterborough.

Curtis, J.T. (1970) The modification of mid-latitude grasslands and forests by man. In: Thomas, W.L. (ed.) *Man's Role in Changing the Face of the Earth.* The University of Chicago Press, Chicago, pp. 721–736.

Dabrowski, M.J. (1959) Late-glacial and Holocene history of Białowieza Primeforest. Part I., Białowieza National Park. *Acta Societatis Botanicorum Poloniae* 28, 197–248.

Dagenbach, H. (1981) Der Speierling, ein seltener Baum in unseren Wäldern und Obstgärten. *Allgemeine Forstzeitschrift* 36, 214–217.

Dannecker, K. (1955) Laubwaldbewirtschaftung im Sinne der Plenteridee. *Schweizerische Zeitschrift für Forstwesen* 106, 291–302.

Dansgaard, W., White, J.W.C. and Johnsen, S.J. (1989) The abrupt termination of the Younger Dryas climate event. *Nature* 339, 532–534.

Darby, H.C. (1970) The clearing of the woodland in Europe. In: Thoman, W.L. Jr (ed.) *Man's Rôle in Changing the Face of the Earth.* The University of Chicago Press, Chicago, London, pp. 183–216.

Darby, H.C. (1976) Domesday England. In: Darby, H.C. (ed.) *A New Historical Geography of England before 1600.* Cambridge University Press, Cambridge, pp. 39–74.

Darley-Hill, S. and Johnson, W.C. (1981) Acorn dispersal by the blue jay (*Cyanocitta cristata*). *Oecologia* 50, 231–232.

Darlington, A. (1974) The galls on oak. In: Morris, M.G. and Perring, F.H. (eds) *The*

British Oak. Its History and Natural History. The Botanical Society of the British Isles, E.W. Classey, Berkshire, pp. 298–311.

Dau, J.H.C. (1829) *Neues Handbuch über den Torf dessen Natur, Entstehung und Wiedererzeugung.* Leipzig.

Davidson, I. (1989) Escaped domestic animals and the introduction of agriculture to Spain. In: Clutton-Brock, J. (ed.) *The Walking Larder. Patterns of Domestication, Pastoralism, and Predation.* Unwin Hyman, London, pp. 59–71.

Davies, W.J. and Pigott, C.D. (1982) Shade tolerance of forest trees. *National Environment Research Council News* 7, 17–18.

Davis, M.B. (1963) On the theory of pollen analysis. *American Journal of Science* 261, 897–912.

Davis, M.B. (1967a) Late-glacial climate in northern United States: A comparison of New England and the Great Lakes region. In: Cushing, E.J. and Wright, H.E. (eds) *Quaternary Ecology.* Yale University Press, New Haven, Connecticut, pp. 11–43.

Davis, M.B. (1967b) Pollen accumulation rates at Rogers Lake, Connecticut, during the Late- and Post-glacial time. *Review of Palaeobotany and Palynology* 2, 219–230.

Davis, M.B. (1984) Holocene vegetational history of the eastern United States. In: Wright, H.E. Jr (ed.) *Late-Quaternary Environments of the United States.* vol. 2. *The Holocene.* Longman, London, pp. 166–181.

Davis, M.B. and Goodlett, J.C. (1960) Comparison of the present vegetation with pollen-spectra in surface samples from Brownington Pond, Vermont. *Ecology* 41, 346–357.

Davis, S.J.M. (1987) *The Archaeology of Animals.* B.T. Batsford, London.

Davison, S.E. (1981) Tree seedling survivorship at Hutcheson Memorial Forest New Jersey. *William L. Hutcheson Memorial Forest Bulletin* 6, 4–7.

Day, D. (1989) *Vanished Species.* Gallery Books, New York.

Day, G.M. (1953) The Indian as an ecological factor in the northeastern forest. *Ecology* 34, 329–346.

Day, S.P. (1991) Post-glacial vegetational history of the Oxford region. *New Phytologist* 119, 445–470.

Day, S.P. (1993) Woodland origin and 'ancient woodland indicators': a case-study from Sidlings Copse, Oxfordshire, UK. *The Holocene* 3, 45–53.

Degerbøl, M. (1964) Some remarks on Late- and Post-glacial vertebrate fauna and its ecological relations in northern Europe. In: Macfadyen, A. and Newbould, P.J. (eds) *British Ecological Society Jubilee Symposium. London, 28–30 March 1963. Journal of Ecology* 52, 71–85.

Degerbøl, M. and Fredskild, B. (1970) *The Urus (Bos primigenius Bojanus) and neolithic domesticated cattle (Bos taurus domesticus Linné) in Denmark. With a revision of Bos-remains from the kitchen middens. Zoölogical and palynological investigations.* Biologiske Skrifter, 17. Det Kongelike Danske Videnskabernes Selskap.

Degerbøl, M. and Iversen, J. (1945) The Bison in Denmark. *Danmarks Geologiske Undersøgelse II.* Raekke No. 73 (*Geological Survey of Denmark* No. 73).

Delcourt, P.A. and Delcourt, H.R. (1987a) Late-Quaternary dynamics of temperate forests: applications of palaeoecology to issues of global environmental change. *Quaternary Science Reviews* 6, 129–146.

Delcourt, P.A. and Delcourt, H.R. (1987b) *Long-term Forest Dynamics of the Temperate Zone.* Springer Verlag, New York.

Delcourt, H.R. and Delcourt, P.A. (1991) *Quaternary Ecology. A Paleoecological Perspective.* Chapman & Hall, London.

De Monté Verloren, J.P.H. and Spruit, J.E. (1982) *Hoofdlijnen uit de ontwikkeling der rechterlijke organisatie in de Noordelijke Nederlanden tot de Bataafse omwenteling,* 6th edn. Kluwer, Deventer.

Dengler, A. (1931) Aus den Südosteuropäischen Urwäldern II, Die Ergebnisse eine Probeflächenaufnahme im Buchenurwald Albaniens. *Zeitschrift für Forst- und Jagdwesen* 63, 20–29.

Dengler, A. (1935) *Waldbau auf ökologischer Grundlage,* 2nd edn. Springer, Berlin.

Dengler, A. (1990) *Waldbau auf ökologischer Grundlage,* Zweiter band. In: Röhrig, E. and Gussone, H.A. (eds) *Baumartenwahl, Bestandesbegründung und Bestandespflege,* 6th edn. Verlag Paul Parey, Hamburg.

Dengler, A. (1992) *Waldbau auf ökologischer Grundlage,* Erster Band. In: Röhrig, E. and Gussone, H.A. (eds). *Der Wald als Vegetationsform und seine Bedeutung für den Menschen,* 6th edn. Verlag Paul Parey, Hamburg.

Den Uyl, D. (1945) Farm woodlands should not be grazed. *Journal of Forestry* 43, 729–732.

Den Uyl, D. (1962) The Central region. In: Barrett, J.W. (ed.) *Regional Silviculture of the United States.* The Ronald Press Company, New York, pp. 137–177.

Den Uyl, D., Diller, O.D. and Day, R.K. (1938) The development of natural reproduction in previously grazed farmwoods. *Purdue University Agriculture Experiment Station Bulletin* 431, 1–28.

Derkman, G.F.M. and Koop, H.G.J.M. (1977) *Struktuur en verjonging van een oerbos.* Praktijkverslag Natuurbehoud en Natuurbeheer. Landbouwhogeschool, Wageningen, LH/NG, projectnr. P2, Wageningen.

Diamond, J.M. (1989) Historic extinction: a Rosette Stone for understanding prehistoric extinctions. In: Martin, P.S. and Klein, R.G (eds) *Quaternary Extinctions. A Prehistoric Revolution.* The University of Arizona Press, Tucson, pp. 824–862.

Dierschke, H. (1974) Saumgesellschaften im Vegetations- und Standortsgefälle an Waldrändern. *Scripta Geobotanica.* (Göttingen) 6, 3–246.

Dietrich, H., Müller, S. and Schlenker, G. (1970) *Urwald von morgen.* Ulmer Verlag, Stuttgart.

Dietsch, M.-F. (1996) Gathered fruits and cultivated plants at Bercy (Paris), a Neolithic village in a fluvial context. *Vegetation History and Archaeobotany* 5, 89–97.

Diez, Chr. (1989) Der Waldkirschbaum. Porträt einer Baumart. *Wald und Holz* 70, 780–795.

Dimbleby, G.W. (1984) Anthropogenic changes from neolithic through medieval times. *New Phytologist* 98, 57–72.

Dister, E. (1980) Geobotanische Untersuchungen in der Hessischen Rheinaue als Grundlage für die Naturschutzarbeit. Thesis, University of Göttingen.

Dister, E. (1985) Zur Struktur und Dynamik alter Hartholzauenwälder (*Querco-ulmetum* Issl. 24) am nördlichen Oberrhein. *Verhandlungen der Zoologisch-Botanischen Gesellschaft in Österreich* 123, 13–32.

Dister, E. and Drescher, A. (1987) Zur Struktur, Dynamik und Ökologie lang überschwemmter Hartholzauenwälder an der unteren March (Niederösterreich). *Verhandlungen der Gesellschaft für Ökologie* (Graz 1985) 15, 295–302.

Dodd, J.R. and Stanton, R.J. Jr (1990) *Paleoecology. Concepts and Applications.* 2nd edn. John Wiley & Sons, New York.

Dodge, S.L. (1997) Successional trends in a mixed oak forest on High Mountain, New Jersey. *Journal of the Torrey Botanical Society* 124, 312–317.

Dohrenbusch, A. (1987) Kann die 'relative Beleuchtungsstärke' die Lichtverhältnisse im Wald zuverlässig charakterisieren? *Forstarchiv* 58, 24–27.

Doing-Kraft, H. and Westhoff, V. (1958) De plaats van de beuk (*Fagus sylvatica*) in het midden- en west Europese bos. *Jaarboek Nederlandse Dendrologische Vereniging* 2, 226–254.

Dolman, P. and Sutherland, W. (1991) Historical clues to conservation. *New Scientist* 129, 40–44.

Domet, P. (1873) *Histoire de la forêt de Fontainebleau*. Hachette, Paris, Laffitte reprints, Marseille (1979).

Don, P. (1985) *Zuid-Holland. Kunstreisboek*. P.N. van Kampen & Zoon, Amsterdam.

van der Donck, A. (1655) In: O'Donnell (ed.) *A Description of the New Netherlands*. (Translation in English). Syracuse University Press, Syracuse.

van der Donck, A. (1655) *Beschryvinghe van Nievv Nederlandt*. Evert Nieuwenhof, Amsterdam.

Dornbusch, P. (1988) Bestockungsprofile in Dauerbeobachtungsflächen im Biosphärenreservat Mittere Elbe, DDR. *Archiv für Naturschutz und Landschaftsforschung* 28, 245–263.

Drent, R.H. and Prins, H.H.T. (1987) The herbivore as prisoner of its food supply. In: van Andel, J., Bakker, J.P. and Snaydon, R.W. (eds) *Disturbance in Grasslands. Causes, Effects and Processes*. Dr W. Junk Publishers, Dordrecht, pp.131–147.

Duby, G. (1968) *Rural Economy and Country Life in the Medieval West* (translated from the French). Edward Arnold Publishers.

Dupré, S., Thiébaut, S. and Teissier du Cros, E. (1986) Morphologie et architecture des jeunes hêtres (*Fagus sylvatica* L.). Influence du milieu, variabilité génétique. *Annales des Sciences Forestières (Paris)* 43, 85–102.

Ebeling, K. and Hanstein, U. (1988) Eichenkulturen unter Kiefernaltholzschirm. *Forst und Holz* 43, 463–467.

Edlin, H.L. (1964) A modern sylva or a discourse of forest trees. 9. Limes – Tilia spp. *Quarterly Journal of Forestry* 58, 135–141.

Edwards, K.J. (1982) Man, space and the woodland edge; speculations on the detection and interpretation of human impact in pollen profiles. In: Bell, M. and Limbrey, S. (eds) *Archaeological Aspects of Woodland Ecology. Symposia of the Association for Environmental Archaeology*, No. 2. BAR International Series 146. Oxford, pp. 5–22.

Edwards, K.J. (1983) Quaternary palynology: consideration of a discipline. *Progress in Physical Geography* 7, 113–125.

Edwards, K.J. (1985) The anthropogenic factor in vegetational history. In: Edwards, K.J. and Warren, W.P. (eds) *The Quaternary History of Ireland*. Academic Press, London, pp. 187–220.

Edwards, K.J. (1993) Models of mid-Holocene forest farming for north-west Europe. In: Chambers, F.M. (ed.) *Climate Change and Human Impact on the Landscape*. Chapman & Hall, London, pp. 133–145.

Edwards, P.J. and Gillman, M.P. (1987) Herbivores and plant succession. In: Gray, A.J., Crawley, M.J. and Edwards, P.J. (eds) *Colonization, Succession and Stability*. Blackwell Scientific Publications, Oxford, pp. 295–314.

Edwards, K.J. and MacDonald, G.M. (1991) Holocene palynology: II. Human influence and vegetation change. *Progress in Physical Geography* 15, 364–391.

Eglar, F.E. (1954) Vegetation science concepts. I. Initial floristic composition, a factor in old-field vegetation development. *Vegetatio* 4, 412–417.

Ehrenfeld, J.G. (1980) Understory response to canopy gaps of varying size in a mature oak forest. *Bulletin of the Torrey Botanical Club* 107, 29–41.

Eichhorn (1927) Waldbauliche Erfahrungen in den Hardtwaldungen des unteren Rheintales. *Allgemeine Forst- und Jagdzeitung* 103, 169–185.

Eichwald, E. (1830) *Naturhistorische Skizze von Lithauen, Volhynien und Podolien in geognostisch-mineralogischer, botanischer und zoölogischer Hinsicht*. Joseph Zawadski, Wilna.

von Eickstedt, F. (1959) Huten und Hutebuchen; ein Characteristikum der Rhön. *Allgemeine Forstzeitschrift* 14, 126.

Eijgenraam, F. (1992) Kalenders van het hout. NRC Handelsblad, 19 maart. *Wetenschap en Onderwijs*, 1–2.

Eisenhut, G. (1957) Blühen, Fruchten und Keimen in der Gattung *Tilia*. Thesis, University of München.

Ekstam, U. and Sjörgen, E. (1973) Studies on past and present changes in deciduous forest vegetation on Öland. *Zoon*, Uppsala (Suppl. 1) 123–135.

Elerie, J.N.H. (1993) Cultuurhistorie en de ecologie van een veldcomplex op de Hondsrug. In: Elerie, J.N.H. (ed.) *Landschapsgeschiedenis van de Strubben/Kniphorstbos. Archeologische en historisch-ecologische studies van een natuurgebied op de Hondsrug*. Van Dijk en Foorthuis REGIO-PRoject, Groningen, III, pp. 79–165.

Ellenberg, H. (1954) Steppenheide und Waldweide. Ein vegetationskundlicher Beitrag zur Siedlungs- und Landschaftsgeschichte. *Erdkunde* 8, 188–194.

Ellenberg, H. (1986) *Vegetation Mitteleuropas mit den Alpen in ökologischer Sicht*. Vierte, verbesserte Auflage. Verlag Eugen Ulmer, Stuttgart.

Ellenberg, H. (1988) *Vegetation Ecology of Central Europe*, 4th edn. Cambridge University Press, Cambridge.

Ellenberg, H. (1990) *Bauernhaus und Landschaft in ökologischer und historischer Sicht*. Verlag Eugen Ulmer, Stuttgart.

Endres, G. (1929a) Die Eichen des Spessarts. *Forstwissenschaftliches Centralblatt* 73, 149–157.

Endres, G. (1929b) Die Eichen des Spessarts. II. Die gegenwärtig vorhandenen Eichenbestände. *Forstwissenschaftliches Centralblatt* 73, 208–216.

Endres, G. (1929c) Die Eichen des Spessarts. III. Preissteigerung während der letzten 100 Jahre. *Forstwissenschaftliches Centralblatt* 73, 229–240.

Endres, G. (1929d) Die Eichen des Spessarts. IV. Masten. *Forstwissenschaftliches Centralblatt* 73, 277–289.

Endres, G. (1929e) Die Eichen des Spessarts. VI. Die Pflege. *Forstwissenschaftliches Centralblatt* 73, 316–327.

Endres, M. (1888) *Die Waldbenutzung von 13. bis Ende des 18. Jahrhunderts. Ein Beitrag zur Geschichte der Forstpolitik*. Verlag der H. Laupp'schen Buchhandlund, Tübingen.

Erdtman, G. (1931) The boreal hazel forests and the theory of pollen statistics. *Journal of Ecology* 19, 158–163.

Erteld, W. (1963) Über die Wachstumsentwichlung der Linde. *Archiv für Forstwesen* 12, 1152–1158.

van Es, W.A. (1994a) Volksverhuizingen en continuïteit. In: van Es, W.A. and Hessing, W.A.M. (eds) *Romeinen, Friezen en Franken in het hart van Nederland*. ROB, Matrijs, Amersfoort, pp. 64–81.

van Es, W.A. (1994b) Friezen, Franken en Vikingen. In: van Es, W.A. and Hessing,

W.A.M. (eds) *Romeinen, Friezen en Franken in het hart van Nederland.* ROB, Matrijs, Amersfoort, pp. 82–119.

Escherich, G. (1917) In den Jagdgründen des Tzaren. 1. Der Wildstand einst und jetzt. Bialowies in deutscher Verwaltung. *Hrsg. v.d. Militärforstverwaltung Bialowies.* Zweites Heft. Berlin, Paul Parey, pp. 192–218.

Escherich, G. (1927) *Im Urwald.* Verlag von Georg Stilke, Berlin.

Evans, J. (1982) Silviculture of oak and beech in northern France: observations and current trends. *Quarterly Journal of Forestry* 76, 75–82.

Evans, J. (1992) Coppice forestry: an overview. In: Buckley, G.P. (ed.) *Ecology and Management of Coppice Woodland.* Chapman & Hall, London, pp. 18–27.

Evans, J.G. (1993) The influence of human communities on the English chalklands form the Mesolithic to the Iron Age: the molluscan evidence. In: Chambers, G.M. (ed.) *Climate Change and Human Impact on the Landscape.* Chapman & Hall, London, pp. 147–156.

Evans, J.G., Limbrey, S. and Cleere, H. (eds) (1975) *The Effect of Man on the Landscape: The Highland Zone.* Research Report no. 11. The Council for British Archaeology, London.

Evans, P. (1975) The intimate relationship: an hypothesis concerning pre-Neolithic land use. In: Evans, J.G., Limbrey, S. and Cleere, H. (eds) *The Effect of Man on the Landscape: The Highland Zone.* Research Report no. 11. The Council for British Archaeology, London, pp. 43–48.

Fabricius (1879) Die rheinischen Auenwaldungen. *Allgemeine Forst- und Jagdzeitung* 55, 84–88.

Fabricius, L. (1929) Forstliche Versuche VII. Neue Versuche zur Feststellung des Einflusses von Wurzelwettbewerb und Lichtentzug des Schirmstandes auf den Jungwuchs. *Forstwissenschaftliches Centralblatt* 13, 477–506.

Faegri, K. and Iversen, J. (1989) In: Faegri, K., Kaland, P.E. and Krzywinsky, K. (eds) *Textbook of Pollen Analyses,* 4th edn. John Wiley & Sons, Chichester.

Falinski, J.B. (1966) Antropogeniczna roslinnosc Puszczy Bialowieskiej jako wynik synantropizacji naturalnego kompleksu lesnego – Végetation anthropogène de la Grande Forêt de Białowieza comme un résultat de la synanthropisation du territoire silvestre naturel. Dissertationes Universitatis Varsoviensis (in Polish with French summary).

Falinski, J.B. (1976) Windwürfe als Faktor der Differenzierung und der Veränderung des Urwaldbiotopes im Licht der Forschungen auf Dauerflächen. *Phytocoenosis* 5, 85–108.

Falinski, J.B. (1977) Białowieza Primeval Forest. *Phytocoenosis* 6, 133–148.

Falinski, J.B. (1986) *Vegetation dynamics in temperate lowland primeval forests.* Ecological studies in Białowieza forest, *Geobotany* 8. Dr W. Junk Publishers, Dordrecht.

Falinski, J.B. (1988) Succession, regeneration and fluctuation in the Białowieza Forest (NE Poland). *Vegetatio* 77, 115–128.

Falinski, J.B. (1993) Aims of nature conservation and scientific functions of reserves. In: Broekmeyer, M.E.A., Vos, W. and Koop, H. (eds) *European Forest Reserves. Proceedings of the European Forest Reserves Workshop, 6–8 May 1992, Wageningen, The Netherlands.* Pudoc Scientific Publishers, Wageningen, pp. 49–53.

Fenner, M. (1987) Seed characteristics in relation to succession. In: Gray, A.J. (ed.) *Colonization, Succession and Stability.* Blackwell Scientific Publications, Oxford, pp. 103–114.

Fenton, E.W. (1948) Some factors affecting the natural regeneration of oak in certain parts of south-east Scotland. *Transactions of the Botanical Society of Edinburgh* 34, 213–232.

Finegan, B. (1984) Forest succession. *Nature* 312, 109–184.

Firbas, F. (1934) Über die Bestimmung der Walddichte und der Vegetation Waldloser Gebiete mit Hilfe der Pollenanalyse. *Planta* 22, 109–146.

Firbas, F. (1935) Die Vegetationsentwicklung des Mitteleuropäischen Spätglacials. *Bibliotheca Botanica* 112, 1–68.

Firbas, F. (1949) *Spät- und nacheiszeitliche Waldgeschichte Mitteleuropas nördlich der Alpen. Erster Band: Allgemeine Waldgeschichte.* Verlag von Gustav Fischer, Jena.

Firbas, F. (1952) *Spät- und nacheiszeitliche Waldgeschichte Mitteleuropas nördlich der Alpen. Zweiter band: Waldgeschichte der einzelnen Landschaften.* Verlag von Gustav Fischer, Jena.

Fitter, A.H. and Jennings, R.D. (1975) The effects of sheep grazing on the growth and survival of seedling junipers (*Juniperus communis* L.). *Journal of Applied Ecology* 12, 637–642.

Fleder, W. (1976) Waldbausysteme im Spessart. *Allgemeine Forstzeitschrift* 31, 737–740.

Fleder, W. (1988) Zur Eichenwirtschaft im Spessart. *Allgemeine Forstzeitschrift* 43, 735–737.

Flexner, S.B. and Hauck, L.C. (1983) *The Random House Dictionary of the English Language,* 2nd edn, unabridged. Random House, New York.

Flörcke, E. (1967) Vegetation und Wild bei Sababurg im Reinhardswald in Vergangenheit und Gegenwart. *Geobotanische Mitteilungen* 48. R. Knapp, Giessen.

Flower, N. (1977) An historical and ecological study of enclosed and unenclosed Woods in the New Forest, Hampshire. MSc thesis, King's College, University of London.

Flower, N. (1980) The management history and structure of unenclosed woods in the New Forest, Hampshire. *Journal of Biogeography* 7, 311–328.

Focke, O. (1871) Ein Stück deutschen Urwaldes. *Österreichische Botanische Zeitschrift* 21, 310–315.

Forbes, A.C. (1902) On the regeneration and formation of woods from seed naturally or artificially sown. *Transactions of the English Arboricultural Society* 5, 239–270.

Foster, D.R. (1992) Land-use history (1730–1990) and vegetation dynamics in central New England, USA. *Journal of Ecology* 80, 753–772.

Fowels, H.A. (1965) *Silvics of Forest Trees of the United States.* US Forest Service, United States Department of Agriculture, Washington, DC.

Frank, F. (1939) Die Nachzucht der Eiche im Badischen Frankenland. *Forst- und Jagdzeitung* 115, 173–196.

Freiherr von Berlepsch, B. (1979) Die Linde im südlichen Vogelsberg. *Allgemeine Forstzeitung* 34, 845–846.

Freist-Dorr, M. (1992) Das Einzelbaumwachstum in langfristig beobachteten Mischbestandsversuchen, dargestellt am Beispiel der Eichen-Buchen-Versuchsfläche Waldbrunn 105. *Forstwissenschaftliches Centralblatt* 111, 106–116.

Frenzel, B. (1983) Die Vegetationsgeschichte Süddeutschlands im Eiszeitalter. In: Müller-Beck, H.J. (ed.) *Urgeschichte in Baden Württemberg.* Konrad Theiss, Verlag, Stuttgart, pp. 91–166.

Fricke, O. (1982) Die Entwicklung von Eichen-Jungwüchsen und -Jungbeständen mit gleichalten Mischbaumarten. Thesis University of Göttingen, Göttingen.

Fricke, O. (1986) Gleichzeitig oder später – wie sollten die Mischbaumarten in die Eichen eingebracht werden? *Forst- und Holzwirt* 41, 7–11.

Fricke, O., Kürschner, K. and Röhrig, E. (1980) Unterbau in einem Stieleichenbestand. *Forstarchiv* 51, 228–232.

Friederici, G. (1930) Der Grad der Durchdringbarkeit Nord-Amerikas im Zeitalter der Entdeckungen und ersten Durchforschung des Kontinents durch die Europäer. *Petermanns Mitteilungen Erganzungsheft* 209, 216–229.

Fröhlich, J. (1930) Der südosteuropäische Urwald und seine Überführung in Wirtschaftswald. *Centralblatt für das Gesamte Forstwesen* 56, 49–65.

Fröhlich, J. (1954) *Urwaldpraxis. 40 jährige Erfahrungen und Lehren.* Neumann Verlag, Radebuil and Berlin.

Frost, I. and Rydan, H. (1997a) Effects of competition, grazing and cotyledon nutrient supply on growth of *Quercus robur* seedlings. *Oikos* 79, 53–58.

Frost, I. and Rydan, H. (1997b) Spatial pattern and size distribution of the animal-dispersed tree *Quercus robur* in two spruce-dominated forests. In: Frost, I. *Dispersal and Establishment of Quercus robur. Importance of Cotyledons, Browsing and Competition.* Part II, PhD thesis, Uppsala University, Uppsala, pp. 2–21.

Fry, G.L.A. (1991) Conservation in agricultural ecosystems. In: Spellenberg, I.F., Goldsmith, F.B. and Morris, M.G. (eds) *The Scientific Management of Temperate Communities for Conservation.* Blackwell Scientific Publications, Oxford, pp. 415–443.

Fuhlendorf, S.D. and Smeins, F.E. (1997) Long-term vegetation dynamics mediated by herbivores, weather and fire in a *Juniperus–Quercus* savanna. *Journal of Vegetation Science* 8, 819–828.

Fuller, R.J. (1992) Effects of coppice management on woodland breeding birds. In: Buckley, G.P. (ed.) *Ecology and Management of Coppice Woodlands.* Chapman & Hall, London, pp. 169–192.

Gaasbeek, F., Kooiman, M., Olde Meierink, B. and Blijdenstein, R. (eds) (1991) *Wijk bij Duurstede. Geschiedenis en architectuur. Monumenten inventarisatie provincie Utrecht.* Kerkebosch, Zeist.

Gaillard, M.-J., Birks, H.J.B., Emanuelsson, U. and Berglund, B.E. (1992) Modern pollen/land-use relationships as an aid in the reconstruction of past land-uses and cultural landscapes: an example from south Sweden. *Vegetation History and Archaeobotany* 1, 3–17.

Gaillard, M.-J., Birks, H.J.B., Emanuelsson, U., Karlsson, S., Lagerås, P. and Olausson, D. (1994) Application of modern pollen/land-use relationships to the interpretation of pollen diagrams; reconstructions of land-use history in south Sweden, 3000–0 BP. *Review of Palaeobotany and Palynology* 82, 47–73.

Gaillard, M.-J., Birks, H.J.B., Ihse, M. and Runberg, S. (1998) Pollen/landscape calibrations based on modern pollen assemblages from surface-sediments samples and landscape mapping – a pilot study in south Sweden. In: Gaillard, M.-J. and Berglund, B.E. (eds) Quantification of land surfaces cleared of forest during the Holocene–modern pollen/vegetation/landscape relationships as an aid to the interpretation of pollen data. *Paläoklimaforschung/Palaeoclimate Research* 27, 31–52.

Gammon, A.D., Rudolph, V.J. and Arend, J.L. (1960) Regeneration following clearcutting of oak during a seed year. *Journal of Forestry* 58, 711–715.

Garcia, D., Zamora, R., Gómez, J.M. and Hódar, J.A. (1999) Bird rejection of unhealthy fruits reinforces mutualism between juniper and its avian dispersers. *Oikos* 85, 536–544.

Gardner, G. (1975) Light and the growth of ash. In: Evans, G.C., Rackham, O. and Bainbridge, R. (eds) *Light as an Ecological Factor II.* Blackwell Scientific Publications, Oxford, pp. 557–563.

Gayer, K. (1886) *Der gemischte Wald, seine Begründung und Pflege, insbesondere durch Horst- und Gruppenwirtschaft.* Paul Parey, Berlin.

Gayer, K. (1895) *Über den Femelschlagbetrieb und seine Ausgestaltung in Bayern.* Berlin.

van Geel, B., Bohncke, S.J.P. and Dee, H. (1980/1981) A palaeoecological study of an upper Lateglacial and Holocene sequence from "De Borchert", the Netherlands. *Review of Palaeobotany and Palynology* 31, 348–367.

van Geel B., Coope, G.R. and van der Hammen, T. (1989) Palaeoecology and stratigraphy of the Lateglacial type section at Usselo (The Netherlands). *Review of Palaeobotany and Palynology* 60, 25–129.

Geerts, G. and Heestermans, H. (1995) *Van Dale; groot woordenboek der Nederlandse taal.* 12th edn. Van Dale Lexicografie, Utrecht, Antwerpen.

Geiser, R. (1983) Die Tierwelt der Weidelandschaften. In: *Schutz von Trockenbiotopen: Trockenrasen, Triften und Hutungen.* Akademie Für Naturschutz und Landschaftspflege (ANL), Laufen/Salzach, pp. 55–65.

Geiser, R. (1992) *Auch ohne Homo sapiens wäre Mitteleuropa von Natur aus eine halboffene Weidelandschaft.* Laufener Seminararbeit, Akademie für Naturschutz und Landschaftspflege (ANL), Laufen/Salzach, 2/92, pp. 22–34.

Geist, V. (1991) Phantom subspecies: the Wood Bison *Bison bison 'athabascae'* Rhoads 1897 is not a valid taxon, but an ecotype. *Arctic* 44, 283–300.

Geist, V. (1992) Endangered species and the law. *Nature* 357, 274–276.

Genssler, H. (1980) Naturwaldzellen in Nordrhein-Westfalen. *Nederlands Bosbouw Tijdschrift* 52, 104–112.

Genthe, F. (1918) Die Geschichte des Wisents in Europa, Bialowies in deutscher Verwaltung. *Hrsg. v.d. Militärforstverwaltung Bialowies.* Paul Parey, Drittes Heft. Berlin, pp. 119–140.

Gerken, B. and Meyer, C (1996) *Wo lebten Pflanzen und Tiere in der Naturlandschaft und der frühen Kulturlandschaft Europas?* Natur und Kulturlandschaft 1, Höxter.

Gerken, B. and Meyer, C. (1997) *Vom Waldinnensaum zur Hecke.* Natur und Kulturlandschaft 2, Höxter.

Gilbert, J.M. (1979) *Hunting and Hunting Reserves in Medieval Scotland.* John Donald Publishers, Edinburgh.

Girling, M.A. and Greig, J.R.A. (1977) Palaeoecological investigations of a site at Hampstead Heath, London. *Nature* 268, 45–47.

Girling, M.A. and Greig, J.R.A. (1985) A first fossil record for *Scolytus scolytus* (F.) (elm bark beetle): its occurrence in elm decline deposits from London and the implications for Neolithic elm disease. *Journal of Archaeological Science* 12, 347–351.

Glavac, V. (1968) Über Eichen-Hainbruchenwälder Kroatiens. *Feddes Repertorium* 79, 115–138.

Glavac, V. (1969) Über die Stieleichen-Auenwälder der Save-Niederung. *Schriftenreihe für Vegetationskunde* 4, 103–108.

Gleason, H.A. (1926) The individualistic concept of the plant association. *Torrey Botanical Club Bulletin* 53, 7–26.

Godwin, H. (1929) The sub-climax and deflected succession. *Journal of Ecology* 17, 144–147.

Godwin, H. (1934a) Pollen analysis. An outline of the problems and potentialities of the method. Part I. Technique and interpretation. *New Phytologist* 33, 278–305.

Godwin, H. (1934b) Pollen analysis. An outline of the problems and potentialities of the method. Part II. General applications of pollen analysis. *New Phytologist* 33, 325–358.

Godwin, H. (1944) Neolithic forest clearance. *Nature* 153, 511–512.

Godwin, H. (1975a) *The History of the British Flora,* 2nd edn. Cambridge University Press, Cambridge.

Godwin, H. (1975b) The history of the natural forests of Britain: establishment, dominance and destruction. *Philosophical Transactions of the Royal Society of London* Series B 271, 47–67.

Godwin, H. and Deacon, J. (1974) Flandrian history of oak in the British Isles. In: Morris, M.G. and Perring, F.H. (eds) *The British Oak. Its History and Natural History.* The Botanical Society of the British Isles. Pendragon Press, Cambridge, pp. 51–61.

Godwin, H. and Tallantire, P.A. (1951) Studies in the Post-glacial history of British vegetation XII, Hockham Mere, Norfolk. *Journal of Ecology* 39, 285–307.

Göransson, H. (1986) Man and the forests of nemoral broad-leaved trees during the Stone Age. *Striae* 24, 143–152.

Gordon, I.J. (1988) Facilitation of red deer grazing by cattle and its impact on red deer performance. *Journal of Applied Ecology* 25, 1–10.

Gordon, R.B. (1969) The natural vegetation of Ohio in pioneer days. *Bulletin of the Ohio Biological Survey,* New Series, vol. III (2), 1–109.

Goriup, P.D., Batten, L.A. and Norton, J.A. (eds) (1991) The conservation of lowland dry grassland birds in Europe. *Proceedings of an International Seminar held at the University of Reading, 20–22 March 1991.* Joint Nature Committee, Peterborough.

Gorter, H.P. (1986) *Ruimte voor natuur.* Vereniging tot Behoud van Natuurmonumenten in Nederland, 's-Graveland.

Gothe, H. (1949) Forstmeister Friedrich Jäger und sein Wirken im Forstamt Schlitz. *Forstwissenschaftliches Centralblatt* December, 761–769.

Götmark, F. (1992) Naturalness as an evaluation criterion in nature conservation: a response to Anderson. *Conservation Biology* 6, 455–458.

Gottwald, H. (1985) Kirschbaum; ein Klassiker unter den Möbelhölzern. *Holz Aktuell* 5, 15–29.

Gould, S.J. (1965) Is uniformitarianism necessary? *American Journal of Science* 263, 223–228.

Gould, S.J. (1989) *Wonderful Life. The Burgess Shale and the Nature of History.* W.W. Norton & Co., New York.

Gove, P.B. (1986) *Webster's Third New International Dictionary of the English Language,* unabridged. Merriam-Webster, Springfield, Massachusetts.

Gradman, R. (1901) Das mitteleuropäische Landschaftsbild nach seiner geschichtlichen Entwicklung. *Geographische Zeitschrift* 7, 361–447.

Graham, R.W. (1986) Response of mammalian communities to environmental changes during the late Quaternary. In: Diamond, J. and Case, T.J. (eds) *Community Ecology.* Harper & Row, New York, pp. 300–313.

Grant, M. (1973) *Tacitus. The Annals of Imperial Rome.* Penguin Books, Harmondsworth, UK.

Grand-Mesnil, M.N. (1982) À propos des réserves biologiques. Questions d'histoire. *Bulletin Association des Amis de Forêt de Fontainebleau* 1, 5–13.

Graumlich, L.J. (1993) High resolution pollen analyses provides new perspectives on catastrophic elm decline. *Trends in Ecology and Evolution* 8, 387–388.

Green, B.H. (1989) Conservation in cultural landscapes. In: Western, D. and Pearl, M.C. (eds) *Conservation for the Twenty-first Century.* Oxford University Press, Oxford, pp. 182–198.

Greig, J. (1982) Past and present lime woods of Europe. In: Bell, M. and Limbrey, S. (eds) *Archaeological Aspects of Woodland Ecology. Symposia of the Association for Environmental Archaeology*. No. 2 BAR International Series 146, Oxford, pp. 23–54.

Greig, J. (1992) The deforestation of London. *Review of Palaeobotany and Palynology* 73, 71–86.

Greig-Smith, P. (1982) A.S. Watt, FRS: A biographical note. In: Newman. E.I. (ed.) *The Plant Community as a Working Mechanism*. Special Publication number 1 of the British Ecological Society. *Produced as a Tribute to A.S. Watt*. Blackwell Scientific Publications, Oxford, pp. 9–10.

Grigson, C. (1978) The Late Glacial and Early Flandrian ungulates of England and Wales; an interim review. In: Limbrey, S. and Evans, J.G. (eds) *The Effect of Man on the Landscape: The Lowland Zone*. BA Research Report 21, London, pp. 46–56.

Grime, J.P. (1974) Vegetation classification by reference to strategies. *Nature* 250, 26–31.

Grime, J.P. (1977) Evidence for the existence of three primary strategies in plants and its relevance to ecological and evolutionary theory. *American Naturalist* 111, 1169–1194.

Grime, J.P. (1979) *Plant Strategies and Vegetation Processes*. John Wiley & Sons, Chichester.

Grimm, E.C. (1988) Data analysis and display. In: Huntley, B. and Webb, T. III (eds) *Vegetation History. Section I: Background and Methods. Handbook of Vegetation Science*, vol. 7. Kluwer Academic Publishers, Dordrecht, pp. 43–76.

Groenman-van Waateringe, W. (1968) The elm decline and the first appearance of *Plantago major. Vegetatio* 15, 292–296.

Groenman-van Waateringe, W. (1983) The early agricultural utilization of the Irish landscape: the last word on the elm decline? In: Reeves-Smyth, T. and Hamond, F. (eds) *Landscape Archeology in Ireland*. British Archaeological Reports, Oxford, BS 116, pp. 217–232.

Groenman-van Waateringe, W. (1988a) Palynologie of plaggen soils on the Veluwe, central Netherlands. In: Groenman-van Waateringe, W. and Robinson, M. (eds) *Man-made Soils Symposia of the Association for Environmental Archaeology* No. 6 BAR International Series 410, Oxford, pp. 55–61.

Groenman-van Waateringe, W. (1988b) New trends in palynoarchaeology in Northwest Europe or the frantic search for local pollen data. In: Webb, R.E. (ed.) *Recent Developments in Environmental Analysis in Old and New World Archaeology*. BAR International Series 416, Oxford.

Groenman-van Waateringe, W. (1993) The effects of grazing on the pollen production of grasses. *Vegetation History and Archaeobotany* 2, 157–162.

Groot Bruinderink, G.W.T.A., Hazebroek, E. and van der Voet, H. (1994) Diet and condition of wild boar, *Sus scrofa scrofa*, without supplementary feeding. *Journal of Zoology* 233, 631–648.

Groot Bruinderink, G.W.T.A., Hazebroek, E. and van der Voet, H. (1997) Wroeten door het wilde zwijn en de gevolgen voor bodem en bosverjonging. In: Wieren, S.E., van Groot Bruinderink, G.W.T.A., Jorritsma, I.T.M. and Kuiters, A.T. (eds) *Hoefdieren in het boslandschap*. Backhuys Publishers, Leiden, pp. 131–145.

Groot Bruinderink, G.T.W.A., Baveco, J.M., Cornelissen, P., Kramer, K., Kuiters, A.T., Lammertsma, D.R., Prins, H.H.T., Roder, F., de Vulink, J.Th., Wieren, S.E., van

Wigbels, V. and Wijdeveen, S. (1999) Dynamische Interacties tussen Hoefdieren en Vegetatie in de Oostvaardersplassen. IBN-rapport nr. 436.

Gross, H. (1933) Die Traubeneiche (*Quercus sessiliflora* Salisb.) in Ostpreussen. *Zeitschrift für Forst- und Jagdwesen* 65, 144–152.

Grossmann, H. (1927) *Die Waldweide in der Schweiz.* Promotionsarbeit, Zürich.

Grubb, P.J. (1977) The maintenance of species-richness in plant communities: the importance of the regeneration niche. *Biological Review* 52, 107–145.

Grubb, P.J. (1985) Plant populations and vegetation in relation to habitat, disturbance and competition: problems of generalization. In: White, J. (ed.) *The Population Structure of Vegetation.* Dr W. Junk Publishers, Dordrecht, pp. 595–621.

Grubb, P.J. (1987) Global trends in species-richness in terrestrial vegetation: a view from the northern hemisphere. In: Giller, P.S. (ed.) *Organization of Communities. Past and Present.* Blackwell Scientific Publications, Oxford, pp. 99–108.

Grubb, P.J., Kelly, D. and Mitchley, J. (1982) The control of relative abundance in communities of herbaceous plants. In: Newman, E.I. (ed.) *The Plant Community as a Working Mechanism.* Blackwell Scientific Publications, Oxford, pp. 79–97.

Grubb, P.J., Lee, W.G., Kollmann, J. and Wilson, J.B. (1996) Interaction of irradiance and soil nutrient supply on growth of seedlings of ten European tall-shrub species. *Journal of Ecology* 84, 827–840.

Guillet, B. and Robin, A.M. (1972). Interprétation de datations par le ^{14}C d'horizons Bh de deux podzols humo-ferrugineux, l'un formé sous callune, l'autre sous hêtraie. *Comptes Rendus Academie des Sciences* (Paris) 274, 2859–2862.

Guintard, G. and Tardy, F. (1994) Les Bovins de l'Île Amsterdam. Un example d'isolement génétique. In: Bailly, L. and Cohën, A.-S. (eds). *Aurochs. Le retour. Aurochs, vaches et autres bovins de la préhistoire à nos jours.* Centre Jurassien du Patrimoine, Lons-le-Saunier, pp. 203–209.

Guliver, R. (1998) What were woods like in the seventeenth century? Examples from the Helmsley Estate, northeast Yorkshire, UK. In: Kirby, K.J. and Watkins, C. (eds) *The Ecological History of European Forests.* CAB International, Wallingford, UK, pp. 135–153.

Gurnell, J. (1993) Tree seed production and food conditions for rodents in an oak wood in southern England. *Forestry* 66, 291–315.

Haan, M.J.M. De (1999) *Beede bosch ende haghe.* Roelofarendsveen.

Habets, J. (1891) *Limburgse wijsdommen. Dorpscostumen en gewoonten, bevattende voornamelijk Bank- Laat- en Boschrechten. Oude Vaderlandsche Rechtsbronnen.* Martinus Nijhoff, 's-Gravenhage.

Hadfield, M. (1974) The oak and its legends. In: Morris, M.G. and Perring, F.H. (eds) *British Oak. Its History and Natural History.* The Botanical Society of the British Isles, E.W. Classey, Berkshire, pp. 123–129.

Hall, S.J. (1989) Running wild. Part 1 and 2. *The Ark*, January, 12–15; February, 46–49.

Hall, S.J. and Moore, C.F. (1987) The feral cattle of Swona, Orkney Islands. *Genetic Resources Information* 6, 2–7.

Hammen, Th. van der (1952) Late-glacial flora and periglacial phenomena in The Netherlands. *Leidse Geologische Mededelingen* 17, 71–184.

Hampicke, U. (1978) Agriculture and conservation. Ecological and social aspects. *Agriculture and Environment* 4, 25–42.

Hanby, J.P. and Bygott, J.D. (1979) Population changes in lions and other predators. In:

Sinclair, A.R.E. and Norton-Griffiths, M. (eds) *Serengeti. Dynamics of an Ecosystem*. The University of Chicago Press, Chicago pp. 249–262.

Hard, G. (1972) Wald gegen Driesch. Das Vorrücken des Waldes auf Flachen junger 'Sozialbrache'. *Bericht zur Deutschen Landeskunde* 46, 49–80.

Hard, G. (1975) Vegetationsdynamik und Verwaldungsprozesse auf den Brachflächen Mitteleuropas. *Erde* 106, 243–276.

Hard, G. (1976) Vegetationsentwicklung auf Brachflächen. In: Bierhals, E. von Gehle, L. Hard, G. and Nohl, W. (eds) *Brachflächen in der Landschaft*. KTBL-Schriften-Vertrieb im Landwirtschaftsverlag, Münster-Hiltrup, KTBL-Schrift 195, pp. 243–276.

Harding, P.T. and Rose, F. (1986) *Pasture-Woodlands in Lowland Britain. A Review of their Importance for Wildlife Conservation*. Natural Environment Research Council. Institute of Terrestrial Ecology, Huntingdon, UK.

Harmer, R. (1990) Relation of shoot growth phases in seedling oak to development of the tap root, lateral roots and fine root tips. *New Phytologist* 115, 23–27.

Harmer, R. (1994a) Natural regeneration of broadleaved trees in Britain. I. Historical aspects. *Forestry* 67, 179–188.

Harmer, R. (1994b) Natural regeneration of broadleaved trees in Britain. 2. Seed production and predation. *Forestry* 67, 275–286.

Harmer, R. (1995) Natural regeneration of broadleaved trees in Britain. 3. Germination and establishment. *Forestry* 68, 1–9.

Harris, E. and Harris, J. (1991) *Wildlife Conservation in Managed Woodlands and Forests*. Basil Blackwell, Oxford.

Harrison, J.S. and Werner, P.A. (1982) Colonization by oak seedlings into a heterogeneous successional habitat. *Canadian Journal of Botany* 62, 559–563.

Hart, G.E. (1966) *Royal Forest. A History of Dean's Woods as Producers of Timber*. Clarendon Press, Oxford.

Hartig, G.L. (1791) *Anweisung zur Holzzucht für Förster*. Marburg.

Hausrath, H. (1898) *Forstgeschichte der rechtsrheinischen Theile des ehemaligen Bisthums Speyer*. Julius Springer, Berlin.

Hausrath, H. (1928) Beiträge zur Geschichte des Nieder- und Mittelwaldes in Deutschland. *Allgemeine Forst- und Jagdzeitung* 104, 345–348.

Hausrath, H. (1982) *Geschichte des deutschen Waldbaus. Von seinen Anfängen bis 1850*. Hochschulverlag Freiburg (Breisgau).

Hedemann, O. (1939) *L'histoire de la Forêt de Białowieza (jusqu'a 1798)*. Wiktor Hartman, Warzawa.

van Hees, A.F.M. (1997) Growth and morphology of pedunculate oak (*Quercus robur* L.) and beech (*Fagus sylvatica* L.) seedlings in relation to shading and drought. *Annales des Sciences Forestières* 54, 9–18.

Helmer, W., Litjens, G. and Overmars, W. (1995) Levende natuur in een nieuw cultuurlandschap. *De Levende Natuur* 96, 182–187.

Heptner, V.G., Nasimovic, A.A. and Bannikov, A.G. (1966) Band 1. Paarhufer und Unpaarhufer. In: Heptner, V.G. and Naumov, N.P. (eds) *Die Säugetiere der Sowjetunion*. Gustav Fischer, Jena,

Herckenrath, C.R.C. and Dory, A. (1990) *Wolters' Woordenboek Frans-Nederlands*, 16th edn. Wolters Noordhoff, Groningen.

Herrera, C.M. (1984) Seed dispersal and fitness determinants in wild roses. Combined effects of hawthorn, birds, mice, and browsing ungulates. *Oecologia* 63, 386–393.

Herrmann, H. (1915) Die Eichelmastnutzung einst und jetzt. *Forstwissenschaftliches Centralblatt* 37, 51–60.

Hesmer, H. (1930) Zur Frage des Aufbaues und der Verjüngung europäischer Urwälder. *Forstarchiv* 6, 265–274.

Hesmer, H. (1932) Die Entwicklung der Wälder des nordwestdeutschen Flachlandes. *Zeitschrirft für Forst- und Jagdwesen* 64, 577–607.

Hesmer, H. (1958) *Wald- und Forstwirtschaft in Nordrhein-Westfalen*. Hannover.

Hesmer, H. (1960) Unterbauversuche mit Winterlinde, Buche und Hainbuche in verschiedenen Verbänden unter Stieleichenstangenholz. *Forstarchiv* 31, 185–192.

Hesmer, H. (1966) Ökologisches und waldbauliches Verhalten der Winterlinde. *Landwirtschaft- Angewandte Wissenschaft* 123, 17–19.

Hesmer, H. and Günther, K.-H. (1966) Kulturversuche und Aufforstungserfahrungen auf Pseudogleyböden des Kottenforstes. *Forstarchiv* 37, 1–26.

Hesmer, H. and Schroeder, F.-G. (1963) *Waldzusammensetzung und Waldbehandlung im Niedersächsischen Tiefland westlich der Weser und in der Münsterschen Bucht bis zum Ende des 18 Jahrhunderts. Forstgeschichtlicher Beitrag zur Klärung der natürlichen Holzartenzusammensetzung und ihrer künstlichen Veränderungen bis in die frühe Waldbauzeit*. Decheniana, Beiheft 11, pp. 1–304.

Hess, E. (1937) Die Bewirtschaftung der Eichenwälder von Blois in Frankreich. *Sweizerische Zeitschrift für Forstwesen* 88, 91–100.

Heukels, H. and van der Meijden, R. (1983) *Flora van Nederland*. Wolters-Noordhoff, Groningen.

Heybroek, H.M. (1984) Bosbeheer ten behoeve van natuurwaarden. *Nederlands Bosbouwtijdschrift* 56, 229–239.

Heymann, P. and Dautzenberg, H. (1988) Wildapfel und Wildbirne. *Forst und Holz* 43, 483–486.

Hibbs, D.E. (1983) Forty years of forest succession in central New England. *Ecology* 64, 1394–1401.

Hicks, S. (1998) Fields, boreal forests and forests clearings as recorded by modern pollen deposition. In: Gaillard, M.-J. and Berglund, B.E. (eds) Quantification of land surfaces cleared of forest during the Holocene-Modern: pollen/vegetation/landscape relationships as an aid to the interpretation of pollen data. *Paläoklimaforschung/Palaeoclimate Research* 27, 53–66.

Hilf, R.B. (1938) *Wald und Weidwerk in Geschichte und Gegenwart*. Erster Teil. *Der Wald*. Akademische Verlaggesellschaft Athenaieon, Potsdam.

Hill, M.O., Evans, D.F. and Bell, S.A. (1992) Long-term effects of excluding sheep from hill pastures in North Wales. *Journal of Ecology* 80, 1–13.

Hill, S.D. (1985) Influence of large herbivores on small rodents in the New Forest, Hampshire. PhD thesis, University of Southampton, UK.

Hillegers, H.P.M. (1986) Kalkgraslanden. *Natuur en Techniek* 54, 264–275.

Hillegers, H. (1994) 'Op de bres voor de jeneverbes'. Herintroductie van de jeneverbes in Zuid-Limburg. *Natuurhistorisch Maandblad* 83, 175–178.

Hillegers, H. and Lejeune, M. (1994) Herstel biologische diversiteit op de St. Pietersberg. *Natuurhistorisch Maandblad* 183, 179–188.

Hjelle, K.L. (1998) Herb pollen representation in surface moss samples from mown meadows and pastures in western Norway. *Vegetation History and Archaeobotany* 7, 79–96.

von Hobe, J.H. (1805) *Freymüthige Gedanken über verschiedene Fehler bey dem*

Forsthaushalt, insbesondere über die Viehude in den Holzungen, deren Abstellung und Einschränkung. Thal-Ehrenbreitstein, in der Gehraschen Hofbuchhandlung.

Hocker, R. (1979) Die Winterlinde im Kottenforst. *Allgemeine Forstzeitschrift* 34, 842–844.

Hodges, J.D. and Gardiner, E.S. (1993) Ecology and physiology of oak regeneration. In: Loftis, D.L. and McGee, C.E. (eds) *Oak Regeneration: Serious Problems, Practical Recommendations. Symposium Proceedings, Knoxville, Tennessee September 8–10, 1992.* Southeastern Forest Experiment Station, Asheville, pp. 54–65.

Hodgson, B. (1994) Buffalo back home on the range. *National Geographic* 186, 64–89.

Hoffmann, C. (1895) Über den Eichenschälwald-Betrieb in Bosniën. *Österreichische Vierteljahres-Schrift für Forstwesen* 3, 226–234.

Hoffmann, G. (1966) Verlauf der Tiefendurchwurzelung und Feinwurzelbildung bei einigen Baumarten. *Archiv für Forstwesen* 15, 825–856.

Hoffmann, G. (1967) Wurzel- und Sprosswachstumsperiodik der Jungpflanzen von *Quercus robur* L. im Freiland und unter Schattenbelastung. *Archiv für Forstwesen* 16, 745–749.

Hofmann, R.R. (1973) *The Ruminant Stomach: Stomach Structure and Feeding Habits of East African Game Ruminants.* East African Literature Bureau, Nairobi, Kenya.

Hofmann, R.R. (1976) Zur adaptiven Differenzierung der Wiederkäuer: Untersuchungsergebnisse auf der Basis der vergleichenden funktionellen Anatomie des Verdauungstrakts. *Praktische Tierärtzt* 57, 351–358.

Hofmann, R.R. (1985) Digestive physiology of the deer. Their morphophysiological specialisation and adaptation. *The Royal Society of New Zealand Bulletin* 22, 393–407.

Hofmann, R.R. (1989) Evolutionary steps of ecophysiological adaptation and diversification of ruminants: a comparative view of their digestive system. *Oecologia* 78, 443–457.

Holeksa, J. (1993) Gap size differentiation and the area of forest reserve. In: Broekmeyer, M.E.A., Vos, W. and Koop, H. (eds) *European Forest Reserves. Proceedings of the European Forest Reserves Workshop 6–8 May 1992, Wageningen, The Netherlands.* Pudoc Scientific Publishers, Wageningen, pp. 159–165.

Holmes, G.D. (1975) History of forestry and forest management. *Philosophical Transactions of the Royal Society of London* Series B 271, 69–80.

Holmes, W. (1989) *Grass. Its Production and Utilization,* 2nd edn. The British Grassland Society, Blackwell Scientific Publications, Oxford.

Hondong, H., Langner, S. and Coch, T. (1993) *Untersuchungen zum Naturschutz an Waldrändern.* Bristol-Schriftenreihe, Band 2, Bristol-Stiftung, Ruth und Herbert UHL – Forschungsstelle für Natur- und Umweltschutz.

Hooke, D. (1998a) Medieval forests and parks in southern and central England. In: Watkins, C. (ed.) *European Woods and Forests. Studies in Cultural History.* CAB International, Wallingford, UK, pp. 19–32.

Hooke, D. (1998b) *The Landscapes of Anglo-Saxon England.* Leicester University Press, London.

Horn, H.S. (1975) Forest succession. *Scientific American* 232, 90–98.

Horton, A., Keen, D.H., Field, M.H., Robinson, J.E., Coope, G.R., Currant, A.P., Graham, D.K., Green, C.P. and Phillips, L.M. (1992) The Hoxnian Interglacial deposits at Woodston, Petersborough. *Philosophical Transactions of the Royal Society of London* Series B 338, 131–164.

Host, G.E., Pregitzer, K.S., Ramm, C.W., Hart, J.B. and Cleland, D.T. (1987) Landformmediated differences in successional pathways among upland forest ecosystems in northwestern lower Mitchigan. *Forest Science* 33, 445–457.

Hötker, H. (ed.) (1991) *Waders Breeding on Wet Grasslands.* Wader Study Group Bulletin Number 61, Supplement April. Joint Nature Conservation Committee, Peterborough.

Hough, A.F. and Forbes, R.D. (1943) The ecology and silvics of forests in the high plateaus of Pennsylvania. *Ecological Monographs* 13, 299–320.

Housley, R.A. (1991) AMS dates from the Late Glacial and early Postglacial in north-west Europe: a review. In: Barton, N., Roberts, A.J. and Roe, D.A. (eds). *The Late Glacial in North-west Europe: Human Adaptation and Environmental Change at the End of the Pleistocene.* CBA Research Report no. 77, Oxford, pp. 25–39.

Huault, M.F. (1976) La végétation au Pléistocène supérieur et au début de l'Holocène dans le Nord. In: de Lumley, H. (ed.) *La Préhistoire Française.* Tome 1. *Les civilations paléolithiques et mésolithiques de la France.* CNRS, Paris, pp. 539–541.

Huddle, J.A. and Pallardy, S.G. (1996) Effects of long-term annual and periodic burning on tree survival and growth in a Missouri Ozark oak–hickory forest. *Forest Ecology and Management* 82, 1–9.

Huijser, M.P., Vulink, J.T. and Zijlstra, M. (1996) *Begrazing in de Oostvaardersplassen. Effecten op de vegetatie-structuur en het terreingebruik van grote herbivoren en ganzen.* Ministerie van Verkeer en Waterstaat. Dorectoraat-Generaal Rijkswaterstaat. Directie IJsselmeer polders. Lelystad.

de Hullu, P.C. (1995) Natuurontwikkeling bij Staatsbosbeheer. *De Levende Natuur* 5, 141–147.

van Hulst, R. (1979a) On the dynamics of vegetation: succession in model communities. *Vegetatio* 39, 85–96.

van Hulst, R. (1979b) On the dynamics of vegetation: Markov chains as models of succession. *Vegetatio* 40, 3–14.

van Hulst, R. (1980) Vegetation dynamics or ecosystem dynamics: dynamic sufficiency in succession theory. *Vegetatio* 43, 147–151.

Huntley, B. (1986) European Post-glacial vegetation history: a new perspective. In: Ouellet, H. (ed.) *Acta XIX Congressus Internationalis Ornithologici* Vol. I. Natural Museum of Natural Sciences, University of Ottowa Press, pp. 1061–1077.

Huntley, B. (1988) Europe. Vegetation history. In: Huntley, B. and Webb, T. III (eds) *Vegetation History. Section III: Glacial and Holocene Vegetation History 220 ky to Present. Handbook of Vegetation Science,* vol. 7. Kluwer Academic Publishers, Dordrecht, pp. 341–382.

Huntley, B. (1989) Historical lessons for the future. In: Spellenberg, I.F., Goldsmith, F.B. and Morris, M.G. (eds) *The Scientific Management of Temperate Communities for Conservation.* Blackwell Scientific Publications, Oxford, pp. 473–503.

Huntley, B. and Birks, H.J.B. (1983) *An Atlas of Past and Present Pollen Maps of Europe: 0–13,000 Years Ago.* Cambridge University Press, Cambridge.

Huntley, B. and Webb, T. III (1989) Migration: species' response to climatic variations caused by changes in the earth's orbit. *Journal of Biogeography* 16, 5–19.

Huss, J. and Stephani, A. (1978) Lassen sich angekommene Buchennatur-verjüngungen durch frühzeitige Auflichtung, durch Düngung oder Unkraut-beämpfung rascher aus der Gefahrenzone bringen? *Allgemeine Forst- und Jagdzeitung* 149, 133–145.

Huston, M. (1979) A general hypothesis of species diversity. *American Naturalist* 113, 81–101.

Hyde, H.A. and Williams, D.A. (1945) Pollen of lime (*Tilia* spp.) *Nature* 155, 457.

Hytteborn, H. (1986) Methods of forest dynamics research. In: Fanta, I. (ed.) *Forest Dynamics Research in Western and Central Europe*. Pudoc, Wageningen, pp. 17–31.

Iben, B. (1993) Schweinehirte – ein ehrenwerter Beruf! Über die Transhumanz von Schweinen, Hirten und Eichenwäldern. *Tierärztliche Umschau* 48, 765–773.

Iversen, J. (1941) Land occupation in Denmark's Stone Age. a pollen-analytical study of the influence of farmer culture on the vegetational development. *Danmarks Geologiske Undersøgelse, II*. Raekke nr. 66 (*Geological Survey of Denmark* No. 66).

Iversen, J. (1954) The late-glacial flora of Denmark and its relation to climate and soil. *Danmarks Geologiske Undersøgelse* 2, 80 (*Geological Survey of Denmark* No. 80), 87–119.

Iversen, J. (1956) Forest clearance in the Stone Age. *Scientific American* 194, 36–41.

Iversen, J. (1958) Pollenanalytischer Nachweis des Reliktencharacters eines Jütischen-Mischwaldes. *Veröffentlichungen des Geobotanischen Instituts Rübel, Zurich* 33, 137–144.

Iversen, J. (1960) Problems of the early Post-Glacial forest development in Denmark. *Danmarks Geologiske Undersøgelse, IV*. Raekke Bd. 4, nr 3 (*Geological Survey of Denmark. IV* Series Vol. 4, No. 3).

Iversen, I. (1964) Retrogressive vegetational succession in the Post-Glacial. *Journal of Ecology* 52, 59–70.

Iversen, I. (1969) Retrogressive development of a forest ecosystem demonstrated by pollen diagrams from fossil mor. *Oikos* (Suppl.) 12, 35–49.

Iversen, J. (1973) The development of Denmark's nature since the last Glacial, *Danmarks Geologiske Undersøgelse, V*. Raekke nr. 7-c (*Geological Survey of Denmark. V*. Series No. 7-c).

Jackson, S.T. and Wong, A. (1994) Using forest patchiness to determine pollen source areas of closed-canopy assemblages. *Journal of Ecology* 82, 89–99.

Jacobi, R.M. (1978) Population and landscape in Mesolithic lowland Britain. In: Limbrey, S. and Evans, J.G. (eds) *The Effect of Man on the Landscape: the Lowland Zone*. CBA Research Report 21, London, pp. 75–85.

Jacobson, G.L. Jr and Bradshaw, R.H.W. (1981) The selection of sites for paleovegetational studies. *Quaternary Research* 16, 80–96.

Jahn, G. (1984) Eichenmischwälder in Nordwestdeutschland: naturnah oder antropogen? *Phytocoenologia* 12, 363–372.

Jahn, G. (1987) Zur Frage der Eichenmischwaldgesellschaften in nordwestdeutschen Flachland. *Forstarchiv* 58, 154–163 and 194–200.

Jahn, G. (1991) Temperate deciduous forests of Europe. In: Röhrig, R. and Ulrich, B. (eds) *Temperate Deciduous Forests. Ecosystems of the World*, 7. Elsevier, Amsterdam, pp. 377–503.

Jahn, G. and Raben, G. (1982) Über den Einfluss der Bewirtschaftung auf Struktur und Dynamik der Wälder. In: Dierschke, H. (ed.) *Struktur und Dynamik von Wäldern Berichte der Internationale Symposium der Internationalen Verein für Vegetationskunde*. J. Cramer, Vaduz, pp. 717–734.

Jakubowska-Gabara, J. (1996) Decline of *Potentillo albae–Quercetum* Libb. 1933 phytocoenoses in Poland. *Vegetatio* 124, 45–59.

Jakucs, P. (1959) Mikroklimaverhältnisse der Flaumeichen-Buschwälder in Ungarn. *Acta Agronomia Academiae Scientiarum Hungaricae* 9, 209–236.

Jakucs, P. (1961) Die Flaumeichen-Buschwälder in der Tschechoslowakei. *Veröffentlichungen. Geobotanisches Institut Eidgenössische Technische Hochschule Stiftung Rübel in Zürich* 36, 91–118.

Jakucs, P. (1969) Die Sprosskolonien und ihre Bedeutung in der dynamischen Vegetationsentwicklung (Polycormonsukzession). *Acta Botanica Croatica* (Zagreb) 28, 161–170.

Jakucs, P. (1972) *Dynamische Verbindung der Wälder und Rasen. Quantitative und qualitative Untersuchungen über die synökologischen, phytozönologischen und strukturellen Verhältnisse der Waldsäume.* Akadémica Kiadó, Ungarische Akademie der Wissenschaften, Budapest.

Jakucs, P. and Jurko, A. (1967) *Querco-petraeae Carpinetum Waldsteinietosum.* Eine neue Subassoziation aus dem Slowakischen und Ungarischen Karstgebiet. *Biologia* (Bratislava) 22, 321–335.

Janis, C. (1975) The evolutionary strategy of the Equidae and the origins of rumen and digestion. *Evolution* 30, 757–774.

Jansen, J.C.G.M. and van de Westeringh, W. (1983) Dat ging over zijn hout. Overmatig gebruik van bossen in het zuiden van Limburg van de Hoge Middeleeuwen tot in de 20ste eeuw. In: *Studies over de sociaal-economische geschiedenis van Limburg.* Jaarboek van het Sociaal Historisch Centrum voor Limburg, Van Gorcum, Assen, 28, pp. 19–63.

Janssen, C.R. (1973) Local and regional pollen deposition. In: Birks, H.J.B. and West, R.G. (eds) *Quaternary Plant Ecology. 14th Symposium of the British Ecological Society, University of Cambridge, 28–30 March 1972.* Blackwell Scientific Publishers, Oxford, pp. 31–42.

Janssen, C.R. (1974) *Verkenningen in de Palynologie.* Oosthoek, Scheltema & Holkema, Utrecht.

Janssen, C.R. (1981) On the reconstruction of past vegetations by pollen analysis; a review. In: *Proceedings of the IVth International Palynoecogical Conference, Lucknow* C84, pp. 197–210.

Jarman, P.J. and Sinclair, A.R.E. (1979) Feeding strategy and the pattern and resource partitioning in ungulates. In: Sinclair, A.R.E. and Norton-Griffiths, M. (eds) *Serengeti. Dynamics of an Ecosystem.* The University of Chicago Press, Chicago pp. 130–163.

Jarvis, P.G. (1963) The effects of acorn size and provenance on the growth of seedlings of sessile oak. *Quarterly Journal of Forestry* 52, 545–571.

Jarvis, P.G. (1964) The adaptability to light intensity of seedlings of *Quercus petraea* (Matt.) Liebl. *Journal of Ecology* 52, 545–571.

Jedrzejewska, B., Okarma, H., Jedrzejewski, W. and Milkowski, L. (1994) Effects of exploitation and protection on forest structure, ungulate density and wolf predation in Białowieza Primeval Forest, Poland. *Journal of Applied Ecology* 31, 664–676.

Jedrzejewska, B., Jedrzejewski, H., Bunevich, N., Milkowski, L. and Krasinski, A. (1997) Factors shaping the population densities and increase rates of ungulates in Białowieza Primeval Forest (Poland and Belarus) in the 19th and 20th centuries. *Acta Theriologica* 42, 399–451.

Jenik, J. (1986) Forest succession: theoretical concepts. In: Fanta, J. (ed.) *Forest Dynamics Research in Western and Central Europe.* Pudoc, Wageningen, pp. 7–16.

Jennersten, O., Loman, J., Møller, A.P., Robertson, J. and Widén, B. (1992) Conservation biology in agricultural habitat islands. In: Hansson, L. (ed.) *Ecological Principles of Nature Conservation.* Elsevier Applied Science, London, pp. 394–424.

Jensen, T.S. (1985) Seed–seed predator interactions of European beech, *Fagus silvatica* and forest rodents, *Clethrionomys glareolus* and *Apodemus flavicollis. Oikos* 44, 149–156.

Jensen, T.S. and Nielsen, O.F. (1986) Rodents as seed dispersers in a heath–oak wood succession. *Oecologia* 70, 214–221.

Joffre, R., Vacher, J. and De Los Llanos, C. (1988) The dehesa: an agrosilvopastoral system of the Mediterranean region with special reference to the Sierra Morena of Spain. In: Nair, P.K.R. (ed.) *Agroforestry Systems in the Tropics.* Kluwer Academic Publishers, Dordrecht, pp. 427–453.

Johnson, P.S. (1992) Oak overstory-reproduction relations in two xeric ecosystems in Michigan. *Forest Ecology and Management* 48, 233–248.

Johnson, W.C. and Adkinson, C.S. (1985) Dispersal of beech nuts by blue jays in fragmented landscapes. *The American Midland Naturalist* 113, 319–324.

Johnson, W.C. and Webb, T. III (1989) The role of blue jays (*Cyanocitta cristata* L.) in the postglacial dispersal of fagaceous trees in eastern North America. *Journal of Biogeography* 16, 561–571.

Jokela, J.J. and Sawtelle, R.A. (1985) Origin of oak stands on the Springfield plain: a lesson on oak regeneration. In: Dawson, J.O. and Magerus, K.A. (eds) *Proceedings of the Fifth Central Hardwood Forest Conference.* SAF Pub. 85–05, Department Forestry, University of Illinois, Urbana-Champaign, pp. 181–188.

Jonassen, H. (1950) Recent pollen sedimentation and Jutland heath diagrams. *Dansk Botanisk Arkiv, Dansk Botanisk Forening* 13, 7.

Jones, E.W. (1945) The structure and repreduction of the virgin forest of the North Temperate Zone. *New Phytologist* 44, 130–148.

Jones, E.W. (1959) Biological flora of the British Isles, *Quercus* L. *Journal of Ecology* 47, 169–222.

Jones, M. (1998) The rise, decline and extinction of spring wood management in south-west Yorkshire. In: Watkins, C. (ed.) *European Woods and Forests. Studies in Cultural History.* CAB International, Wallingford, UK, pp. 55–71.

Jonsson, L. (1993) *Vogels van Europa, Noord-Afrika en het Midden-Oosten,* 3rd edn. Tirion, Baarn.

Jorritsma, I.T.M., Mohren, G.M.J., Hees, A.F.M. and van Seigers, G. (1997) Bosontwikkeling in Aanwezigheid van hoefdieren: een modelbenadering. In: Wieren, S.E. van, Groot Bruinderink, G.W.T.A., Jorritsma, I.T.M. and Kuiters, A.T. (eds) *Hoefdieren in het boslandschap.* Backhuys Publishers, Leiden, pp. 165–191.

Jorritsma, I.T.M., Hees, A.F.M. and van Mohren, G.M.J. (1999) Forest development in relation to ungulate grazing: a modelling approach. *Forest Ecology and Management* 120, 23–34.

Kahlke, H.D. (1994) *Die Eiszeit.* Urania, Leipzig.

Kalicki, T. and Krapiec, M. (1995) Problems of dating alluvium using buried subfossil tree trunks: lessons from the 'black oaks' of the Vistula Valley, Central Europe. *The Holocene* 5, 243–250.

Kalis, A.J. (1983) Die menschliche Beeinflussung der Vegetationsverhältnisse auf der Aldenhovener Platte (Rheinland) während der vergangenen 2000 Jahre. *Archäologie in den Rheinischen Lössbörden. Beiträge zur Siedlungsgeschichte im Rheinland, Sonderdruck.* Rheinland Verlag, Köln, pp. 331–345.

Kalis, A.J. (1988) Zur Umwelt des frühneolitischen Menschen: ein Beitrag der Pollenanalyse. Der Prähistorische Mensch und seine Umwelt. *Festschrift für Udelgard Körber-Grohne. Forschungen und Berichte zur Vor- und Frühgeschichte in Baden-Württemberg* 31 Stuttgart, pp. 125–137.

Kalis, A.J. and Meurers-Balke, J. (1988) Wirkungen neolithischer Wirtschaftsweisen in Pollendiagrammen. *Archäologische Informationen* 11, 39–53.

Kalis, A.J. and Bunnik, F.P.M. (1990) Holozäne Vegetationsgeschichte in der westlichen niederrheinischen Bucht. In: Schirmer, W. (ed.) *Rheingeschichte zwischen Mosel und Maas.* Dengna Führer 1, 266–272.

Kalis, A.J. and Zimmermann, A. (1988) An integrative model for the use of different landscapes in Linearbandkeramik times. In: Bintliff, L., Davidson, D.A. and Grant, E.G. (eds) *Conceptual Issues in Environmental Archeology.* Edinburgh University Press, pp. 145–152.

Karcev, G. (1903) *Beloveskaja Pusca. [Belovezhskaya Primeval Forest. Historical description, contemporary game management, and monarchial hunts in the forest].* A. Marks, St Petersburg [in Russian].

Kaspers, H. (1957) *Comitatus nemoris. Die Waldgrafschaft zwischen Maas und Rhein.* Beiträge zur Geschichte des Dürener Landes, Band 7, Düren und Aachen.

Kausch von Schmeling, W. (1985) Der Europäische Kirschbaum. Geschichte und Gegenwart. *Holz Aktuell* 5, 7–13.

Keiper, J. (1916) Die Linde im Pfälzerwald und den übrigen Waldgebieten der Pfalz. *Forstwissenschaftliches Centralblatt* 38, 222–237.

Kerner, A. (1929) *Das Pflanzenleben der Donauländer,* 2nd edn. F. Vierhapper, Innsbruck.

Kidwell, C.S. (1992) Systems of knowledge. In: Josephy, A.M. Jr (ed.) *America in 1492. The World of the Indian Peoples Before the Arrival of Columbus.* Alfred A. Knopf, New York, pp. 369–403.

Kiess, R. (1998) The word 'Forst/forest' as an indicator of fiscal property and possible consequences for the history of Western European forests. In: Watkins, C. (ed.) *European Woods and Forests: Studies in Cultural History.* CAB International, Wallingford, UK, pp. 11–18.

Kingsland, S.E. (1991) Foundational papers. Defining ecology as a science. In: Real, L. and Brown, J.H. (eds) *Foundations of Ecology.* Classic papers with commentaries. The University of Chicago Press, Chicago, pp. 1–13.

Kingsolver, J.G. and Pain, R.T. (1991) Conversational biology and ecological debate. In: Real, L. and Brown, J.H. (eds) *Foundations of Ecology.* Classic Papers with commentaries. The University of Chicago Press, Chicago, pp. 309–317.

Kinnaird, J.W. (1974) Effect of site conditions on the regeneration of birch (*Betula pendula* Roth. and *B. pubescens* Ehrh.). *Journal of Ecology* 62, 467–472.

Klein, J.P., Pont, B., Faton, J.M. and Knibiely, P. (1993) The network of river system nature reserves in France and the preservation of alluvial forests. In: Broekmeyer, M.E.A., Vos, W. and Koop, H. (eds) *European Forest Reserves. Proceedings of the European Forest Reserves Workshop 6–8 May 1992, Wageningen, The Netherlands.* Pudoc Scientific Publishers, Wageningen, pp. 91–96.

Kleinschmit, J. (1998) Die wildbirne – Baum des Jahres 1998. *Forst und Holz* 53, 35–39.

Klepac, D. (1981) Les forêts de chêne en Slavonie. *Revue Forestière Française* 33 (numéro spécial), 86–104.

Klika, J. (1954) The influence of pasturing on the phytocenosis of the Slovak Karst. *Festschrift für Erwin Aichinger Sonderfolde der Schriftenreihe Angewandte Pflanzensociologie* Band I. Wien, 1235–1237.

Knapp, A.K., Blair, J.M., Briggs, J.M., Collins, S.L., Hartnett, D.C., Johnson, L.C. and Towne, E.G. (1999) The Keystone role of bison in North American tallgrass prairie. *Bioscience* 49, 39–50.

Knapp, E. (1971) *Wiesen und Weiden. Eine Grünlandlehre.* Paul Parey, Berlin/Hamburg.

Koenen, M.J. and Drewes, J.B. (1987) *Wolters' Woordenboek Nederlands, Koenen.* Wolters-Noordhoff, Groningen.

von Koenigswald, W. (1983) Die Säugetierfauna des süddeutschen Pleistozäns. In: Müller-Beck, H.J. (ed.) *Urgeschichte in Baden-Württemberg.* Konrad Theiss Verlag, Stuttgart, pp. 167–216.

Kollmann, J. (1992) Gebüschenwicklung in Halbtrockenrasen des Kaiserstuhls. *Natur und Landschaft* 67, 20–26.

Kollmann, J. and Schill, H.-P. (1996) Spatial patterns of dispersal, seed predation and germination during colonization of abandoned grassland by *Quercus petraea* and *Corylus avellana. Vegetatio* 125, 193–205.

Kolstrup, E. (1980) Climate and stratigraphy in northwestern Europe between 30,000 and 13,000 BP with special reference to the Netherlands. *Mededelingen Rijks Geologische Dienst* 32–15, 181–253.

Kolstrup, E. (1991) Palaeoenvironmental developments during the Late Glacial of the Weichselian. In: Barton, N., Roberts, A.J. and Roe, D.A. (eds) *The Late Glacial in North-west Europe: Human Adaptation and Environmental Change at the End of the Pleistocene.* CBA Research Report No. 77. Council for British Archaeology, London, pp. 1–6.

Königsson, L.-K. (1968) The Holocene history of the Great Alvar of Öland. *Acta Phytogeographica Seucica* 55, 1–172.

Koop, H. (1981) *Vegetatiestructuur en dynamiek van twee natuurlijke bossen: het Neuenburger en Hasbrucher Urwald.* Pudoc Centrum voor Landbouwpublicaties en Landbouwdocumentatie, Wageningen.

Koop, H. (1982) Waldverjüngung Sukzessionsdynamik und kleinstandörtliche Differenzierung infolge spontaner Waldentwicklung. In: Dierschke, H. (ed.) *Struktur und Dynamik von Wäldern.* Berichte der Internationale Verein für Vegetationskunde. J. Cramer, Vaduz, pp. 235–273.

Koop, H. (1986) Omvormingsbeheer naar natuurlijk bos: een paradox? *Nederlands Bosbouwtijdschrift* 55, 51–56.

Koop, H. (1987) Vegetative reproduction of trees of some European natural forests. *Vegetatio* 72, 103–110.

Koop, H. (1989) *Forest Dynamics. Silvi-Star: A Comprehensive Monitoring System.* Springer Verlag, Berlin.

Koop, H. and Hilgen, P. (1987) Forest dynamics and regeneration mosaic shifts in unexploited beech (*Fagus sylvatica*) stands at Fontainebleau (France). *Forest Ecology and Management* 20, 135–150.

Koop, H. and Siebel, H.N. (1993) Conversion management towards more natural forests: evaluation and recommendations. In: Broekmeyer, M.E.A., Vos, W. and Koop, H. (eds) *European Forest Reserves. Proceedings of the European Forest Reserves Workshop, 6–8 May 1992, Wageningen, The Netherlands.* Pudoc Scientific Publishers, Wageningen, pp. 199–204.

Kornas, J. (1983) Man's impact upon the flora and vegetation in Central Europe. In: Holzner, W., Werger, M.J.A. and Ikusima, I. (eds) *Man's Impact on Vegetation.* Dr W. Junk Publishers, The Hague pp. 277–286.

Korpel, S. (1982) Degree of equilibrium and dynamical changes of the forest on example of natural forests of Slovakia. *Acta Facultatis Forestalis* 24, 9–31.

Korpel, S. (1987) Dynamics of the structure and development of natural beech forests in Slovakia. *Acta Facultatis Forestalis* 29, 59–85.

Korpel, S. (1995) *Die Urwälder der Westkarpaten.* Gustav Fischer Verlag, Stuttgart.

Korstian, C.F. (1927) Factors controlling germination and early survival in oaks. PhD thesis, Yale University, New Haven, Connecticut.

Koss, H. (1982) Verbreitung, ökologishe Ansprüche und waldbauliche Verwendung der Winterlinde (*Tilia cordata* Mill.). *Forst- und Holzwirt* 37, 381–385.

Koster, E.A. (1988) Ancient and modern cold-climate aeolien sand deposition: a review. *Journal of Quaternary Science* 3, 69–83.

Kraft, G. (1894) Zur Erziehung der Eiche, mit besonderer Rücksicht auf den Spessart. *Zeitschrift für Forst- und Jagdwesen* 26, 389–405.

Krahl-Urban, J. (1959) *Die Eichen. Forstliche Monographie der Traubeneiche und der Stieleiche.* Paul Parey, Berlin.

Krasinska, M. and Krasinski, Z.A. (1998) Het succes van de wisent in Białowieza. *Nieuwe Wildernis* 4, 16–21.

Krasinski, Z.A. (1978) Dynamics and structure of the European Bison population in the Białowieza Primeval Forest. *Acta Theoriologica* 23, 3–48.

Krasinski, Z.A. (1993) *Bison. A Relict of Ancient Times.* Białowieza National Park.

Krause, E.H.L. (1892a) Die Heide. Beitrag zur Geschichte des Pflanzenwuchses in Nordwesteuropa. *Engleis Botanisches Jahrbuch* 14, 517–539.

Krause, E.H.L. (1892b) Urkundliche Nachrichten über Bäume und Nutzpflanzen des Gebiets der brandenburgischen Flora. *Verhandlungen des Botanischen Vereins der Provinz Brandenburg und die Angrenzende Laender* 33, 75–87.

Krebs, C.J. (1972) *Ecology. The Experimental Analysis of Distribution and Abundance*, 3rd edn. Harper & Row, New York.

Krüger, U. (1999) Das niederländische Beispiel: Die 'Oostvaardersplassen' – ein Vogelschutzgebiet mit Großherbivoren als Landschaftsgestaltern. *Natur und Landschaft* 74, 428–435.

Kuiters, A.T., Slim, P.A. and van Hees, A.F.M. (1997) Spontane Bosverjonging en hoefdieren. In: van Wieren, S.E., Groot Bruinderink, G.W.T.A., Jorritsma, I.T.M. and Kuiters, A.T. (eds) *Hoefdieren in het boslandschap.* Backhuys Publishers, Leiden, pp. 99–129.

Kuper, J.H. (1994) Sustainable development of Scots pine forests. PhD thesis, Wageningen Agricultural University Papers 94–2, Agricultural University Wageningen, Wageningen.

Kwasnitschka, K. (1965) Das Naturschutzgebiet 'Unterhölzerwald'. *Mitteilungen des badischen Landesvereines für Naturkunde und Naturschutz e.V., Freiburg im Breisgau* 8, 725–728.

Kwiatkowska, A.J. (1986) Reconstruction of the old range and the present-day boundary of a *Potentillo albae–Quercetum* (Libb.) 1933. Phytocoenosis in the Białowieza Primeval Forest landscape. *Ekologia Polska* 34, 31–45.

Kwiatkowska, A.J. and Wyszomirski, T. (1988) Decline of *Potentillo albae–Quercetum* phytocoenoses associated with the invasion of *Carpinus betulus. Vegetatio* 75, 49–55.

Kwiatkowska, A.J. and Wyszomirski, T. (1990) Species deletion in *Potentillo albae–Quercetum* phytocoenoses reversed by the removal of *Carpinus betulus. Vegetatio* 87, 115–126.

Kwiatkowska, A.J., Spalik, K., Michalak, E., Palinska, A. and Panufnik, D. (1997) Influence of the size and density of *Carpinus betulus* on the spatial distribution and rate of deletion of forest-floor species in thermophilous oak forest. *Plant Ecology* 129, 1–10.

Landolt, E. (1866) *Der Wald, seine Verjüngung, Pflege und Benutzung.* Schweizerischen Forstverein, Zürich.

Lanier, L. (1988) Weeding and cleaning of broadleaved high forest. In: Savill, P.S. (ed.) National Hardwoods Programme. *Report of the Eighth Meeting and Second Meeting of the Uneven-Aged Silviculture Group 7 January 1988.* OFI Occasional Papers nr. 37. Oxford Forestry Institute, University of Oxford, pp. 52–77.

van der Lans, H. and Poortinga, G. (1985) *Natuurbos in Nederland. Een uitdaging.* IVN, Amsterdam.

Latalowa, M. (1992) The last 1500 years on Wolin Island (N.W. Poland) in the light of palaeobotanical studies. *Review of Palaeobotany and Palynology* 73, 213–226.

Lautenschlager, D. (1917) Die forstlichen Verhältnisse des Bialowieser Urwaldes. In: *Bialowies: Bialowies in deutscher Verwaltung.* Hrsg. v.d. Militärforstverwaltung pp. 1–5, 1917–1919.

Lauwerier, R.C.G.M. (1988) *Animals in Roman times in the Dutch Eastern River Area.* Rijksdienst voor het Oudheidkundig Bodemonderzoek ROB. Nederlandse oudheden 12. Oostelijk rivierengebied 1. (ROB) Amersfoort, SDU, 's-Gravenhage.

van Lear, D.H. and Watt, J.M. (1993) The role of fire in oak regeneration. In: Loftis, D.L. and McGee, C.E. (eds) *Oak Regeneration: Serious Problems, Practical Recommendations. Symposium Proceedings Knoxville, Tennessee September 8–10, 1992.* Southeastern Forest Experiment Station, Asheville, pp. 66–78.

Lebreton, P. (1990) Histoire de la grande faune mammifère des forêt françaises depuis 2000 ans. Thèse pour le Doctorat vétérinaire, diplôme d'état, Nantes.

Le Duc, M.G. and Havill, D.C. (1998) Competition between *Quercus petraea* and *Carpinus betulus* in an ancient wood in England: seedling survivorship. *Journal of Vegetation Science* 9, 873–880.

Leemans, R. (1991a) Canopy gaps and establishment patterns of spruce (*Picea abies* (L.) Karst.) in two old-growth coniferous forests in central Sweden. *Vegetatio* 93, 157–165.

Leemans, R. (1991b) Sensitivity analysis of a forest succession model. *Ecological Modelling* 53, 247–262.

van Leeuwarden, W. and Janssen, C.R. (1987) Differences between valley and upland vegetation development in eastern Noord-Brabant, the Netherlands, during the Late Glacial and Early Holocene. *Review of Palaeobotany and Palynology* 52, 170–204.

van Leeuwen, C.G. (1966) Het beheer van natuurreservaten op struktuur-oecologische grondslag. *Gorteria* 3, 16–28.

Leibundgut, H. (1945) Waldbauliche Untersuchungen über den Aufbau von Plenterwäldern. *Mitteilungen Schweizerische Anstalt für das Forstliche Versuchswesen* 24, 219–296.

Leibundgut, H. (1959) Über Zweck und Methodik der Struktur und Zuwachsanalyse von Urwäldern. *Schweizerische Zeitschrift für Forstwesen* 110, 111–124.

Leibundgut, H. (1978) Über die Dynamik europäischer Urwälder. *Allgemeine Forstzeitschrift* 33, 686–690.

Leibundgut, H. (1984a) *Die Waldpflege.* Dritte überarbeitete und ergäntzte Auflage unter Mitverwendung von 'Auslesedurchforstung als Erziehungsbetrieb höchster Wertleistung' von Walter Schädelin. Paul Haupt, Bern.

Leibundgut, H. (1984b) *Die natürliche Waldverjüngung.* Zweite, überarbeitete und erweiterte Auflage. Paul Haupt, Bern.

Leibundgut, H. (1993) *Europäische Urwälder. Wegweiser zur naturnahen Waldwirtschaft.* Paul Haupt, Bern.

Leimbach, W. (1948) Zur Waldsteppenfrage in der Sowjetunion. *Erdkunde* 2, 238–256.

Leitner, L.A. and Jackson, M.T. (1981) Presettlement forests of the unglaciated portion of southern Illinois. *The American Naturalist* 105, 290–304.

Lemdahl, G. (1985) Fossil insect faunas from Late-glacial deposits in Scania (South Sweden). *Ecologia Mediterranea* Tome XI (Fascimile 1), 185–191.

Lemée, G. (1966) Sur l'intérêt écologiques des réserves biologiques de la forêt de Fontainebleau. *Bulletin Societe Botanique de France* 113, 305–323.

Lemée, G. (1978) La hêtraie naturelle de Fontainebleau. In: Lamotte, M. and Boresliène, F (eds) *Problèmes d'ecologie: structure et functionnement des écosystemes terrestres.* Masson, Paris, pp. 75–128.

Lemée, G. (1981) Contribution à l'histoire des landes de la forêt de Fontainebleau d'après l'analyse pollinique des sols. *Bulletin Societe Botanique de France* 128, Lettres Bot. 3, 189–200.

Lemée, G. (1985) Rôle des arbres intolérants à l'ombrage dans la dynamique d'une hêtraie naturelle (forêt de Fontainebleau). *Oecologia Plantarum* 6, 3–20.

Lemée, G. (1987) Les populations de chênes (*Quercus petraea* Liebl.) des réserves biologiques de La Tillaie et du Gros Fouteau en forêt de Fontainebleau: structure, démographie et évolution. *Revue d'Ecologie* 42, 329–355.

Lemée, G., Faille, A. and Pontailler, J.Y. (1986) Dynamique de cicatrisation des ouvertures naturelles dans des reserves biologiques de la forêt de Fontainebleau (Environs de Paris). In: Fanta, J. (ed.) *Forest Dynamics Research in Western and Central Europe. Proceedings of the Workshop held 17–20 September 1985 in Wageningen, The Netherlands.* IUFRO-subject group SI 01–00 Ecosystems. Pudoc, Wageningen, pp. 170–183.

Lemée, G., Faille, A., Pontailler, J.Y. and Roger, J.M. (1992) Hurricanes and regeneration in a natural beech forest. In: Teller, A., Malthy, P. and Jeffers, J.N.R. (eds) *Responses of Forest Ecosystems to Environmental Changes.* Elsevier Applied Science, London, pp. 987–988.

Lieckfeld, C.-P. (1991) Blanco's Schecks Sind Gedeckt. Sanien: Privatinitiative. In: *Naturerbe Europa. Alte Welt-Neue Chancen.* Pro Futura, Munich, pp. 16–35.

Ligtendag, W.A. (1995) *De Wolden en het water. De landschaps- en waterstaatsontwikkeling in het lege land ten oosten van de stad Groningen vanaf de volle Middeleeuwen tot ca. 1870.* Regio en Landschapsstudies nr.2. Stichting Historisch Onderzoek en Beleid. REGIO-PRoject Uitgevers, Groningen.

Lindbladh, M. (1999) The influence of former land-use on vegetation and biodiversity in the boreo-nemoral zone of Sweden. *Ecography* 22, 485–498.

Lindbladh, M. and Bradshaw, R. (1995) The development and demise of a Medieval forest-meadow system at Linnaeus' birthplace in southern Sweden: implications for conservation and forest history. *Vegetation History and Archaeobotany* 4, 153–160.

Lindbladh, M. and Bradshaw, R. (1998) The origin of present forest composition and pattern in southern Sweden. *Journal of Biogeography* 25, 463–477.

Lindquist, B. (1938) Dalby Söderskog. En skånsk lövskog i forntid ock nutid. Mit einer deutschen Zusammenfassung. *Acta Phytogeographica Suecica* 10, 1–273.

Linhart, Y.B. and Whelan, R.J. (1980) Woodland regeneration in relation to grazing and fencing in Coad Gorswen, North Wales. *Journal of Applied Ecology* 17, 827–840.

Linnard, W. (1980) Coppicing, lopping, and natural high forest in medieval Wales. *Quaterly Journal of Forestry* 74, 225–228.

Linnman, G. (1978) Some aspects of the colonization of *Corylus avellana* L. in North West Europe during Early Flandrian Times. *Striae* 14, 72–75.

Litt, T. (1992) Fresh investigations into the natural and anthropogenically influenced vegetation of the earlier Holocene in the Elbe-Saale region, Central Germany. *Vegetation History and Archaeobotany* 1, 69–74.

Lödl, J., Mayer, H. and Pitterle, A. (1977) Das Eichen-Naturschutzgebiet Rohrberg im Hochspessart. *Forstwissenschaftliches Centralblatt* 96, 294–312.

Loftas, T. (ed.) (1995) *Bronnen voor ons Bestaan. Een atlas van voedsel en landbouw.* FAO 1945–1995. Voedsel en Landbouworganisatie van de Verenigde Naties, Rome.

Loftis, D.L. (1983) Regenerating southern Appalachian mixed hardwood stands with the shelterwood method. *Southern Journal of Applied Forestry* 7, 212–217.

Lohmeyer, W. and Bohn, U. (1973) Wildsträucher-Sprosskolonien (Polycormone) und ihre Bedeutung für die Vegetationsentwicklung auf brachgefallenem Grünland. *Natur und Landschaft* 48, 75–79.

Löhrl, H. (1970) Der Tannenhäher (*Nucifraga caryocatactus*) beim Sammeln und Knacken von Nüßchen der Zirbelkiefer (*Pinus cembra*). *Anzeiger der Ornithologischen Gesellschaft in Bayern* 9, 185–196.

Londo, G. (1990) Conservation and management of semi-natural grasslands in northwestern Europe. In: Bohnand, U. and Neuhäust, R. (eds) *Vegetation and Flora of Temperate Zones.* SPB Academic Publishers, The Hague, pp. 67–77.

Londo, G. (1991) *Natuurtechnisch bosbeheer. Natuurbeheer in Nederland* deel 4. Pudoc, Wageningen.

Lorimer, C.G. (1984) Development of the red maple understory in northeastern oak forests. *Forest Science* 30, 3–22.

Lorimer, C.G. (1993) Causes of the oak regeneration problem. In: Loftis, D.L. and McGee, C.E. (eds) *Oak Regeneration: Serious Problems, Practical Recommendations. Symposium Proceedings Knoxville, Tennessee, September 8–10, 1992.* Southeastern Forest Experiment Station. Asheville, pp. 14–39.

Lorimer, C.G., Chapman, J.W. and Lambert, W.D. (1994) Tall understory vegetation as a factor in the poor development of oak seedlings beneath mature stands. *Journal of Ecology* 82, 227–237.

Loucks, O.L. (1970) Evolution of diversity, efficiency and community stability. *American Zoologist* 10, 17–25.

Louwe Kooijmans, L.P. (1985) *Sporen in het land. De Nederlandse delta in de prehistorie.* Meulenhoff Informatief, Amsterdam.

Louwe Kooijmans, L.P. (1987) Neolithic settlement and subsistence in the wetlands of the Rhine-Meuse Delta of the Netherlands. In: Coles, J.M. and Lawson, A.J. (eds) *European Wetlands in Prehistory.* Clarendon Press, Oxford, pp. 227–252.

Lüdi, W. (1934) Zur Frage des Waldklimaxes in der Nordschweiz. *Bericht über das geobotanische Forschungsinstitut Rübel in Zürich für das Jahr 1934* 13, 143–147.

von Lüpke, B.V. (1982) Versuche zur Einbringung von Lärche und Eiche in Buchenbestände. *Schriften aus der Forstlichen Fakultät der Universität Göttingen und der Niedersächsischen Forstlichen Versuchsanstalt* 74, 1–123.

von Lüpke, B.V. (1987) Einflüsse von Altholzüberschirmung und Bodenvegetation auf das Wachstum junger Buchen und Eichen. *Forstarchiv* 58, 18–24.

von Lüpke, B.V. (1989) Die Esche; wertvolle Baum im Buchenwald. *Allgemeine Forstzeitschrift* 44, 1040–1042.

von Lüpke, B.V. and Hauskeller-Bullerjahn, K. (1999) Kahlschlagfreire Waldbau: wird die Eiche an den Rand gedrängt? *Forst und Holz* 18, 563–568.

Lutgerink, R.H.P., Swertz, Ch.A. and Janssen, C.R. (1989) Regional pollen assemblages versus landscape regions in the monts du Forez, Massif Central, France. *Pollen et Spores* 31, 45–60.

Lutz, H.J. (1930) The vegetation of Heart's Content, a virgin forest in northwestern Pennsylvania. *Ecology* 11, 2–29

Lyr, H., Hoffmann, G. and Engel, W. (1965) Über den Einfluss unterschiedlicher Beschattung auf die Stoffproduktion von Jungpflanzen einiger Waldbäume (II. Mitteilung). *Flora* (Jena) 155, 305–330.

van der Maarel, E. (1971) Plant species diversity in relation to management. In: Duffey, E. and Watt, A.S. (eds) *The Scientific Management of Animal and Plant Communities for Conservation. Symposia of the British Ecological Society* 11, 45–63.

Madsen, P. (1994) Growth and survival of *Fagus sylvatica* seedlings in relation to light intensity and soil water content. *Scandinavian Journal of Forest Research* 9, 316–322.

Madsen, P. (1995) Effects of soil water content, fertilization, light, weed competition and seedbed type on natural regeneration of beech (*Fagus sylvatica*). *Forest Ecology and Management* 72, 251–264.

Magri, D. (1995) Some questions on the late-Holocene vegetation of Europe. *The Holocene* 5, 354–360.

Malmer, N., Lindgren, K. and Persson, S. (1978) Vegetational succession in a south-Swedish deciduous wood. *Vegetatio* 36, 17–29.

Mannion, A.M. (1992) *Global Environmental Change. A Natural and Cultural Environmental History.* Longman Scientific & Technical, Essex.

Mantel, K. (1968) Die Anfänge der Waldpflege und Forstkultur im Mittelalter unter der Einwirkung der lokalen Waldordnung in Deutschland. *Forstwissenschaftliches Centralblatt* 67, 75–100.

Mantel, K. (1980) *Forstgeschichte des 16. Jahrhunderts unter dem Einfluß der Forstordnungen und Noe Meurers.* Paul Parey, Hamburg.

Mantel, K. (1990) *Wald und Forst in der Geschichte.* M. und H. Schaper, Alfeld-Hannover.

Manten, A.A. (1967) Lennaert von Post and the foundations of modern palynology. *Review of Palaeobotany and Palynology* 1, 11–22.

Mantyk, A. (1957) Mehr Beachtung unseren Linden! *Forst- und Jagdzeitung* 7, 84–86.

Markgraf, F. (1927) An den Grenzen des Mittelmeergebiets. *Feddes Repertorium* Beih, 45.

Markgraf, F. (1931) Aus den südosteuropäischen Urwäldern. *Forst- und Jagdwesen* 63, 1–19.

Markgraf, F. and Dengler, A. (1931) Aus den südeuropäischen Urwälder. *Zeitschrift für Forst- und Jagdwesen* 63, 1–31.

Marks, J.B. (1942) Land use and plant succession in Coon Valley, Wisconsin. *Ecological Monographs* 12, 113–133.

Marziani, G. and Tacchini, G. (1996) Palaeoecological and palaeoethnological analysis of botanical macrofossils found at the Neolithic site Rivaltella ca'Romensini, northern Italy. *Vegetation History and Archaeobotany* 5, 131–136.

Matlack, G.R. (1994) Vegetation dynamics of the forest edge – trends in space and successional time. *Journal of Ecology* 82, 113–123.

Matthiae, P.E. and Stearns, F. (1981) Mammals in forest islands in southeastern

Wisconsin. In: Burgess, R.L. and Sharpe, D.M. (eds) *Forest Island Dynamics in Man-dominated Landscapes.* Springer Verlag, New York, pp. 55–66.

Mattingly, H. (1986) *Tacitus. The Agricola and the Germania.* Translated with an introduction by H. Mattingly, translation revised by S.A. Handford. Penguin Classics, Harmondsworth.

Matuszkiewicz, A. (1977) Der Thermophile Eichenwald in NO-Polen als anthropo-zoogene Gesellschaft. In: Tüxen, R. (ed.) *Vegetation und Fauna. Berichte der Internationalen Symposien der Internationalen Verein für Vegatationskunde.* J. Cramer, Valduz, pp. 527–540.

Mauve, K. (1931) Ueber Bestandesaufbau, Zuwachsverhältnisse und Verjüngung im galizischen Karpaten Urwald. *Mitteilungen aus Forstwirtschaft und Forstwissenschaft* 2, 257–311.

May, H. (1981) *Die 10 ökologischen Wald-Wild-Geboten für naturnahen Waldbau und naturnahe Jagdwirtschaft.* Waldbau-Institut. Universität für Bodenkultur, Wien.

May, T. (1993) Beeinflussten Grosssäuger die Waldvegetation der pleistozänen Warmzeiten Mitteleuropas? Ein Diskussionsbeitrag. *Natur und Museum* 123, 157–170.

Mayer, H. (1975) Der Einfluss des Schalenwildes auf die Vergüngung und Erhaltung von Naturwaldreservaten. *Forstwissenschaftliches Centralblatt* 94, 209–224.

Mayer, H. (1976) Zur Wiederherstellung und Erhaltung eines ökologischen Gleichgewichtes zwischen Wald und Wild im Gebirge. *16. IUFRO World Conference*, Oslo, 23–28.

Mayer, H. (1992) *Waldbau auf soziologisch-ökologischer Grundlage*, 4., teilweise neu bearbeitete Auflage. Gustav Fischer, Stuttgart.

Mayer, H., Neumann, M. and Sommer, H.G. (1980) Bestandsaufbau und Verjüngungsdynamik unter dem Einfluß natürlicher Wilddichten im kroatischen Urwaldreservat Corkova Uvala/Plitvicer Seen. *Schweizerische Zeitschrift für Forstwesen* 131, 45–70.

Mayer, H. and Neumann, M. (1981) Struktureller und entwicklungsdynamischer Vergleich der Fichten-Tannen-Buchen-Urwälder Rothwald/Niederösterreich und Corkova Uvala/Kroatien. *Forstwissenschaftliches Centralblatt*, 100, 111–132.

Mayer, H. and Reimoser, F. (1978) Die Auswirkungen des Ulmensterbens im Buchen-Naturwaldreservat Dobra (Niederösterreichisches Waldviertel). *Forstwissenschaftliches Centralblatt* 97, 314–321.

Mayer, H. and Tichy, K. (1979) Das Eichen-Naturschutzgebiet Johannser Kogel im Lainzer Tiergarten, Wienerwald. *Centralblatt für das gesamte Forstwesen* 4, 193–226.

Mayer-Wegelin, H. (1943) *Aufforstungen in der ukrainischen Steppe.* Mittlg. d. H.G. Akad. d. Deutsch. Forstw., Frankfurt am Main 3, pp. 1–34.

McAndrews, J.H. (1965) Postglacial history of prairie, savanna, and forest in northwestern Minnesota. *Memoirs of the Torrey Botanical Club* 22, 1–72.

McCarthy, B.C. (1994) Experimental studies of hickory recruitment in a wooded hedgerow and forest. *Bulletin of the Torrey Botanical Club* 121, 240–250.

McCarthy, B.C. and Bailey, D.R. (1996) Composition, structure, and disturbance history of Crabtree Woods: an old-growth forest of western Maryland. *Bulletin of the Torrey Botanical Club* 123, 350–365.

McCook, L.J. (1994) Understanding ecological community succession: causal models and theories, a review. *Vegetatio* 110, 115–147.

McCracken, D.I. and Bignal, E.M. (eds) (1995) Farming on the edge: the nature of traditional farmland in Europe. *Proceedings of the Fourth European Forum on Nature Conservation and Pastoralism 2–4 November 1994, Trujillo, Spain.* Joint Nature Conservation Committee, Peterborough.

McCune, B. and Cottam, G. (1985) The successional status of a southern Wisconsin oak woods. *Ecology* 66, 1270–1278.

McDonald, J.N. (1981) *North American Bison. Their Classification and Evolution.* University of California Press, Berkeley.

McGee, C.E. (1981) *Responses of Overtopped White Oak to Release.* Research Note SO-273, Southern Forest Experiment Station, New Orleans.

McGee, C.E. (1984) *Heavy Mortality and Succession in a Virgin Mixed Mesophytic Forest.* Research Paper SO-209, Southern Forest Experiment Station, New Orleans.

McHugh, T. (1972) *The Time of the Buffalo.* University of Nebraska Press, Lincoln.

McNaughton, S.J. (1979) Grassland-herbivore dynamics. In: Sinclair, A.R.E. and Norton-Griffiths, M. (eds) *Serengeti. Dynamics of an Ecosystem.* The University of Chicago Press, Chicago, pp. 46–81.

Medwecka-Kornás, A. (1977) Ecological problems in the conservation of plant communities, with special reference to Central Europe. *Environmental Conservation* 4, 27–33.

Meiggs, R. (1982) *Trees and Timber in the Ancient Mediterranean World.* Clarendon Press, Oxford.

Meiggs, R. (1989) *Farm Forestry in the Ancient Mediterranean.* Social Forestry Network. Network Paper 8b Odi. Agricultural Administration Unit, Regent's College, London.

Mellanby, K. (1968) The effects of some mammals and birds on regeneration of oak. *Journal of Applied Ecology* 5, 359–366.

Mellars, P. (1975) Ungulate populations, economic patterns, and the Mesolithic landscape. In: Evans, J.G., Limbrey, S. and Cleere, H. (eds) *The Effect of Man on the Landschape, the Highland Zone.* Research Report No. 11, The Council for British Archaeology. pp. 49–56.

Mellars, P. (1976) Fire ecology, animal populations and man: a study of some ecological relationships in prehistory. *Proceedings of the Prehistoric Society* 42, 15–45.

Meusel, H. (1951/52) Die Eichen-Mischwälder des Mitteldeutschen Trockengebietes. *Wissenschaftliches Zeitschrift der Martin Luther Universität Halle-Wittenberg* 1, 49–72.

Meyer, K.A. (1931) Geschichtliches von den Eichen in der Schweiz. *Mitteilungen der Schweizerischen Centralanstalt für das forstlichen Versuchswesen* 16, 231–452.

Meyer, K.A. (1941) Holzartenwechsel und frühere Verbreitung der Eiche in der Westschweiz; Kanton Waadt: Vom Jura zum Jorat. *Mitteilungen der Schweizerischen Centralanstalt für das forstlichen Versuchswesen* 22, 63–141.

Miedema, T. and Gevers, E. (1990) Het vlotte al eeuwen. Reusachtige Rijnvlotten brachten Duits hout naar Dordrecht. *Het Houtblad,* September, 55–87.

Mikan, C.J., Orwig, D.A. and Abrams, M.D. (1994) Age structure and successional dynamics of a presettlement-origin chestnut oak forest in the Pennsylvania Piedmont. *Bulletin of the Torrey Botanical Club* 121, 13–23.

Milchunas, D.G., Lauenroth, W.K. and Burke, I.C. (1998) Livestock grazing: animal and plant biodiversity of shortgrass steppe and the relationship to ecosystem function. *Oikos* 83, 65–74.

Miles, J. (1987) Vegetation succession: past and present perceptions. In: Gray, A.J.,

Grawley, M.J. and Edwards, P.J. (eds) *Colonization, Succession and Stability*. Blackwell Scientific Publications, Oxford, pp. 1–29.

Miles, K. and Kinnaird, J.W. (1979a) The establishment and regeneration of birch, juniper and Scots pine in the Scottish highlands. *Scottish Forestry* 33, 102–107.

Miles, K. and Kinnaird, J.W. (1979b) Grazing: with particular reference to birch, juniper and Scots pine in the Scottish Highlands. *Scottish Forestry* 33, 280–289.

Mitchell, F.J.G. (1998) The investigation of long-term succession in temperate woodland using fine spatial resolution pollen analysis. In: Kirby, K.J. and Watkins, C. (eds) *The Ecological History of European Forests*. CAB International, Wallingford, UK, pp. 213–223.

Mitchell, F.J.G. and Cole, E. (1998) Reconstruction of long-term successional dynamics of temperate woodland in Białowieza Forest, Poland. *Journal of Ecology* 86, 1042–1059.

Mitchell, F.J.G. and Kirby, K.J. (1990) The impact of large herbivores on the conservation of semi-natural woods in British uplands. *Forestry* 63, 333–353.

Mlinsek, D. (1993) Forestry and society-oriented research on the history of virgin forests and their future needs. In: Broekmeyer, M.E.A., Vos, W. and Koop, H. (eds) *European Forest Reserves. Proceedings of the European Forest Reserves Workshop, 6–8 May 1992, Wageningen, The Netherlands*. Pudoc Scientific Publishers, Wageningen, pp. 29–33.

Moe, D. and Rackham, O. (1992) Pollarding and a possible explanation of the neolithic elmfall. *Vegetation History and Archaeobotany* 1, 63–68.

Monk, C.D. (1961a) The vegetation of the William L. Hutcheson Memorial Forest, New Jersey. *Bulletin of the Torrey Botanical Club* 88, 156–166.

Monk, C.D. (1961b) Past and present influences on reproduction in the William L. Hutcheson Memorial Forest, New Jersey. *Bulletin of the Torrey Botanical Club* 88, 167–175.

Moore, P.D. (1987) Chalk grasslands in the ice age. *Nature* 329, 388–389.

Moore, P.D. and Webb, J.A. (1978) *An Illustrated Guide to Pollen Analysis*. Hodder and Stoughton, London.

Morey, H.F. (1936) A comparison of two virgin forests in north-western Pennsylvania. *Ecology* 17, 43–55.

Morgan, R.K. (1987a) Composition, structure and regeneration characteristics of the open woodlands of the New Forest, Hampshire. *Journal of Biogeography* 14, 423–438.

Morgan, R.K. (1987b) An evaluation of the impact of anthropogenic pressures on woodland regeneration in the New Forest, Hampshire. *Journal of Biogeography* 14, 439–450.

Morgan, R.K. (1991) The role of protective understory in the regeneration system of a heavily browsed woodland. *Vegetatio* 92, 119–132.

Morosow, G. (1928) *Die Lehre vom Walde* (Translation in German). Neumann, Neudamm.

Morris, M.G. (1967) Differences between the invertebrate faunas of grazed and ungrazed chalk grassland. Responses of some phytophagous insects to cessation of grazing. *Journal of Applied Ecology* 4, 459–474.

Morris, M.G. (1974) Oak as a habitat for insect life. In: Morris, M.G. and Perring, F.H. (eds) *The British Oak, Its History and Natural History*. The Botanical Society of the British Isles, E.W. Classey, Berkshire, pp. 274–297.

Morzadec-Kerfourn, M.-T. (1976) La végétation au Pléistocène supérieur et au début de l'Holocène en Armorique. In: de Lumley, H. (ed.) *La Préhistoire Française* Tome I. *Les civilisations paléolithiques et mesolithiques de la France.* CNRS, Paris, pp. 531–533.

Moss, C.E. (1910) The fundamental units of vegetation: historical development of the concepts of the plant association and the plant formation. *New Phytologist* 9, 18–53.

Moss, C.E. (1913) *Vegetation of the Peak District.* Cambridge University Press, Cambridge.

Moss, C.E., Rankin, W.M. and Tansley, A.G. (1910) The woodlands of England. *New Phytologist* 9, 113–149.

Muller, F. and Renkema, E.H. (1995) *Beknopt Latijns-Nederlands Woordenboek,* 12th edn. Wolters-Noordhoff, Groningen.

Müller, H. (1953) *Zur spät- und nacheiszeitlichen Vegetationsgeschichte des mitteldeutschen Trockengebietes.* Nova Acta Leopoldina. Abhandlungen der deutschen Akademie der Naturforscher (Leopoldina) zu Halle/Saale, No. 10, Band 16.

Müller, I. (1947) Der pollenanalytische Nachweis der Menschlichen Besiedlung im Federsee- und Bodenseegebiet. *Planta* 35, 70–87.

Müller, K.M. (1929) *Aufbau, Wuchs und Verjüngung der Südosteuropäischen Urwälder.* M & H, Hannover.

Müller, T. (1962) Die Saumgesellschaften der Klasse *Trifolio-Geranieta sanguinei. Mitteilungen. Floristisch-Soziologische Arbeitsgemeinschaft N.F.* 9, 94–140.

Müller-Kroehling, S. and Schmidt, O. (1999) Zu: Megaherbivoren-Theorie. Mit großen Säugern gegen kleine Bäume? *Allgemeine Forst Zeitschrift/Der Wald* 20, 1072–1073.

Müller-Schneider, P. (1977) Über die Rolle der Tiere bei der Samenverbreitung. In: Tüxen, R. (ed.) *Vegetation und Fauna. Berichte der Internationalen Symposien der Internationalen Verein für Vegetationskunde.* J. Cramer, Vaduz, pp. 119–130.

Musall, H. (1969) *Die Entwicklung der Kulturlandschaft der Rheinniederung zwischen Karlsruhe und Speyer vom Ende des 16. bis zum Ende des 19. Jahrhunderts.* Heidelberger Geografische Arbeiten, 22. Geographischen Institut der Universität Heidelberg, Heidelberg.

Myster, R.W. (1993) Tree invasion and establishment in old fields at Hutcheson Memorial Forest. *The Botanical Review* 59, 251–272.

Nabokov, P. and Snow, D. (1992) Farmers of the woodlands. In: Josephy, A.M. Jr (ed.) *America in 1492. The World of the Indian Peoples Before the Arrival of Columbus.* Alfred A. Knopf, New York, pp. 119–145.

Namvar, K. and Spethmann, W. (1985a). Waldbaumarten aus der Gattung *Ulmus* (Ulme, Rüster). *Allgemeine Forstzeitschrift* 40, 1220–1225.

Namvar, K. and Spethmann, W. (1985b). Die Baumarten der Gattung *Sorbus:* Vogelbeere, Mehlbeere, Elsbeere und Speierling. *Allgemeine Forstzeitschrift* 40, 937–943.

Namvar, K. and Spethmann, W. (1986a) Die heimischen Waldbaumarten der Gattung 'Tilia' (Linde). *Allgemeine Forstzeitschrift* 41, 42–44.

Namvar, K. and Spethmann, W. (1986b) Die Wild- oder Holzbirne (*Pyrus pyraster*). *Allgemeine Forstzeitschrift* 41, 520–522.

Neemann, G. and Stickan, W. (1992) Carbohydrate partitioning and storage in beech saplings of a mature stand's understory; studies of carbon balance in a montane beech forest (*Fagus sylvatica* L.) in the Solling area, FRG. In: Teller, A., Mathy, P. and Jeffers, N.J.R. (eds) *Responses of Forest Ecosystems to Environmental Changes.* Elsevier Applied Science, London, pp. 635–636.

Neuweiler, E. (1905) *Die Prähistorischen Pflanzenreste Mitteleuropas, mit besonderer Berücksichtigung der Schweizerischen Funde.* Verlag von Albert Raustein, Zürich.

Newbold, A.J. and Goldsmith, F.B. with an addendum on birch by Harding, J.S. (1981) *The Regeneration of Oak and Beech: a Literature Review.* Discussion Papers in Conservation. No. 33. University College London, London.

Niering, W.A. and Goodwin, R.H. (1974) Creation of relatively stable shrublands: arresting 'succession' on rights-of-way and pastureland. *Ecology* 55, 784–795.

Nietsch, H. (1927) Mittelauropäischer Urwald. *Zeitschrift der Gesellschaft für Erdkunde zu Berlin* 1–16.

Nietsch, H. (1928) Die Eiche in der indogermanischen Vorzeit. *Mannus* 20, 44–53.

Nietsch, H. (1935) *Steppenheide oder Eichenwald? Eine urlandschaftkundliche Untersuchung zum Verständnis der vorgeschichtlichen Siedlung in Mitteleuropa.* Uschmann, Weimar.

Nietsch, H. (1939) *Wald und Siedlung im vorgeschichtlichen Mitteleuropa.* Mannus-Bücherei 64, Rabitzsch Verlag, Leipzig.

Nilsson, S.G. (1985) Ecological and evolutionary interactions between reproduction of beech *Fagus sylvatica* and eating animals. *Oikos* 44, 157–164.

Nilsson, S.G. (1992) Forests in the temperate-boreal transition – natural and man-made features. In: Hanssen, L. (ed.) *Conservation Ecology Series: Principles, Practices and Management.* Elsevier Applied Science, London, pp. 373–393.

Nitzschke, H. (1932) Der Neuenburger Uwrwald bei Bockhorn in Oldenburg. *Vegetationsbilder* 23(6/7). Gustav Fischer, Jena.

Noffke, J. (1989) Sorgfältige Pflege und starke Durchforstung für die Vogelkirsche. *Allgemeine Forstzeitschrift* 44, 1034–1036.

Norland, E.R. and Hix, D.M. (1996) Composition and structure of a chronosequence of young, mixed-species forests in southeastern Ohio, USA. *Vegetatio* 125, 11–30.

Nowacki, G.J. and Abrams, M.D. (1992) Community, edaphic and historical analysis of mixed oak forests of the Ridge and Valley Province in central Pennsylvania. *Canadian Journal of Forest Research* 22, 790–800.

Nowacki, G.J. and Abrams, M.D. (1994) Forest composition, structure, and disturbance history of the Alan Seeger Natural Area, Huntington County, Pennsylvania. *Bulletin of the Torrey Botanical Club* 121, 277–291.

Nüsslein, H. (1978) Eichenmasten im Spessart und ihre Ausnutzung. *Allgemeine Forstzeitschrift* 33, 667–668.

Oberdorfer, E. (ed.) (1983) *Süddeutsche Pflanzengesellschaften.* Teil III., *Wirtschaftswiesen und Unkrautgesellschaften.* Zweite stark bearbeitete Auflage. Gustav Fischer Verlag, Stuttgart.

Oberdorfer, E. (ed.) (1992a) Süddeutsche Pflanzengesellschaften. Teil IV., *Wälder und Gebüsche.* A. Textband. Zweite stark bearbeitete Auflage. Gustav Fischer Verlag, Jena.

Oberdorfer, E. (ed.) (1992b) *Süddeutsche Pflanzengesellschaften.* Teil IV., *Wälder und Gebüsche.* B. Tabellenband. Zweite stark bearbeitete Auflage. Gustav Fischer Verlag, Jena.

O'Connell, M. (1986) Pollenanalytische Untersuchungen zur Vegetations- und Siedlungsgeschichte aus dem Lengener Moor, Friesland. *Probleme der Küstenforschung* 16, 171–193, Hildesheim.

O'Donnell (ed.) (1968) *Adriaen van der Donck (1656). A Description of the New Netherlands.* Syracuse University Press, Syracuse.

Ogden, J.G. III (1961) Forest history of Martha's Vineyard, Massachusetts. I. Modern and pre-colonial forests. *The American Midland Naturalist* 66, 417–430.

Oksanen, L. (1990) Predation, herbivory, and plant strategies along gradients of primary production. In: Grace, J.B. and Tilman, D. (eds) *Perspectives on Plant Competition*. Academic Press, London, pp. 445–474.

Olberg, A. (1957) Beiträge zum Problem der Kieferverjüngung. *Schriftenreihe der Forstliche Fakultät der Universität Göttingen und Mitteilungen der Niedersächsischen Fortslichen Versuchsanstalt* 18, 1–96.

Oldeman, R.A.A. (1990) *Forests: Elements of Sylvology*. Springer, Berlin.

Oliver, C.D. and Stephens (1977) Reconstruction of a mixed-species forest in central New England. *Ecology* 58, 562–572.

Ollf, H., Vera, F.W.M., Bokdam, J., Bakker, E.S., Gleichman, J.M., de Maeyer, K. and Smit, R. (1999) Shifting mosaics in grazed woodlands driven by the alternation of plant facilitation and competition. *Plant Biology* 1, 127–137.

Oosterbaan, A. and van Hees, A.F.M. (1989) *Resultaten van een lichtingsproef in een Beuken-Wintereikenbos*. Rapport nr. 551. De Dorschkamp, Wageningen.

Orwig, D.A. and Abrams, M.D. (1994) Land-use history (1720–1992), composition, and dynamics of oak–pine forests within the Piedmont and Coastal Plain of northern Virginia. *Canadian Journal of Forest Research* 24, 1216–1225.

O'Sullivan, P.E., Oldfield, F. and Batterbee, R.W. (1973) Preliminary studies of Lough Neagh sediments. I. Stratigraphy, chronology and pollen analyses. In: Birks, H.J.B. and West, R.G. (eds) *Quaternary Plant Ecology. The 14th symposium of the British Ecological Society, University of Cambridge, 28–30 March 1972*. Blackwell Scientific Publications, Oxford, pp. 267–279.

Otto, H. (1987) Zum waldbaulichen Verhalten der Vogelkirsche. *Forst- und Holzwirt* 42, 44–45.

Otto, S.F. (1780) *Forstbeschreibungsprotokoll*. Handgeschiebenes Original im Landesarchiv Oldenburg, Forst-, Jagd-, und Fischereiachen.

Ouden, J.B. den (1992) Floodplain forest reserve Ranspuk – Structure and dynamics of a Fraxino pannonicae-Ulmetum in South Moravia, CSFR. MSc thesis, Wageningen Agricultural University, Wageningen.

Ovington, J.D. (1965) *Woodlands*. The English Universities Press, London.

Ovington, J.D. and McRae, C. (1960) The growth of seedlings of *Quercus petraea*. *Journal of Ecology* 48, 549–555.

Packham, J.R. and Harding, D.J.L. (1982) *Ecology of Woodland Processes*. Edward Arnold, London.

Paczoski, J. (1928) *La flore de la Forêt de Białowieza*. V-me Excurs. Phytogéograph. Internat. (V.I.P.E.), Varsovie.

Paczoski, J. (1930) *Lasy Bialowiezy (Die Waldtypen von Białowieza)*. Panstowa Rada Ochrony Przyrody. Monografje Naukowe, nr. 1, Poznan.

Page, R.I. (1972) *Life in Anglo-Saxon England*. Batsford, London.

Paice, J.P. (1974) The ecological history of Grizedale Forest, Cumbria, with particular reference to *Tilia cordata* (Mill.). MSc thesis, University of Lancaster.

Parker, G.R., Leopold, D.J. and Eichenberger, J.K. (1985) Tree dynamics in an old-growth, deciduous forest. *Forest Ecology and Managment* 11, 31–57.

Parker Pearson, M. (1993) *Bronze Age Britain*. English Heritage and B.T. Batsford, London.

Peckham, W.D. (1925) *Thirteen Costumals of the Sussex Manors of the Bishop of Chichester*. Sussex Record Society (SRS), Vol. 31, W. Heffer & Sons, Cambridge.

Peglar, S.M. (1993a) The mid-holocene *Ulmus*-decline at Diss Mere, Norfolk, UK: a year-by-year pollen stratigraphy from annual laminations. *The Holocene* 3, 1–13.

Peglar, S.M. (1993b) The development of the cultural landscape around Diss-Mere, Norfolk, UK, during the past 7000 Years. *Review of Palaeobotany and Palynology* 76, 1–47.

Peglar, S.M. and Birks, H.J.B. (1993) The mid-Holocene *Ulmus* fall at Diss Mere, South-East England – disease and human impact? *Vegetation History and Archaeobotany* 2, 61–68.

Peltier, A., Toezet, M.-C., Armengaud, C. and Ponge, J.-F. (1997) Establishment of *Fagus sylvatica* and *Fraxinus excelsior* in an old-growth beech forest. *Journal of Vegetation Science* 8, 13–20.

Penistan, M.J. (1974) Growing oak. In: Morris, M.G. and Perring, F.H. (eds) *The British Oak, Its History and Natural History*. The Botanical Society of the British Isles, E.W. Classey Ltd, Berkshire, pp. 98–112.

Pennington, W. (Tutin, T.G.) (1970) Vegetation history in the North-West of England: A regional synthesis. In: Walker, D. and West, R.G. (eds) *Studies in the Vegetational History of the British Isles*. Cambridge University Press, Cambridge, pp. 41–79.

Pennington, W. (1975) A chronostratigraphic comparison of Late-Weichselien and Late-Devensian subdivisions, illustrated by two radiocarbon-dated profiles from western Britain. *Boreas* 4, 157–171.

Pennington, W. (1977) The Late Devensian flora and vegetation of Britain. *Philosophical Transactions of the Royal Society of London* Series B, 280, 247–271.

Perlin, J. (1991) *A Forest Journey: The Role of Wood in the Development of Civilization*. Harvard University Press, Cambridge, Massachusetts.

Perry, I. and Moore, P.D. (1987) Dutch elm disease as an analogue of Neolithic elm decline. *Nature* 326, 72–73.

Persson, S. (1974) Vegetation development after the exclusion of grazing cattle in a meadow in the south of Sweden. PhD thesis. Department of Plant Ecology University of Lund, Lund.

Persson, S. (1980) Succession in a south Swedish deciduous wood: a numerical aproach. *Vegetatio* 43, 103–122.

Peterken, G.F. (1981) *Woodland Conservation and Management*. Chapman and Hall, London.

Peterken, G.F. (1991) Ecological issues in the management of woodland nature reserves. In: Spellenberg, I.F., Goldsmith, F.B. and Morris, M.G. (eds) *The Scientific Management of Temperate Communities for Conservation*. Blackwell Scientific Publications, Oxford, pp. 245–272.

Peterken, G.F. (1992) Coppices in the lowland landscape. In: Buckley, G.P. (ed.) *Ecology and Management of Coppice Woodlands*. Chapman & Hall, London, pp. 3–17.

Peterken, G.F. (1993) Long-term studies in forest nature reserves. In: Broekmeyer, M.E.A., Vos, W. and Koop, H. (eds) *European Forest Reserves. Proceedings of the European Forest Reserves Workshop, 6–8 May 1992, Wageningen, The Netherlands*. Pudoc Scientific Publishers, Wageningen, pp. 35–48.

Peterken, G.F. (1996) *Natural Woodland. Ecology and Conservation in Northern Temperate Regions*. Cambridge University Press, Cambridge.

Peterken, G.F. and Game, M. (1984) Historical factors affecting the number and distribution of vascular plants in the woodlands of central Lincolnshire. *Journal of Ecology* 72, 155–182.

Peterken, G.F. and Tubbs, C.R. (1965) Woodland regeneration in the New Forest Hampshire, since 1650. *Journal of Applied Ecology* 2, 159–170.

Peters, K. (1992) *Begrazing door runderen gedurende de laatste eeuwen in het woud van Białowieza (N.O. Polen/ W. Wit-Rusland).* Landbouwuniversiteit Wageningen, Vakgroep Natuurbeheer, verslag no. 3022, Wageningen.

Peterson, C.J. and Picket, S.T.A. (1995) Forest reorganization: a case study in an old-growth forest catastrophic blowdown. *Ecology* 76, 763–774.

Pickett, S.T.A. (1980) Non-equilibrium co-existence of plants. *Bulletin of the Torrey Botanical Club* 107, 238–248.

Pickett, S.T.A. and White, P.S. (1985) Patch dynamics: a synthesis. In: Pickett, S.T.A. and White, P.S. (eds) *The Ecology of Natural Disturbance and Patch Dynamics.* Academic Press, Orlando, Florida, pp. 371–384.

Pietzarka, U. and Roloff, A. (1993) Dynamische Waldrandgestaltung. Ein Modell zur Strukturverbesserung von Wald aussenrändern. *Natur und Landschaft* 68, 555–560.

Pigott, C.D. (1975) Natural regeneration of *Tilia cordata* in relation to forest-structure in the forest of Białowieza, Poland. *Philosophical Transactions of the Royal Society of London* Series B 270, 151–179.

Pigott, C.D. (1981) The status, ecology and conservation of *Tilia platyphyllos* in Britain. In: Synge, H. (ed.) *The Biological Aspects of Rare Plant Conservation.* John Wiley & Sons, Chichester, pp. 305–317.

Pigott, C.D. (1983) Regeneration of oak–birch woodland following exclusion of sheep. *Journal of Ecology* 71, 629–646.

Pigott, C.D. (1985) Selective damage to tree-seedlings by bank voles (*Clethrionomys glareolus*). *Oecologia* 67, 367–371.

Pigott, C.D. (1988) The ecology and silviculture of limes (*Tilia* spp.). In: Savill, P.S. (ed.) *O.F.I. Occasional Papers,* No 37. *National Hardwoods Programme. Report of the Eighth Meeting and Second Meeting of the un-even Aged Silviculture Group 7 January 1988.* Oxford Forestry Institute, University of Oxford, pp. 27–32.

Pigott, C.D. (1991) Biological flora of the British Isles. *Tilia Cordata* Miller. *Journal of Ecology* 79, 1147–1207.

Pilcher, J.R., Baillie, M.G.L., Schmidt, B. and Becker, B. (1984) A 7272-year tree-ring chronology for western Europe. *Nature* 312, 150–152.

Planchais, N. (1976) La végétation pendant le Post-Glaciaire: aspects de la végétation holocène dans les plaines françaises. In: Guilaine, J. (ed.) *La Préhistoire françaises.* Tome II. *Les civilisations néolithiques et protohistoriques de la France.* CNRS, Paris, pp. 35–42.

Platt, W.J. and Strong, D.R. (1989) Gaps in forest ecology. *Ecology* 70, 535.

Pockberger, J. (1963) Die Linden. Ein Beitrag zur Bereicherung des mitteleuropäischen Waldbildes. *Centralblatt für das gesamte Forstwesen* 80, 99–123.

Pockberger, J. (1967) *Die Verbreitung der Linde, insbesondere in Oberösterreich.* Mitteilungen der Forstlichen Bundes-Versuchsanstalt, Wien.

Polak, B. (1959) Palynology of the 'Uddeler meer': A contribution to our knowledge of the vegetation and the agriculture in the northern part of the Veluwe in prehistoric and early historic times. *Acta Botanica Neerlandica* 9, 547–571.

Ponel, P. and Coope, G.R. (1990) Late glacial and early Flandrian coleoptera from La Taphanel, Massif Central, France: climatic and ecological implications. *Journal of Quarternary Science* 5, 235–249.

Ponge, J.-F. and Ferdy, J.-B. (1997) Growth of *Fagus sylvatica* saplings in an old growth forest as affected by soil and light conditions. *Journal of Vegetation Science* 8, 789–796.

Pontailler, J.-Y., Faille, A. and Lemée, G. (1997) Storms drive successional dynamics in natural forests: a case study in Fontainebleau forest (France). *Forest Ecology and Management* 98, 1–15.

von Post, L. (1916) Forest tree pollen in south Swedish peat bog deposits. (Om skogstradspollen i sydvenska torfmosselager foljder (foredragsreferat). Geologiska Foereningen in Stockholm, *Four handlingar* 38, 384–434. Translation by Margaret Bryan Davis and Knut Faegri with an introduction by Knut Faegri and Johs. Iversen. *Pollen et Spores* (1967) 9, 378–401. In: Real, L.A. and Brown, J.H. (eds) *Foundations of Ecology*. Classic Papers with commentaries. The University of Chicago Press, Chicago, pp. 456–482.

Pott, R. (1983) Geschichte der Hude- und Schneitelwirtschaft in Nordwestdeutschlands und ihre Auswirkung auf die Vegetation. *Oldenburger Jahrbuch* 83, 357–384.

Pott, R. (1992a) Entwicklung der Kulturlandschaft Nordwestdeutschlands unter dem Einfluss des Menschen. *Zeitschrift der Universität Hannover, Hochschulgem* 19, 3–48.

Pott, R. (1992b) *Die Pflanzengesellschaften Deutschlands*. Eugen Ulmer, Stuttgart.

Pott, R. (1993) *Farbatlas Waldlandschaften. Ausgewählte Waldtypen und Waldgesellschaften unter dem Einfluss des Menschen*. Eugen Ulmer, Stuttgart.

Pott, R. and Hüppe, J. (1991) *Die Hudenlandschaften Nordwestdeutschlands*. Westfälisches Museum für Naturkunde, Landschafsverband Westfalen-Lippe. Veröffentlichung der Arbeitsgemeinschaft für Biol.-ökol. Landesforschung, *ABÖL*, nr. 89, Münster.

Prentice, I.C. (1986) Forest-composition calibration of pollen data. In: Berglund, B.E. (ed.) *Handbook of Holocene Palaeoecology and Palaeohydrology*. John Wiley & Sons, Chichester, pp. 799–816.

Prentice, I.C. and Leemans, R. (1990) Pattern and process and the dynamics of forest structure: a simulation approach. *Journal of Ecology* 78, 340–355.

Price, T.D. (1987) The Mesolithic of Western Europe. *Journal of World Prehistory* 1, 225–305.

Prins, H.H.T. (1998) Origins and development of grassland communities in northwestern Europe. In: Wallis de Vries, M.F., Bakker, J.P. and van Wieren, S.E. (1998) *Grazing and Conservation Management*. Kluwer Academic Publishers, Dordrecht, pp. 55–105.

Prins, H.H.T. and van der Jeugd, H.P. (1993) Herbivore population crashes and woodland structure in East Africa. *Journal of Ecology* 81, 305–314.

van Prooije, L.A. (1990) *De invoer van Rijns hout per vlot 1650–1795*. Economisch en Sociaal-Historisch Jaarboek 53. Amsterdam, Nederlands Economisch-Historisch Archief, pp. 30–79.

van Prooije, L.A. (1992a) *Dordrecht als centrum van de Rijnse houthandel in de 17de en 18de eeuw*. Economisch en Sociaal-Historisch Jaarboek 55. Amsterdam, Nederlands Economisch-Historisch Archief, pp. 143–158.

van Prooije, L.A. (1992b) De houtvlotterij en Dordrecht in de 17de en de 18de eeuw. *Kwartaal en Teken* 18, 14–24.

van Prooije, L.A. (1992c) De verwerking en de distributie vanuit Dordrecht van hout in de 17de en de 18de eeuw. *Kwartaal en Teken* 18, 7–16.

Prusa, E. (1982) Kurzgefasste Ergebnisse von Untersuchungen einiger Urwald

Bestände in Böhmen und Mähren. In: Mayer, H. (ed.) *Urwald-Symposium IUFRO-Gruppe Urwald. Waldbau-Institut.* Universität für Bodenkultur, Wien, pp. 81–100.

Prusa, E. (1985) Urwald Lanzhot. *Die Böhmischen und Mährischen Urwälder. Ihre Struktur und Ökologie.* Verlag der Tscheckoslowakische Akademie der Wissenschaften, Prag, pp. 38–105.

Pruski, W. (1963) Ein Regenerationsversuch des Tarpans in Polen. *Zeitschrift für Tierzüchtung und Züchtungsbiologie* 79, 1–30.

Pryor, S.N. (1985) The silviculture of wild cherry or gean (*Prunus avium* L.). *Quarterly Journal of Forestry* 79, 95–109.

Pryor, S.N. (1988) *The Silviculture and Yield of Wild Cherry.* Forestry Commission Bulletin 75, London.

Pucek, Z. (1984) What to do with the European bison, now saved from extinction? *Acta Zoologica Fennica* 172, 187–190.

Puster, D. (1924) Auenwirtschaft. *Forstwissenschaftliches Centralblatt* 68, 448–460.

Putman, R.J. (1986) *Grazing in Temperate Ecosystems: Large Herbivores and the Ecology of the New Forest.* Croom Helm, London.

Putman, R.J. (1996a) Ungulates in temperate forest ecosystems: perspectives and recommendations for future research. *Forest Ecology and Management* 88, 205–214.

Putman, R.J. (1996b) *Competition and Resource Partitioning in Temperate Ungulate Assemblages.* Chapman & Hall, London.

Putman, R.J., Edwards, P.J., Mann, J.C.E., How, R.C. and Hill, S.D. (1989) Vegetational and faunal changes in an area of heavily grazed woodland following relief of grazing. *Biological Conservation* 47, 13–22.

Pyne, S.J. (1982) *Fire in America. A Cultural History of Wildland and Rural Fire.* Princeton University Press, Princeton, New Jersey.

Raben, G. (1980) Geschichtliche Betrachtung der Waldwirtschaftung im Naturwaldreservat Priorteich und deren Einfluss auf den heutigen Bestand. Diplomarbeit. Institut für Waldbau der Universität Göttingen, Göttingen.

Rackham, O. (1975) *Hayley Wood. Its History and Ecology.* Cambridgeshire and Isle of Ely Naturalists' Trust, Cambridge.

Rackham, O. (1976) *Trees and Woodland in the British Landscape.* Archaeology in the Field Series. J.M. Dent & Sons, London.

Rackham, O. (1980) *Ancient Woodland. Its History, Vegetation and Uses in England.* Edward Arnold, London.

Rackham, O. (1992) Mixtures, mosaics and clones: the distribution of trees within European Woods and forests. In: Canell, M.G.R., Malcolm, D.C. and Robertse, P.A. (eds) *The Ecology of Mixed-Species Stands of Trees.* Special Publication number 11 of the British Ecological Society. Blackwell Scientific Publications, Oxford, pp. 1–20.

Rackham, O. (1993) *The History of the Countryside. The Classic History of Britain's Landscape, Flora and Fauna.* J.M. Dent, London.

Rackham, O. (1998) Savanna in Europe. In: Kirby, K.J. and Watkins, C. (eds) *The Ecological History of European Forests.* CAB International, Wallingford, UK, pp. 1–24.

Ralska-Jasiewiczowa, M. and van Geel, B. (1992) Early human disturbance of the natural environment recorded in annually laminated sediments of Lake Gosciaz, central Poland. *Vegetation History and Palaeobotany* 1, 33–42.

Ranney, J.W. and Bruner, M.C. (1981) The importance of edge in the structure and dynamics of forest islands. In: Burgess, R.L. and Sharpe, D.M. (eds) *Forest Island Dynamics in Man-Dominated Landscapes*. Springer, New York, pp. 67–95.

Raup, H.M. (1937) Recent changes of climate and vegetation in southern New England and adjacent New York. *Journal of Arnold Arbor* 18, 79–117.

Raup, H.M. (1964) Some problems in ecological theory and their relation to conservation. *Journal of Ecology* 52 (Suppl.), 19–28.

Raus, D. (1986) Die Stieleiche in ihrem slawonischen Optimum. *Allgemeine Forstzeitschrift* 41, 761–762.

Real, L. and Brown, J.H. (eds) (1991) *Foundations of Ecology*. Classic papers with commentaries. The University of Chicago Press, Chicago.

Reed, J.L. (1954) *The Forests of France*. Faber and Faber, London.

Regnell, M., Gaillard, M.-J., Bartholin, T.S. and Karsten, P. (1995) Reconstruction of environment and history of plant use during the late Mesolithic (Ertebølle culture) at the inlands settlement of Bökeberg III, Southern Sweden. *Vegetation History and Archaeobotany* 4, 67–61.

Reif, A., Jolitz, Th., Münch, D. and Bücking, W. (1998) Sukzession vom Eichen-Hainbuchen-Wald zum Ahorn-Wald- Prozesse der Naturverjüngung im Bannwald 'Bechtaler Wald' bei Kenzingen, Südbaden. *Allgemeine Forst- und Jagdzeitung* 170, 67–74.

Reimers, N.F. (1958) The reforestation of burns and forest tracts devasted by silkworms in the mountain cedar–pine taiga of cisbaikal and the role of vertebrate animals in this process. *Byull. Moip, Otdel. Biol* 63, 49–56.

Reinhold, F. (1949) Zusammensetzung und Aufbau eines natürlichen Eichen-Buchenwaldes auf der Baar bei Donaueschingen. *Forstwissenschaftliches Centralblatt* 68, 691–698.

Remling, F.X. (1852) *Urkundenbuch zur Geschichte der Bischöfe zu Speijer.* Band I. Ältere Urkunden. Neudruck der Ausgabe Mainz, 1852. Scientia Verlag, Aalen (1970).

Remling, F.X. (1853) *Urkundenbuch zur Geschichte der Bishöfe zu Speijer.* Band II. Jünger Urkunden. Neudruck der Ausgabe Mainz, 1853. Scientia Verlag, Aalen (1970).

Remmert, H. (1991) The mosaic-cycle concept of ecosystems. An overview. In: Remmert, H. (ed.) *The Mosaic-Cycle Concept of Ecosystems*. Springer, Berlin, pp. 11–21.

Rempe, H. (1937) Untersuchungen über die Verbreitung des Blütenstaube durch die Luftströmungen. *Planta* 27, 93–147.

Rennie, R. (1810) Essays on the natural history of peat moss (cited in Clements, 1916).

Reynolds, H.W., Glaholt, R.D. and Hawley, A.W.L. (1982) Bison (*Bison bison*). In: Chapman, L.J.A. and Feldhammer, G.A. (eds) *Wild Mammals of North America: Biology, Management, and Economics*. Johns Hopkins University Press, Baltimore, Maryland, pp. 972–1007.

Roberts, N. (1989) *The Holocene. An Environmental History*. Basil Blackwell, Oxford.

Rodenberg, L. (1988) Das naturschutzgebiet in Gösslunda. In: Sjögren, E. (ed.) *Plant Cover on the Limestone Alvar of Öland. Ecology, Sociology, and Taxonomy. Acta Phytogeographica Suecica* 76, 73–86.

Rodenwaldt, U. (1951) Reviergeschichte als eine Grundlage der Waldbauplanung. Die Entwicklung der gräfl. Erbach-Arbachsen Waldungen in Odenwald. *Forstwissenschaftliches Centralblatt* 70, 469–514.

Rodwell, J.S. (ed.) (1991) *British Plant Communities. vol. I. Woodlands and Scrub*. Cambridge University Press, Cambridge.

Roger, C.A. (1993) Application of aeropalynological principles in palaeoecology. *Review of Palaeobotany and Palynology* 79, 133–140.

Röhle, H. (1984) Ertragskundliche Merkmale von Stieleichen-Mischbeständen auf grundwasserbeeinflussten Standorten in den Auewaldgebieten Südbayerns. *Forstwissenschaftliches Centralblatt* 103, 330–349.

Röhrig, E. (1967) Wachstum junger Laubholzpflanzen bei unterschiedlichen Lichtverhältnissen. *Allgemeine Forst- und Jagdzeitung* 138, 224–239.

Röhrig, E. (1991) Vegetation structure and forest succession. In: Röhrig, E. and Ulrich, B. (eds) *Temperate Deciduous Forests. Ecosystems of the World 7.* Elsevier, Amsterdam.

Röös, M. (1990) Zum Wachstum der Vogelkirsche (*Prunus avium*) in Nordrhein-Westfalen und angrenzenden Gebieten. Dissertation, Forstliche Fakultät Universität Göttingen.

Rösch, M. (1992) Human impact as registered in the pollen record: some results from the western Lake Constance region, Southern Germany. *Vegetation History and Archaeobotany* 1, 101–109.

Rose, F. (1974) The epiphytes of oak. In: Morris, M.G. and Perring, F.H. (eds) *The British Oak. Its History and Natural History.* The Botanical Society of the British Isles, E.W. Classey, Berkshire, pp. 250–273.

Rose, F. (1992) Temperate forest management: its effects on bryophyte and lichen floras and habitats. In: Bates, J.W. and Farmar, A.M. (eds) *Bryophytes and Lichens in a Changing Environment.* Clarendon Press, Oxford, pp. 211–233.

Rose, F. and James, P.W. (1974) Regional studies on the British lichen flora. I. The corticolous and lignicolous species of the New Forest, Hampshire. *Lichenologist* 6, 1–72.

Rosén, E. (1982) Vegetation development and sheep grazing in limestone grasslands of south Öland, Sweden. *Acta Phytogeographica Suecica* 72, 1–72.

Rosén, E. (1988) Shrub expansion in alvar grassland on Öland. In: Sjögren, E. (ed.) *Plant Cover on the Limestone Alvar of Öland. Ecology, Sociology and Taxonomy. Acta Phytogeographica Suecica* 76, 76–100.

Ross, M.S., Sharik, T.L. and Smith, D.W. (1986) Oak regeneration after clear felling in southwest Virginia. *Forest Science* 32, 157–169.

Rostlund, E. (1960) The geographic range of the historic bison in the southeast. *Annals of the Association of American Geographers* 50, 395–407.

Rousset, O. and Lepart, J. (1999) Shrub facilitation of *Quercus humilis* regeneration in succession on calcareous grasslands. *Journal of Vegetation Science* 10, 493–502.

Rowley-Conwy, R. (1982) Forest grazing and clearance in temperate Europe with special reference to Denmark: an archaeological view. In: Bell, M. and Limbrey, S. (eds) *Archaeological Aspects of Woodland Ecology. Symposia of the Association for Environmental Archaeology*, No. 2 BAR. International Series 146, Oxford, pp. 199–215.

Rübel, E.A. (1914) Heath and steppe, macchia and garigue. *Journal of Ecology* 2, 232–237.

Rubner, K. (1920) Die waldbaulichen Folgerungen des Urwaldes. *Naturwissenschaftliche Zeitschrift für Forst- und Landwirtschaft* 18, 201–214.

Rubner, K. (1924) *Die pflanzengeographischen Grundlagen des Waldbaus.* Verlag von J. Neumann, Neudamm.

Rubner, K. (1934) *Die Pflanzengeographischen Grundlagen des Waldbaus*, 3rd edn. J. Neumann, Neudamm.

Rubner, H. (1960) *Die Hainbuche in Mittel- und Westeuropa. Untersuchungen über ihre ursprünglichen Standorte und ihre Förderung durch die Mittelwaldwirtschaft.* Forschungen zur Deutschen Landes- und Volkskunde 121. Bundesanstalt für Landeskunde und Raumforschung, Bad Godesberg.

Ruffner, C.M. and Abrams, M.D. (1998a) Relating land-use history and climate to the dendrology of a 326-year-old *Quercus prinus* talus slope forest. *Canadian Journal of Forest Research* 28, 347–358.

Ruffner, C.M. and Abrams, M.D. (1998b) Lightning strikes and resultant fires from archival (1912–1917) and current (1960–1997) information in Pennsylvania. *Journal of the Torrey Botanical Society* 125, 249–252.

Rühl, A. (1968) Lindenmischwälder im südlichen Nordwestdeutschland. *Allgemeine Forst- und Jagdzeitung* 139, 118–130.

Rümelin (1926) Eichenschnitt in Kulturen. *Allgemeine Forst- und Jagdzeitung* 87, 359–361.

Runkle, J.R. (1981) Gap regeneration in some old-growth forests of the eastern United States. *Ecology* 62, 1041–1051.

Runkle, J.R. (1982) Patterns of disturbance in some old-growth mesic forests in eastern North America. *Ecology* 63, 1533–1546.

Runkle, J.R. (1985) Disturbance regimes in temperate forests. In: Pickett, S.T.A. and White, P.S. (eds) *The Ecology of Natural Disturbance and Patch Dynamics.* Academic Press, Orlando, Florida, 17–33.

Runkle, J.R. (1989) Synchrony of regeneration, gaps, and latitudinal differences in tree species diversity. *Ecology* 70, 546–547.

Russell, E.W.B. (1981) Vegetation of northern New Jersey before European settlement. *The American Midland Naturalist* 105, 1–12.

Russell, E.W.B. (1983) Indian-set fires in the forests of the northeastern United States. *Ecology* 64, 78–88.

Rymer, L. (1978) The use of uniformitarianism and analogy in palaeoecology, particularly pollen analysis. In: Walker, D. and Guppy, J.C. (eds) *Biology and Quaternary Environments.* Australian Academy of Sciences, Canberra, pp. 245–258.

Salisbury, E.J. (1918) The ecology of scrub in Hertfordshire. A study in colonization. *Transactions of the Hertfordshire National History Society and Field Club* 17, 53–64.

Sanderson, J.L. (1958) The autecology of *Corylus avellana* (L.) in the neighbourhood of Sheffield with special reference to its regeneration. PhD thesis, The University of Sheffield, Sheffield.

Savill, P.S. (1991) *The Silviculture of Trees used in British Forestry.* CAB International, Wallingford, UK.

Scaife, R.G. (1982) Late-Devensian and early flandrian vegetation changes in southern England. In: Bell, M. and Limbrey, S. (eds) *Archaeological Aspects of Woodland Ecology. Symposia of the Association for Environmental Archaeology,* No. 2 BAR. International Servies 146. Oxford, pp. 57–74.

Schalk, P.H. (1990) Prunus in bos en landschap in Nederland. *Nederlands Bosbouw Tijdschrift* 62, 144–151.

Schama, S. (1995) *Landschap en herinnering.* Uitgeverij Contact, Amsterdam/Antwerpen.

Scheiner, S.M. (1993) Introduction: theories, hypotheses, and statistics. In: Scheiner, S.M. and Gurevitch, J. (eds) *Design and Analysis of Ecological Experiments.* Chapman & Hall, New York, pp. 1–13.

Schenck, C. A. (1924) Der Waldbau des Urwaldes. *Allgemeine Forst- und Jagd-Zeitung* 100, 377–388.

Schepers, F. (1993) De broedvogels van Koningsteen in 1991 en 1992. *Natuurhistorisch Maandblad* 82, 245–251 (with English summary).

Scherf, B.D. (ed.) (1995) *World Watch List for Domestic Animal Diversity*, 2nd edn. FAO, UNEP. Food and Agriculture Organization of the United Nations, Rome.

Scholz, H. (1975) Grassland evolution in Europe. *Taxon* 24, 81–90.

Schott, C. (1934) Kanadische Biberwiesen. Ein beitrag zur Frage der Wiesenbildung. *Zeitschrift der Gesellschaft für Erdkunde zu Berlin* 370–374.

Schreiber, K-F. (1993) Standordsabhängige Entwicklung von Sträuchern und Bäumen im Sukzessionsverlauf von Brachgefallenem Grünland in Südwestdeutschland. *Phytocoenologia* 23, 539–560.

Schröder, W. (1974) Über einige Fragen der Ökologie der Cerviden im Walde. *Forstwissenschafliches Centralblatt* 93, 121–127.

Schrötter, H. (1964) Die Hainbuche; eine Gefahr für die Naturverjüngung der Rotbuche. *Sozialistische Forstwirtschaft* 282–283.

Schubart, W. (1966) *Die Entwicklung des Laubwaldes als Wirtschaftswald zwischen Elbe, Saale und Weser.* Aus dem Walde. Mitteilungen aus der Niedersächsischen Landesforstverwaltung 14.

Schuster, L. (1950) Über den Sammeltrieb des Eichelhähers (*Garrulus glandarius*). *Vogelwelt* 71, 9–17.

Schwabe, A. and Kratochwil, A. (1986) Zur Verbreitung und Individualgeschichte von Weidbuchen im Schwarzwald. *Abhandlungen Landesmuseum für Naturkunde zu Münster in Westfalen* 48, 21–54, Münster.

Schwabe, A. and Kratochwil, A. (1987) *Weidhuchen im Schwarzwald und ihre Entstehung durch Vebliss des Weideviehs.* Beihefte zu den Veröffentlichungen für Naturschutz und Landschafspflege in Baden-Württemburg 49, Karlsruhe.

Schwappach, A. (1916) Zur Entwicklung der Mischbestände von Eiche und Buche. *Forst- und Jagdwesen* 38, 615–623.

Scot, E.L. (1915) A study of pasture trees and shrubbery. *Bulletin of the Torrey Botanical Club* 42, 451–461.

Scott, G.H. (1963) Uniformitarianism, the uniformity of nature, and Paleoecology. *New Zealand Journal of Geology and Geophysics* 6, 510–527.

Seagrief, S.C. (1960) Pollen diagrams from Southern England: Cranes Moor, Hampshire. *New Phytologist* 59, 73–83.

Seeger, M. (1930) Erfahrungen über die Eiche in der Rheinebene bei Emmendingen (Baden). *Allgemeine Forst- und Jagdzeitung* 106, 201–219.

Seeger, M. (1938) Erfahrungen met der Eichenwirtschaft in Baden. *Mitteilungen aus Forstwirtschaft und Forstwissenschaft* 9, 657–690.

Semken, H.A. (1983) Holocene mammalian biography and climatic change in the eastern and central United States. In: Wright, H.E. Jr (ed.) *Late-Quaternary Environments in the United States*, Vol. 2. *The Holocene*. Longman, London, pp. 182–207.

Shaw, M.W. (1968a) Factors affecting the natural regeneration of sessile oak (*Quercus petraea*) in North Wales. I. A preliminary study of acorn production, viability and losses. *Journal of Ecology* 56, 565–583.

Shaw, M.W. (1968b) Factors affecting the natural regeneration of sessile oak (*Quercus petraea*) in North Wales. II. Acorn losses and germination under field conditions. *Journal of Ecology* 56, 647–660.

Shaw, M.W. (1974) The reproductive characteristics of oak. In: Morris, M.G. and Perring, F.H. (eds) *The British Oak, Its History and Natural History*. The Botanical Society of the British Isles, E.W. Classey Ltd, Berkshire, pp. 162–181.

Shea, J.H. (1983) Twelve fallacies of uniformitarianism. *Geology* 10, 455–460.

Sheail, J. (1980) *Historical Ecology: the Documentary Evidence*. Natural Research Council, Institute of Terrestrial Ecology, Cambridge.

Shugart, H.H. (1984) *A Theory of Forest Dynamics*. Springer Verlag, New York.

Shugart, H.H. and Seagle, S.W. (1985) Modelling forest landscapes and the rôle of disturbance in ecosystems and communities. In: Pickett, S.T.A. and White, P.S. (eds) *The Ecology of Natural Disturbance and Patch Dynamics*. Academic Press, Orlando, Florida, pp. 353–368.

Shugart, H.H. and Urban, D.L. (1989) Factors affecting the relative abundance of forest tree species. In: Grubb, P.J. and Whittaker, J.B. (eds) *Towards a More Exact Ecology*. Blackwell Scientific, Oxford, pp. 249–273.

Shugart, H.H. and West, D.C. (1977) Development of an Appalachian deciduous forest succession model and its application to assessment of the impact of the chestnut blight. *Journal of Environmental Management* 5, 161–170.

Shugart, H.H. and West, D.C. (1980) Forest succession models. *Bioscience* 30, 308–313.

Shugart, H.H. and West, D.C. (1981) Long-term dynamics of forest ecosystems. *American Scientist* 69, 647–652.

Siebel, H.N. and Bijlsma, R.J. (1998) *Patroonontwikkeling en begrazing in boslandschappen: New Forest en Fontainebleau als referenties*. IBN-rapport 357. IBN-DLO, Instituut voor Bos- en Natuuronderzoek.

Simmons, I.G. (1993) Vegetation change during the Mesolithic in the British Isles; some amplifications. In: Chambers, F.M. (ed.) *Climate Change and Human Impact at the Landscape*. Chapman & Hall, London, pp. 109–118.

Simmons, I.G. and Innes, J.B. (1987) Mid-Holocene adaptations and later Mesolithic forest disturbance in northern England. *Journal of Archaeological Science* 14, 385–403.

Simmons, I.G., Dimble, G.W. and Grigson, C. (1984) The Mesolithic. In: Simmons, I.G. and Tooley, M.J. (eds) *The Environment in British Prehistory*. Duckworth, London, pp. 82–124.

Simmons, J. (1992). Mid-Holocene woodland history in upland England. In: Teller, A., Maltley, P. and Jeffers, J.N.R. (eds) *Responses of Forest Ecosystems to Environmental Changes*. Elsevier, London, pp. 567–568.

Simpson, G. (1998) English cathedrals as sources of forest and woodland history. In: Watkins, C. (ed.) *European Woods and Forests: Studies in Cultural History*. CAB International, Wallingford, UK, pp. 39–53.

Sims, R.E. (1973) The anthropogenic factor in East Anglian vegetational history: an approach using A.P.F. techniques. In: Birks, H.J.B. and West, R.G. (eds) *Quaternary Plant Ecology. The 14th Symposium of the British Ecological Society, University of Cambridge, 28–30 March 1972*. Blackwell Scientific Publications, Oxford, p. ix.

Sinclair, A.R.E. (1979a) Dynamics of the Serengeti ecosystem: process and pattern. In: Sinclair, A.R.E. and Norton-Griffiths, M. (eds) *Serengeti. Dynamics of an Ecosystem*. The University of Chicago Press, Chicago, pp. 1–30.

Sinclair, A.R.E. (1979b) The eruption of the ruminants. In: Sinclair, A.R.E. and

Norton-Griffiths, M. (eds) *Serengeti. Dynamics of an Ecosystem*. The University of Chicago Press, Chicago, pp. 82–103.

Sissingh, G. (1983) Betekenis en gevolgen van menselijke ingrepen voor de samenstelling en instandhouding van bossen, speciaal onder Nederlandse omstandigheden. In: Goor, C.P. (ed.) *Ecologie en gebruik van bossen*. Pudoc, Wageningen, pp. 41–50.

Sjögren, E. (1988) Studies of vegetation on Öland; changes and development during a century. In: Sjögren, E. (ed.) *Plant Cover on the Limestone Alvar of Öland. Ecology, Sociology and Taxonomy. Acta Phytogeographica Suecica* 76, 5–8.

Skeen, J.N. (1976) Regeneration and survival of woody species in a naturally-created forest opening. *Bulletin of the Torrey Botanical Club* 6, 259–265.

Slicher van Bath, B. (1987) *De agrarische geschiedenis van Europa 500–1850*, 6th edn. Aula, Spectrum, Utrecht.

Sloet, J.J.S. Baron (1911) *Gelderse markerechten. Oud-vaderlandse rechtsbronnen*. Werken der Vereeniging tot Uitgaaf der Bronnen van het Oud-Vaderlands Recht, Utrecht. Tweede reeks no. 12. Martinus Nijhoff, 's-Gravenhage.

Sloet, J.J.S. Baron (1913) *Gelderse markerechten. Oud-vaderlandse rechtsbronnen*. Werken der Vereeniging tot Uitgaaf der Bronnen van het Oud-Vaderlands Recht, Utrecht. Tweede reeks no. 15. Martinus Nijhoff, 's-Gravenhage.

Smith, A.G. (1958) Two lacustrine deposits in the south of the English Lake District. *New Phytologist* 57, 363–386.

Smith, A.G. (1970) The influence of mesolithic and neolithic man on British vegetation: discussion. In: Walker, D. and West, R.G. (eds) *Studies in the Vegetational History of the British Isles*. Cambridge University Press, Cambridge, pp. 81–96.

Smith, C. (1993) Regenerating oaks in the Central Appalachians. In: Loftis, D.L. and McGee, C.E. (eds) *Oak Regeneration: Serious Problems, Practical Recommendations. Symposium Proceedings, Knoxville, Tennessee, September 8–10, 1992*. Southeastern Forest Experiment Station. Asheville, pp. 211–221.

Smith, C.J. (1980) *Ecology of the English Chalk*. Academic Press, London.

Smith, D.M. (1962) The forest of the United States. In: Barrett, J.W. (ed.) *Regional Silviculture of the United States*. The Ronald Press Company, New York, pp. 3–29.

Snow, B. and Snow, D. (1988) *Birds and Berries. A Study of an Ecological Interaction*. T. and A.D. Poyser, Calton.

van Soest, P.J. (1982) *Nutritional Ecology of the Ruminant. Ruminant Metabolism, Nutritional Strategies, the Cellulolytic Fermentation and the Chemistry of Forages and Plant Fibers*. O & B Books, Corvallis, Oregon.

Söffner, W. (1982) *Über die Grosssäugerfauna Mitteleuropas im Postglazial. Ein Beitrag zur Kenntnis der Beziehungen zwischen Wild und Vegetation. Zulassungsarbeit*. Institut für Botanik der Universität Hohenheim.

Sonnesson, L.K. (1994) Growth and survival after cotyledon removal in *Quercus robur* seedlings, grown in different natural soil types. *Oikos* 69, 65–70.

Sorensen, A.E. (1981) Interactions between birds and fruit in a temperate woodland. *Oecologia* 50, 242–249.

Speidel, G. (1975) Schalenwildbestände und Leistungsfähigkeit des Waldes als Problem der Forst- und Holzwirtschaft aus der sicht der Forstökonomie. *Allgemeine Forstzeitschrift* 30, 247–250.

Spethmann, W. and Namvar, K. (1985) Der Bergahorn und die Gattung Acer. *Allgemeine Forstzeitschrift* 40, 1126–1131.

Spiecker, M. and Spiecker, H. (1988) Erziehung von Kirschenwertholz. *Allgemeine Forstzeitschrift* 43, 562–565.

Spies, T.A. and Franklin, J.F. (1989) Gap characteristics and vegetation response in coniferous forests of the Pacific Northwest. *Ecology* 70, 543–545.

Spurr, S.H. (1952) Origin of the concept of forest succession. *Ecology* 33, 426–427.

Spurr, H.S. (1956) Natural restocking of forests following the 1938 hurricane in central New England. *Ecology* 33, 426–427.

Stagg, D.J. (1987) Ships and timber. In: *Explore the New Forest*. Forestry Commission, HMSO, London, pp. 22–25.

Stamper, P. (1988) Woods and parks. In: Astill, G. and Grant, A. (eds) *The Countryside of Medieval England*. Blackwell Scientific Publications, Oxford, pp. 128–148.

Steele, R.C. (1974) Variations in oakwoods. In: Morris, M.G. and Perring, F.H. (eds) *The British Oak, Its History and Natural History*. The Botanical Society of the British Isles, E.W. Classey, Berkshire, pp. 130–140.

Steenstrup, J.J.S. (1841) *Geognostisk-geologisk Undersögelse af Skovmoserne Vidnesdam- og Lillemose i det nordlige Sjælland, ledsaget af sammenligende Bemærkninger, hentede fra Danmarks Skov- Kjær- og Lyngmoser Ialmindelighed*. Copenhagen.

Steinbrenner, E.C. (1951) Effect of grazing on floristic composition and soil properties of farm woodlands in southern Wisconsin. *Journal of Forestry* 12, 906–910.

Stoks, F.C.M. (1994) *Van Dale Handwoordenboek Duits-Nederlands*, 2nd edn. Van Dale Lexicografie, Utrecht/Antwerpen.

Stover, M.E. and Marks, P.L. (1998) Successional vegetation on abandoned cultivated and pastured land in Tompkins County, New York. *Journal of the Torrey Botanical Society*, 125, 150–164.

Street, M. (1991) Bedburg-Königshoven: A Pre-Boreal Mesolithic site in the Lower Rhineland (Germany). In: Barton, N., Roberts, A.J. and Roe, D.A. (eds) *The Late Glacial in North-west Europe: Human Adaptation and Environmental Change at the End of the Pleistocene*. CBA Research Report No. 77. Council for British Archaeology, London, pp. 256–270.

Streitz, H. (1967) Bestockungswandel in Laubwaldgesellschaften des Rhein-Main-Tieflandes und der Hessischen Rheinebene. Dissertation Forstlichen Fakultät der Georg-August-Universität zu Göttingen in Hannover, Münden.

Stuart, A.J. (1982) *Pleistocene Vertebrates in the British Isles*. Longman, London.

Stuart, A.J. (1991) Mammalian extinctions in the late Pleistocene of northern Eurasia and North America. *Biological Review* 66, 453–562.

Sugita, S. (1994) Pollen representation of vegetation in Quaternary sediments: theory and method in patchy vegetation. *Journal of Ecology* 82, 881–897.

Sugita, S. (1998) Modelling pollen representation of vegetation. In: Gaillard, M.-J. and Berglund, B.E. (eds) Quantification of land surfaces cleared of forest during the Holocene-Modern pollen/vegetation/landscape relationships as an aid to the interpretation of pollen data. *Paläoklimaforschung/Palaeoclimate Research* 27, 1–16.

Sugita, S., Gaillard, M.-J. and Broström, A. (1999) Landscape openness and pollen records: a simulation approach. *The Holocene* 9, 409–421.

Sugita, S., MacDonald,G. and Larsen, C.P.S. (1997) Reconstruction of fire disturbance and forest succession from fossil pollen in lake sediments: potential and limitations. In: Clark, J.S., Cashier, H., Goldmmer, J.G. and Stocks, B.J. (eds) *Sediment Records of Biomass Burning and Global Change*. Springer, Berlin, pp. 387–412.

Sukopp, H. (1972) Wandel von Flora und Vegetation in Mittelauropa unter dem Einfluss des Menschen. *Berichte über Landwirtschaft* 50, 112–139.

Sulser, J.S. (1971) Twenty years of change in the Hutcheson Memorial Forest. *William L. Hutcheson Memorial Forest Bulletin* 2, 15–25.

Suner, A. and Röhrig, E. (1980) Die Entwicklung der Buchennaturverjüngung in Abhängigkeit von der Auflichtung des Altbestandes. *Forstarchiv* 51, 145–149.

Sutter, E. and Amann, F. (1953) Wie weit fliegen vorratssammelnde Tannenhähe? *Ornithologische Beobachter* 50, 89–90.

Swaine, M.D. and Whitmore, T.C. (1988) On the definition of ecological species groups in tropical rain forests. *Vegetatio* 75, 81–86.

Swanberg, P.O. (1951) Food storage, territory and song in the thick-billed nutcracker. *Proceedings of the 10th International Ornithological Congress, 1950, Uppsala*, pp. 545–554.

Swart, G. (1953) Naturschutz im Reinhardswald. *Natur und Landschaft* 28, 66–69.

Szafer, W. (1968) The ure-ox, extinct in Europe since the seventeenth century: an early attempt at conservation that failed. *Biological Conservation* 1, 45–47.

Tacitus (AD 98) *The Agricola and the Germania* (transl. with introduction by Mattingly, H.). Translation revised by Handford, S.A. (1970). Penguin Classics, Harmondsworth.

Tack, G., van den Bremt, P. and Hermy, M. (1993) *Bossen van Vlaanderen: een historische ecologie*. Davidsfonds, Leuven.

Tallis, J.H. (1991) *Plant Community History. Long-term Changes in Plant Distribution and Diversity*. Chapman & Hall, London.

Tamboer-van den Heuvel, G. and Janssen, C.R. (1976) Recent pollen assemblages from the crest region of the Vosges mountains (France). *Review of Palaeobotany and Palynology* 21, 219–240.

Tangermann (1932) Zur Frage der Eichenverjüngungsverfahrens im Lehrrevier Freienwalde. *Zeitschrift für Forst- und Jagdwesen* 14, 321–226.

Tanouchi, H. and Yamamoto, S. (1995) Structure and regeneration of canopy species in an old-growth evergreen broad-leaved forest in Aya district, southwestern Japan. *Vegetatio* 117, 51–60.

Tansley, A.G. (ed.) (1911) *Types of British Vegetation*. Cambridge University Press, Cambridge.

Tansley, A.G. (1916) The development of vegetation. Review of Clements' 'Plant succession', 1916. *Journal of Ecology* 4, 198–204.

Tansley, A.G. (1922) Studies on the vegetation of the English chalk II. Early stages of redevelopment of woody vegetation on chalk grassland. *Journal of Ecology* 10, 168–177.

Tansley, A.G. (1935) The use and abuse of vegetational concepts and terms. *Ecology* 16, 284–307.

Tansley, A.G. (1949) *The British Islands and their Vegetation*, Vols 1 and 2, 2nd edn. Cambridge University Press, Cambridge.

Tansley, A.G. (1953) *The British Islands and their Vegetation*, Vols 1 and 2, 3rd edn. Cambridge University Press, Cambridge.

Tanton, M.T. (1959) Acorn destruction potential of small mammals and birds in British woodlands. *Quarterly Journal of Forestry* 59, 1–5.

Tapper, P.-G. (1992) Demography of persistent juveniles in *Fraxinus excelsior*. *Ecography* 15, 385–392.

Tapper, P.-G. (1993) The replacement of *Alnus glutinosa* by *Fraximus excelsior* during succession related to regenerative differences. *Ecography* 16, 212–218.

Tauber, H. (1965) Differential pollen dispersion and the interpretation of pollen diagrams. With a contribution to the interpretation of the elm fall. *Danmarks Geologiske Undersøgelse II* Raekke, nr. 89. *(Geological Survey of Denmark. II. Series, No. 89).*

Tauber, H. (1967) Differential pollen dispersion and filtration. In: Cushing, E.J. and Wright, H.E. (eds) *Quaternary Paleoecology*, Vol. 7. *Proceedings of the VII Congress of the International Association for Quaternary Research.* Yale University Press, New Haven, Connecticut, pp. 131–141.

Tendron, G. (1983) *La fôret de Fontainebleau. De l'ecology à la sylviculture.* Office National des Fôrets, Fontainebleau.

Thill, A. (1975) Contribution à l'étude du frêne, de l'érable sycamore et du merisier (*Fraxinus excelsior* L., *Acer pseudoplatanus* L., *Prunus avium* L.). *Bulletin de la Societé Royale Forestière de Belgique* 82, 1–12.

Thompson, D.Q. and Smith, R.H. (1970) The forest primeval in the Northwest – a great myth? *Proceedings of the Annual Tall Timbers Fire Ecology Conference* 10, 255–265.

Tichy, K.J. (1978) Analyse der Waldstruktur des Naturschutzgebietes Johannser Kogel im Lainzer Tiergarten. Diplomarbeit, forst- und holzwirtschaftlichen Studienrichtung der Universität für Bodenkultur in Wien.

Tilman, D. (1985) The resource-ratio hypothesis of plant succession. *American Naturalist* 125, 827–852.

Tilman, D. (1990) Constraints and tradeoffs: toward a predictive theory of competition and succession. *Oikos* 58, 3–15.

Tinner, W., Conedera, M., Ammann, B., Gäggeler, H.W., Gedye, S., Jones, R. and Sägesser, B. (1998) Pollen and charcoal in lake sediments compared with historically documented forest fires in southern Switzerland since AD 1920. *The Holocene* 8, 31–42.

Tinner, W., Hubschmid, P., Wehrli, M., Ammann, B. and Conedera, M. (1999) Long-term fire ecology and dynamics in southern Switzerland. *Journal of Ecology* 87, 273–289.

Tipping, R., Buchanan, J., Davies, A. and Tisdall, E. (1999) Woodland biodiversity, palaeo-human ecology and some implications for conservation management. *Journal of Biogeography* 26, 33–43.

Tittensor, R.M. (1978) A history of the Mens: a Sussex woodland common. *Sussex Archaeological Collections* 116, 347–374.

van Tol, G. (1982) Verjonging en beheer loofbossen. Verslag KNBV-excursie naar Frankrijk van 7–11 juni 1982. *Nederlands Bosbouw Tijdschrift* 54, 332–336.

Trautmann, W. (1969) Zur Geschichte des Eichen-Hainbuchenwaldes im Münsterland auf Grund pollenanalytischer Untersuchungen. *Schriftenreihe für Vegetationskunde* 4, 109–129.

Treiber, R. (1997) Vegetationsdynamik unter dem Einfluß des Wildschweins (*Sus scrofa* L.) am Beispiel bodensaurer Trockenrasen der elsässischen Harth. *Zeitschrift für Ökologie und Naturschutz* 6, 83–95.

Trepp, W. (1947) *Der Lindenmischwald (Tilieto-Aspenulatum Taurinae) des Schweizerischen voralpinen Föhn- und Seenbezirkes, seine pflanzensoziologische und forstliche Bedeutung.* Beiträge zur geobotanischen Landsaufnahme der Schweiz 27.

Trier, J. (1952) *Holz. Etymologien aus dem Niederwald.* Münstersche Forschungen 6, Böhlau Verlag, Münster/Köln.

Trier, J. (1963) *Venus: Etymologien um das Futterlaub.* Münstersche Forschungen 15, Böhlau Verlag, Münster/Köln.

Tripathi, R.S. and Khan, M.L. (1990) Effects of seed weight and microsite characteristics on germination and seedling fitness in two species of *Quercus* in a subtropical wet hill forest. *Oikos* 57, 289–296.

Troels-Smith, J. (1954) *Ertebøllekultur- Bondekultur. Resultater af de sidste 10 års undersøgelser i Åmosen.* Årb. f. Nordisk Oldkyndighed og. hist. 1953.

Troels-Smith, J. (1955) *Pollenanalytische Untersuchungen zu einigen schweizerischen Pfahlbauproblemen. Das Pfahlbau-problem.* Schaffhausen, 1954.

Troels-Smith, J. (1956) Neolithic period in Switzerland and Denmark. *Science* 124, 876–879.

Troels-Smith, J. (1960) Ivy, mistletoe and elm: climate indicator – fodder plants. *Danmarks Geologiske Undersøgelse IV.* Raekke 4, nr. 4 (*Geological Survey of Denmark* Series 4, No. 4), 1–32.

Truett, J. (1996) Bison and elk in the American southwest: in search of the pristine. *Environmental Management* 20, 195–206.

Tschadek, G. (1933) Die Waldentwicklung Mitteleuropas im Lichte der pollenanlytischen Ergebnisse und der urgeschichtlichen Forschung. *Centralblatt für das gesamte Forstwesen* 59, 206–213.

Tschermak (Cermak), L. (1910) Einiges über den Urwald von waldbaulichen Gesichtspunkten. *Centralblatt für das gesamte Forstwesen* 36, 340–370.

Tsouvalis-Gerber, J. (1998) Making the invisible visible: ancient woodlands, British forest policy and the social construction of reality. In: Watkins, C. (ed.) *European Woods and Forests: Studies in Cultural History.* CAB International, Wallingford, UK, pp. 215–229.

Tubbs, C.R. (1964) Early encoppicements in the New Forest. *Forestry* 37, 95–105.

Tubbs, C.R. (1988) *The New Forest. A Natural History.* The New Naturalist, Collins, London.

Tubbs, C.R. and Tubbs, J.M. (1985) Buzzards *Buteo buteo* and land use in the New Forest, Hampshire, England. *Biological Conservation* 31, 41–65.

Turbang, J. (1954) Contribution à l'étude de la régéneration naturelle du Chêne en Lorraine Belge. *Bulletin Institut Agronomique et Stations de Recherches de Gembloux* (Belgium) 22, 90–133.

Turcek, F.J. (1966) Über das Wiederauffinden von im Boden versteckten Samen durch Tannen- und Eichelhäher. *Waldhygiene* 6, 215–217.

Turcek, F.J. and Kelso, L. (1968) Ecological aspects of food transportation and storage in the Corvidae. *Communications in Behavioral Biology* Part A, 1, 277–297.

Turner, J. (1962) The *Tilia* decline: an anthropogenic interpretation. *New Phytologist* 61, 328–341.

von Tüxen, R. (1932) Ist die Buche die 'Naturmutter des deutschen Waldes?' *Forstarchiv* 8, 27–32.

von Tüxen, R. (1952) Hecken und Gebüsche. *Mitteilungen Geographische Gesellschaft in Hamburg* 50, 85–117.

von Tüxen, R. (1956) Die heutige potentielle natürliche Vegetation als Gegenstand der Vegetationskartierung. *Angewandte Pflanzensoziologie* 13, 1–42.

von Tüxen, R. (1974) Synchronologie einzelner Vegetationseinheiten in Europa. In:

Knapp, R. (ed.) *Vegetation Dynamics*. Part. VIII. *Handbook of Vegetation Science*. Dr W. Junk Publishers, The Hague, pp. 267–292.

van Twist, M. (1995) Verbale vernieuwing. Aantekeningen over de kunst van bestuurskunde. PhD thesis, Erasmus University, Rotterdam.

Vanselow, K. (1926) *Die Waldbautechniek im Spessart*. Eine historisch-kritische Untersuchung ihrer Epochen. Verlag von Julius Springer, Berlin.

Vanselow, K. (1949) *Theorie und Praxis der natürlichen Verjüngung im Wirtschaftswald*. Neumann Verlag, Berlin.

Vanselow, K. (1957) Die Verjüngungsformen. Entstehung-Entwicklung-Sinn und Wert. *Allgemeine Forstzeitschrift* 12, 205–208.

Vaupell, C. (1857) *Bögens Indvandring i de danske Skove*. Forlagt af C. A. Reitzels Bo og Arvinger, Copenhagen.

Vedel, H. (1961) Natural regeneration in juniper. *Proceedings. Botanical Society of the British Isles* 4, 146–148.

van Veen, P.A.F. and van der Sijs, N. (1990 and 1991) *Etymologisch woordenboek. De herkomst van onze woorden*. Van Dale Lexicografie, Utrecht/Antwerpen.

van de Veen, H.E. (1979) Food selections and habitat use of the red deer *(Cervus elaphus)*. PhD thesis, Rijksuniversiteit Groningen, Groningen.

van de Veen, H.E. (1985) Natuurontwikkelingsbeleid en bosbegrazing. *Landschap* 2 (1), 14–28.

van de Veen, H.E. and van Wieren, S.E. (1980) *Van grote grazers, kieskeurige fijnproevers en opportunistische gelegenheidsvreters; over het gebruik van grote herbivoren bij de ontwikkeling en duurzame instandhouding van natuurwaarden*. Rapport 80/11, Instituut voor Milieuvraagstukken, Amsterdam.

Vera, F.W.M. (1986) Grote Plantenetende zoogdieren. Voor natuur in Nederland nog steeds tweederangs elementen? *Huid en Haar* 5, 214–228.

Vera, F.W.M. (1988) *De Oostvaardersplassen. Van spontane natuuruitbarsting tot gerichte natuurontwikkeling*. IVN/Grasduinen. Oberon, Haarlem.

Vera, F.W.M. (1989) Epiloog: Natuurontwikkeling, uiteraard ook met zoogdieren! *Huid en Haar* 8, 123–157.

Vera, F. (1998) Das Multi-Spezies-Projekt Oostvaardersplassen. In: Cornelius, R., Hofmann, R.R. and Lindner, U. (eds) *Extensive Haltung robuster Haustierrassen, Wildtiermanagement, Multi-Spezies-Projekte – Neue Wege in Naturschutz und Landschaftspflege?* Institut für Zoo- und Wildtierforschung (IZW) im Forschungsverband Berlin e.V., Berlin, pp. 108–115.

Vera, F.W.M (1999) Ohne Pferd und Rind wird die Eiche nicht überleben. In: Gerken, B. and Görner, M. (eds) *Europäische Landschaftsentwicklung mit großen Weidetieren – Geschichte, Modelle und Perspektiven*. Natur und Kulturlandschaft (Höxter/Jena), 3 pp. 404–425.

Vereshchagin, N.K. and Baryshnikov, G.F. (1989) Quaternary mammalian extinctions in northern Eurasia. In: Martin, P.S. and Klein, R.G. (eds) *Quaternary Extinctions. A Prehistoric Revolution*. The University of Arizona Press, Tucson, pp. 483–516.

Vitousek, P.M. and Walker, L.R. (1987) Colonization, succession and resource availability; ecosystem-level interactions. In: Gray, A.J., Crawley, M.J. and Edwards, P.J. (eds) *Colonization, Succession, and Stability*. Blackwell Scientific Publications, Oxford, pp. 207–223.

Vogl, R.J. (1980) The ecological factors that produce perturbation-dependent ecosystems. In: Cairns, J. Jr (ed.) *The Recovery Process in Damaged Ecosystems*. Ann Arbor Science, Ann Arbor, Michigan, pp. 63–94.

Volf, J. (1979) Der Tarpan und das polnische "Konik". *Zeitschrift des Kölner Zoo* 21, 119–123.

Voloshcuk, I. (1993) The virgin forests and reserves in Slovakia. In: Broekmeyer, M.E.A., Vos, W. and Koop, H. (eds) *European Forest Reserves. Proceedings of the European Forest Reserves Workshop, 6–8 May 1992, Wageningen, The Netherlands.* Pudoc Scientific Publishers, Wageningen, pp. 69–74.

Voous, K.H. (1986) *Roofvogels en Uilen van Europa.* Nederlandse Vereniging tot Bescherming van Vogels; Wereld Natuur Fonds. E.J. Brill/W. Backhuis, Leiden.

de Vries, J. (1970) *Etymologisch woordenboek. Waar komen onze woorden vandaan?* Het Spectrum, Utrecht/Antwerpen.

de Vries, J. and de Tollenaere, F. (1997) *Etymologisch Woordenboek.* Het Spectrum, Utrecht.

Vulink, J.T and Drost, H.J. (1991a) Nutritional characteristics of cattle forage in the eutrophic nature reserve Oostvaardersplassen, Netherlands. *Netherlands Journal of Agricultural Science* 39, 263–271.

Vulink, J.T and Drost, H.J. (1991b) A causal analysis of diet composition in free ranging cattle in reed-dominated vegetation. *Oecologia* 88, 167–172.

Vulink, J.T and van Eerden M.R. (1998) Hydrological conditions and herbivory as key operators for ecosystem development in Dutch artificial wetlands. In: Wallis de Vries, M., Bakker, J.P. and van Wieren, S.E. (eds) *Grazing and Conservation Mangement.* Kluwer Academic Publishers, Dordrecht, pp. 217–252.

Vullmer, H. and Hanstein, U. (1995) Der Beitrag des Eichelhähers zur Eichenverjüngung in einem naturnah bewirtschafteten Wald in der Lünerburger Heide. *Forst und Holz* 50, 643–646.

Wagner, I. (1998) Artenschutz bei Wildapfel. *Forst und Holz* 53, 40–43.

Wagner, I. (1999) Schutz und Nutzen von Wildobst – Probleme bei der direkten Nutzung von Wildobstrelikten. *Forstarchiv* 70, 23–27.

Wahrig, G. (1980) *Brockhaus Wahrig. Deutsches Wörterbuch in sechs Bänden.* Erster Band A – BT. Deutsche Verlags-Anstalt, Stuttgart.

Walker, D. (1990) Purpose and method in Quaternary palynology. *Review of Palaeobotany and Palynology* 64, 13–27.

Walker, D. and Singh, G. (1993) Earliest palynological records of human impact on the world's vegetation. In: Chambers, F.M. (ed.) *Climate Change and Human Impact on the Landscape.* Chapman & Hall, London, pp. 101–118.

Walker, M.J.C. (1993) Holocene (Flandrian) vegetation change and human activity in the Carneddau area of upland mid-Wales. In: Chambers, F.M. (ed.) *Climate Change and Human Impact on the Landscape.* Chapman & Hall, London, pp. 169–183.

Waller, M.P. (1993) Flandrian vegetational history of south-eastern England. Pollen data from Pannel Bridge, East Sussex. *New Phytologist* 124, 345–369.

Waller, M.P. with an appendix by Marlow, A.D. (1994) Flandrian vegetational history of south-eastern England. Stratigraphy of the Brede valley and pollen data from Brede Bridge. *New Phytologist* 126, 369–392.

Wallis de Vries, M.F. (1994) Foraging in a landscape mosaic. Diet selection and performance of free-ranging cattle in heathland and riverine grassland. PhD thesis, Agricultural University Wageningen, Wageningen.

Wallis de Vries, M.F. (1995) Large herbivores and the design of large-scale nature reserves in western Europe. *Conservation Biology* 9, 25–33.

Wallis de Vries, M. (1998) Large herbivores as key factors for nature conservation. In:

Wallis de Vries, M., Bakker, J.P. and van Wieren, S.E. (eds) *Grazing and Conservation Managment.* Kluwer Academic Publishers, Dordrecht, pp. 1–20.

Walter, H. (1954) Klimax und zonale Vegetation. *Angewandte Pflanzensociologie* 1 *Festschrift für Erwin Aichiner* 144–150.

Walter, H. (1974) *Die Vegetation Osteuropas, Nord und Centralasiens.* Gustav Fischer, Stuttgart.

Walter, H. and Breckle, S.W. (1983) *Ökologie der Erde.* Band 1, *Grundlagen.* Zweite Auflage. Gustav Fischer, Stuttgart.

Walter, H. and Breckle, S.W. (1994) *Ökologie der Erde.* Band 3. *Spezielle Ökologie der Gemässigten und Arktischen Zonen Euro-Nordasiens,* 2nd edn. Gustav Fischer, Stuttgart.

Ward, J.S. and Parker, G.R. (1989) Spatial dispersion of woody regeneration in an old-growth forest. *Ecology* 70, 1279–1285.

Ward, L.K. (1981) The demography, fauna and conservation of *Juniperus communis* in Britain. In: Synge, H. (ed.) *The Biological Aspects of Rare Plant Conservation.* John Wiley & Sons, Chichester, pp. 319–329.

Ward, R.T. (1961) Some aspects of the regeneration habits of the American beech. *Ecology* 42, 828–832.

Wardle, P. (1959) The regeneration of *Fraxinus exelsior* in woods with a field layer of *Mercurialis perennis. Journal of Ecology* 47, 483–497.

Ware, S.A. (1970) Southern mixed hardwood forest in the Virginia coastal plain. *Ecology* 51, 921–924.

Warming, E. assisted by Vahl, M. (1909) *Ecology of Plants: an Introduction to the Study of Plant Communities.* Clarendon Press, Oxford.

Warren, M.S. and Thomas, J.A. (1992) Butterfly responses to coppicing. In: Buckley, G.P. (ed.) *Ecology and Management of Coppice Woodland.* Chapman & Hall, London, pp. 249–270.

Wartena, R. (1968) Vier eeuwen bosbeheer in Gelderland 1400–1800. *Tijdschrift Koninklijke Nederlandse Heidemij* 79, 33–40, 94–100, 182–191.

Watkins, C. (1990) *Britain's Ancient Woodland. Woodland Management and Conservation.* Davis & Chasler, London.

Watt, A.S. (1919) On the causes of failure of natural regeneration in British oakwoods. *Journal of Ecology* 7, 173–203.

Watt, A.S. (1923) On the ecology of the British beech woods with special reference to their regeneration. Failure of natural regeneration of the beech. *Journal of Ecology* 11, 1–48.

Watt, A.S. (1924) On the ecology of British beech woods with special reference to their regeneration. Part II. The development and structure of beech communities on the Sussex Downs. *Journal of Ecology* 12, 145–204.

Watt, A.S. (1925) On the ecology of British beech woods with special reference to their regeneration. Part II, sections II and III. The development and structure of beech communities. *Journal of Ecology* 13, 27–73.

Watt, A.S. (1926) Yew communities of the South Downs. *Journal of Ecology* 14, 282–316.

Watt, A.S. (1934a) The vegetation of the Chiltern Hills with special reference to the beechwoods and their seral relationship. Part I. *Journal of Ecology* 22, 230–270.

Watt, A.S. (1934b) The vegetation of the Chiltern Hills with special reference to the beechwoods and their seral relationship. Part II. *Journal of Ecology* 22, 445–507.

Watt, A.S. (1944) Ecological principles involved in the practice of forestry. *Journal of Ecology* 32, 96–104.

Watt, A.S. (1947) Pattern and process in the plant community. *Journal of Ecology* 35, 1–22.

Watt, A.S. (1957) The effect of excluding rabbits from Grassland B (Mesobrometum) in Breckland. *Journal of Ecology* 45, 861–878.

Watts, W.A. (1979) Late Quaternary vegetation of Central Appalachia and the New Jersey coastal plain. *Ecological Monographs* 49, 427–469.

Watts, W.A. (1983) Vegetational history of the eastern United States 25,000 to 10,000 years ago. In: Wright, H.E. (ed.) *Late-Quaternary Environments of the United States. The Late Pleistocene.* Longman, London, pp. 294–310.

Watts, W.A. (1985) Quaternary vegetation cycles. In: Edwards, K.J. and Warren, W.P. (eds) *The Quaternary History of Ireland.* Academic Press, London, pp. 155–185.

Webb, T., Cushing, E.J. and Wright, H.E. (1984) Holocene changes in the vegetation of the Midwest. In: Wright, H.E. (ed.) *Late-Quaternary Environments of the United States. The Holocene.* Longman, London, pp. 142–165.

Webb, T. III (1988) Eastern North America. In: Huntley, B. and Webb, T. III (eds) *Vegetation History. Section III: Glacial and Holocene Vegetation History –20 Ky to Present. Handbook of Vegetation Science*, vol. 7. Kluwer Academic Publishers, Dordrecht, pp. 633–672.

Weeda, E.J., Westra, R., Westra, Ch. and Westra, T. (1985) *Nederlandse oecologische flora; wilde planten en hun relaties.* 1. IVN, Amsterdam.

Weeda, E.J., Westra, R., Westra, Ch. and Westra, T. (1987) *Nederlandse oecologische flora; wilde planten en hun relaties.* 2. IVN, Amsterdam.

Weeda, E.J., Westra, R., Westra, Ch. and Westra, T. (1988) *Nederlandse oecologische flora; wilde planten en hun relaties*, 3. IVN, Amsterdam.

Wehage (1930) Deutsche Urwälder. Beiträge zur Geschichte und Beschreibung dreier urwaldähnlicher Waldungen im Landesteil Oldenburg. *Mitteilungen Deutsche Dendrologische Gesellschaft* 42, 249–260.

Weimann, H.J. (1977) Art und Höhe der Wildschäden im Wald. *Allgemeine Forstzeitschrift* 32, 106–109.

Weimann, K. (1911) *Die Mark- und Walderbengenossenschaften des Niederrheins.* Untersuchungen zur Deutschen Staats- und Rechtsgeschichte 106 Heft. M & H, Marcus, Breslau.

van der Werf, S. (1991) *Bosgemeenschappen. Natuurbeheer in Nederland*, Vol. 5. Pudoc, Wageningen.

Werner, P.A. and Harbeck, A.L. (1982) The pattern of tree seedling establishment relative to staghorn sumac cover in Michigan old fields. *The American Midland Naturalist* 108, 124–132.

West, D.C., Shugart, H.H. and Botkin, D.B. (1980) *Forest Succession. Concepts and Application.* Springer Verlag, New York.

Westhoff, V. (1952) The management of nature reserves in densely populated countries considered from a botanical viewpoint. In: *Proceedings and Papers of the Technical Meeting of the International Union for the Protection of Nature.* The Hague/Brussels, pp. 77–82.

Westhoff, V. (1971) The dynamic structure of plant communities in relation to the objectives of conservation. In: Duffey, E. and Watt, A.S. (eds) *The Scientific Management of Animal and Plant Communities for Conservation. Symposia of the British Ecological Society* 11, pp. 3–14.

Westhoff, V. (1976) Het zichzelf handhaven van bos in de gematigde luchtstreken. *Nederlands Bosbouwkundig Tijdschrift* 48, 58–65.

Westhoff, V. (1983a) Het zichzelf handhaven van bos in de gematigde luchtstreken. In: Goor, C.P. (ed.) *Ecologie en gebruik van bossen.* Pudoc, Wageningen, pp. 12–20.

Westhoff, V. (1983b) Man's attitude towards vegetation. In: Holzner, W., Werger, M.J.A. and Ikusima, I. (eds) *Man's Impact on Vegetation.* Dr W. Junk Publishers, The Hague; *Geobotany* 5, 7–24.

Westhoff, V. and den Held, A.J. (1975) *Plantengemeenschappen in Nederland.* Thieme en Cie, Zutphen.

Westhoff, V., Bakker, P.A., van Leeuwen, C.G. and van der Voo, E.E. (1970) *Wilde Planten. Flora en vegetatie in onze natuurgebieden,* Vol. 1. Vereniging tot Behoud van Natuurmonumenten in Nederland, 's-Graveland.

Westhoff, V., Bakker, P.A., van Leeuwen, C.G. and van der Voo, E.E. (1971) *Wilde Planten. Flora en vegetatie in onze natuurgebieden.* Vol. 2: *het lage land.* Vereniging tot Behoud van Natuurmonumenten in Nederland, s'-Gravenland.

Westhoff, V., Bakker, P.A., van Leeuwen, C.G., van der Voo, E.E. and Zonneveld, I.S. (1973) *Wilde Planten. Flora en vegetatie in onze natuurgebieden.* Vol. 3: *de hogere gronden.* Vereniging tot Behoud van Natuurmonumenten in Nederland, s'-Graveland.

Whitmore, T.C. (1982) On pattern and process in forests. In: Newman, E.I. (ed.) *The Plant Communities as a Working Mechanism.* Blackwell Scientific Publications, Oxford, pp. 45–59.

Whitmore, T.C. (1989) Canopy gaps and the two major groups of forest trees. *Ecology* 70, 536–538.

Whitney, G.G. (1987) An ecological history of the great lakes forest of Michigan. *Journal of Ecology* 75, 667–684.

Whitney, G.C. and Davis, W.C. (1986) From primitive woods to cultivated woodlots: Thoreau and the forest history Concord, Massachusetts. *Journal of Forest History* 30, 70–81.

Whitney, G.G. and Runkle, J.R. (1981) Edge versus age effects in the development of a beech–maple forest. *Oikos* 37, 377–381.

Whitney, G.G. and Somerlot, W.J. (1985) A case study of woodland continuity and change in the American midwest. *Biological Conservation* 31, 265–287.

Whittaker, R.H. (1953) A consideration of climax theory: the climax as a population and pattern. *Ecological Monographs* 23, 41–78. In: Golley, F.B. (ed.) *Ecological Succession. Benchmark Papers in Ecology* 5 (1977). Dowden Hutchinson & Ross, Stroudsburg, pp. 240–277.

Whittaker, R.H. (1977) Animal effects on plant species diversity. In: Tüxen, R. (ed.) *Vegetation und Fauna. Berichte der Internationalen Symposium der Internationalen Vereinigung für Vegetationskunde.* Cramer, Vaduz, pp. 409–425.

Whittaker, R.H. and Woodwell, G.M. (1969) Structure, production and diversity of the oak–pine forest at Brookhaven, New York. *Journal of Ecology* 57, 155–174.

Wiecko, E. (1963) The Białowieza Forest in 1795–1918. *Kwartalnik Historii Kultury Materialnej* 11, 345–352. (In Polish, with English summary.)

Wiedemann, E. (1931) Eichen-Buchen-Mischbestände. *Zeitschrift für Forst- und Jagdwesen* 63, 614–638.

Wiegers, J. and van Geel, B. (1983) The bryophyte *Tortella flavovirens* (Bruch) Broth. in late glacial sediments from Usselo (the Netherlands) and its significance as a palaeo-environmental indicator. *Acta Botanica Neerlandica* 32, 431–436.

van Wieren, S.E. (1985) Het jaar van de Schotse Hooglande. *Huid en Haar* 4, 19–30.

van Wieren, S.E. (1988) *Runderen in het bos. Begrazingsproef met Schotse Hooglandrunderen in het natuurgebied de Imbos.* Eindrapport. Instituut voor Milieuvraagstukken. V.U. Boekhandel/Uitgeverij, Amsterdam.

van Wieren, S.E. (1991) The management of populations of large mammals. In: Spellenberg, I.F., Goldsmith, F.B. and Morris, M.G. (eds) *The Scientific Management of Temperate Communities for Conservation.* Blackwell, Oxford, pp. 103–127.

van Wieren, S.E. (1996) Browsers and grazers: foraging strategies in ruminants, Chapter 8. PhD thesis, Agricultural University Wageningen, Wageningen.

van Wieren, S.E. and Borgesius, J.J. (1988) *Evaluatie van bosbegrazingsprojecten in Nederland.* RIN-rapport 88/63. Rijksinstituut voor Natuurbeheer, Arnhem.

van Wieren, S.E. and Kuiters, A.T. (1997) Hoefdieren in het boslandschap van de hogere zandgronden: evaluatie en perspectieven. In: van Wieren, S.E., Groot Bruinderink, G.W.T.A., Jorritsma, I.T.M. and Kuiters, A.T. (eds) *Hoefdieren in het boslandschap.* Backhuys Publishers, Leiden, pp. 193–208.

van Wieren, S.E. and Wallis de Vries, M. (1998) Ecologisch profiel van de wisent. *Nieuwe Wildernis* 4, 4–7.

van Wijk, N. (1949) *Franck's etymologisch woordenboek der Nederlandse taal.* Martinus Nijhoff, 's-Gravenhage.

van Wijk, W. and Allebas, J. (1995) De houthandel in Dordrecht. In: *Dordt in de kaart gekeken.* Dienst Kunsten, Gemeentearchief Dordrecht, Waanders Uitgevers, Zwolle 114, pp. 37–59,

Williams, M. (1982) Marshland and waste. In: Cantor, L. (ed.) *The English Medieval Landscape.* Croom Helm, London, pp. 86–125.

Williams, M. (1992) *Americans and their Forests. A Historical Geography.* Cambridge University Press, Cambridge.

Willis, K.J. (1993) How old is ancient woodland? *Trends in Ecology and Evolution* 8, 427–428.

Wilmanns, O. (1989) Zur Entwicklung von Trespenrasen im letzten halben Jahrhundert: Einblick-Ausblick-Rückblick, das Beispiel des Kaiserstuhls. *Düsseldorfer Geobotanischen Kolloqien* 6, 3–17.

Wiltshire, E.J. and Edwards, J. (1993) Mesolithic, early Neolithic, and later prehistoric impacts on vegetation at a riverine site at Derbyshire, England. In: Chambers, F.M. (ed.) *Climate Change and Human Impact on the Landscape.* Chapman & Hall, London, pp. 157–168.

Wistendahl, W.A. (1975) Buffalo beats, a relict prairie within a southeastern Ohio forest. *Bulletin of the Torrey Botanical Club* 102, 178–186.

Wolf, G. (1982) Beobachtungen zur Entwicklung von Baumsämlingen im Eichen-Hainbuchen und Eichen-Buchenwald. In: Dierschke, H. (ed.) *Struktur und Dynamik von Wäldern.* Berichte der Internationale Symposium der Internationale Verein für Vegetationskunde. J. Cramer, Valduz, pp. 475–494.

Wolf, G. (1988) Dauerflächen-Beobachtungen in Naturwaldzellen der Niederrheinischen Bucht. Veränderungen in der Feldschicht. *Natur und Landschaft* 63, 167–172.

Wolf, R. (1984) Heiden im Kreis Ludwigsburg. *Beiträge zu den Veröffentlichungen für Naturschutz und Landschaftspflege Baden-Württenberg* 35, 1–76, Karlssruhe.

Wolfe, M.L. and von Berg, F.C. (1988) Deer and forestry in Germany. Half a century after Aldo Leopold. *Journal of Forestry* 86, 25–31.

Wolkinger, F. and Plank, S. (1981) *Dry Grasslands of Europe*. European Committee for the Conservation of Nature and Natural Resources, Council of Europe. Nature and Environment Series no. 21. Strasbourg.

Woolsey T.S. Jr and Greeley, W.B. (1920) *Studies in French Forestry*. John Wiley & Sons, New York.

van der Woud, A. (1987) *Het Lege Land. De ruimttelijke orde van Nederland 1798–1848*. Meulenhoff Informatief, Amsterdam.

Wright, H.E. Jr (1971) Late quaternary vegetational history of North America. In: Turekian (ed.) *The Late Cenozoic Glacial Ages*. Yale University Press, New Haven, Connecticut, pp. 425–464.

Wright, H.E. (1984) Introduction. In: Wright, H.E. (ed.) *Late-Quaternary Environments of the United States. The Holocene*. Longman, London, pp. xi–xvii.

Young, P. T. (1994) Natural die-offs of large mammals: implications for conservation. *Conservation Biology* 8, 410–418.

van Zeist, W. (1959) Studies on the post-boreal vegetational history of south-eastern Drenthe (Netherlands). *Acta Botanica Neerlandica* 8, 156–184.

van Zeist, W. (1964) A palaeobotanical study of some bogs in western Brittany (Finistere), France. *Palaeohistoria* 10, 157–180.

van Zeist, W. and van der Spoel-Walvius, M.R. (1980) A palynological study of the late-glacial and the postglacial in the Paris Basin. *Palaeohistoria* 12, 67–109.

Ziegenhagen, B. and Kausch, W. (1995) Productivity of young shaded oaks (*Quercus robur* L.) as corresponding to shoot morphology and leaf anatomy. *Forest Ecology and Management* 72, 97–108.

Zimmermann, A. (1982) Naturnahe Traubeneichenwälder und ihre Kontaktgesellschaften im Rennfeldgebiet bei Bruck a.d. Mur (Steiermark). *Urwald Symposium IUFRO Gruppe Urwald*. Waldbau Institut Universität für Bodenkultur, Wien, pp. 121–126.

Zoller, H. and Haas, J.N. (1995) War Mitteleuropa ursprünglich eine halboffene Weidelandschaft oder von geschlossenen Wäldern bedeckt? *Schweizerische Zeitschrift für Forstwesen* 5, 321–354.

Zukrigl, K. (1991) Succession and regeneration in the natural forests in Central Europe. *Geobios* 18, 202–208.

Zukrigl, K., Eckhart, G. and Nather, J. (1963) *Standortskundliche und Waldbaukundliche Untersuchungen in Urwaldresten der niederÖsterreichischen Kalkalpen*. Mitteilungen Forstliche Bundes-Versuchsanstalt 62.

Zvelebil, M. (1986a) Introduction: the scope of the present volume. In: Zvelebil, M. (ed.) *Hunters in Transition: Mesolithic Societies of Temperate Eurasia and their Transition to Farming*. Cambridge University Press, Cambridge, pp. 1–4.

Zvelebil, M. (1986b) Mesolithic prelude and neolithic revolution. In: Zvelebil, M. (ed.) *Hunters in Transition: Mesolithic Societies of Temperate Eurasia and their Transition to Farming*. Cambridge University Press, Cambridge, pp. 5–15.

Zvelebil, M. (1986c) Mesolithic societies and the transition to farming: problems of time, scale and organisation. In: Zvelebil, M. (ed.) *Hunters in Transition: Mesolithic Societies of Temperate Eurasia and their Transition to Farming*. Cambridge University Press, Cambridge, pp. 167–188.

Zvelebil, M. and Rowley-Conwy, P. (1986) Foragers and farmers in Atlantic Europe. In: Zvelebil, M. (ed.) *Hunters in Transition: Mesolithic Societies of Temperate Eurasia and their Transition to Farming*. Cambridge University Press, Cambridge, pp. 67–93.

Index

Acker (acker, Acker, aecker, aker, Aecer) as
concept,
as concept 123–127, 129–130, 138,
174–175, 183
trees removed from 175
see also forestis as concept; oak; pannage;
wald as concept; weide as concept
Acorns 20, 21
dispersal by jays 47, 96, 300–307, 360,
376–377
predation 20, 21, 47, 304
see also acker as concept
Acre *see* acker as concept
Actualism 62
Agriculture,
as analogy of biotopes of wild species
356
enriching nature 5–6, 37
impoverishing biodiversity 379–382
invention of 379
in connection with wilderness 102–104,
106–111, 113–115, 123–127,
129–131, 133, 135–138, 140,
142–144, 164–168, 170–176,
370–372, 381–398
intensification propagated by the physiocrats 173–175
introduction in prehistory 78–85, 98
for preserving biodiversity 6, 370
see also acker as concept; elm pollen
decline; fodder; foliage; forestis

as concept; Landnam theory;
livestock; nature conservation;
pannage; pasture; regulations;
Wald as concept; wood-pasture
Alternative hypothesis 8–9, 378
see also null hypothesis
Alterphase 29
see also Leibundgut's cyclical model of
forest regeneration
Ash (*Fraxinus excelsior*)
colonization after last Ice Age 68–70
as coppice 132
establishment in closed forest 22–24,
44, 241, 274, 332–333, 365
in forestis and Wald 110, 133, 142
growth of seedlings with different
amounts of daylight 332–333
growth of seedlings compared with oak
333
in primeval vegetation 3,
white ash (*F. americana*) 39–40, 277,
278
in wood-pasture 22, 28, 54, 333
Atlanticum 9, 63, 65, 70, 77–79, 94
see also prehistoric eras
Aurochs (*Bos primigenius*)
feeding strategy 53, 55, 347–348,
350–356
last refuges 246, 250, 255
role in the wilderness 273, 348–350,
354–356, 358, 367–368, 370